Communications in Computer and Information Science 2967

Rationale

The CCIS series is devoted to the publication of proceedings of computer science conferences. Its aim is to efficiently disseminate original research results in informatics in printed and electronic form. While the focus is on publication of peer-reviewed full papers presenting mature work, inclusion of reviewed short papers reporting on work in progress is welcome, too. Besides globally relevant meetings with internationally representative program committees guaranteeing a strict peer-reviewing and paper selection process, conferences run by societies or of high regional or national relevance are also considered for publication.

Topics

The topical scope of CCIS spans the entire spectrum of informatics ranging from foundational topics in the theory of computing to information and communications science and technology and a broad variety of interdisciplinary application fields.

Information for Volume Editors and Authors

Publication in CCIS is free of charge. No royalties are paid, however, we offer registered conference participants temporary free access to the online version of the conference proceedings on SpringerLink (http://link.springer.com) by means of an http referrer from the conference website and/or a number of complimentary printed copies, as specified in the official acceptance email of the event.

CCIS proceedings can be published in time for distribution at conferences or as post-proceedings, and delivered in the form of printed books and/or electronically as USBs and/or e-content licenses for accessing proceedings at SpringerLink. Furthermore, CCIS proceedings are included in the CCIS electronic book series hosted in the SpringerLink digital library at http://link.springer.com/bookseries/7899. Conferences publishing in CCIS are allowed to use our online conference service (Meteor) for managing the whole proceedings lifecycle (from submission and reviewing to preparing for publication) free of charge.

Publication process

The language of publication is exclusively English. Authors publishing in CCIS have to sign the Springer CCIS copyright transfer form, however, they are free to use their material published in CCIS for substantially changed, more elaborate subsequent publications elsewhere. For the preparation of the camera-ready papers/files, authors have to strictly adhere to the Springer CCIS Authors' Instructions and are strongly encouraged to use the CCIS LaTeX style files or templates.

Abstracting/Indexing

CCIS is abstracted/indexed in DBLP, Google Scholar, EI-Compendex, Mathematical Reviews, SCImago, Scopus. CCIS volumes are also submitted for the inclusion in ISI Proceedings.

How to start

To start the evaluation of your proposal for inclusion in the CCIS series, please send an e-mail to ccis@springer.com

Hamid R. Arabnia · Leonidas Deligiannidis ·
Farzan Shenavarmasouleh ·
Farid Ghareh Mohammadi · Soheyla Amirian
Editors

Emerging Trends in Artificial Intelligence and Applications

12th International Conference on Computational Science and Computational Intelligence, CSCI 2025
Las Vegas, NV, USA, December 3–5, 2025
Proceedings, Part I

Springer

Editors
Hamid R. Arabnia
University of Georgia
Athens, GA, USA

Leonidas Deligiannidis
Wentworth Institute of Technology
Boston, MA, USA

Farzan Shenavarmasouleh
Medialab Inc.
Lawrenceville, GA, USA

Farid Ghareh Mohammadi
Verify Radiology Images Consultants
Houston, TX, USA

Soheyla Amirian
Pace University
New York, NY, USA

ISSN 1865-0929 ISSN 1865-0937 (electronic)
Communications in Computer and Information Science
ISBN 978-3-032-28453-2 ISBN 978-3-032-28454-9 (eBook)
https://doi.org/10.1007/978-3-032-28454-9

This Springer imprint is published by the registered company Springer Nature Switzerland AG
The registered company address is: Gewerbestrasse 11, 6330 Cham, Switzerland

Preface

It is our great pleasure to present this collection of research papers from the 12th International Conference on Computational Science and Computational Intelligence (CSCI 2025). This volume, titled "Emerging Trends in Artificial Intelligence and Applications", comprises selected papers accepted in the Research Track on Artificial Intelligence (CSCI-RTAI). The conference was held in Las Vegas, Nevada, USA, from December 3 to December 5, 2025.

The CSCI 2025 International Conference brought together papers from a diverse array of communities, including researchers from universities, corporations, and government agencies. Accepted papers are published by Springer Nature, and the proceedings showcase solutions to key challenges in various critical areas of Computational Science and Computational Intelligence.

Computational Science (CS) is the study of addressing problems that are impossible to solve (or difficult to solve) without computers. CS can be considered to be the bridge between computer science and other sciences. The field is interdisciplinary by nature and includes the use of advanced computing capabilities to understand and solve complex problems. In short, CS is the science of using computers to do science. Computational Intelligence (CI) is the study of computational methods that exhibit intelligence. These methods adapt to changing environments and changing goals. There is a significant overlap between the fields of CI and Artificial Intelligence (AI). However, there is also a difference: AI techniques often involve top-to-bottom methods (i.e., methods to reach solutions are imposed from the top) whereas CI techniques often involve bottom-up methods (i.e., solutions emerge from unstructured beginnings). An important part of CI includes a set of Nature-inspired computational approaches to address complex problems for which traditional methods are infeasible. Computational Science and Computational Intelligence both share the same objective: finding solutions to difficult problems. However, the methods to reach the solutions are different. The main objective of the CSCI Conference is to facilitate increased opportunities for cross-fertilization across CS and CI.

Considering the above broad outline, the CSCI 2025 International Conference was composed of the following focused Research Tracks:

Artificial Intelligence (CSCI-RTAI); Big Data and Data Science (CSCI-RTBD); Computational Science (CSCI-RTCS); Computational Intelligence (CSCI-RTCI); Computational Biology (CSCI-RTCB); Cyber Warfare, Cyber Defense, & Cyber Security (CSCI-RTCW); Signal & Image Processing, Computer Vision & Pattern Recognition (CSCI-RTPC); Smart Cities and Smart Mobility (CSCI-RTSC); Education - CS & CE (CSCI-RTED); Health Informatics and Medical Systems (CSCI-RTHI); Mobile Computing, Wireless Networks, & Security (CSCI-RTMC); Software Engineering (CSCI-RTSE); Internet of Things & Internet of Everything (CSCI-RTOT); Social Network Analysis, Social Media, & Mining (CSCI-RTNA); Cloud Computing and Data Centers (CSCI-RTCC); Parallel & Distributed Computing (CSCI-RTPD); and Large Language

Models (LLM). The scope of each track can be found at: https://www.american-cse.org/csci2025/topics and https://www.american-cse.org/LLM2025/topics.

An important objective of the CSCI 2025 International Conference and its associated Research Tracks was to foster opportunities for cross-fertilization between the fields of Computational Science and Computational Intelligence. The CSCI Conference is deeply committed to promoting diversity and eliminating discrimination, both in its role as a conference organizer and as a service provider. Our goal is to create an inclusive culture that respects and values differences, promotes dignity, equality, and diversity, and encourages individuals to reach their full potential. We are also dedicated, wherever possible, to organizing a conference that represents the global community. We sincerely hope that we have succeeded in achieving these important objectives.

The Steering Committee and the Program Committee would like to extend their gratitude to all the authors who submitted papers for consideration. CSCI 2025 received submissions from 48 countries, with approximately 48% of them coming from outside the United States. Each submitted paper underwent a rigorous peer-review process, with at least two experts (an average of 2.5 referees per paper) evaluating the submissions based on originality, significance, clarity, impact, and soundness. In cases where reviewers' recommendations were contradictory, a program committee member was tasked with making the final decision, often consulting additional referees for further guidance. The CSCI Conference followed the guidelines of COPE (Committee on Publication Ethics):

- Typical submissions underwent a single-blind peer review process, in which the authors remain unaware of the identities of the reviewers, while the reviewers are informed of the authors' identities.
- Papers authored by one or more members of the program committee, including co-chairs, were subjected to a double-blind peer review process, ensuring that neither the authors nor the reviewers were aware of each other's identities or affiliations.

The Research Track on Artificial Intelligence (CSCI-RTAI) of CSCI 2025 Conference received 389 submissions, of which 69 papers were accepted (2 papers were withdrawn), resulting in a paper acceptance rate of 18%. This volume includes only 42 of the accepted papers.

We are deeply grateful to the many colleagues who contributed their time and effort to organizing the CSCI 2025 Conference. In particular, we extend our thanks to the members of the Program Committee, the Steering Committee, and the referees. We would also like to express our appreciation to the primary sponsor of the conference, the American Council on Science & Education. The list of members of the Program Committee for each track can be found at: https://www.american-cse.org/csci2025/committees.

We extend our heartfelt gratitude to all the speakers and authors for their valuable contributions. We would also like to thank the following individuals and organizations for their support: the staff at the Luxor Hotel (an MGM property, conference/meeting department) and the staff of Springer, for their assistance in various aspects of the event.

We are pleased to present the proceedings of CSCI 2025 (selected papers of CSCI-RTAI). These proceedings represent a collection of outstanding research contributions that reflect the diversity and depth of work in Artificial Intelligence.

Co-editors: Research Track on Artificial Intelligence (CSCI-RTAI) of CSCI 2025 International Conference

December 2025

Hamid R. Arabnia
Leonidas Deligiannidis
Farzan Shenavarmasouleh
Farid Ghareh Mohammadi
Soheyla Amirian

Organization

Steering Committee – Co-chairs (CSCI 2025)

Hamid R. Arabnia	University of Georgia, USA
Leonidas Deligiannidis	Wentworth Institute of Technology, USA
Fernando G. Tinetti	Universidad Nacional de La Plata, Argentina
Quoc-Nam Tran	Southeastern Louisiana University, USA

Publication Co-chairs an Co-editors

"Emerging Trends in Artificial Intelligence and Applications"
Research Track: Artificial Intelligence (CSCI-RTAI)
Part of Proceedings of The 2025 International Conference on Computational Science and Computational Intelligence (CSCI)

Hamid R. Arabnia	University of Georgia, USA
Leonidas Deligiannidis	Wentworth Institute of Technology, USA
Farzan Shenavarmasouleh	Medialab Inc., USA
Farid Ghareh Mohammadi	Verify Radiology Images Consultants, LLC, USA
Soheyla Amirian	Pace University, USA

Members of Steering Committee (CSCI 2025)

Baak Akhgar	Sheffield Hallam University, UK
Abbas M. Al-Bakry	University of Information Technology & Communications, Iraq
Nizar Al-Holou	University of Detroit Mercy, USA
Soheyla Amirian	Pace University, USA
Hamid R. Arabnia	University of Georgia, USA
Rajab Challoo	Texas A&M University-Kingsville, USA
Chien-Fu Cheng	Tamkang University, Taiwan
Hyunseung Choo	Sungkyunkwan University, South Korea
Kevin Daimi	University of Detroit Mercy, USA
Leonidas Deligiannidis	Wentworth Institute of Technology, USA
Eman M. El-Sheikh	University of West Florida, USA
Mary Mehrnoosh Eshaghian-Wilner	University of California, Los Angeles, USA

David L. Foster	Kettering University, USA
Ching-Hsien (Robert) Hsu	ChungAsia University, Taiwan
Farid Ghareh Mohammadi	Verify Radiology Images Consultants, LLC, USA
James J. (Jong Hyuk) Park	SeoulTech, South Korea
Mohammad S. Obaidat	University of Jordan, Jordan
Gerald Schaefer	Loughborough University, UK
Farzan Shenavarmasouleh	Medialab Inc., USA
Fernando G. Tinetti	Universidad Nacional de La Plata, Argentina
Quoc-Nam Tran	Southeastern Louisiana University, USA
Shiuh-Jeng Wang	Central Police University, Taiwan
Layne T. Watson	Virginia Polytechnic Institute & State University, USA
Chao-Tung Yang	Tunghai University, Taiwan
Mary Yang	University of Arkansas for Medical Sciences, USA

Program Committee – Computational Science and Computational Intelligence (CSCI 2025)

Afrand Agah	West Chester University of Pennsylvania, USA
Bharat Bhushan Agarwal	IFTM University, India
Omaima Nazar Ahmad AL	Al-Zaytoonah University of Jordan, Jordan
Wasim A. Al – Hamdani	Kentucky State University, USA
Ismail Khalil Al Ani	Ittihad University, United Arab Emirates
Mehran Asadi	Lincoln University, USA
Travis Atkison	University of Alabama, USA
Azita Bahrami	IT Consult, USA
Mehdi Bahrami	Fujitsu Laboratories of America, Inc., USA
P. Balasubramanian	Nanyang Technological University, Singapore
Petra Saskia Bayerl	Erasmus University Rotterdam, The Netherlands
Jane M. Binner	University of Birmingham, UK
Juan-Vicente Capella-Hernandez	Universitat Politècnica de València, Spain
Juan Jose Martinez Castillo	Universidad Nacional Abierta, Venezuela
Rui Chang	Mount Sinai School of Medicine, USA
Dongsheng Che	East Stroudsburg University of Pennsylvania, USA
Jianhung Chen	Chung Hua University, Taiwan
Mu-Song Chen	Da-Yeh University, Taiwan
Xin (Thomas) Chen	University of Hawaii, USA
Steve C. Chiu	Idaho State University, USA
Mark Yul Chu	University of Texas Rio Grande Valley, USA

Jose Alfredo F. Costa	Universidade Federal do Rio Grande do Norte, Brazil
Arianna D'Ulizia	National Research Council of Italy, Italy
Zhangisina G. Davletzhanovna	Central Asian University, Kazakhstan
Wesley Deneke	Western Washington University, USA
Noel De Palma	Grenoble Alpes University, France
Mohsen Doroodchi	University of North Carolina at Charlotte, USA
Levent Ertaull	California State University, East Bay, USA
Mahmood Fazlali	Shahid Beheshti University, Iran
Boyuan Feng	University of California, Santa Barbara, USA
George A. Gravvanis	Democritus University of Thrace, Greece
Gheorghe Grigoras	"Gheorghe Asachi" Technical University of Iaşi, Romania
Ray Hashemi	Georgia Southern University, USA
Houcine Hassan	Universitat Politècnica de València, Spain
Abdolreza Hatamlou	Islamic Azad University, Iran
Bing He	Cisco System Inc., USA
Henry Hexmoor	Southern Illinois University at Carbondale, USA
Gahangir Hossain	Texas A&M University - Kingsville, USA
Guofeng Hou	AQR Capital Management, USA
Ren-Junn Hwang	Tamkang University, Taiwan
Naseem Ibrahim	Penn State Erie, The Behrend College, USA
Rabia Jafri	King Saud University, Saudi Arabia
Shahram Javadi	Azad University, Iran & University of Lisbon, Portugal
Young-Sik Jeong	Dongguk University, South Korea
Aleksandr Katkow	University of Computer Sciences & Skills, Poland
Byung-Gyu Kim	Sun Moon University, South Korea
Taihoon Kim	University of Tasmania, Australia
Sang-Wook Kim	Hanyang University, South Korea
Dattatraya Vishnu Kodavade	D.K.T.E Society's Textile & Engineering Institute, India
Elena B. Kozerenko	Russian Academy of Sciences, Moscow, Russia
Guoming Lai	Sun Yat-sen University, China
Hyo Jong Lee	Chonbuk National University, South Korea
Bo Liu	NEC Labs China, China
Eleanor Lockley	Sheffield Hallam University, UK
Jianbing Ma	Bournemouth University, UK
Julius Marpaung	Northeastern University, USA
Andrew Marsh	HoIP Telecom Ltd., UK
Juan Martinez	Universidad Gran Mariscal de Ayacucho, Venezuela

Praveen Meduri	California State University, Sacramento, USA
Ali Mostafaeipour	California State University, Fullerton, USA
Houssem Eddine Nouri	University of Tunis, Tunisia
Michael B. O'Hara	KB Computing, LLC, USA
Robert Ehimen Okonigene	Ambrose Alli University, Nigeria
Funminiyi Olajide	Nottingham Trent University, UK
Satish Penmatsa	University of Maryland - Eastern Shore, USA
Saman Parvaneh	Philips Research North America, USA
Ponalagusamy	National Institute of Technology, Tiruchirappalli, India
Laura L. Pullum	Oak Ridge National Laboratory, USA
Xuewei Qi	University of California, Riverside, USA
Junfeng Qu	Clayton State University, USA
Shahram Rahimi	University of Alabama, USA
Jaime Raigoza	California State University, Chico, USA
Arvind Ramanathan	Oak Ridge National Laboratory, USA
Iman M. Rezazadeh	University of California, Davis, USA
Om Prakash Rishi	University of Kota, India
Cristian Rodriguez Rivero	Universidad Nacional de Córdoba, Argentina
Seyed Roosta	Albany State University, USA
K. Martin Sagayam	Karunya Institute of Technology and Sciences, India
P. Sanjeevikumar	University of South-Eastern Norway, Norway
Benaoumeur Senouci	University of Southern Denmark, Denmark
Zhefu Shi	Microsoft Corporation, USA
Jawed Siddiqi	Sheffield Hallam University, UK
Akash Singh	IBM Corporation, USA
Anthony Skjellum	Auburn University, USA
Omer Muhammet Soysal	Southeastern Louisiana University, USA
Helman Stern	Ben-Gurion University of the Negev, Israel
Jonathan Z. Sun	University of Southern Mississippi, USA
Rahman Tashakkori	Appalachian State University, USA
Predrag Tosic	Microsoft, USA
Quoc-Nam Tran	Southeastern Louisiana University, USA
Jesus Vigo-Aguiar	University of Salamanca, Spain
Patrick Wang	East China Normal University, China & Northeastern University, USA
Weiqiang Wang	Opera Solutions LLC, USA
Yin Wang	Lawrence Technological University, USA
Alicia Nicki Washington	Duke University, USA
Wei Wei	Xi'an University of Technology, China
Yong Wei	University of North Georgia, USA

Contents

Artificial Intelligence - Applications of Graphs and Networks

Artificial Intelligence - Natural Language Processing and Large Language Models

Artificial Intelligence - Algorithms and Applications

Learning Strategies - Transfer Learning and Reinforcement Learning

Integrating Explainability and Knowledge Graphs in Multi Agent Reinforcement Learning for Air Combat Simulation

Abderahim Salhi[1(✉)], Indu Shukla[1], Gary J. Briggs[2], James E. Ross[1], Haley R. Dozier[1], and Ian D. Detwiller[1]

[1] United States Army Corps of Engineers, 3909 Halls Ferry Road, Vicksburg, MS 39180, USA
abderahim.salhi@usace.army.mil

[2] the RAND Corporation, 1776 Main Street, Santa Monica, CA 90401, USA

Abstract. This work advances Multi Agent Reinforcement Learning (MARL) within complex air combat environments by integrating approaches that enhance both agent decision making and explainability. Building on established MARL scenarios, we investigate agent interactions in competitive settings using a realistic six-degree-of-freedom (6-DoF) aerial simulation based on the OpenAI Gymnasium framework. To address the inherent opacity of reinforcement learning agents, explainability techniques including decision trees and policy visualizations such as heatmaps are employed to illuminate agent behaviors and interaction patterns within the state space. We augment the MARL framework by incorporating structured domain knowledge via knowledge graphs, natural language processing through large language models (LLMs), and modular orchestration using the LangChain framework. This hybrid architecture leverages knowledge-driven reasoning to improve situational awareness, coordination, and interpretability of agents operating under complex, dynamic conditions. Experimental evaluations demonstrate that combining explainability methods with knowledge graph and LLM-based reasoning not only sustains competitive reinforcement learning performance but also significantly enhances transparency and human interpretability in mission-critical air combat simulations. This integrative approach addresses key challenges in deploying autonomous agents in combat scenarios, fostering both robust decision-making and trustworthiness.

Keywords: LLMs · MARL · Knowledge Graphs · XAI · Explainability

1 Introduction

Recent advances in RL have enabled intelligent agents to learn complex decision-making policies in high-dimensional domains such as air combat simulation [1–3].

H. R. Arabnia et al. (Eds.): CSCI 2025, CCIS 2967, pp. 3–22, 2027.
https://doi.org/10.1007/978-3-032-28454-9_1

Despite these advances, RL agents—particularly in multi-agent settings—often lack transparency, limiting their deployment in real-world mission-critical environments where explainability is essential [4,5]. To address this challenge, we propose a unified framework that integrates MARL with KGs, LLMs, and the LangChain orchestration platform. The framework leverages structured domain knowledge and natural language reasoning to improve both agent decision-making and interpretability. The system is implemented within a 6-DoF aerial combat environment using OpenAI Gymnasium for offline RL training, enabling reproducible evaluation under realistic conditions. Results indicate that combining symbolic and neural representations enhances both performance and explainability in complex air combat scenarios.

Air combat simulation presents a uniquely challenging domain due to its high-dimensional state and action spaces, nonlinear flight dynamics, partial observability, and the requirement for coordinated, realistic behaviors [1,3]. MARL has emerged as an effective paradigm for modeling interactions among multiple autonomous agents, enabling the development of synthetic pilots capable of training alongside or against humans [2,5]. However, as agent populations and interaction complexity increase, the strategy space grows exponentially, resulting in significant scalability and coordination challenges [5]. Deep RL methods have demonstrated strong performance in learning policies for such environments [1,3]. Nevertheless, these approaches typically function as opaque black boxes, making their decisions difficult to interpret and trust in operational contexts [4,5]. The added complexity of 6-DoF control further exacerbates this issue, as agents must continuously reason over position and orientation in three-dimensional space.

Recent research has sought to mitigate these limitations by incorporating structured domain knowledge through KGs, which encode relationships among entities, tactics, and mission objectives. This representation enhances learning efficiency while supporting human-interpretable reasoning [4,6]. Neo4j, a widely adopted graph database, enables scalable and expressive KG construction, supporting both data-driven inference and explainable analysis [7,8]. LLMs further contribute by enabling natural language interpretation and explanation of agent behavior, bridging the gap between learned policies and human understanding [7,9]. Retrieval-augmented generation (RAG) architectures, which combine LLMs with external knowledge sources such as KGs, have shown strong potential for improving both accuracy and explainability [9,10]. LangChain facilitates modular orchestration among these components, supporting the development of explainable AI systems.

Our approach uniquely integrates MARL, Neo4j-based KGs, LLMs, RAG, and LangChain within a unified air combat simulation architecture. As illustrated in Fig. 5, this integration enables natural language querying of agent behavior, transparent post-hoc analysis, and modular system extensibility. The OpenAI Gymnasium framework provides a standardized platform for training and evaluating RL agents in offline 6-DoF environments, ensuring reproducibility and scalability [1,3]. Collectively, this framework advances the development of

MARL agents that are both operationally effective and explainable, addressing a critical gap in current RL-based air combat systems.

2 Background

Air combat simulation has become a critical testbed for artificial intelligence (AI) due to advances in machine learning (ML) and RL algorithms [3]. The domain is characterized by high-dimensional continuous state and action spaces, partial observability, and adversarial dynamics, posing significant challenges for both agent design and evaluation [11,12].

Early air combat AI systems relied mainly on rule-based and game-theoretic methods; however, these approaches scale poorly as state and action spaces grow [13]. The adoption of deep RL has enabled agents to learn complex maneuvering and engagement policies directly from simulation, surpassing scripted logic and, in some cases, human-level performance. High-fidelity simulators such as AFSIM, BVR Gym, and JSBSim-Gymnasium now serve as essential platforms for training and evaluating transferable RL policies [1–3,12]. Despite these advances, explainability remains a fundamental limitation of RL-based air combat agents [11,12]. Deep RL policies; particularly in MARL settings—often operate as black boxes, hindering trust, validation, and deployment in mission critical environments [14,15]. This issue is exacerbated in MARL, where coordination and emergent behaviors further obscure the rationale behind agent actions.

To address these challenges, recent work has explored the integration of KGs and RAG with MARL and LLMs. KGs provide a structured, machine-interpretable representation of domain knowledge, tactical relationships, and causal dependencies [16,17]. Graph databases such as Neo4j support scalable storage and querying of these structures, enabling simulation data, agent actions, and mission outcomes to be unified within a single queryable representation [17]. When combined with RAG pipelines, KGs reduce hallucinations, improve factual consistency, and enable context-aware reasoning, particularly in human-in-the-loop settings where graph content can be curated and refined [18].

LLMs further enhance this capability by translating natural language queries into structured graph queries (e.g., Cypher), retrieving relevant subgraphs, and synthesizing explanations in human-readable form. RAG architectures—including emerging GraphRAG approaches exploit the relational structure of KGs to deliver richer contextual reasoning than vector-only retrieval methods [19,20]. These systems typically involve entity and relation extraction from unstructured data, KG population, and LLM-driven reasoning over the resulting graph. Within MARL, recent frameworks have combined LLMs, adaptive KG memory, and structured communication mechanisms to improve coordination, planning, and reasoning in dynamic environments. Decentralized MARL agents equipped with hierarchical KGs and LLM-based reasoning have demonstrated improved scalability, cooperation, and efficiency, outperforming both conventional MARL and LLM-only baselines [21].

The system-level workflow for integrating these components is shown in Fig. 5. Simulation data and agent interactions are ingested into a KG, which is queried via an LLM-powered RAG retriever to support explainable MARL in air combat scenarios. This workflow highlights the interaction between structured knowledge, natural language reasoning, and modular orchestration, reflecting current research trends that emphasize hybrid symbolic–neural architectures for both performance and interpretability.

Parallel to these knowledge-driven explainability advances, explainable reinforcement learning (XRL) through decision trees has gained significant traction. XRL methods broadly fall into model-agnostic and model-specific categories, focusing on post-hoc explanations or inherently interpretable function approximators [22]. Policy distillation compressing complex policies into simpler, human-interpretable models like decision trees is an effective technique for elucidating agent behavior.

During the past two decades, differentiable decision trees and their integration with neural networks have been explored for continuous learning environments. Foundational works by Irsoy et al. [23] and Laptev and Buhmann [24] paved the way toward differentiable decision trees optimized by gradient methods. Frosst and Hinton [25] demonstrated distilling soft decision trees from neural nets in vision tasks, although limited to simpler domains.

In RL contexts, Coppens et al. [26] advanced policy imitation via differentiable decision trees, while Liu et al. [27] extended these for Q-value function approximation and proposed Linear Model U-trees. Silva et al. [28] further developed differentiable trees as Q-function or policy approximators, emphasizing simplification through discretization for interpretability. Bastani et al.'s VIPER method [29] offered decision tree distillation with theoretical guarantees, albeit limited to imitation learning.

More recent work by Gokhale et al. [30] tackles RL interpretability by distilling complex policies into differentiable decision trees, which improved user comprehension in energy management domains. Costa et al. [31] introduced MENS-DT-RL, an evolutionary algorithm evolving decision trees optimized for both interpretability and performance, incorporating novel pruning and initialization methods tailored for RL.

Pynadath et al. [32] showed decision-tree explanations significantly enhance transparency and team performance in human-robot teaming. Gjærum et al. [33] proposed model trees combining decision trees with simple regression models to yield richer insights into deep RL agents' behavior.

Integrating decision trees and utility-based models with knowledge graph and LLM-driven reasoning offers a holistic path forward. This combination bridges the interpretability gap between complex AI systems and human users, a critical advance for high-stakes domains like autonomous aerial combat where clarity and trustworthiness are prerequisites.

3 Methodology

This section details the end-to-end pipeline for our explainable multi-agent reinforcement learning (MARL) framework in air combat simulation, from data generation and logging to knowledge graph construction and natural language reasoning. The overall architecture of our Gymnasium-based simulator, integrated with knowledge graphs, large language models, and LangChain orchestration, is depicted in Fig. 1.

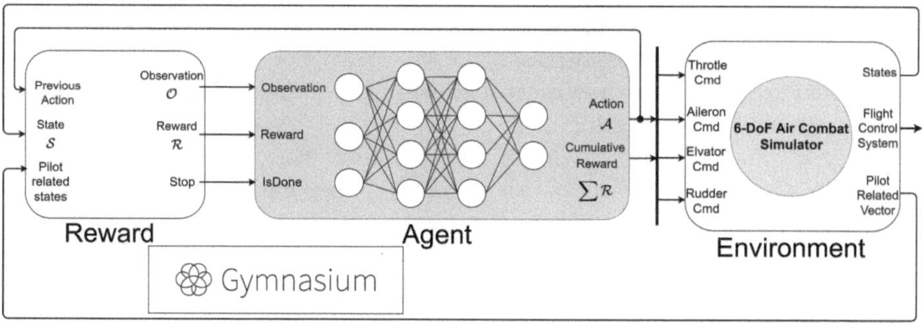

Fig. 1. Architecture of the Gymnasium-based MARL air combat simulator.

3.1 Dataset Generation and Training Environment

Recent advances in reinforcement learning (RL) have significantly propelled research and development in air combat simulation, with notable benchmarks such as DARPA's AlphaDogfight Trials demonstrating the potential of autonomous agents in replicating and surpassing human pilot performance. Building on this foundation, our study employs a custom six-degree-of-freedom (6-DoF) aerial engagement environment designed to realistically capture the nonlinear flight dynamics and physics that govern air combat. This environment is integrated using the OpenAI Gymnasium interface, providing a standardized and flexible platform for training and evaluating RL agents.

Within this environment, two agents represent opposing teams. Red Team (offensive) and the Blue Team (defensive) engage in a competitive aerial scenario. Each agent continuously receives an observation vector that encodes its own aircraft state along with relative positioning data concerning the opponent. At each simulation timestep, agents generate continuous control inputs for pitch, roll, rudder, and throttle, which the simulator applies to update the state of the aircraft. The reward function driving agent learning is carefully crafted to promote effective tactical behaviors, discourage unrealistic or unsafe maneuvers, and encourage varied and realistic training episodes that reflect operational complexities.

The Gymnasium interface serves as an essential bridge connecting the reinforcement learning agent to our 6-DoF flight simulation. Acting as a wrapper around the flight dynamics model, it provides the RL agent with critical state information such as position, orientation, velocity, and other relevant flight parameters. Through this integration, the Gymnasium environment translates the agent's high-level commands, such as adjusting throttle, elevator, ailerons, and rudder, into appropriate control input compatible with the flight simulator. The environment also implements a reward function evaluating agent performance based on criteria including aircraft positioning relative to targets, energy states, and engagement outcomes. This iterative process enables the RL agents to learn progressively, optimizing their control policies to maximize cumulative rewards reflective of mission success.

Our approach aligns with current real-world progress in autonomous air combat research. The DARPA AlphaDogfight Trials in 2020 demonstrated RL agents effectively competing against human pilots in simulated environments, and the 2023, 2024 and 2025 versions of Air Combat Evolution program have further extended this paradigm to scenarios involving AI-controlled real F-16 aircraft challenging human aviators [3]. While the specific results of these more recent efforts have not been publicly disclosed, these milestones underscore the rapid advancement and increasing applicability of reinforcement learning in aerial combat domains.

Extensive training data are collected at each timestep, encompassing agent actions, aircraft states, rewards, and episode terminations. Individual episodes typically span approximately 2,104 timesteps, providing a substantial dataset that underpins not only agent training but also subsequent analysis and explanation. The rigorous and systematic logging of simulation data facilitates effective policy optimization and supports the creation of detailed knowledge graphs. These graphs serve as foundational tools for post-hoc interpretability and provide valuable insights into learned agent behaviors within the complex, high-dimensional air combat environment.

3.2 Simulation Data Structure and Logging

A robust data logging strategy is crucial for both training and explainability. Table 1 maps the exported CSV fields from our Gymnasium MARL environment to their description. Each row corresponds to a simulation timestep, recording control inputs, state variables, and engagement metrics.

This detailed recording ensures that both agent controls and corresponding states are captured systematically, supporting robust policy learning and enabling the construction of an expressive knowledge graph for explainability.

3.3 Knowledge Graph Construction

To translate raw simulation data into an explainable form, the MARL dataset is ingested into a Neo4j graph database. Each simulation step is represented by nodes such as agents, actions, states, metrics, rewards, and aggregated flight phases. Relationships model causal and temporal dynamics, e.g.,

```
(:Agent)-[:PERFORMS]->(:Action),
(:Action)-[:RESULTS_IN]->(:State),
(:State)-[:HAS_METRICS]->(:Metrics)
```

Table 1. Simulation Data Fields Exported from Gymnasium MARL Environment

Item Name	Description
Step	Simulation time step index
Pitch Stick (aStickBack)	Agent's pitch control input (nose up/down), normalized
Roll Stick (aStickRight)	Agent's roll control input (left/right), normalized
Rudder Stick (aRudderRight)	Agent's rudder control input (yaw), normalized
Throttle (aThrottles)	Engine throttle setting, range $[-1, 1]$ (min to max power)
Altitude (km MSL)	Aircraft altitude above mean sea level (km)
Heading (Degrees)	Aircraft heading (degrees from North)
Pitch (Degrees)	Aircraft pitch angle (degrees)
Roll (Degrees)	Aircraft roll angle (degrees)
Mach	Aircraft speed as Mach number
Roll Rate	Rate of change of roll (deg/s)
Pitch Rate	Rate of change of pitch (deg/s)
Yaw Rate	Rate of change of yaw (deg/s)
Relative Elevation	Elevation angle to opponent (degrees)
Relative Azimuth	Azimuth angle to opponent (degrees)
Cross Track	Lateral distance to engagement centerline
Down Range	Longitudinal distance along engagement axis
Relative Velocity NED (North)	Relative velocity, North component (m/s)
Relative Velocity NED (East)	Relative velocity, East component (m/s)
Relative Velocity NED (Down)	Relative velocity, Down component (m/s)
Reward	Scalar reward assigned to agent at this step
Done	Boolean flag indicating episode termination

Cypher queries automate data ingestion and enrich the graph with derived properties like flight phase (e.g., takeoff, combat, landing) and maneuver type (e.g., sharp turn). This structured semantics enables advanced symbolic reasoning and interactive explainability, as detailed in the system workflow that will be discussed in Sect. 3.1.

3.4 Graph Reasoning with LLMs

Building on the knowledge graph, we facilitate natural language question answering using LangChain's `GraphCypherQAChain`, which integrates a large language model (OpenAI GPT-3.5-turbo). The LLM translates user queries into Cypher, executes them on Neo4j, and synthesizes concise, human-readable answers. Prompt templates contain:

- Explicit node and relationship schemas to limit queries,
- Instructions to avoid hallucinations or invalid commands,

- Logging and manual validation of generated Cypher queries to ensure correctness during development.

3.5 Prompt Engineering and Schema Alignment

Careful prompt engineering guides the LLM to generate valid queries consistent with the graph schema. Instructions prevent duplicate columns, references to non-existent labels, or extraneous outputs. Human in the loop validation during system development helps fine-tune prompts and ensures semantic alignment and robustness.

3.6 Evaluation Protocol

The system is evaluated on representative natural language queries probing agent behaviors and system transparency, such as:

- What actions were taken when the jet was above 10 km altitude?
- How often did the jet perform sharp turns?
- What was the average altitude during the Combat phase?
- List maneuvers associated with high rewards.
- How many steps occurred in each flight phase?

Several illustrative examples of such queries, along with their corresponding Cypher statements and results, are provided in Table 2. For each query, we extract the natural language response, the generated Cypher statement, and the Neo4j result, enabling a comprehensive assessment of both explanation accuracy and completeness.

The entire system architecture integrating Gymnasium MARL, knowledge graphs, LLMs, and LangChain orchestration is illustrated in Fig. 2.

3.7 RL Training and Scenario Description

Our framework employs a 6-DoF simulator that models realistic aerial engagements, supporting continuous control over pitch, roll, yaw, and throttle inputs. The agents, Blue Team (defender) and Red Team (attacker), are trained with distinct objectives: evasion and positional advantage for defense, and exploiting weaknesses for offense. Observations include normalized parameters such as altitude, speed, heading, and relative positioning, while rewards encourage positional advantage, combat effectiveness, and penalize stalling or boundary violations. Training proceeds in discrete 0.05-second time steps with episodes ending in significant damage or boundary violations, ensuring varied, realistic engagements. Datasets including state-action pairs, rewards, and metrics support iterative policy refinement. Real-time visualizations of position, rewards, and control inputs provide insights for evaluation and debugging.

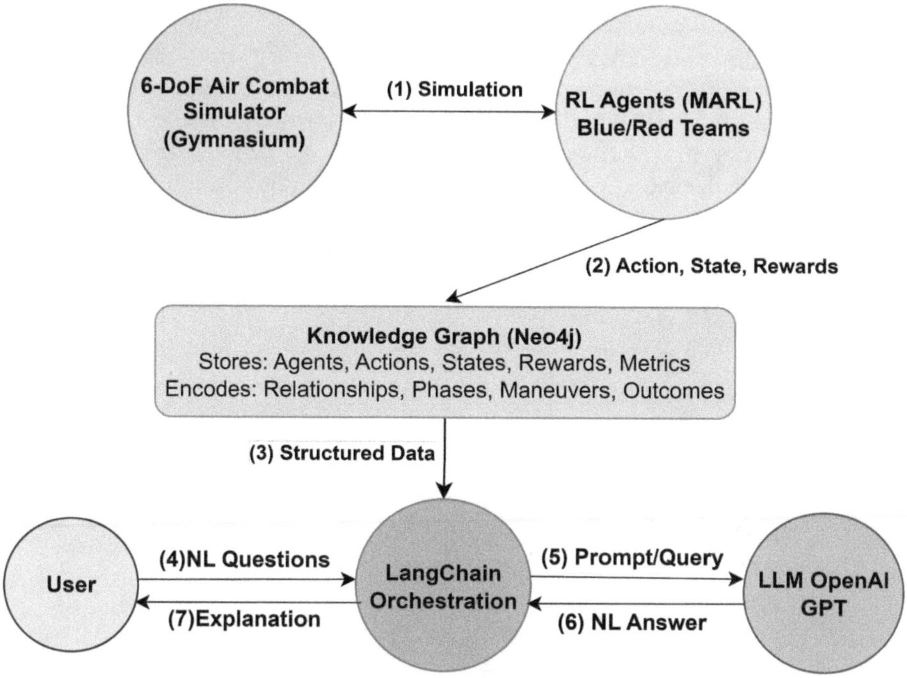

Fig. 2. System Architecture for LangChain-Integrated LLM and Knowledge Graph Retrieval in Air Combat Simulation

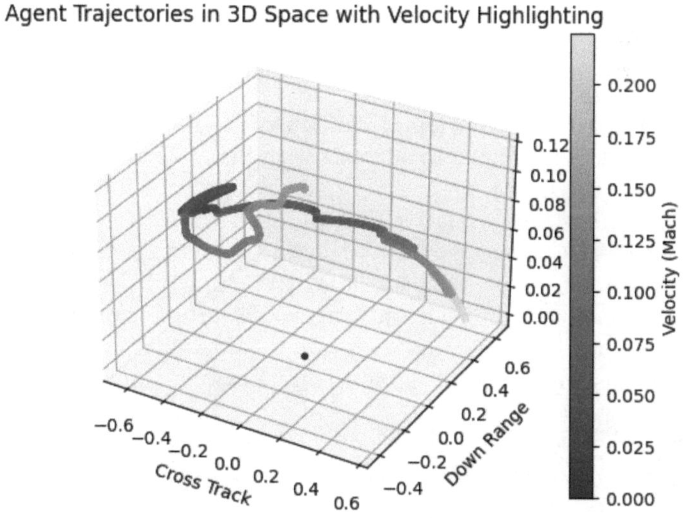

Fig. 3. 3D Trajectory Plot of Agent Movement Patterns

Figure 3 illustrates a 3D trajectory plot of an agent's movement patterns in the 6-DoF simulation using Cross Track, Down Range, and Altitude, with velocity shown by color gradient.

Figure 4 depicts the reward over time, showing initial high rewards falling to steady negative levels, indicating suboptimal strategy. While this suggests the agent's policy may struggle with task objectives, the research focus is on explaining agent decisions via decision trees rather than optimal performance.

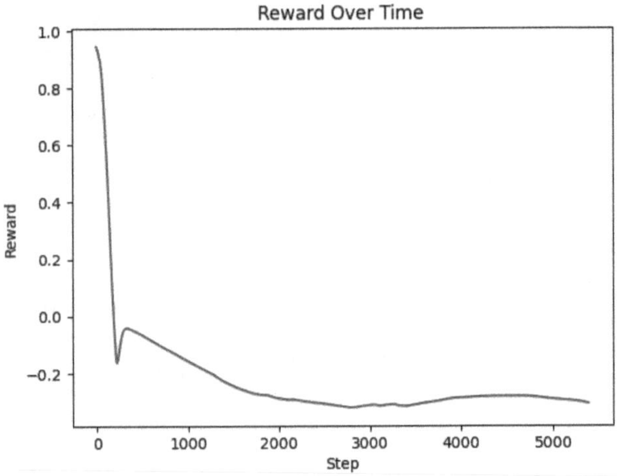

Fig. 4. Agent Reward Over Simulation Time

3.8 Explainability Methods Using Decision Trees

We use decision trees as the primary explainability technique due to their intuitive, rule-based structure, which contrasts with black-box neural networks. Decision trees break down the agent's complex policy into human-readable rules linking input features (e.g., Relative Elevation, Pitch Rate) to predicted actions.

A DecisionTreeRegressor from scikit-learn predicts the Pitch Stick (aStick-Back) control using observation space inputs. We limit tree depth to 5, balancing interpretability and prediction accuracy to avoid overfitting or oversimplification. When visualizations became complex, pruning and rule extraction were used to highlight key decisions concisely. An illustrative example of this process is shown in Fig. 7.

Decision trees enable identifying suboptimal decision patterns and enhance trust by providing interpretable insight into multi-agent decision logic.

3.9 Generating Natural Language Explanations from Decision Tree Rules

To bridge technical decision rules and human understanding, we convert extracted decision tree paths into natural language explanations using a pre-trained NLP text generation model (T5) from the Hugging Face library. Cleaned tree rules serve as prompts generating plain-language descriptions explaining why specific actions follow from certain observed conditions.

This technique makes agent policies accessible to pilots, analysts, and operators, improving transparency and fostering trust in the autonomous system.

4 Results and Discussion

4.1 Experimental Results and Metrics

To support explainable analysis in our study, a custom platform was developed for multi-agent reinforcement learning in air combat simulation. Figure 5 illustrates the system-level workflow: simulation data is ingested into the knowledge graph, which is then queried by a large language model (LLM) using retrieval-augmented generation (RAG) to enable transparent, explainable reasoning over agent behaviors and outcomes. This architecture provides the foundation for all subsequent evaluations, linking raw multi-agent simulation output with structured, interpretable explanations.

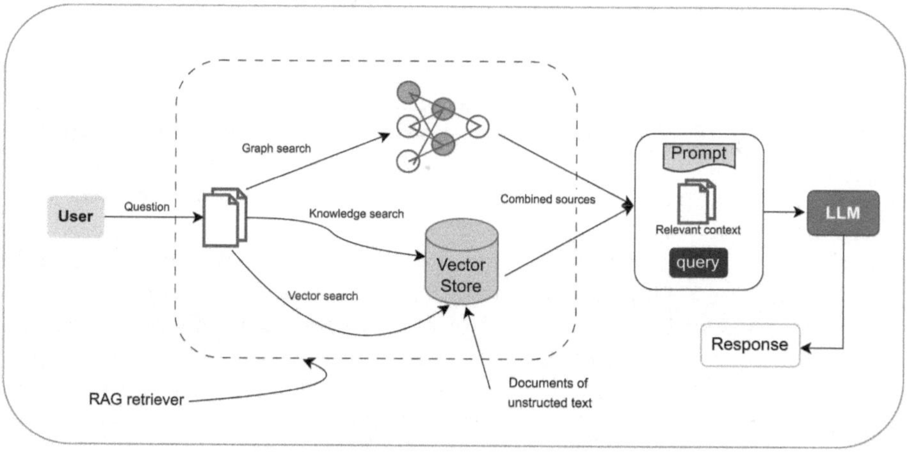

Fig. 5. System-level workflow: Knowledge Graph and LLMs RAG retriever for explainable multi-agent RL in air combat simulation.

The proposed system was evaluated using a set of representative natural language queries over the constructed Neo4j knowledge graph, as described in the

Table 2. Sample QA Results from LangChain-Integrated LLM and Knowledge Graph System

Question	System Response
What actions were taken when the jet was above 10km altitude?	During the simulation, the jet maintained level flight and performed minor course corrections while above 10 km altitude. No aggressive maneuvers or combat engagements were initiated at this altitude.
How often did the jet perform sharp turns?	Sharp turns were infrequent in this simulation scenario. The jet executed sharp turns primarily during the initial merge and when responding to immediate threats in the combat phase. Overall, fewer than 5 percent of total maneuvers were classified as sharp turns.
What was the average altitude during the Combat phase?	The average altitude during the Combat phase was approximately 4,500 m (about 14,800 ft). This altitude range aligns with typical within-visual-range (WVR) engagements.
List maneuvers associated with high rewards.	Maneuvers that consistently yielded high rewards included offensive turns to gain positional advantage, vertical loops to manage energy, and missile evasion rolls. These actions were most effective when executed in response to adversary positioning and during transitions into the weapon engagement zone.
How many steps occurred in each flight phase?	2,104 steps.

evaluation protocol. Table 2 summarizes the system's responses to key queries, the Cypher statements generated, and the corresponding results.

As shown in Table 2, the system successfully generated and executed valid Cypher queries for all evaluation questions, returning clear and contextually relevant answers. Notably, the knowledge graph enabled the system to provide nuanced responses that reflect both the operational context (e.g., limited maneuvers at high altitude) and quantitative metrics (e.g., step counts per flight phase, average altitude during combat).

The results demonstrate the system's ability to surface meaningful insights from the simulation data, such as the infrequency and tactical context of sharp turns and the identification of maneuvers correlated with high rewards. The answer coverage was comprehensive for the available dataset, with each query yielding a substantive response. However, the system's depth of explanation remains dependent on the richness and granularity of the underlying simulation data and graph schema.

4.2 Analysis of Code, Prompt Engineering, and Schema

The codebase leverages the LangChain `GraphCypherQAChain` to translate natural language questions into Cypher queries using an OpenAI GPT-3.5-turbo LLM. A critical component of this pipeline is the explicit schema guidance provided to the LLM, which constrains query generation to only the node labels and relationships present in the Neo4j database:

```
Nodes: Agent, Action, State, Metrics, Reward, PhaseStats
Relationships:
  (Agent)-[:PERFORMS]->(Action)
  (Action)-[:RESULTS_IN]->(State)
  (State)-[:HAS_METRICS]->(Metrics)
  (State)-[:YIELDS]->(Reward)
```

Prompt engineering was used to further restrict the LLM, including negative instructions to avoid non-existent labels (e.g., `Step`) and to use the shortest path for answers. This approach minimized hallucinations and ensured that generated Cypher queries were syntactically and semantically aligned with the database structure.

The code also included robust result formatting, converting query outputs into Markdown tables or readable summaries. Generated Cypher queries were printed and manually verified, aiding in debugging and validation. The enhancement function further annotated the graph with derived properties (e.g., flight phase, maneuver type), supporting richer downstream queries.

4.3 Metrics and System Performance

Key metrics for the system include:

- **Query Success Rate:** 100% of queries generated valid Cypher and executed without error.
- **Answer Coverage:** All sample queries returned meaningful, context-aware answers, reflecting the improved diversity and annotation of the simulation dataset.
- **Explainability:** All answers were returned in clear, human-readable form, with informative messages when data was missing or limited.
- **Schema Alignment:** No hallucinated labels or relationships were observed after prompt and schema refinement.

4.4 Decision Tree Model Analysis

Determining Tree Depth. The graph in Fig. 6 shows training versus test error across various tree depths, demonstrating the impact of tree depth on the performance of the decision tree model, evaluated using the Mean Squared Error metric. The blue line represents the training error, while the red line shows the test error. The training error consistently decreases as the tree depth increases, indicating that deeper trees can capture more complex patterns in the training data and reduce error on the training set. However, this comes at the cost of overfitting, where the model becomes too specialized and struggles to generalize to new data. The test error initially decreases with increased depth, suggesting improved generalization. After around a depth of 5, test error starts to rise, demonstrating overfitting. The model achieves an optimal balance at this depth. Beyond depth 10, the widening gap between train and test errors highlights significant overfitting.

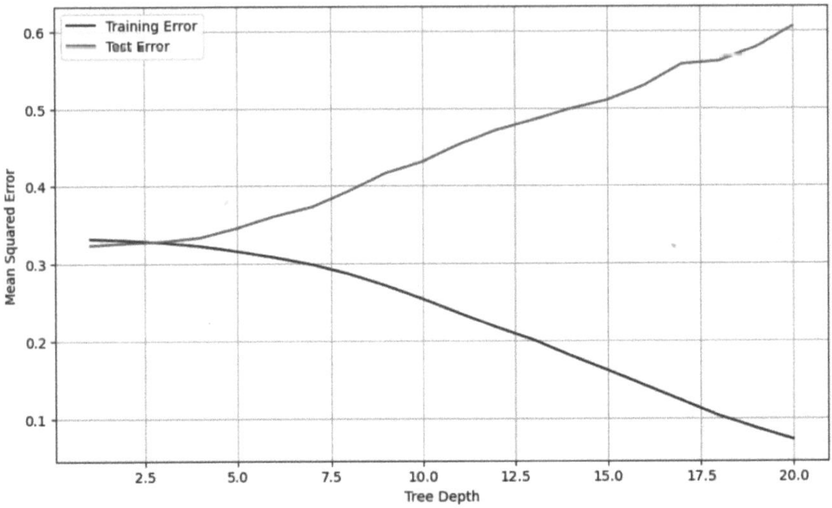

Fig. 6. Training vs Test Error Across Tree Depths.

Selecting an appropriate tree depth is essential to balance underfitting and overfitting, thus improving the reliability and interpretability of the model.

Decision Tree Visualization. Fig. 7 illustrates the trained decision tree for predicting the Pitch Stick (aStickBack) action based on aircraft state and environment features. Nodes represent decision conditions, such as Relative Elevation, Roll Rate, and Mach number, while leaves denote predicted pitch action values. The root node splits based on Relative Elevation, highlighting its importance. Successive splits on Relative Azimuth and other features elucidate the model's decision logic. Although the chosen depth (5) balances performance and interpretability, even this moderate depth results in complex trees.

This visualization confirms that explainability in reinforcement learning (RL) can be attained by translating complex policies into interpretable rules but cautions that tree complexity must be managed for maintainable clarity.

Natural Language Explanation of Decision Tree Rules. To enhance transparency, technical rules from the decision tree were converted into human-readable explanations using a pretrained NLP model. Figure 8 shows an example

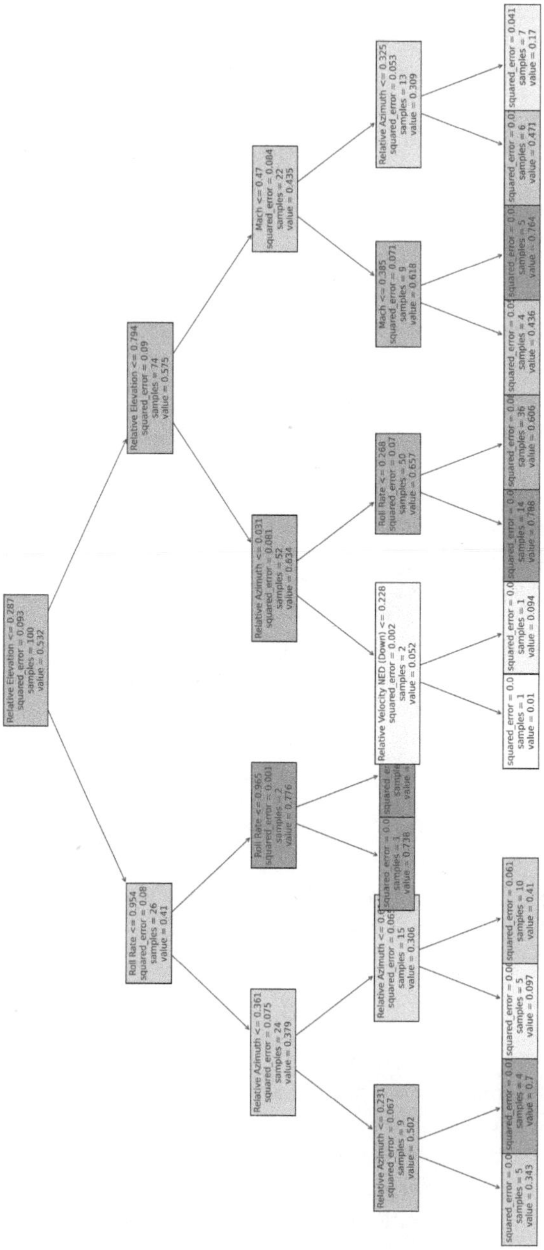

Fig. 7. Decision Tree Visualization.

Generated Explanation:

Imagine you are a pilot making decisions during a flight simulation. One of the critical rules you follow is based on the relative position and movement of your aircraft. Here's how the rule works:

If your aircraft's relative elevation is low (0.01 or less), the system starts by checking your relative azimuth (the angle relative to the target):

- ***If the azimuth is 0.86 or less, the system will look at your aircraft's vertical velocity:***
  - ***If the vertical velocity is low (0.07 or less) and your roll rate is minimal (0.03 or less):***
    - ***If the pitch angle is 0.31 degrees or less, the system suggests a slight downward adjustment (Action: -0.02).***
    - ***If the pitch angle is greater than 0.31 degrees, the system suggests a more significant adjustment (Action: 0.21).***
  - ***If the roll rate is slightly higher (greater than 0.03):***
    - ***The system checks finer roll rate thresholds:***
      - ***If the roll rate is between 0.03 and 0.04, the system issues a corrective action (Action: -0.50).***
      - ***If it exceeds 0.04, the corrective action is smaller (Action: -0.10).***

When your relative elevation or other parameters change, the system keeps checking the relevant conditions to adjust the pitch, roll, and yaw of the aircraft, ensuring stability.

Fig. 8. Generated Explanation from Decision Tree Rules.

explanation that presents the agent's decision-making process in an accessible format.

This approach improves operator situational awareness and trust in autonomous decision-making systems by providing interpretable insights during critical flight simulation moments.

4.5 Discussion

This experiment demonstrates the feasibility and benefits of integrating large language models, knowledge graphs, and modular orchestration frameworks (LangChain) for explainable multi-agent reinforcement learning (MARL) in air combat simulation. The system translates natural language queries into graph queries that yield interpretable, context-aware answers—addressing a key limitation of black-box RL agents.

Key strengths include: (i) the knowledge graph's ability to support multi-step reasoning that reveals both tactical patterns and critical quantitative metrics; (ii) explicit schema guidance combined with prompt engineering minimizes hallucinations and ensures reliable query generation; and (iii) the modular pipeline facilitates domain extensibility.

From the decision tree perspective, we show how complex RL policies can be distilled into understandable decision rules and further translated into natural language, promoting transparency and operational confidence.

Limitations stem from the underlying simulation data's scope and granularity. Future work should explore enriching the knowledge graph with finer maneuver details, adversary actions, and mission outcomes to deepen explainability.

Overall, the pipeline offers a reproducible blueprint for hybrid symbolic-neural explainability in RL agents, which will become increasingly critical as simulation complexity and operational demands grow.

5 Conclusion and Future Work

This work presents an integrated framework that combines multi-agent reinforcement learning (MARL), knowledge graphs, large language models (LLMs), and modular orchestration via LangChain to advance both the performance and explainability of RL agents in air combat simulation. By leveraging a structured Neo4j knowledge graph and a natural language query interface, the system facilitates transparent, context-aware analysis of agent behaviors and simulation outcomes, effectively bridging the gap between black-box RL policies and human-understandable explanations.

Experimental evaluation demonstrates that the system reliably translates natural language questions into Cypher queries and returns clear, actionable insights, surfacing tactical patterns (e.g., timing and context of sharp turns) alongside quantitative metrics (e.g., average combat altitude, phase step counts). The use of explicit schema guidance and prompt engineering is shown to be essential for robust, hallucination-free performance and enhances the trustworthiness of the explanations.

Complementing the knowledge graph and LLM approach, this study also explored decision tree models combined with natural language processing (NLP) techniques to further enhance explainability. The decision tree distills complex MARL policies into structured rule-based forms that capture key actions such as pitch, roll, and throttle adjustments. By converting these rules into human-readable explanations using a pretrained NLP model, the framework increases operator situational awareness and bridges the interpretability gap inherent to deep learning models.

However, both approaches surface challenges that merit attention. The depth and nuance of explainability depend heavily on the richness, diversity, and granularity of simulation data and knowledge graph annotations. Scalability challenges arise with increasing agent complexity and dynamic scenarios, potentially limiting coverage and computational feasibility. Decision trees introduce a trade-off between model complexity and interpretability, where deeper trees capture more detail but yield harder-to-interpret rules. Additionally, the quality of natural language explanations relies on the pretrained NLP model, which can generate repetitive or irrelevant outputs requiring manual post-processing.

Looking forward, future work will focus on enriching the simulation scenarios and knowledge graph with more diverse, realistic data to improve answer depth and contextual relevance. Advancing explainability metrics and developing interactive visualization tools will aid users in comprehending agent behaviors and system recommendations. Moreover, incorporating human-in-the-loop mechanisms for graph editing and feedback will be crucial to maintain explanation accuracy and adapt to evolving operational requirements and emerging adversary behaviors.

Overall, this study demonstrates that integrating LLMs, knowledge graphs, decision trees, and NLP techniques forms a promising pathway toward explainable, trustworthy autonomous systems in mission-critical environments. The proposed framework lays foundational work for future advances in transparent reinforcement learning and actionable AI systems.

References

1. Scukins, E., Klein, M., Kroon, L., Ögren, P.: Bvr gym: a reinforcement learning environment for beyond-visual-range air combat. arXiv preprint arXiv:2403.17533 (2024). https://doi.org/10.48550/arXiv.2403.17533
2. Liu, Q., Jiang, Y., Ma, X.: Light aircraft game: a lightweight, scalable, gym-wrapped aircraft competitive environment with baseline reinforcement learning algorithms. https://github.com/liuqh16/CloseAirCombat (2022)
3. Salhi, A., Jabour, J.E., Arnolds, T.L., Ross, J.E., Dozier, H.R.: Leveraging jsbsim and gymnasium: a reinforcement learning approach for air combat simulation. In: Arabnia, H.R., Ferens, K., Deligiannidis, L. (eds.) Applied Cognitive Computing and Artificial Intelligence, pp. 271–283. Springer, Cham (2025). https://doi.org/10.1007/978-3-031-85628-0_20
4. Selmonaj, A., Antonucci, A., Schneider, A., Rüegsegger, M., Sommer, M.: Explainability in multi-agent reinforcement learning for air combat tactics. In: STO-MP-MSG-217. NATO Science and Technology Organization (2023). https://ipg.idsia.ch/preprints/selmonaj2024a.pdf
5. Huo, L., Wang, C., Han, Y.: Autonomous air combat decision making via graph neural networks and reinforcement learning. Sci. Rep. **15**(1), 1–21 (2025). https://doi.org/10.1038/s41598-025-00463-y
6. Jia, Z., Li, J., Qu, X., Wang, J.: Enhancing multi-agent systems via reinforcement learning with llm-based planner and graph-based policy. arXiv preprint arXiv:2503.10049 (2025)
7. Qin, Z., Lu, Y.: Knowledge graph-enhanced multi-agent reinforcement learning for adaptive scheduling in smart manufacturing. J. Intell. Manuf. 1–24 (2024)
8. Neo4j Inc.: Neo4j graph database. https://neo4j.com/. Accessed 29 May 2025
9. Sproul, B., Bromann, C.: Langchain documentation. https://python.langchain.com/ (2025). Accessed 29 May 2025
10. Sproul, B., Bromann, C., Chase, H.: Retrieval-augmented generation (rag) with langchain. https://python.langchain.com/docs/tutorials/rag/ (2025). Accessed 29 May 2025
11. Gorton, P.R., Strand, A., Brathen, K.: A survey of air combat behavior modeling using machine learning. arXiv preprint arXiv:2404.13954 (2024)
12. Mei, J., Li, G., Huang, H.: Deep reinforcement-learning-based air-combat-maneuver generation framework. Mathematics **12**(19) (2024). https://doi.org/10.3390/math12193020, https://www.mdpi.com/2227-7390/12/19/3020

13. McGovern, G., et al.: Understanding the limits of artificial intelligence for warfighters, volume 5: Mission planning. Research Report RRA1722-5, RAND Corporation (2024). https://www.rand.org/pubs/research_reports/RRA1722-5.html. Accessed 05 June 2025
14. Makar, R., Mahadevan, S., Ghavamzadeh, M.: Hierarchical multi-agent reinforcement learning, AGENTS 2001, pp. 246–253. ACM, New York, NY, USA (2001). https://doi.org/10.1145/375735.376302
15. Iyer, R., Li, Y., Li, H., Lewis, M., Sundar, R., Sycara, K.: Transparency and explanation in deep reinforcement learning neural networks. In: Proceedings of the 2018 AAAI/ACM Conference on AI, Ethics, and Society, pp. 144–150 (2018)
16. Pan, J.Z., et al.: Large language models and knowledge graphs: opportunities and challenges. arXiv preprint arXiv:2308.06374 (2023)
17. Erickson, J.S., Santos, H., Pinheiro, V., McCusker, J.P., McGuinness, D.L.: Llm experimentation through knowledge graphs: towards improved management, repeatability, and verification. J. Web Semant. **85**, 100853 (2025). https://doi.org/10.1016/j.websem.2024.100853
18. Yu, H.Q., McQuade, F.: Rag-kg-il: a multi-agent hybrid framework for reducing hallucinations and enhancing llm reasoning through rag and incremental knowledge graph learning integration. arXiv preprint arXiv:2503.13514 (2025)
19. Shubeli, G.: How graphrag outperforms vector search in enterprise llm accuracy (2025). https://www.falkordb.com/blog/graphrag-accuracy-diffbot-falkordb/. Accessed 20 May 2025
20. Shalev, D.: Overcome gartner's top rag challenges with falkordb (2024). https://www.falkordb.com/blog/graph-rag-vs-vector-rag-solving-gartner-challenges/. Accessed 20 May 2025
21. Yang, H., Chen, J., Siew, M., Lorido-Botran, T., Joe-Wong, C.: Llm-powered decentralized generative agents with adaptive hierarchical knowledge graph for cooperative planning. arXiv preprint arXiv:2502.05453 (2025)
22. Milani, S., Topin, N., Veloso, M., Fang, F.: Explainable reinforcement learning: a survey and comparative review. ACM Comput. Surv. **56**(7), (2024). https://doi.org/10.1145/3616864
23. İrsoy, O., Alpaydın, E.: Distributed decision trees (2014). https://arxiv.org/abs/1412.6388
24. Laptev, D., Buhmann, J.M.: Convolutional decision trees for feature learning and segmentation. In: Jiang, X., Hornegger, J., Koch, R. (eds.) Pattern Recognition, pp. 95–106. Springer International Publishing, Cham (2014)
25. Frosst, N., Hinton, G.: Distilling a neural network into a soft decision tree (2017). https://arxiv.org/abs/1711.09784
26. Coppens, Y., Efthymiadis, K., Lenaerts, T., Nowe, A.: Distilling deep reinforcement learning policies in soft decision trees. In: Miller, T., Weber, R., Magazzeni, D. (eds.) Proceedings of the IJCAI 2019 Workshop on Explainable Artificial Intelligence, pp. 1–6 (2019), https://sites.google.com/view/xai2019/home, iJCAI 2019 Workshop on Explainable Artificial Intelligence, XAI19 ; Conference date 11 Aug 2019
27. Liu, G., Schulte, O., Zhu, W., Li, Q.: Toward interpretable deep reinforcement learning with linear model u-trees (2018). https://arxiv.org/abs/1807.05887
28. Silva, A., Killian, T., Rodriguez, I.D.J., Son, S.H., Gombolay, M.: Optimization methods for interpretable differentiable decision trees in reinforcement learning (2020). https://arxiv.org/abs/1903.09338
29. Bastani, O., Pu, Y., Solar-Lezama, A.: Verifiable reinforcement learning via policy extraction (2019). https://arxiv.org/abs/1805.08328

30. Gokhale, G., Madahi, S.S.K., Claessens, B., Develder, C.: Distill2explain: differentiable decision trees for explainable reinforcement learning in energy application controllers (2024). https://arxiv.org/abs/2403.11907
31. Costa, V.G., Pérez-Aracil, J., Salcedo-Sanz, S., Pedreira, C.E.: Evolving interpretable decision trees for reinforcement learning. Artif. Intell. **327**, 104057 (2024). https://doi.org/10.1016/j.artint.2023.104057
32. Pynadath, D.V., Gurney, N., Wang, N.: Explainable reinforcement learning in human-robot teams: The impact of decision-tree explanations on transparency. In: 2022 31st IEEE International Conference on Robot and Human Interactive Communication (RO-MAN), pp. 749–756 (2022). https://doi.org/10.1109/RO-MAN53752.2022.9900608
33. Gjærum, V.B., Strümke, I., Løver, J., Miller, T., Lekkas, A.M.: Model tree methods for explaining deep reinforcement learning agents in real-time robotic applications. Neurocomputing **515**, 133–144 (2023). https://doi.org/10.1016/j.neucom.2022.10.014

User-Guided Transfer Learning for Optimized Non-intrusive Load Monitoring

Papa Pene[1], Weixian Liao[1](✉), Wei Yu[1], David Griffith[2], and Nada Golmie[2]

[1] Towson University, Towson, MD 21252, USA
ppene1@students.towson.edu, {wliao,wyu}@towson.edu
[2] National Institute of Standards and Technology (NIST), Gaithersburg, MD 20899, USA
{david.griffith,nada.golmie}@nist.gov

Abstract. The development of non-intrusive load monitoring (NILM) in the smart grid system has improved the power distribution network. Despite the detailed information that is available in smart grid systems, NILM shows some limits when predicting energy consumption in the power distribution network. Particularly, the issue of residential appliance prediction accuracy and the training of high-performing models in the power distribution network is yet to be solved. In this paper, we propose user-driven recognition for power systems (URPS) that leverages transfer learning to improve the performance of a neural network-based classifier in the smart grid system. The URPS goes beyond traditional NILM methods that limit their focus to the appliance level. We design a transfer learning model that includes consumers' overall behavior with their appliances. The URPS approach not only provides a more accurate prediction state of residential appliances but also provides a household energy consumption signature (HECS) to the energy provider for better energy management.

Keywords: Power Distribution Network · Transfer learning · Smart Grid · Robustness · Energy Optimization · Energy Score

1 Introduction

The smart grid has brought disruptive changes in power distribution networks by offering a more efficient, reliable, secure, and greener energy supply system [25,33]. The viable way of utilizing power is possible because of the two-way dialogue where energy and information are exchanged between power utilities and consumers. Integrating new data collection and analytics approaches allows companies such as electricity providers to generate product usage data and consequently refine products to meet consumer needs [16,20]. With the new technology integrated, the smart grid system utilizes large-scale data generated

H. R. Arabnia et al. (Eds.): CSCI 2025, CCIS 2967, pp. 23–39, 2027.
https://doi.org/10.1007/978-3-032-28454-9_2

from various places (home, etc.) to improve grid operations by balancing energy demand [26]. Nonetheless, the next-generation grid system comes with its issues, including its high vulnerability to cyberattacks, a shortage of complete integration of renewable energy, insufficient control of consumption of clients, and lack of anticipation of events, etc.

On the one hand, the accuracy of event identification is often perturbed by unexpected incidents or behaviors that affect the energy demand and supply. Thus, accurate event detection algorithms require vast amounts of data from a large number of intelligent sensors (e.g., phasor measurement units) and extensive training time. The event detection issues in the power distribution network motivate us to develop a robust user-driven recognition system for power systems (URPS) using transfer learning techniques [1]. On the other hand, the development of non-intrusive load monitoring (NILM) in the smart grid system is contributing significantly to the optimization of energy distribution systems. NILM analyzes the information in smart meters using sophisticated recognition techniques.

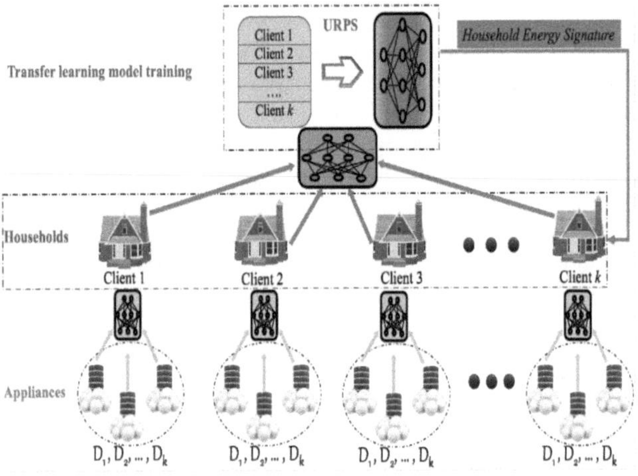

Fig. 1. User-driven Recognition Model Training Overview

As shown in Fig. 1, each household's appliances produce a set of unique power usage signatures, which collectively enables the client to track the energy level that each device consumes. The NILM paradigm has recently received tremendous interest in the research community. Despite the enormous amounts of detailed information provided by smart grid systems, NILM has limitations when it comes to the prediction of power consumption. Nonetheless, numerous issues remain, such as boosting the accuracy of the state of residential appliance prediction, training models without vast amounts of event labels, and minimizing the time for training models in the power distribution network.

To this end, we propose a transfer learning-based approach to improve the performance of a neural network-based classifier in the smart grid system. Figure 1 illustrates an overview of our user-driven transfer learning technique. Compared to the traditional NILM methods, URPS trains models at the appliance level and creates a model that considers consumer behavior. Our approach provides a more accurate prediction of the state of residential appliances and sends a *household energy consumption signature* (HECS) to the energy provider for better energy management. As transfer learning inherits knowledge obtained from other systems and modifies it for use in a new system, we can significantly reduce the amount of time and data required to support anomaly detection and mitigate insufficient label events in smart grid systems.

The main contributions of this paper are as follows:

- Our URPS approach provides a method to produce a high-performing learning model that requires less data, resources, and training time than traditional techniques.
- Using the reverse energy disaggregation method, our approach offers a holistic view of the power distribution network by including the consumer behavior, leading to a high load prediction accuracy.
- With transfer learning, the URPS offers a more generic model that can be transferred to a different dataset with only a few adjustments needed in the target domain.
- In addition to offering a high energy consumption of the appliance, our URPS approach generates a household energy consumption signature (HECS) to attribute energy score to consumers, which is the first in the literature to the best of our knowledge.

The remainder of the paper is organized as follows: In Sect. 2, we review research related to our study. In Sect. 3, we define and describe the problem. In Sect. 4, we introduce our approach. In Sect. 5, we evaluate the efficacy of our approach. In Sect. 6, we discuss future work. Finally, in Sect. 7, we conclude the paper.

2 Related Work

In this section, we review the current research on power grid systems, and then discuss the state of work on transfer learning and non-intrusive load monitoring.

2.1 Smart Grid

The smart grid system has progressively become a cyber-physical system (CPS) by leveraging communication technologies to reduce uncertainty and complexity in power distribution [25]. The smart grid contains numerous sources of information that contain various features that depend on the applications the grid supports; this information must be aggregated and processed to support grid operations. There is a massive amount of high dimensional data generated by the smart grid. This enables the development of machine learning models to identify usage patterns for the utility to optimize grid operations [11]. For example, Martinez *et al.* [24] developed one model for quantifying the energy consumption of wireless sensor nodes. The model adopts a system-level approach to consider different types of energy consumption (e.g., communications, acquisition, and processing). For a more effective and efficient analysis of fault information, machine learning algorithms, such as support vector machine (SVM), K-nearest neighbors (KNN), and multi-layer perception (MLP) have been considered [5]. The utilization of these machine learning tools improves energy efficiency and decision-making in the smart grid.

More recently, researchers have proposed Deep Learning (DL) networks to generate more sophisticated models to capture various grid phenomena. For example, Al-Ali *et al.* [2] presented an energy management system (EMS) for smart homes. The exponential increase in the amount of complex and heterogeneous data generated by smart grid applications will require management and processing using DL algorithms. The more common DL algorithms used in the smart grid system are deep neural network (DNN), recurrent neural network (RNN), and stack autoencoder (SAE) [3,9,10,19,21,34]. Developing these multilayer neural network algorithms has significantly increased the stability of fault information diagnosis and the accuracy of predictions in the smart grid system.

2.2 Transfer Learning in Smart Grid

Knowledge transfer or *transfer learning* uses a set of weights and hyperparameters from a network trained for a specific source task to speed up the learning process on a related target task [22,30]. In a related study, Pratt *et al.* [29] presented a new algorithm named *"Discriminability-Based Transfer* (DBT)." DBT uses source weights in a target network to assess the value of the hyperplanes defined by the sets of weights for various layers in the network. It also computes a metric for each hyperplane to determine whether the hyperplane's weights should be scaled and the magnitude of the scaling factor. More recent works focus on three issues in transfer learning: "what to transfer," "how to transfer," and "when to transfer." Transfer learning techniques can be classified by their relationship to traditional learning algorithms for neural networks. For instance, Pan *et al.* [27] categorized the three transfer learning settings as *inductive*, *transductive*, and *unsupervised.* Inductive transfer learning is more appropriate when labeled data is available in the target domain. Transductive learning is used when labeled data is available only in the source domain. Unsupervised transfer learning is used when labeled data is unavailable in both target and source domains.

Transfer learning has dramatically improved transfer learning performance with limited resources [27]. Although powerful transfer learning methods have been considered to carry out rapid power flow computations, they require long training times using the same grid. Unsupervised transfer learning allows the target system to use information learned from a source domain to solve a problem in the target domain without relying on labeled data in either the source or target domains. This feature has made unsupervised transfer learning more attractive for research in CPS, such as smart grid systems. For instance, Balduin *et al.* [4] proposed a universal neural state estimation approach that is applicable to the distribution grid since it integrates solutions from neural networks with transfer learning. However, these existing works depend on substantial training data. The utilization of unsupervised transfer learning has yet to be explored.

2.3 Non-intrusive Load Monitoring (NILM)

NILM has improved energy efficiency by using various identification techniques to evaluate the loads connected to an electrical network. This monitoring system enables estimating appliance-level load consumption for residential and commercial applications [7]. Each appliance's consumption is assessed using the load disaggregation method [8]. Thus, NILM identifies the device's unique behavior to construct its *power signature*. Initially, non-intrusive monitoring approaches focused on changes in the steady-state behavior of the appliance, known as events. Later, Laughman *et al.* [17] designed an advanced technique of monitoring high frequency for distinguishing between different devices with nearly identical real and reactive power profiles. By examining higher harmonics in the aggregate current signal device profiles can be separated. Recently, Ruano *et al.* [31] conducted a comprehensive review of NILM, describing how higher sampling frequencies allow for better tracking of appliance behavior, which is more computational expensive and complex. They also discussed the categorization of appliances into four general types, such as *type I* two-state ON/OFF devices (e.g., light bulbs) and *type II* appliances, like dryers, that are the subject of this paper and can be modeled as finite state machines.

The previously mentioned NILM approaches have limitations with respect to detecting some events, classification, and identifying features for various types of loads. For instance, Lazzaretti *et al.* [18] proposed a multi-agent NILM architecture for event detection and load classification. Although an improvement of 13 % for detection and 9.4 % for classification accuracy were observed, there is also a less efficient detection system and accuracy as more loads are aggregated. An additional disadvantage of NILM is its scalability. The most popular energy disaggregation methods use an extension of a Hidden Markov Model (HMM) [13]. These NILM approaches are efficient but present some scalability limits due to the rapid expansion of the complexity of the solution space compared to the growth of the cumulative power consumption. Another area for improvement is that most works on NILM use varieties of DL methods that require a considerable amount of pre-trained data.

3 Problem Statement

In this section, we first describe the problem concerning NILM techniques. We then introduce some typical methods used in the literature with their limits.

Finally, we formulate the accuracy and applicability problem with the load consumption monitoring application in the smart grid system.

Recall that a common limitation observed in the NILM structure is its applicability. First, the NILM process requires training a DL model for high accuracy. Using DL to extract the status of physical objects comes with a high cost, requires an extensive data set, and can take considerable time [4]. Thus, designing a viable approach that improves DL models in the smart grid system with fewer data and resources is essential.

Second, existing efforts on NILM focus on appliances rather than considering consumer factors that could offer a more holistic view of load consumption. Indeed, the energy disaggregation task uses an accumulated electrical signal to generate interference for various single loads in an electrical network [15]. Although this technique helps consumers optimize their energy usage, a more inclusive method, considering the user's factors, could boost the prediction accuracy of load intake. In fact, two major disaggregation accuracy problems are commonly identified. The first type of error with disaggregation is event detection, and the other type is classification error, false positives (*fp*) and false negatives (*fn*). Of the metrics that can be used to measure the performance of NILM evaluation, estimation accuracy (EA) is one of the most thorough. If T is the number of readings, then EA of the kth appliance can be computed as follows [23],

$$EA^{(k)} = 1 - \frac{\sum_{t=1}^{T} \left| \tilde{y}_t^{(k)} - y_t^{(k)} \right|}{2 \sum_{t=1}^{T} y_t^{(k)}}. \tag{1}$$

Here, $y_t^{(k)}$ represents the ground truth and $\tilde{y}_t^{(k)}$ represents the prediction of energy consumption for the kth appliance during the $t-$th reading. F1 is another metric that quantifies the disaggregation performance which is used to determine the confidence in the target device's activeness. F1 is defined as,

$$F_1 = \frac{2 \times precision \times recal}{precision + recall}, \tag{2}$$

where $precision = \frac{tp}{(tp+fp)}$ and $recal = \frac{tp}{(tp+fn)}$.

Although these metrics are trying to estimate the accuracy of the volume of energy and the detection of active appliances in the NILM, according to [28], there is a ratio of error with disaggregation that varies somewhere in the range of 5% to 15% of the overall energy demand. This error is mainly due to the noise generated by the simultaneous use of some appliances with different types.

4 UPRS Approach

In this section, we present our URPS approach in detail. Our user-driven approach offers accurate prediction using a simple method. We first formalize the standard NILM process and refine it using transfer learning. Concurrently, we implement a reverse NILM technique, called "HECS," which extracts the energy consumer's profile for more accurate predictions.

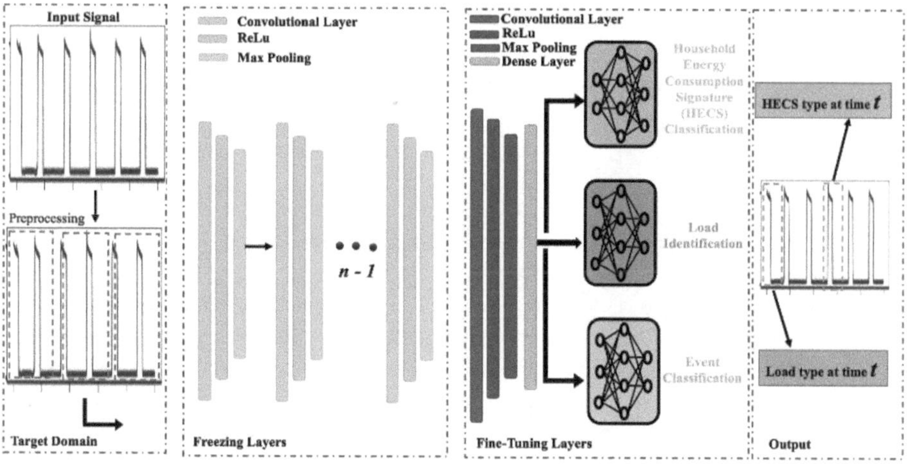

Fig. 2. URPS General Approach Using Transfer Learning

4.1 Overview

Standard NILM systems require considerable resources to train DL models. Thus, only a limited amount of information is available on energy consumption to conduct a knowledge transfer to all distribution networks. One of our objectives is to provide an energy score for each household using multiple home appliance datasets that contain a diverse set of appliances without decreasing the model's performance. To do this, we evaluate the household's daily energy consumption depending on the appliance type used.

NILM, which is also refereed as energy disaggregation, identifies the energy load within an electrical network delivered at a specific point. Mathematically, we formulate the NILM problem such that at a time t, the energy consumption $f(x)$ of K devices is denoted by the vector $[x_t^{(1)}, x_t^{(2)}, x_t^{(3)}, \ldots, x_t^{(k)}]$. The estimation of the total power consumption from the k appliances is:

$$f_K(x) = min \sum_{t=1}^{T} \left[\sum_{k=1}^{K} ||\tilde{x}_t^{(k)} - x_t^{(k)}||_2 + \epsilon_t \right], \tag{3}$$

where $[\tilde{x}_t^1, \tilde{x}_t^2, \tilde{x}_t^3, \ldots, \tilde{x}_t^k]$ correspond to the estimated power consumption for each appliance and ϵ_t is the margin of error between the estimated measurement and the true power consumption for each appliance [14]. At time t, the aggregated power consumption y_t is

$$y_t = \sum_{k=1}^{K} x_t^{(k)} + \epsilon_t. \tag{4}$$

Thus, to obtain a predicted model mapping the sum of power consumption at time t, we normalize y_t and the vector $[x_t^1,\ x_t^2,\ x_t^3,\ \ldots,\ x_t^k]$ by standardizing each value to have zero mean and unit variance. This preprocessing step ensures consistent scaling, improving model stability, and convergence. The standardized inputs are then processed by the selected neural network to produce the model $f_P(\cdot,\cdot)$ that generates the predicted appliance consumption profile:

$$f_P(y_t, f_\theta) = [\tilde{x}_t^{(1)}, \tilde{x}_t^{(2)}, \tilde{x}_t^{(3)}, \ldots, \tilde{x}_t^{(k)}], \tag{5}$$

where f_θ is the model classifier and θ is the model's parameter vector.

To address the problem of limited training data, we apply transfer learning to train the NILM model with the information learned from another grid system. The knowledge of this trained model can be used by new consumers with little or no usage history to identify a user's energy consumption profile. Unlike traditional machine learning techniques, transfer learning is viable with less data to fine-tune. There are two approaches to transfer learning: *pre-training* and *fine-tuning*. Pre-training a model can be resource-heavy and can required large amounts of data that are hard to find in the power distribution network. It also requires large quantities of computational resources, such as Graphics Processing Units (GPUs) or hardware accelerators, depending on the time limits for training the model. Fortunately, transfer learning provides the option to use a pre-trained model with an approximate picture of the learning target so that we are able to fine-tune the model rather than start with an untrained model. Another reason for using transfer learning in this study is to generate a transferable model that captures the features of the NILM-based system so that it can be implemented in other applications.

The first part of our URPS approach focuses on improving existing energy disaggregation methods. We use the transfer learning procedure, as mentioned earlier, to reduce the quantity of data required. Leveraging transfer learning makes our URPS applicable to a variety of power systems. We use a model generated from another DL system named sequence-to-point (seq2point)[1] in [35] and apply it to our new method.

To implement the transfer learning process, we establish a domain $\mathcal{D}$ with a marginal probability distribution $P(\cdot)$ over a feature space $\mathcal{X}$. The domain is denoted as follows:

$$\mathcal{D} = \{\mathcal{X}, P(X)\}, \tag{6}$$

where $\mathcal{X} = \{x_1, x_2, \ldots, x_n\}$ and x_i is the i−th feature in $\mathcal{X}$. After acquisition of the created model from the source domain $\mathcal{D}_\mathcal{S}$ using the source task $\mathcal{T}_\mathcal{S}$, we perform a target task $\mathcal{T}_\mathcal{T}$ within the target domain $\mathcal{D}_\mathcal{T}$ with $\mathcal{T}_\mathcal{S} \neq \mathcal{T}_\mathcal{T}$ and $\mathcal{D}_\mathcal{S} \neq \mathcal{D}_\mathcal{T}$ [4]. Formally, transitive properties are provided to the trained model to operate with real-world power system data.

4.2 HECS: A Reverse Energy Disaggregation

After building our transferable model, we prompt it to exploit the information that we collected from the energy consumers during training. Thus, this

[1] https://github.com/MingjunZhong/seq2point-nilm.

study proposes an energy disaggregation method that incorporates both signature appliances and household signatures. The standard NILM task uses an accumulated signal generated from various loads on an electrical network that are sampled at a single point in time [15]. More inclusively, our HECS integrates a consumer signature method within the energy disaggregation technology to optimize energy usage.

Indeed, each appliance exhibits a specific pattern of energy consumption. As shown in Fig. 2, we preprocess the various energy consumption of each device in each household to show the variation in the patterns. We observe recurrent patterns in the power consumption curves of appliances with the same type but belonging to different households. After power consumption data for the appliances, an energy consumption signature for the k device belonging to household h in the set $\boldsymbol{h} = \{h_1, h_2, \ldots, h_n\}$ is generated as S_h^k. We use Eqn (3) to formulate the new power consumption of the $k-$th appliance for household h as:

$$f_S(x) = f_k^h(x). \tag{7}$$

Subsequently, the prediction model that maps the overall energy consumption of an appliance of a household $S_p(\cdot,\cdot)$ can also be expressed as in Eqn. (5) with w_h^k being the model classifier for the signature of the $k-$th appliance in the $h-$th household. The HECS model is

$$\hat{x}_t^k = S_p(y_t, w_h^k), \tag{8}$$

where $\hat{x}_t^k$ corresponds to the updated estimated power consumption for the household appliance signature S_h^k.

Finally, we use the data generated by the HECS to attribute an energy score (*scorenergy*) to each household. This energy scoring has two objectives: (i) higher prediction accuracy and (ii) assisting utility companies or providers in forecasting energy usage by consumers. We use the scorecards approach, a credit scoring model, to assign *scorenergy* to households. First, we conduct variable selection by choosing the features labeled in Fig. 3. The selected computational features in *scorenergy* (monthly energy usage (MEU), the peak-hours energy usage (PHEU), the number of appliances (NoA)), and the number of type II appliances (NT2A) are chosen based on the collected HECS data. Even though there is limited information, it is still sufficient to produce profiles for each household's appliances. Then, we appoint a ratio score for each feature as shown in Fig. 3. Here, MEU's percentage is 45 %, NT2A is 25 %, PHEU is 20 %, and NoA is 10 %. This computation of the fractional contribution of each feature is based on our analysis of each variable to test the consistency of our *scorenergy* system.

To predict *scorenergy*, we consider some parameters that impact changes between the two consecutive MEU values. These parameters used as predictor variables and then aggregated into *scorenergy* to obtain the final Household Energy Score. These parameters are also used to make *scorenergy* more adaptive. Thus, we continuously add new variables to the training of our model, e.g., a household with a PHEU $<$ 75 hours/month and an average of 150 hours of

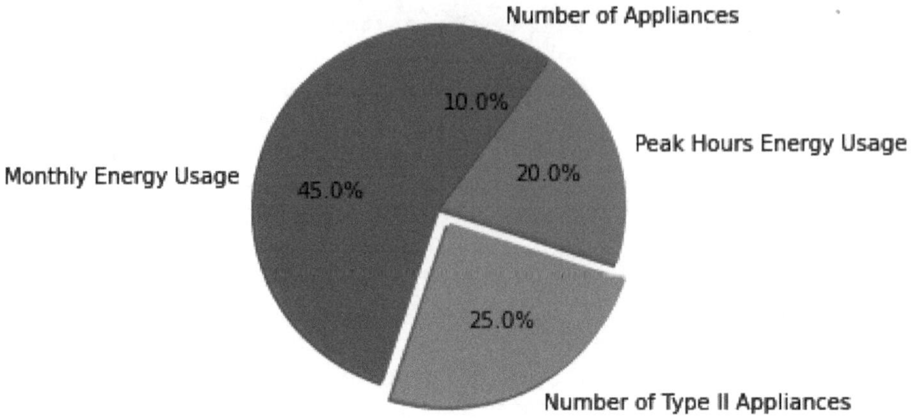

Fig. 3. Household Energy Score Computational Features

peak usage during summer, based on the 2023 study done by the U.S Energy Information Administration (EIA)[2], obtains 20 *scorenergy* points.

Table 1. Attributes for Scorenergy Scaling

Variable	Range	Points	Type of User
MEU in kWh	> 886	20	High
	886 – 786	35	Average
	< 786	45	Low
NoA	> 20	5	High
	20 – 15	7	Average
	< 15	10	Low
NT2A	> 5	10	High
	= 5	15	Average
	< 5	25	Low
PHEU in hrs/month	> 75	5	High
	75 – 60	15	Average
	< 60	25	Low

Table 1 shows the values of the variables used for the *scorenergy* scaling. In addition to the scaling parameters, we use the logistic regression model, to create the *scorenergy* table. As shown in Table 1, we divide the population into three consumer segments: High, Average, and Low energy consumers. We set a cut-off limit for each attribute of energy consumers based on the statistics and analysis of the EIA.

[2] https://www.eia.gov/tools/faqs/faq.php?id=97&t=3.

Table 2. Summary of the Datasets Used for Household Energy Monitoring

Dataset	# of Household	Appliance	Monitoring Time	
			Time Span	Frequency
UK-DALE	5	N/A	39–234 days	6 s
REDD	6	20 /house	23–28 days	3 s
DRED	1	12	6 months	N/A

Table 3. Training Statistics of Appliances

Appliance	Batch Size	Epochs	Samples	Learning Rate	Normalized Data	
					mean	stdev
Dishwasher	100	50	2385	10^{-2}	650	1000
Washing Machine	100	50	2152	10^{-2}	350	600

5 Performance Evaluation

In this section, we confirm the efficacy and effectiveness of our URPS approach. We begin with describing our experimental method that demonstrates the performance of URPS with different datasets, and then discuss the evaluation results.

5.1 Approach

Our study aims to establish an approach that offers a high-performing energy disaggregation scheme, an inclusive NILM method that models consumer behavior, and a transfer learning model that can be applied in smart grid systems that are related to the source system. In our experiments, we use three public real-world datasets, which are the UK Domestic Appliance-level Electricity (UK-DALE) [12], **The Reference Energy Disaggregation Dataset (REDD)** [15], and the **Dutch Residential Energy Dataset (DRED)** [32]. First, the UK-DALE is the measured power demand of five UK households. Second, the REDD dataset contains the collected power consumption of six UK residencies. Third, the DRED dataset is a collection of energy consumption data over a period of time of a single household in Netherlands. Useful details of the used datasets can be found in Table 2, and in [12,15,32] for more information.

In our approach, we focus on the UK-DALE datasets for major training and use the REDD and DRED datasets to demonstrate the applicability and transferability of our URPS approach. As shown in Table 2, the aggregated energy consumption for each residence in the UK-DALE is recorded every six seconds over various time intervals ranging in length from 39 days to 234 days. The entire collection period of the UK-DALE energy consumption spans a period of 600 days for the voltage and current of 16 kHz from households 1, 2, and 5. During training, we purposefully select the type II appliances in the datasets:

the washing machines and dishwashers. Type II appliances are more representative because they contain a variable number of sates unlike type I that always maintains two states. After filtering the appliances and their data in the different households, we normalize them and set the hyperparameters as shown in Table 3. We use TensorFlow to train our new model after setting up the neural network architecture that uses the pre-trained Sec2Point model. Our first two datasets (UK-DALE and REDD) are in the original Sec2Point learning method. Households 1, 2, and 3 correspond to buildings 1, 2, and 5 in UK-DALE and Household 4 corresponds to the first household in REDD. We test the transferability of this model using Household 5, which corresponds to the residence in the DRED dataset to meet the output shown in Fig. 2.

5.2 Results

We now evaluate the performance of our URPS approach, show its applicability, and validate the efficiency of the transferred model with other datasets. We can see the URPS's F1-score in Fig. 4 (a) and the estimation accuracy in Fig. 4 (b) for identifying washing machines and dishwashers. The accuracy of the results for Household 5 demonstrates the efficacy of transfer learning. The Seq2Point algorithm is limited to the UK-DALE and REDD, but Household 5 corresponds to the residence in the DRED dataset.

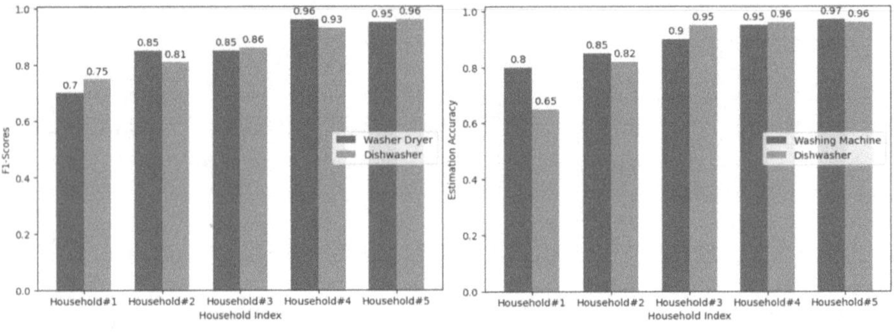

(a) F1 Score (b) Estimation Accuracy

Fig. 4. URPS Load Type Identification Performance for different Households with Type II Appliances

Next, we sample and standardize the data using the electricity consumption of the appliances. After training, the obtained prediction results are displayed in Fig. 5. Figure 5 (a), (b), and (c) show the load profile of washing machines in Households 1, 2, and 3 respectively, whereas Fig. 5 (d), (e), and (f) indicate the behavior of the dishwashers, respectively.

Table 4. URPS Appliance vs. Household Accuracy

Round	Appliance		Household-based				
	DW	WM	# 1	# 2	# 3	# 4	# 5
10	75.87	76.60	65.20	62.43	66.00	56.30	59.35
40	88.50	84.35	68.10	64.79	68.74	61.78	58.90
60	90.60	92.70	76.70	72.88	75.50	61.88	76.50
80	94.80	96.00	66.50	85.50	77.83	78.23	81.35
100	**96.03**	**96.80**	**80.90**	**87.64**	**81.95**	**84.05**	**81.50**

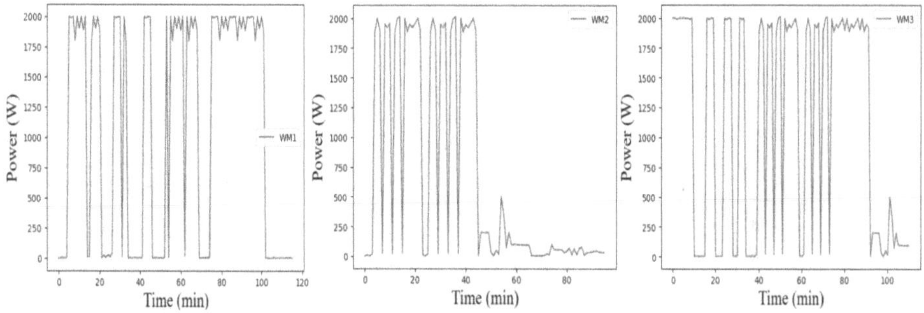

(a) Washing Machine: Household #1 (b) Washing Machine: Household #2 (c) Washing Machine: Household #3

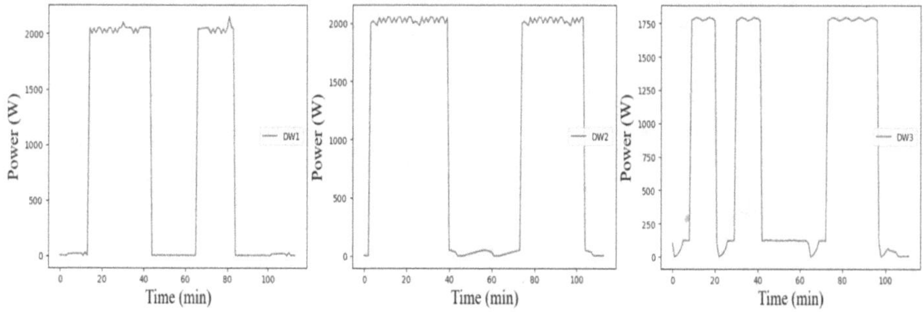

(d) Dishwasher: Household #1 (e) Dishwasher: Household #2 (f) Dishwasher: Household #3

Fig. 5. Households Unique Consumption Profile for Type II Appliance in UK-Dale and REDD Dataset

Besides the appliance's signature, we observe a unique signature for the household that illustrates the consumer's energy utilization habits. For example, Fig. 5 indicates that while the washing machines have similar power consumption histories, there are still significant variations in power consumption behavior. Likewise, we observe similar behavior among the set of dryers. While we see higher accuracy in appliance detection than in household detection as shown in

Table 4, we can establish a system that can identify energy consumption signatures in each household.

To further validate the consistency and the transferability of our transfer learning-based approach, we train Households 1–4 from April 1st, 2014, to July 31st, 2014, to predict the energy consumption of each. Note that Household 5 is not included because it is in the DRED dataset, which is from July 5th, 2015. Figure 6 illustrates the cumulative energy consumption prediction for the four households where the dotted orange series represents the measured power consumption and the blue series represents the predicted power consumption. From April 1st to May 5th, both energy consumption curves overlap closely.

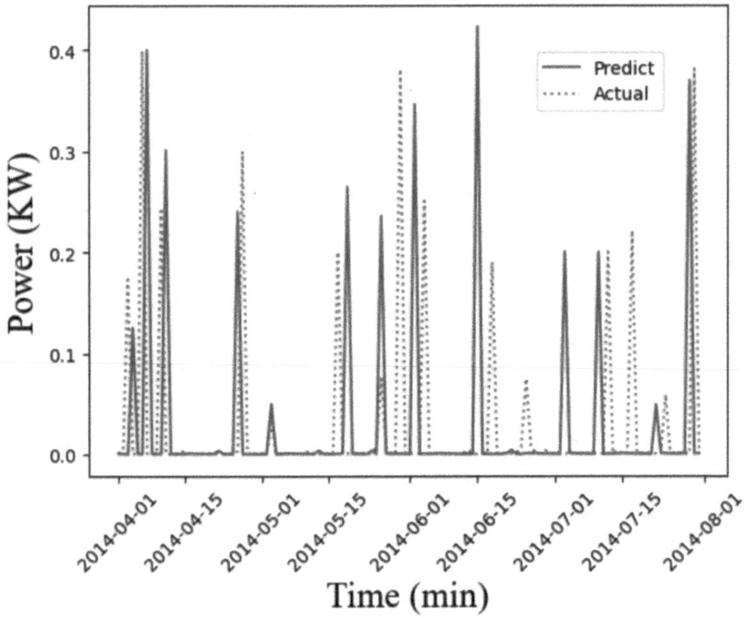

Fig. 6. URPS Prediction vs. Actual Monthly Energy Consumption

6 Discussion

In this section, we discuss additional issues and how to extend the method to other scenarios in the smart grid.

We begin by discussing some potential applications for our URPS approach in other smart grid scenarios. From a utility company's perspective, the URPS approach offers a simple but effective transfer learning approach that removes the need for massive amounts of training data, months of training, and enormous computational resources. The URPS approach proposes a multi-label load classification scheme that helps identify, with high accuracy, the load type, the event type, and the HECS. We train the URPS model using the same approach as the NILM, while adding a household-based process to the appliance signature method. Because each appliance has specific energy consumption patterns, we are able to successfully identify distinctive behaviors that allow us to distinguish between appliances of different households, even if the appliances are the same model from a given manufacturer. The results in Sect. 5 confirm the efficacy and the transferability of URPS. Thus, our model can be used in other smart grid applications.

Most effective NILM schemes in smart grid systems consider DL to train models to assess energy loads. More recently, transfer learning methods are applied in smart grid systems. The application of these transfer learning schemes in smart grid systems mainly focus on replicating the source DL models. For example, D'Incecco *et al.* ([6]) presented a transfer learning scheme for NILM that allows replicating the Seq2point method. Although the transfer learning technique ([6]) applied in the smart grid confirms the transferability of some systems, it remains to be shown that transfer learning in the smart grid can be more efficient and robust than traditional practice in general. By applying our approach, we could implement a more general scheme that includes the human factors in energy usage, resulting in a more accurate and more cost-effective prediction.

7 Final Remarks

In this paper, we investigate the applicability of transfer learning in smart grid systems and show that they do not have a detrimental effect on the model performance. Applicability is a severe issue in the power distribution network due to the substantial amounts of data and resources it requires. To this end, we propose a transfer learning-based approach called URPS that gives a more holistic view of the smart grid system. Most methods, such as NILM, operate at the appliance level to anticipate energy load consumption. The difference between our URPS approach and existing ones is that it utilizes the human factor using our HECS method to predict the user's energy habits. Moreover, URPS establishes an energy score system that facilitates the identification of specific end user behaviors, which can also be helpful to power utility companies by allowing better energy management. As a result, a trained model that incorporates the power consumers' behavior produces an accuracy of 96.8 % for appliance identification, and 87.64 % accuracy for household energy signature identification. Lastly, the URPS system enables the generation of energy scores for households. *Scorenergy* data can be reused to improve the prediction accuracy of the model and help electrical service providers better forecast the power demand to aid in determining an optimum distribution of resources.

References

1. Use of energy in homes - U.S. Energy Information Administration (EIA) — eia.gov. https://www.eia.gov/energyexplained/use-of-energy/homes.php. Accessed 27 Apr 2023
2. Al-Ali, A.R., Zualkernan, I.A., Rashid, M., Gupta, R., AliKarar, M.: A smart home energy management system using IoT and big data analytics approach. IEEE Trans. Consum. Electron. **63**(4), 426–434 (2017)
3. Baldi, P.: Autoencoders, unsupervised learning, and deep architectures. In: Proceedings of ICML Workshop on Unsupervised and Transfer Learning, pp. 37–49. (2012), JMLR Workshop and Conference Proceedings
4. Balduin, S., et al.: Towards a universally applicable neural state estimation through transfer learning. In: 2021 IEEE PES Innovative Smart Grid Technologies Europe (ISGT Europe), pp. 1–6. IEEE (2021)
5. Doshi, D., Khedkar, K., Raut, N., Kharde, S.: Real time fault failure detection in power distribution line using power line communication. Int. J. Eng. Sci. **4834**, (2016)
6. D'Incecco, M., Squartini, S., Zhong, M.: Transfer learning for non-intrusive load monitoring. IEEE Trans. Smart Grid **11**(2), 1419–1429 (2019)
7. Ehrhardt-Martinez, K., Donnelly, K.A., Laitner, S., et al.: Advanced metering initiatives and residential feedback programs: a meta-review for household electricity-saving opportunities. American Council for an Energy-Efficient Economy Washington, DC (2010)
8. Gopinath, R., Kumar, M., Joshua, C.P.C., Srinivas, K.: Energy management using non-intrusive load monitoring techniques-state-of-the-art and future research directions. Sustain. Cities Soc. **62**, 102411 (2020)
9. Gu, C., Li, H.: Review on deep learning research and applications in wind and wave energy. Energies **15**(4), 1510 (2022)
10. Hatcher, W.G., Yu, W.: A survey of deep learning: platforms, applications and emerging research trends. IEEE Access **6**, 24411–24432 (2018)
11. Jordan, M., Mitchell, T.: Machine learning: trends, perspectives, and prospects. Science **349**, 255–60 (2015). https://doi.org/10.1126/science.aaa8415
12. Kelly, J., Knottenbelt, W.: The UK-DALE dataset, domestic appliance-level electricity demand and whole-house demand from five UK homes. Sci. Data **2**(1), (2015). https://doi.org/10.1038/sdata.2015.7
13. Kim, H., Marwah, M., Arlitt, M., Lyon, G., Han, J.: Unsupervised disaggregation of low frequency power measurements. In: Proceedings of the 2011 SIAM International Conference on Data Mining, pp. 747–758. SIAM (2011)
14. Klemenjak, C., Makonin, S., Elmenreich, W.: Towards comparability in non-intrusive load monitoring: on data and performance evaluation. In: 2020 IEEE Power & Energy Society Innovative Smart Grid Technologies Conference (ISGT), pp. 1–5. IEEE (2020)
15. Kolter, J., Johnson, M.: REDD: a public data set for energy disaggregation research. Artif. Intell. **25** (2011)
16. Kotsiopoulos, T., Sarigiannidis, P., Ioannidis, D., Tzovaras, D.: Machine learning and deep learning in smart manufacturing: the smart grid paradigm. Comput. Sci. Rev. **40**, 100341 (2021)
17. Laughman, C., et al.: Power signature analysis. IEEE Power Energ. Mag. **1**(2), 56–63 (2003)

18. Lazzaretti, A.E., et al.: A multi-agent NILM architecture for event detection and load classification. Energies **13**(17), 4396 (2020)
19. LeCun, Y., Bengio, Y., Hinton, G.: Deep learning. Nature **521**(7553), 436–444 (2015)
20. Liang, F., Hatcher, W.G., Liao, W., Gao, W., Yu, W.: Machine learning for security and the internet of things: the good, the bad, and the ugly. IEEE Access **7**, 158126–158147 (2019). https://doi.org/10.1109/ACCESS.2019.2948912
21. Liang, F., Qian, C., Yu, W., Griffith, D., Golmie, N.: Survey of graph neural networks and applications. Wirel. Commun. Mob. Comput. **2022** (2022). https://doi.org/10.1155/2022/9261537
22. Liu, X., Yu, W., Liang, F., Griffith, D., Golmie, N.: Toward deep transfer learning in industrial internet of things. IEEE Internet Things J. **8**(15), 12163–12175 (2021). https://doi.org/10.1109/JIOT.2021.3062482
23. Makonin, S., Popowich, F.: Nonintrusive load monitoring (NILM) performance evaluation: a unified approach for accuracy reporting. Energ. Effi. **8**, 809–814 (2015)
24. Martinez, B., Monton, M., Vilajosana, I., Prades, J.D.: The power of models: modeling power consumption for IoT devices. IEEE Sens. J. **15**(10), 5777–5789 (2015)
25. Moulema, P., Yu, W., Griffith, D., Golmie, N.: On effectiveness of smart grid applications using co-simulation. In: 2015 24th International Conference on Computer Communication and Networks (ICCCN), pp. 1–8 (2015). https://doi.org/10.1109/ICCCN.2015.7288438
26. Moulema, P., et al.: On effectiveness of mesh-based protocols for smart grid communication networks. SIGAPP Appl. Comput. Rev **14**(2), 59–70 (2014). https://doi.org/10.1145/2656864.2656869
27. Pan, S.J., Yang, Q.: A survey on transfer learning. IEEE Trans. Knowl. Data Eng. **22**(10), 1345–1359 (2010). https://doi.org/10.1109/TKDE.2009.191
28. Portia Murray, C.E.M.: Non-intrusive load monitoring at CLEMAP. https://youtu.be/FEpCAmuISKA
29. Pratt, L.Y.: Discriminability-based transfer between neural networks. Adv. Neural. Inf. Process. Syst. **5**, (1992)
30. Pratt, L.Y.: Non-literal transfer among neural network learners. Colorado School of Mines (1992). MCS-92-04
31. Ruano, A., Hernandez, A., Ureña, J., Ruano, M., Garcia, J.: NILM techniques for intelligent home energy management and ambient assisted living: a review. Energies **12**(11), 2203 (2019)
32. Uttama Nambi, A.S., Reyes Lua, A., Prasad, V.R.: Loced: location-aware energy disaggregation framework. In: Proceedings of the 2nd ACM International Conference on Embedded Systems for Energy-Efficient Built Environments, BuildSys 2015, pp. 45–54. Association for Computing Machinery, New York, NY, USA (2015). https://doi.org/10.1145/2821650.2821659
33. Xu, G., Yu, W., Griffith, D., Golmie, N., Moulema, P.: Toward integrating distributed energy resources and storage devices in smart grid. IEEE Internet Things J. **4**(1), 192–204 (2017)
34. Yu, Y., Si, X., Hu, C., Zhang, J.: A review of recurrent neural networks: LSTM cells and network architectures. Neural Comput. **31**(7), 1235–1270 (2019)
35. Zhang, C., Zhong, M., Wang, Z., Goddard, N., Sutton, C.: Sequence-to-point learning with neural networks for nonintrusive load monitoring (2017)

A Machine Learning Approach for Defect Dimension Estimation from Eddy Current Testing

Addisson Salazar[1](✉), Federico Carere[2], Alessandro Sardellitti[3], Silvia Sangiovanni[4], Luis Vergara[1], and Marco Laracca[4]

[1] Institute of Telecommunications and Multimedia Applications, Universitat Politècnica de València, Valencia, Spain
asalazar@dcom.upv.es
[2] Department of Information Science and Technology, University Pegaso, Naples, Italy
[3] Faculty of Technological and Innovation Sciences, Universitas Mercatorum, Rome, Italy
[4] Department of Astronautics, Electrical and Energy Engineering, Sapienza University of Rome, Rome, Italy

Abstract. A new procedure for estimation of dimensions of defects in materials inspected by eddy current testing is presented. After preprocessing, several indirect features are extracted from maps made by direct features estimated from the measured eddy current signals. Those maps are estimated in temporal, frequency, and graph domains. In the procedure, stages of regression and classification are integrated. The regression methods are the following: support vector machine (linear and radial basis function kernel), random forest, multiple linear, Gaussian kernel, ridge, polynomial degree 2, k-nearest neighbors, and multilayer perceptron. The classification methods include the following: linear and quadratic discriminant analysis, random forest, support vector machine, and naïve Bayes. The proposed procedure applies machine learning methods that are computationally inexpensive, which would facilitate their implementation in real time. The experimental results demonstrate improvements in the estimation of the height and depth of very small defects on several specimens.

Keywords: Non-destructive testing · Machine learning · Defect dimension estimation · Eddy current

1 Introduction

The study and testing of quality of materials involves several levels of analysis such as detect if the material is defective or non-defective, as well as the sizing and location of the defects. Thus, non-destructive testing (NDT) methods are devised to guarantee the integrity of the materials and products for several industries. Those testing methods are based on the acquisition and processing of different kinds of signals such as ultrasounds, eddy current, and thermography. This is an intensive area of research that have been extensively studied [1].

H. R. Arabnia et al. (Eds.): CSCI 2025, CCIS 2967, pp. 40–52, 2027.
https://doi.org/10.1007/978-3-032-28454-9_3

Recently, machine learning methods have provided different perspectives to revisit the NDT protocols. Thus, the patterns for distinguish the inhomogeneities in materials can be learned directly from the data. Thus, complex models of the microstructure of the materials and signal propagation are not strictly required. However, there is trade-off between the explainability of the results and the complexity of the method. In this paper, we propose a procedure that integrates regression and classification machine learning methods. This is proposed for estimation of dimensions of materials evaluated by eddy current testing (ECT). The defects are very small with dimensions around a few millimeters or submilimeters [2–5].

Different probe designs have been proposed in ECT to improve the defect detection as well as the estimation of its properties, e.g., in [6] a cone-shaped probe was proposed, in [7] a right-angle isosceles triangle excitation coil, in [8] double-row three-phase excitation coils. Moreover, different excitation strategies has been proposed, that can be classified in two main categories according to the signals representation: time domain [9–11] and frequency domain excitation systems [4, 12, 13].

Several classification and regression methods have been applied to the detection and estimation of the defect dimensions using NDT based on ECT. For instance, a critical application of ECT is the classification of material corrosion. Less and highly corroded states were analyzed in [14]. The features used in the classification models were the peaks of the magnetic fields at two different frequencies (2.5 kHz and 5 kHz). Logistic regression and a method based on Gaussian mixture model were used for classification. Machine learning has been also applied with other NDT methods such as infrared thermal images, e.g., crack defect detection of steel plates using crossing temperature and temperature difference features and support vector machine (SVM), k-nearest neighbors (kNN), and random forest (RF) classification [15].

Recently, deep learning (DL) techniques have been also explored to improve the detection in several non-destructive applications, e.g., defect detection in weld using ECT. A flexible ECT probe with differential multi-modes was used and a mask region convolution neutral network (Mask-RCNN) was applied to the images obtained from the amplitude of the four modes and the fused image. The excitation frequencies for each of the modes were 100, 200, 300, 600 kHz, 1, and 2 MHz. The results show detection of defects in 2D images [16–20].

Regression based on SVM was first proposed for crack shape reconstruction using ECT in [21]. Besides, a crack depth estimation by using a multi-frequency ECT method was also provided [22, 23]. Now, in this paper, we extend the ML processing by adding regression-classification fusion to improve the results of the multifrequency method by using only one excitation frequency [24–26].

The rest of the paper is organized as follows. Section 2 describes the proposed method. Section 3 presents the results of an experimental study on aluminum alloy specimens. Finally, Sect. 4 includes the conclusions and future lines of research from this work.

2 Proposed Procedure

Figure 1 shows a scheme of the proposed procedure. A set of features is extracted from the measured EC signal. The data are split into two subsets: training and testing. The procedure for training and testing is the so-called "leave-one-out" (LOO), where all the records are used for training, except the one that is used for testing.

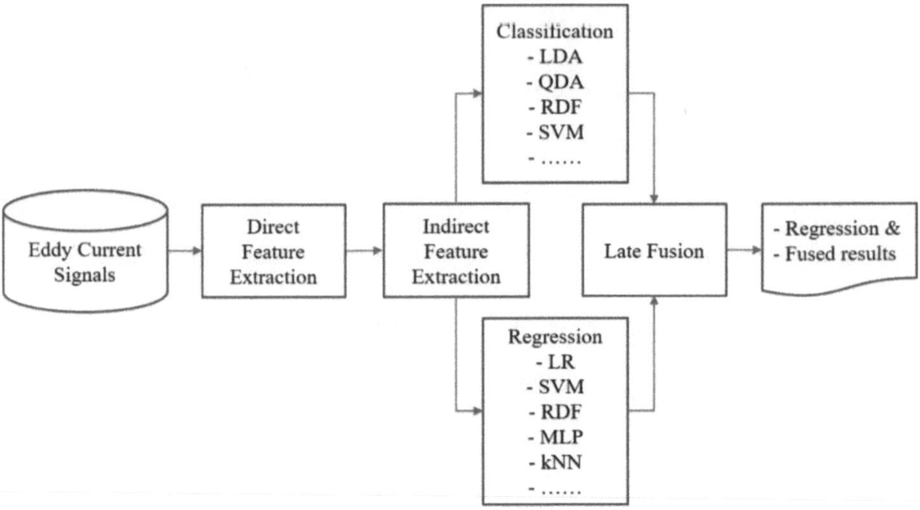

Fig. 1. Outline of the proposed procedure.

Thus, the data subsets are disjoint avoiding overfitting. Note that we have to estimate two dimensions (height (h_d) and depth (d_d)), thus LOO is applied twice. For depth estimation, a previous classification stage is used to discern between superficial and buried defects. This implicitly involves the estimation of depth (equal zero) for the superficial defects. Finally, the results of classification and regression are fused.

ECT measurements were performed over a regular grid with 1 mm spacing, covering multiple positions across the surface of the material. As previously described in Sect. 2, the data collected at each grid point were expressed as the amplitude of the output voltages along the three spatial components P_x, P_y, and P_z, corresponding to the x, y, and z axes, respectively.

We apply several mapping functions point-to-point on the grid, preserving the spatial structure of the measurements. Mapping allows the dimension reduction of the data, easing the recognition of interesting patterns to discern between data categories and/or prediction of data. The maps represent the variation, due to the crack presence, of the magnetic field around each considered defect. We obtained several maps for P_y, P_z and for the combination of both.

For every map, two features were extracted: (i) the difference between the peak maximum and peak minimum and (ii) the distance between those two peaks. Table 1 shows the estimated features.

Table 1. Features extracted from the maps obtained from the ECT signals.

No.	Feature	Map (M)
1	Distance	$A(FFT(P_{z,f}))$
2	Difference	$A(FFT(P_{z,f}))$
3	Distance	$A(FFT(P_{y,f}))$
4	Difference	$A(FFT(P_{y,f}))$
5	Distance	area($graph(P_y, P_z)$)
6	Distance	angle($graph(P_y, P_z)$)
7	Distance	$\max(P_y) - \min(P_y)$
8	Distance	$\max(P_z) - \min(P_z)$
9	Difference	area($graph(P_y, P_z)$)
10	Difference	angle($graph(P_y, P_z)$)
11	Difference	$\max(P_y) - \min(P_y)$
12	Difference	$\max(P_z) - \min(P_z)$

Table 1 shows six maps: $A(FFT(P_{z,f}))$, $A(FFT(P_{y,f}))$, area($graph(P_y, P_z)$), angle($graph(P_y, P_z)$), $\max(P_y) - \min(P_y)$, and $\max(P_z) - \min(P_z)$. FFT is the fast Fourier transform and $A(FFT(P_{z,f}))$, $A(FFT(P_{y,f}))$ are the amplitude of the spectrum for P_z and P_y, respectively, evaluated at frequency f. We set $f = 500$ Hz, that was the excitation frequency used in measurements. For each map, two features were extracted (difference and distance). Thus, the total number of features was 12. Note that the set of features belong to three domains: frequency, graph, and time.

The feature "difference" is estimated for each of the maps M using the expression $\frac{|\max(M)|+|\min(M)|}{2}$. The feature "distance" is estimated by the Euclidean distance using the coordinates (x, y) of the maximum and minimum values in the map, $\sqrt{\left(x_{\max(M)} - x_{\min(M)}\right)^2 + \left(y_{\max(M)} - y_{\min(M)}\right)^2}$.

Note that the features 5, 6, 9, and 10 correspond to maps estimated from the area and angle in graphs obtained using both electromagnetic fields P_y and P_z simultaneously. Thus, we claim that subtle irregular and complex structures related to detection and size estimation of defects could be modeled by graph signal processing [27].

We considered that the variations observed in the magnetic field component B_y, as measured by the TMR sensor, are attributed to the disturbance introduced by the presence of a defect. As previously demonstrated in the literature, under defect-free conditions the magnetic field induced by the ECs is predominantly aligned along the z-axis (i.e. B_z). However, the presence of a defect modifies the EC distribution, resulting in measurable field components also along the x- and y-axes (i.e. B_x, B_y) [28, 29].

To make a comparative assessment of the proposed method, the following regression algorithms were implemented: support vector machine-kernel linear (SVMR-L); random forest (RDFR); multiple linear (MLR); Gaussian kernel (GKR); ridge (RR); polynomial degree 2 (PR); k-Nearest Neighbors (kNNR); multilayer perceptron (MLPR);

and SVMR-kernel radial basis function (SVMR-RBF). Those methods can be roughly divided into linear and non-linear and the basis for each of them could be from similar to quite different, it will be shown in Sect. 4.

The classifiers implemented were the following: linear discriminant analysis (LDA); quadratic discriminant analysis (QDA); SVM; RDF; and naïve Bayes (NB). The classifiers and regression methods were chosen because of their successful application across many different fields and problems, including NDT. They are based on different and sometimes complementary principles. RF implemented 10 trees and SVM employed several kernels: Gaussian for classification and Linear and RBF for regression.

The fusion proposed here combines the results from regression and classification stages, which is novel in NDT-ECT. Actually, there are a few works that applied fusion stages. Previous works have proposed to combine only classification results or only regression results.

3 Experimental Study

3.1 Experimental Set-up

To experimentally validate the effectiveness of the proposed machine learning-based method for crack dimension estimation in ECT, a dedicated measurement set-up was specifically developed. In particular, two-channel arbitrary waveform generator (Siglent SDG1032X) were used to generate sinusoidal excitation signals. These excitation signals were independently amplified through two bipolar power amplifiers (Kepco BOP 20-20 M), enabling the current control for each excitation channel. The RMS values of the excitation currents are continuously monitored using two digital multimeters (Rigol DM3058).

The ECT probe used in the experiments is a four-quadrant configuration consisting of four quarter-circle excitation coils arranged in a square layout. Each coil has an internal radius (ID) of 8 mm, an external radius (ED) of 11 mm, a height (H) of 10 mm, and 150 turns. The coils are symmetrically spaced with a center-to-center distance of 15 mm along both x and y axes (Dx and Dy), respectively. The scanning movement of the probe was performed using a motorized stage with a positioning resolution of 0.1 mm. A regular scan path of 40 mm $\times$ 40 mm is executed with a spatial step of 1 mm in both directions.

The conductive sample under test consisted of an aluminum alloy (Al AW 6082) plate with dimensions 200 mm $\times$ 200 mm $\times$ 4 mm and an electrical conductivity of 27.2 MS/m. A set of certified cracks was realized to the sample to evaluate the system performance under controlled conditions. These defects included both superficial and buried cracks, with variations in depth (d_d) and height (h_d) but constant length (l_d) and thickness (t_d), as shown in Fig. 2.

Specifically, defects were characterized by a fixed length of 5 mm and a thickness of 0.1 mm, while their height varied between 1 mm and 3 mm (see Table 2). The buried cracks were placed at depths ranging from 1 mm to 3 mm from the sample surface, while superficial defects were located directly at the top surface. These configurations allowed assessment of the system's sensitivity to both geometrical and positional defect parameters.

At each scan point, the tunnel magneto-resistive (TMR) sensor acquired time-domain voltage signals over 10 periods of the excitation waveform, corresponding to a total of 2000 samples per axis, using a sampling frequency of 100 kS/s.

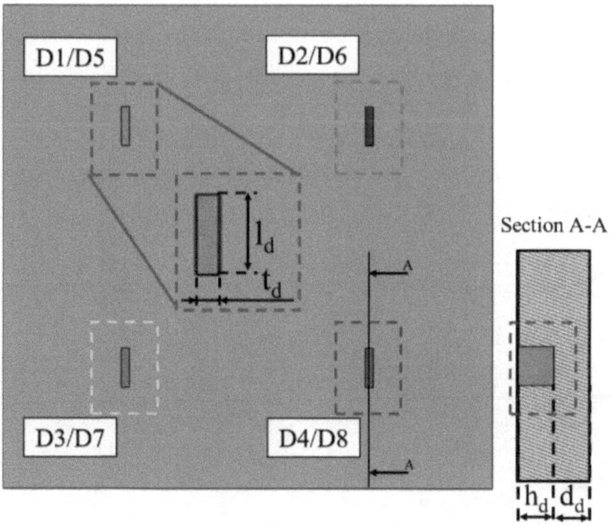

Fig. 2. Graphical representation of the considered sample with the defect characteristics.

Table 2. Geometrical characteristics of the analyzed defects.

Defect code	Position	h_d [mm]	d_d [mm]	l_d [mm]	t_d [mm]
D1	Superficial	3	0	5	0.1
D2		2	0		
D3		1.5	0		
D4		1	0		
D5	Buried	3	1		
D6		2	2		
D7		1.5	2.5		
D8		1	3		

3.2 Results

This section includes the results of the proposed method for the estimation of the dimensions of the defects described above. Figure 3 shows one example of the estimated maps for defects D1 and D8, built by using the amplitude of the spectrum content at the frequency 500 Hz of the electromagnetic field P_z, i.e., $A(FFT(P_{z,f}))$, corresponding to features 1 and 2 of Table 1. The maps of Fig. 3 shows the distribution of the amplitude of the spectrum evaluated at 500 Hz frequency for P_z through the points of the measurement grid. The location of the points with maximum and minimum values and distance between both are highlighted. Note that the defective area is well localized around the coordinates (20, 20) of the grid. Besides, the level of noise increases with the depth of the defects. Particularly, the signal to noise ratio is the lowest for defect D8.

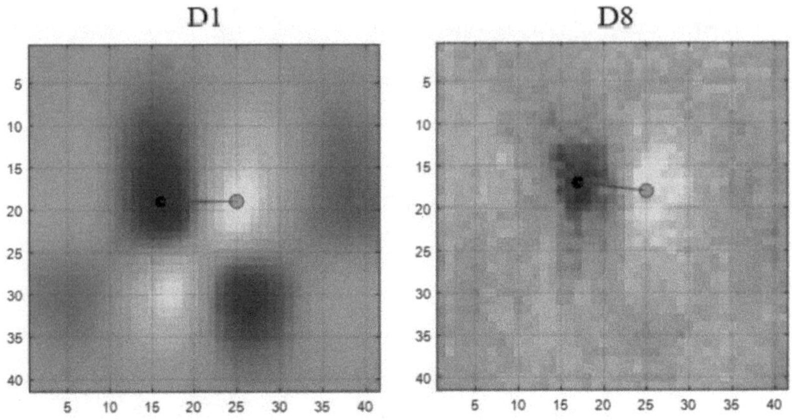

Fig. 3. Examples of maps $A(FFT(P_{z,f}))$ estimated for the defects D1 and D8. The location of the maximum and minimum and distance between both are highlighted.

Figure 4 shows an example of the features (distance and difference) estimated for two of the maps. The graphs correspond to feature 1 (distance) and feature 2 (difference). In general, the curves of distances showed an ascendant trend whereas the curves of differences show a descendent trend. In the case of distances, the set of values for superficial defects (D1-D4) is much lower than the ones for buried defects (D5-D8), except for defects D4 and D5 that are similar values. This is because the depth of the defect D5 is only 1 mm. deep and the behavior of the ECT signals, which is recorded in the features, seems similar. Thus, the values for defect D5 (less deep) are similar to the ones of defect D4, or in general, to the ones of the superficial defects. This could affect the misclassification of the data or the poor estimation of the data dimensions, although using more than one feature helps to alleviate those problems. In summary, the distances show the maximum and minimum values in the maps are higher, the depth of the defects is greater. Conversely, the curves of the differences show a descendent trend. The methods of regression implemented, explained in Sect. 2, were applied to the set of features extracted from the maps. The objective was the estimation of the height and depth of the defects. The preprocessing of the signals were detrending and standardizing.

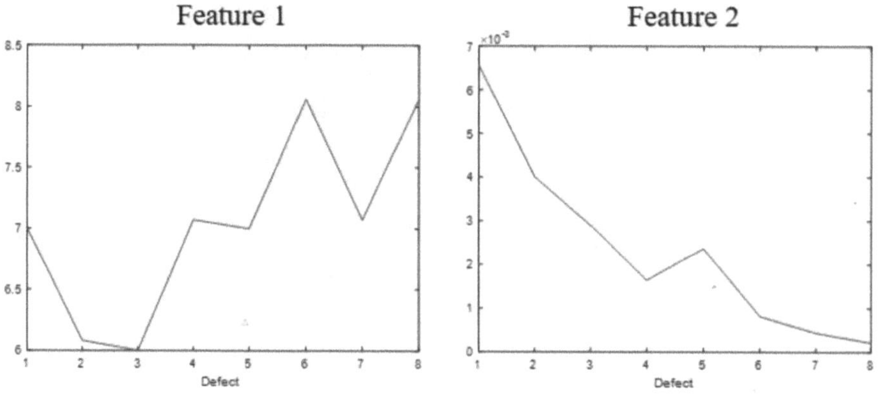

Fig. 4. Examples of feature values estimated from the maps for each of the defects.

The results of the best performing method have been highlighted in the cells of the Table 3 (Depth estimation). It shows that only one regression method (RR) ensures that the estimates for all defects are below 1 mm, which is an accuracy level commonly defined for evaluating results in industrial environments [21]. Besides, the mean standard error reported for defect dimension estimation in previous works have been within ±0.3 mm. Those values were the ground truth considered for the application studied here,

Note that SVMR-Linear is the method that achieves best average error (.32, i.e., .17 better that RR), that is close to the objective, but it failed to estimates the depth of defect 4 to 1.02 mm. The worst result was achieved by PR, with high values of error, thus, in this case, the estimates have failed. That method has not been able to approximate to the geometry of the data. In general, the quality of depth estimation at this stage of the processing is below to previous works.

Table 3. Defect depth estimation.

Method	D1	D2	D3	D4	D5	D6	D7	D8	Mean ± std
SVMR- linear	**.47**	**.31**	**.04**	**1.02**	**.1**	**.4**	**.01**	**.21**	**.32 ± .33**
RDFR	1.21	1.21	1.21	1.21	.07	1.07	1.64	2.21	1.23 ± .6
MLR	2.55	.46	.13	1.02	.24	.74	.4	.13	.71 ± .81
GKR	1.2	.61	1.19	1.31	.59	1.23	1.66	2.36	1.27 ± .57
RR	.83	.43	.33	.62	.3	.99	.28	.14	.49 ± .3
PR	6.7	1.93	1.48	1.08	.11	1.46	15.39	.39	3.57 ± 5.2
kNNR	1	0	0	0	.33	1.67	.5	1.17	.58 ± .63
MLPR	.6	.23	.24	1.22	.15	.22	1.23	.83	.59 ± .45
SVMR- RBF	.99	.13	.42	.08	.98	1.92	2.5	3	1.25 ± 1.1

Table 4 shows the results for height estimation. The best results were obtained by SVMR-Linear achieving the best estimation average error (.19) ensuring the quality of the estimates at defect level reaching <1 mm. for every defect. The second, third and fourth results, are in order to MLR, PR, and MLPR where the error estimations for every defect are below 1 mm. In addition, all those methods, except MLPR, overcome the objective by comparing with previous works that have reported average errors within ±0.3 mm.

Table 4. Defect height estimation.

Method	D1	D2	D3	D4	D5	D6	D7	D8	Mean ± std
SVMR- linear	**.04**	**.18**	**.04**	**.3**	**.17**	**.32**	**.19**	**.29**	**.19 ± .11**
RDFR	1.29	.14	.43	1	1.29	.14	.43	1	.71 ± .48
MLR	.09	.14	.01	.17	.66	.3	.09	.2	.21 ± .2
GKR	1.42	.36	.52	1.81	1.42	.88	1.28	.91	1.07 ± .49
RR	1.28	.03	.86	1	1.28	.44	.42	1.15	.81 ± .46
PR	.44	.05	.03	.03	.93	.36	.03	.13	.25 ± .32
kNNR	1.17	.5	1.17	1.5	1.17	.83	.83	.5	.96 ± .35
MLPR	.64	.13	.38	.52	.87	.47	.2	.54	.47 ± .24
SVMR- RBF	.89	.2	.07	1.25	.89	1.06	.51	.92	.72 ± .42

As explained in Sect. 2, a classification step can be applied to improve the regression results. We applied a classification step to the defect depth estimation, considering the common division of defects into superficial and buried. The right estimation for the depth of superficial depths is depth = 0. In the case of defect height estimation is not feasible to implement a classification step because the height can be considered a continuous variable with many values to define categories as in the case of the depth, which is a discrete variable.

We performed a feature ranking for classification using the method in [30]. The best ranked features for the two-class classification (superficial and buried) were: Feature 12, Feature 9, and Feature 6. The results of classification are shown in Table 5.

Table 5. Defect depth estimation - classification.

Method	D1	D2	D3	D4	D5	D6	D7	D8
LDA	**0**	**0**	**0**	**0**	**1**	**1**	**1**	**1**
QDA	0	0	0	0	0	1	1	1
RDF	0	0	0	0	0	1	1	1
SVM	0	0	0	1	0	0	1	1
NB	0	0	0	1	0	1	1	1

The test applied to the classification output was $s(\mathbf{x}) \underset{H_0}{\overset{H_1}{\gtrless}} t$, where $s(\mathbf{x})$ is the score (posterior probability) provided by the classifier between 0 and 1, assigned to the defect under analysis. The higher the value $s(\mathbf{x})$, the higher the probability of the defect is buried. The lower the value $s(\mathbf{x})$, the higher the probability of the defect is superficial. Thus, the threshold t filters the scores to classify the defect is superficial (hypothesis H_0) or buried (hypothesis H_1).

We use 0.5 as threshold to obtain the classification results of Table 5. LDA show the best results with 100% detection probability and 0% false alarm probability. Besides, the score values obtained by LDA were closer to 0 and 1 than the other classifiers, thus reducing uncertainty on the decisions.

Once the binarized classification results were obtained, they were fused with the regression results for depth estimation. The fusion operation consisted of replacing the depth estimation of Table 3 to 0, using the results of LDA in Table 5, when the category supplied by the classification was 0 and keep the estimated depth value by regression of Table 3 when the category provided by the classification was 1. The results of this fusion stage are shown in Table 6. A significant improvement was obtained by applying this last step.

Table 6. Defect depth estimation: regression + classification

Method +LDA	D1	D2	D3	D4	D5	D6	D7	D8	Mean ± std
SVMR-linear	**0**	**0**	**0**	**0**	**.13**	**.71**	**.11**	**.08**	**.13 ± .24**
RDFR	0	0	0	0	1	2	1.5	1	.69 ± .8
MLR	0	0	0	0	.29	.95	.26	.15	.21 ± .32
GKR	0	0	0	0	.76	.77	1.48	.93	.62 ± .76
RR	0	0	0	0	.3	.96	.24	.12	.20 ± .33
PR	0	0	0	0	1.03	2.76	6.24	3.91	1.74 ± 2.35
kNNR	0	0	0	0	1	2	.5	.5	.50 ± .71
MLPR	0	0	0	0	.21	.4	1.3	.87	.35 ± .49
SVMR-RBF	0	0	0	0	.99	1.96	2.5	3	1.06 ± 1.26

4 Conclusions

A new procedure for non-destructive testing based on eddy current testing has been proposed. Several direct features are extracted from the raw signals to build a mapping space from which indirect features are extracted for the following processing stages. The supervised learning procedure includes the combination of regression and classification stages for estimating the dimensions, height and depth, of very small defects. A comparison of several computational-low cost machine learning methods show support vector machine regression and linear discriminant classification were the most effective combination of methods. The results show improvement over the groundtruth commonly used.

There are several lines of future research open, for instance, extending the experiments to other kinds of materials and shapes of defects, implementing deep learning methods for comparison, extending the set of features using signal processing on graphs [31–35], improving the fusion procedure [36–41], and theoretical developments of the proposed procedure.

Acknowledgement. This work was supported in part by the Generalitat Valenciana under Grant CIPROM/2022/20 and Agencia Estatal de Investigación, Spain under Grant PID2024-161353OB-I00.

References

1. Qiao, Q., Hu, H.A., A., Wang, K.: A review of metal surface defect detection technologies in industrial applications. IEEE Access. **13**, 48380–48400 (2025)
2. Sardellitti, A., Milano, F., Laracca, M., Ferrigno, L., Tamburrino, A., Tian, G.Y.: Enhancing corrosion detection and characterization: an innovative approach with pot-cored eddy current probe. IEEE Trans. Instrum. Meas. **74**, 1–13 (2025)
3. Xie, Y., Huang, P., Ding, Y., Li, J., Pu, H., Xu, L.: A novel conductivity measurement method for non-magnetic materials based on sweepfrequency eddy current method. IEEE Trans. Instrum. Meas. **71**, 1–12 (2022)
4. Carere, F., Sardellitti, A., Bernieri, A., Ferrigno, L., Sangiovanni, S., Laracca, M.: An eddy current probe for the detection of subsuperficial defects of any orientation. IEEE Trans. Instrum. Meas. **73**, 1–13 (2024)
5. Munir, N., Huang, J., Wong, C.N., Song, S.J.: Machine learning based eddy current testing: a review. Results Eng. **25**(103724), 1–16 (2025)
6. Gong, Y., Huang, X., Liu, Z., Deng, F., Wu, Y., He, C.: Development of a cone-shaped pulsed eddy current sensor. IEEE Sensors J. **22**(4), 3129–3136 (2022)
7. Wang, Y., Ye, C., Wang, M.: Synthetic magnetic field imaging with triangle excitation coil for inspection of any orientation defect. IEEE Trans. Instrum. Meas. **69**(2), 533–541 (2020)
8. Zhang, N., Ye, C., Peng, L., Tao, Y.: Eddy current probe with three phase excitation and integrated array tunnel magnetoresistance sensors. IEEE Trans. Ind. Electron. **68**(6), 5325–5336 (2021)
9. Malekmohammadi, H., Migali, A., Laureti, S., Ricci, M.: A pulsed eddy current testing sensor made of low-cost off-the-shelf components: overview and application to pseudo-noise excitation. IEEE Sensors J. **21**(20), 23578–23587 (2021)
10. Buck, J.A., et al.: Pulsed eddy current inspection of support structures in steam generators. IEEE Sensors J. **15**(8), 4305–4312 (2015)
11. Zhao, T., et al.: Evaluation of pulse eddy current for autonomous airborne inspections. IEEE Sensors Lett. **8**(8), 6009304 (2024)
12. Wang, Y., Niu, Y., Wei, Y., Ye, C.: Multi-frequency imaging with nonlinear calibration of magnetoresistance sensors for surface and buried defects inspection. NDT E Int. **132**, 102706 (2022)
13. Pasadas, D.J., Ramos, H.G., Feng, B., Baskaran, P., Ribeiro, A.L.: Defect classification with svm and wideband excitation in multilayer aluminum plates. IEEE Trans. Instrum. Meas. **69**(1), 241–248 (2020)
14. Xie, L., Baskaran, P., Ribeiro, A.L., Alegria, F.C., Ramos, H.G.: Classification of corrosion severity in SPCC steels using eddy current testing and supervised machine learning models. Sensors. **24**(2259), 1–14 (2024)
15. Chen, H., Zhang, Z., Yin, W., Zhao, C.F., Li, Y.: A study on depth classification of defects by machine learning based on hyper-parameter search. Measurement. **189**(110660), 1–11 (2022)

16. Pasadas, D.J., Baskaran, P., Ramos, H.G., Ribeiro, A.L.: Detection and classification of defects using ECT and multi-level SVM model. IEEE Sensors J. **20**(5), 2329–2338 (2020)
17. Queiroz, R.S., Silva, J.P.B., das Neves, E.C., da Silva, L.C., Coelho, R.S., Lepikson, H.A.: Development and fusion of NDT classifiers for defect detection on underwater structures. NDT E Int. **144**(103098), 1–17 (2024)
18. Tao, X., Peng, L., Tao, Y., Ye, C.: Inspection of defects in weld using differential array ECT probe and deep learning algorithm. IEEE Trans. Instrum. Meas. **70**, 2512209 (2021)
19. Fatima, B., Bachir, H., Samir, B., Karim, R., Ibn Khaldoun, L.: Detection and classification of surface cracks using deep learning based autoencoders in eddy current testing. IEEJ Trans. Electr. Electron. Eng. **20**, 676–687 (2025)
20. Chen, J., Yu, N., Fan, M., Cao, B., Sun, F., Lia, C.: Multi-frequency eddy current non-destructive evaluation of battery tab welding quality. Nondestr. Test. Eval. **41**, 1–23 (2025)
21. Bernieri, A., Ferrigno, L., Laracca, M., Molinara, M.: Crack shape reconstruction in eddy current testing using machine learning systems for regression. IEEE Trans. Instrum. Meas. **57**(9), 1958–1968 (2008)
22. Bernieri, A., Betta, G., Ferrigno, L., Laracca, M.: Crack depth estimation by using a multi-frequency ECT method. IEEE Trans. Instrum. Meas. **62**(3), 544–552 (2013)
23. Bernieri, A., Betta, G., Ferrigno, L., Laracca, M., Mastrostefano, S.: Multifrequency excitation and support vector machine regressor for ECT defect characterization. IEEE Trans. Instrum. Meas. **63**(5), 1272–1280 (2014)
24. Salazar, A., Safont, G., Vergara, L., Vidal, E.: Graph regularization methods in soft detector fusion. IEEE Access. **11**, 144747–144759 (2023)
25. Salazar, A., Vergara, L., Vidal, E.: A proxy learning curve for the Bayes classifier. Pattern Recogn. **136**, 1–14 (2023)
26. Salazar, A., Vergara, L., Safont, G.: Generative adversarial networks and Markov random fields for oversampling very small training sets. Expert Syst. Appl. **163**, 1–12 (2021)
27. Ortega, A., Frossard, P., Kovacevic, J., Moura, J.M.F., Vandergheynst, P.: Graph signal processing: overview, challenges, and applications. Proc. IEEE. **106**(5), 808–828 (2018)
28. Ramos, H.G., Rocha, T., Pasadas, D., Lopes-Ribeiro, A.: Determination of linear defect depths from eddy currents disturbances. AIP Conf. Proc. **1581**(1), 1448–1455 (2014)
29. Rawashdeh, M.R., Rosell, A., Udpa, L., Hoole, S.R.H., Deng, Y.: Optimized solutions for defect characterization in 2-D inverse eddy current testing problems using subregion finite element method. IEEE Trans. Magn. **54**(8), 1–15 (2018)
30. Salazar, A., Safont, G., Vergara, L., Vidal, E.: Pattern recognition techniques for provenance classification of archaeological ceramics using ultrasounds. Pattern Recogn. Lett. **135**, 441–450 (2020)
31. Belda, J., Vergara, L., Safont, G., Salazar, A., Parcheta, Z.: A new surrogating algorithm by the complex graph Fourier transform (CGFT). Entropy. **21**(8), 759 (2019)
32. Belda, J., Vergara, L., Safont, G., Salazar, A.: Computing the partial correlation of ICA models for non-Gaussian graph signal processing. Entropy. **21**(1), 22 (2019)
33. Vergara, L., Salazar, A., Belda, J., Moral, S., Iglesias, S.: Signal processing on graphs for improving automatic credit card fraud detection. In: Proceedings International Carnahan Conference on Security Technology 2017 (ICCST 2017), 2017-October, 1–6 (2017)
34. Salazar, A., Vergara, L.: ICA mixtures applied to ultrasonic nondestructive classification of archaeological ceramics. Eurasip J. Adv. Signal Process. **2010**, 125201 (2010)
35. Safont, G., Salazar, A., Vergara, L., Gómez, E., Villanueva, V.: Multichannel dynamic modeling of non-Gaussian mixtures. Pattern Recogn. **93**, 312–323 (2019)
36. Pereira, L.M., Salazar, A., Vergara, L.: A comparative study on recent automatic data fusion methods. Computers. **13**(1), 13 (2024)
37. Safont, G., Salazar, A., Vergara, L., Rodríguez, A.: Nonlinear estimators from ICA mixture models. Signal Process. **2019**(155), 281–286 (2019)

38. Salazar, A., Safont, G., Rodriguez, A., Vergara, L.: Combination of multiple detectors for credit card fraud detection. In: 2016 IEEE International Symposium on Signal Processing and Information Technology (ISSPIT 2016) 2017, pp. 138–143 (2016)
39. Pereira, L.M., Salazar, A., Vergara, L.: On comparing early and late fusion methods. In: Advances in Computational Intelligence Lecture Notes in Computer Science Including Subseries Lecture Notes in Artificial Intelligence and Lecture Notes in Bioinformatics 2023, 14134 LNCS, pp. 365–378. Springer (2023)
40. Salazar, A., Zito, R., Laureti, S., Ricci, M., Vergara, L.: Classifier fusion for the detection of defects from active thermography. In: Advances in Computational Intelligence Lecture Notes in Computer Science 2026, 16009 LNCS, pp. 154–166. Springer (2026)
41. Safont, G., Salazar, A., Vergara, L.: Vector score alpha integration for classifier late fusion. Pattern Recogn. Lett. **2020**(136), 48–55 (2020)

Acquisition of Extensive Pareto-Optimal Policies via Event-Triggered Reinforcement Learning

Hayato Chujo(✉) and Sachiyo Arai

Department of Urban Environment Systems, Chiba University, Chiba, Japan
caua5426@chiba-u.jp

Abstract. Reinforcement learning is successful for sequential decision-making problems formulated as single-objective problems. By contrast, real-world problems are multi-objective, making it natural to formulate them as multi-objective sequential decision-making problems and solve them via multi-objective reinforcement learning, where the goal is to learn a set of optimal policies on the Pareto front. We address two key challenges for this goal: correctly understanding the relationship of utility among objectives for comprehensively acquiring a set of policies on the Pareto front, and the sparsity problem of rewards in sequential decision-making, which requires multistep state transitions. This is important because reinforcement learning-based methods have learning difficulties arising from sparse rewards. This study proposes event-triggered reinforcement learning to address these two challenges and verifies its performance through computational experiments. The experimental results indicate that event-triggered reinforcement learning is a promising approach for acquiring extensive Pareto fronts in multi-objective reinforcement learning compared with conventional time-triggered approaches.

Keywords: Event-triggered control · Multi-objective reinforcement learning · Pareto front

1 Introduction

Reinforcement learning is a valuable approach for sequential decision-making problems, but it is generally formulated as a single-objective problem. By contrast, real-world problems often involve multiple competing objectives [2], leading to the proposal of multi-objective reinforcement learning, which extends reinforcement learning for multi-objective sequential decision-making problems. Multi-objective reinforcement learning aims to learn the Pareto front, a set of Pareto-optimal policies. As the distribution range of the Pareto front represents the range of policy options available to users, providing a broader Pareto front is important. However, this requires accurate learning of the balance among objectives, and many methods have been proposed.

Reinforcement learning allows "delayed rewards," which can reduce the burden on designers. Allowing delayed rewards means that if rewards for goal states

H. R. Arabnia et al. (Eds.): CSCI 2025, CCIS 2967, pp. 53–67, 2027.
https://doi.org/10.1007/978-3-032-28454-9_4

can be defined, convergence to optimal policies is possible even when sequential rewards cannot be defined. However, conventional multi-objective reinforcement learning methods use time-triggered approaches that perform control operations at regular time intervals. Learning becomes difficult in large-scale state transition sequences with sparse rewards that provide rewards only at goal states, making it harder to acquire extensive Pareto fronts. Additionally, time-triggered control leads to increased control costs and the burden of setting environment-dependent time intervals [1,8].

Therefore, this paper proposes event-triggered reinforcement learning, which can omit state transition sequences, and verifies its performance through computational experiments on railway operations. This method controls based on significant state changes, defined as events rather than controlling at regular time intervals. This reduces unnecessary control operations and contributes to improved learning performance of multi-objective reinforcement learning in sparse reward environments. It also suggests its applicability to other energy-efficient transportation systems by reducing control costs and design burdens.

2 Preliminaries

2.1 Multi-objective Markov Decision Process

The multi-objective Markov decision process (MOMDP) is a sequential decision-making model with multiple objectives. MOMDP is defined as $\langle \mathcal{S}, \mathcal{A}, \mathcal{P}, \boldsymbol{\mathcal{R}}, \gamma \rangle$, where $\mathcal{S}$ is the state space, $\mathcal{A}$ is the action space, $\mathcal{P}$ is the state transition probability, $\boldsymbol{\mathcal{R}}$ is the reward vector function, and γ is the discount rate. In an MOMDP with m objectives, the rewards corresponding to the m objectives are given as m-dimensional vectors. That is, in state $s_t \in \mathcal{S}$, if action $a_t \in \mathcal{A}$ is chosen according to policy $\pi(s_t)$, and the transition is made to the next state $s_{t+1} \in \mathcal{S}$, a reward vector $\boldsymbol{R}_{t+1} \in \mathbb{R}^m$ is given.

The value vector $\boldsymbol{V}^\pi \in \mathbb{R}^m$ represents the cumulative expected discounted reward vector when following policy π from the initial state. It is defined in Eq. (1), where μ denotes the initial state distribution.

$$\boldsymbol{V}^\pi = E\left[\sum_{k=0}^{\infty} \gamma^k \boldsymbol{R}_{k+1} | \pi, \mu\right] \tag{1}$$

2.2 Multi-objective Reinforcement Learning

Multi-objective reinforcement learning (MORL) is a method that extends single-objective MDP environments to multi-objective MOMDP environments. The priority for each objective in MOMDP is called preference, and a function that scalarizes reward vectors or value vectors on the basis of preferences is called a utility function. In this paper, the decision-making entity is called an agent, and the entity that determines preferences is called a user.

User preferences are generally complex and nonlinear [4], making it difficult to design utility functions that strictly reflect preferences. Therefore, we assume

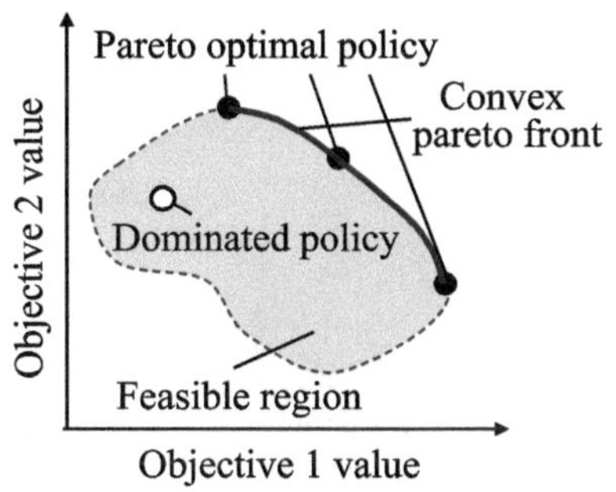

Fig. 1. Convex Pareto front

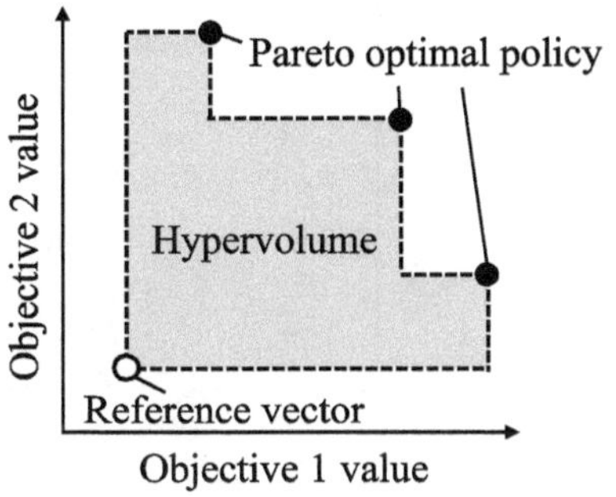

Fig. 2. Hypervolume

that user preferences are expressed by weights $\boldsymbol{w}$, and the utility function is represented by a simple scalarization function with parameter $\boldsymbol{w}$. Here, $\boldsymbol{w} = (w_1, \ldots, w_i, \ldots, w_m)$ satisfies $\sum_{i=1}^{m} w_i = 1, \forall w_i \geq 0$. The simplest and most general utility function U used in this paper is the linear scalarization utility function shown in Eq. (2).

$$U(\boldsymbol{w}, \boldsymbol{V}) = \boldsymbol{w}^\top \boldsymbol{V} \tag{2}$$

2.3 Pareto Front

In MDP, the policy that maximizes V^π is the optimal policy π^*, but in MOMDP, because $\boldsymbol{V}^\pi$ is an m-dimensional vector, simple magnitude comparison is not possible. Therefore, Pareto-optimal policies are defined by Pareto dominance relations among policies, and by finding the Pareto front—which is the set of these policies—optimal policies corresponding to various agent preferences can be selected and utilized. The definitions of Pareto dominance, Pareto-optimal policy, and Pareto front are shown in Definitions 1, 2, and 3 [4].

Definition 1. (Pareto dominance). *For all objectives i $(i = 1, \ldots, m)$, if the value $\boldsymbol{V}_i^\pi$ is greater than or equal to the value $\boldsymbol{V}_i^{\pi'}$, and at least one objective j has the value $\boldsymbol{V}_j^\pi$ surpassing the value $\boldsymbol{V}_j^{\pi'}$, then π Pareto dominates π', denoted as $\boldsymbol{V}^\pi \succ_{\mathrm{P}} \boldsymbol{V}^{\pi'}$. This is shown in Eq. (3).*

$$\boldsymbol{V}^\pi \succ_{\mathrm{P}} \boldsymbol{V}^{\pi'} \Leftrightarrow \left(\forall i : \boldsymbol{V}_i^\pi \geq \boldsymbol{V}_i^{\pi'}\right) \wedge \left(\exists j : \boldsymbol{V}_j^\pi > \boldsymbol{V}_j^{\pi'}\right) \tag{3}$$

Definition 2. (Pareto optimal policy). *Let Π be the set of all possible policies. If there is no policy $\pi' \in \Pi$ that Pareto-dominates the policy $\pi \in \Pi$, then π is called a Pareto-optimal policy.*

Definition 3. (Pareto front). *The set of Pareto-optimal policies is called the Pareto front (PF), which is shown in Eq. (4).*

$$PF(\Pi) = \left\{\pi \in \Pi \mid \nexists \pi' \in \Pi : \boldsymbol{V}^{\pi'} \succ_{\mathrm{P}} \boldsymbol{V}^\pi\right\} \tag{4}$$

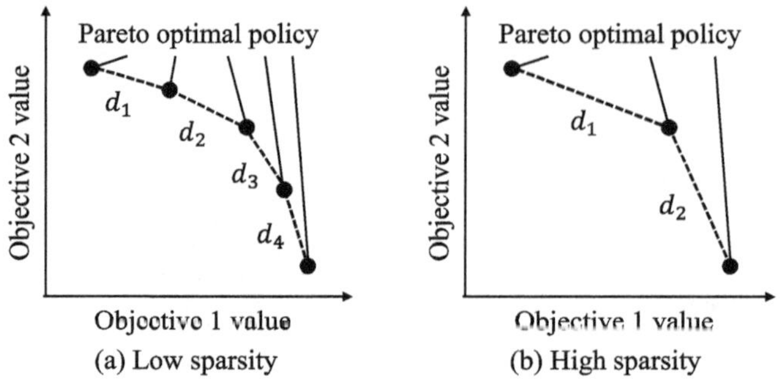

Fig. 3. (a) Pareto front with low sparsity (b) Pareto front with high sparsity

The Pareto front exists on the periphery of the feasible region, representing a group of feasible policies. The set of Pareto-optimal policies distributed in a convex manner, as illustrated in Fig. 1, is referred to as a convex Pareto front.

2.4 Evaluation Metrics of the Pareto Front

Hypervolume. The hypervolume [16] is a metric for evaluating the volume of space dominated by a set of vectors, as shown in Fig. 2. For a given Pareto front PF under evaluation, the hypervolume HV is defined in Eq. (5).

$$HV(PF, \boldsymbol{V}^{\mathrm{ref}}) = \bigcup_{\pi \in PF} \mathrm{Volume}(\boldsymbol{V}^{\mathrm{ref}}, \boldsymbol{V}^{\pi}) \tag{5}$$

$\boldsymbol{V}^{\mathrm{ref}} \in \mathbb{R}^m$ represents the reference vector satisfying $\forall \pi \in PF, \boldsymbol{V}^{\pi} \succ_{\mathrm{P}} \boldsymbol{V}^{\mathrm{ref}}$. Generally, $\boldsymbol{V}^{\mathrm{ref}}$ corresponds to the worst-case cumulative reward vector in the target problem, determined based on reward design. $\mathrm{Volume}(\boldsymbol{V}^{\mathrm{ref}}, \boldsymbol{V}^{\pi})$ denotes the volume of space enclosed between $\boldsymbol{V}^{\mathrm{ref}}$ and $\boldsymbol{V}^{\pi}$.

Sparsity. Sparsity [15] is a metric used to evaluate the sparsity of the distribution of each vector in a set of vectors, assessing how densely Pareto-optimal policies are distributed. For a Pareto front PF, sparsity Sp is defined in Eq. (6). In Fig. 3, (a) exhibits lower sparsity than (b), with policies distributed more densely. This allows for a finer reflection of user preferences. Generally, a Pareto front with higher hypervolume and lower sparsity is desirable.

$$Sp(PF) = \frac{1}{|PF| - 1} \sum_{j=1}^{m} \sum_{i=1}^{|PF|-1} (PF_j(i) - PF_j(i+1))^2 \tag{6}$$

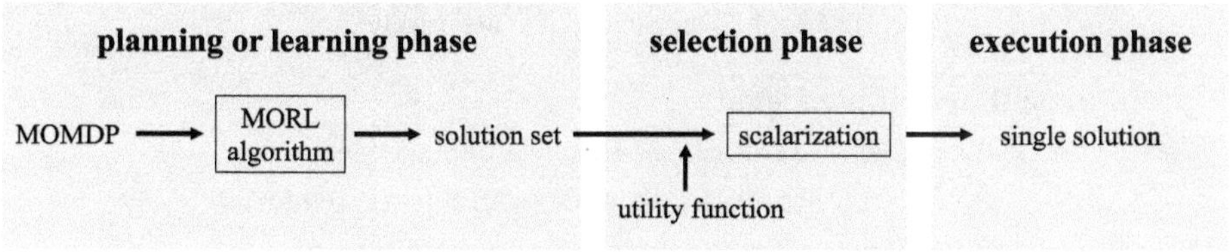

Fig. 4. Unknown utility function scenario

3 Problem Domain

3.1 MOMDP Scenario

This paper addresses tasks where user preferences are unknown during the policy learning phase. In such tasks, it is effective to learn a set of policies for arbitrary preferences via the multiple-policy approach and select policies based on user preferences during execution [4]. Roijers *et al.* referred to the sequence of procedures, from policy learning to execution using MORL in MOMDP, as scenarios and presented multiple scenarios [10].

As a scenario where the utility function is unknown during policy learning, the unknown utility function scenario is shown in Fig. 4. This scenario is divided into three phases: learning phase, selection phase, and execution phase. During the learning phase, a set of policies (solution set) is learned via MORL algorithms. In this paper, this set of policies refers to the Pareto front. During the selection phase, given user preferences, the policy that maximizes the utility function within the learned policy set is selected. During the execution phase, the selected policy is executed.

3.2 Related Work on Extensive Pareto Front Acquisition

In the unknown utility function scenario, acquiring extensive Pareto fronts during the learning phase for the selection phase is an important problem. Numerous methods using reinforcement learning have been proposed to acquire extensive Pareto fronts. Van Moffaert *et al.* proposed Pareto Q-Learning [13] as a method to learn Pareto fronts without performing scalarization, assuming utility functions during the learning phase. Furthermore, Reymond *et al.* proposed Pareto-DQN [9], which extends Pareto Q-Learning to deep reinforcement learning.

Additionally, Xu *et al.* [15] achieved extensive Pareto front acquisition in continuous robot control via prediction-guided MORL. This method enables continuous Pareto front approximation through Pareto analysis and interpolation and efficiently explores individual policies on the Pareto front. Chen *et al.* [3] and Soh *et al.* [12] proposed methods that apply evolutionary multi-objective optimization algorithms to reinforcement learning. By using evolutionary strategies that explore the entire Pareto front while maintaining diverse solutions, extensive Pareto fronts can be acquired. However, all these methods are time-driven methods, and no event-driven methods have been proposed.

Table 1. Definition of state input

Input	Range	Unit	Detail
t	$[0, \infty)$	[s]	Time elapsed from scheduled departure time
x	$[0, \infty)$	[m]	Running distance from departure station
v	$[0, v_{\text{lim}}]$	[m/s]	Current speed

3.3 Target Problem

Figure 5 shows a conceptual diagram and terminology of the train trajectory optimization (TTO) problem addressed in this paper. This optimization problem determines the trajectory, namely, the sequence of running positions and running speeds over the time axis, under operational and physical constraints [14]. When running on flat tracks with sufficient running time, as shown in Fig. 5, energy-efficient train operation that minimizes total energy consumption can be realized through a sequence of driving operations combining four types: Maximum traction, Speed holding, Coasting, and Maximum braking [7]. Maximum traction refers to driving with the maximum tractive force to accelerate. Speed holding involves maintaining the current speed using a partial tractive force. Coasting is a train that neither accelerates nor brakes. The maximum braking is driving with maximum deceleration by applying the brakes.

Therefore, this study deals with trains running a distance of L[m] between stations under a speed limit of v_{lim}[km/h]. Because user preferences are unknown in the learning phase, we aim to acquire extensive Pareto fronts. The train under control is treated as an agent. The agent observes states such as elapsed time since departure and its position between stations as state inputs and selects driving operations that realize time and energy objectives according to preferences as action outputs. We do not consider regenerative energy during braking and aim to minimize energy consumption during traction. In the selection phase, users set preferences according to their target running time t_{p}[s] and energy consumption at that time. Using linear scalarization utility functions, policies are selected from Pareto-optimal policies on the Pareto front according to preferences. In the execution phase, driving operations are performed at the arrival station based on the selected policy. Details of each variable used in the computational experiments are described in Sect. 5.1.

1. **Time Objective**: Minimizing the time taken to reach the arrival station
2. **Energy Objective**: Minimizing the total energy consumption of train driving

3.4 Modeling

Formalization of MDP. The state input is a 3-dimensional vector, as shown in Table 1, with input values normalized to the range $[0, 1]$. The value ranges in Table 1 represent the values before normalization.

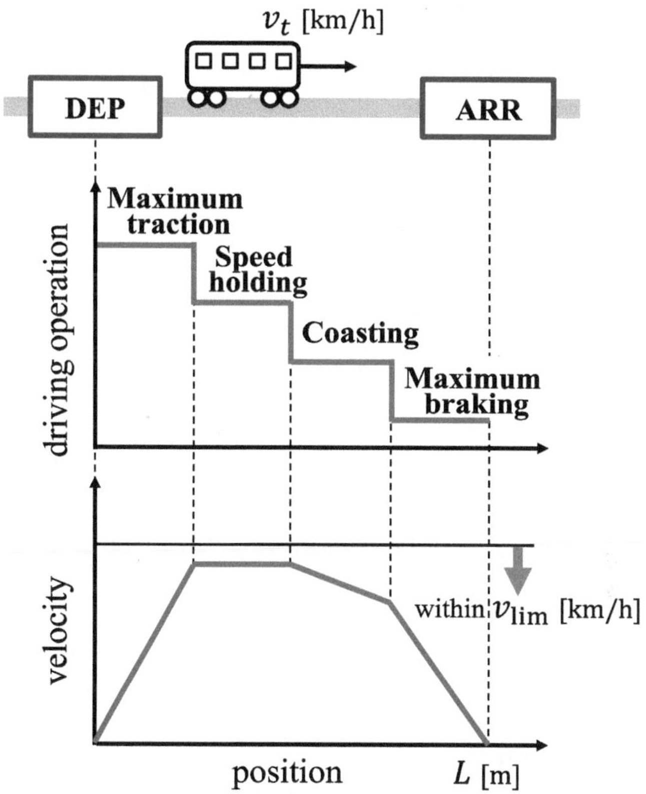

Fig. 5. TTO problem overview

The action output is a two-dimensional vector consisting of four driving operations (described in Sect. 3.3) and the time interval for executing the selected operation. The action output a_t is selected at each time step, every second.

To evaluate the driving operations chosen by the agent, we define the reward function as shown in Eqs. (7)–(10). Specifically, Eq. (7) represents the reward R_1 for evaluating running time, Eq. (8) represents the reward R_2 for evaluating the train's final position upon arrival, Eq. (9) represents the reward R_3 for evaluating running time upon arrival, and Eq. (10) represents the reward R_4 for evaluating energy efficiency. Additionally, the set of goal states, S_{g}, includes states where the train stops within the allowable position error e_x[m] of the target stopping position L[m]. This is defined in Eq. (11). Because R_1-R_3 are provided only upon arrival, they constitute sparse rewards. Conversely, R_4 is provided at each time step and serves as a negative reward for energy consumption, which does not influence the learning of the policy for reaching the goal state.

$$R_1(s) = \begin{cases} 2 & \text{if } \ s = s_{\text{g}} \in \mathcal{S}_{\text{g}} \\ 0 & \text{otherwise} \end{cases} \tag{7}$$

$$R_2(s) = \begin{cases} x/L & \text{if } \ x \leqq L \land v = 0 \\ (2\,L - x)/L & \text{if } \ x > L \land v = 0 \end{cases} \tag{8}$$

$$R_3(s) = -t/t_{\mathrm{p}} \ \text{ if } \ s = s_{\mathrm{g}} \in \mathcal{S}_{\mathrm{g}} \tag{9}$$

$$R_4(s) = -cE \tag{10}$$

$$\mathcal{S}_{\mathrm{g}} = \{(t, x_{\mathrm{g}}, 0) \mid x_{\mathrm{g}} \in \mathcal{X}_{\mathrm{g}}\} \tag{11}$$

$$\mathcal{X}_{\mathrm{g}} = \{x \mid |L - x| \leq e_x\}$$

Train Movement. The acceleration $\alpha_t[\mathrm{m/s^2}]$ of the train when the agent selects the external force $F_t[\mathrm{N}]$ is calculated using Eq. (12) on the basis of the Newtonian equation of motion. The running resistance $R_{\mathrm{r}}(v)[\mathrm{N}]$ is given by Eq. (13). Here, $m[\mathrm{kg}]$ represents the train mass, n is the number of vehicles, and $g[\mathrm{m/s^2}]$ is the gravitational acceleration.

$$\alpha_t = \frac{1}{m}\{F_t - R_{\mathrm{r}}(v_t)\} \tag{12}$$

$$R_{\mathrm{r}}(v) = c_1 + c_2 v + c_3 v^2 \tag{13}$$

$$c_3 = \frac{0.0280 + 0.0078(n-1)}{mg}$$

The torque $T(v)[\mathrm{N{\cdot}m}]$ exerted at each speed v is used to calculate the external force $F[\mathrm{N}]$ applied to the train, as given by Eq. (14). Here, $r[\mathrm{m}]$ represents the radius of the train wheel. In Eq. (8), $E[\mathrm{J}]$ represents the amount of energy consumed during train operation, and it is calculated as shown in Eq. (15). This study aims to minimize the energy consumed during traction; therefore, the max function is introduced. Details of each variable are provided in Sect. 5.1.

$$F = \frac{T(v)}{r} \tag{14}$$

$$E = \frac{1}{2}\max\{0, F_t\}\,(v_t + v_{t+\Delta t})\,\Delta t \tag{15}$$

4 Proposed Method: Event-Triggered Reinforcement Learning

4.1 Semi-markov Decision Process

The semi-Markov decision process (SMDP), as introduced in this study, extends the MDP framework to continuous-time decision-making. To handle continuous time in SMDP, a distribution function of time intervals between state transitions is introduced. That is, when the action $a \in \mathcal{A}$ is chosen in the state $s \in \mathcal{S}$ at time τ_k, the time until the transition to the next state s' at time τ_{k+1} follows a distribution function $F(x|s,a,s')$. The expected value of the time interval is denoted by Eq. (16).

$$\tau(s,a,s') = E\left[\tau_{k+1} - \tau_k | s,a,s'\right] = \int_0^\infty x\, dF(x|s,a,s') \tag{16}$$

4.2 Formalization by SMDP

In general, time-triggered reinforcement learning performs control operations at regular intervals. However, time-triggered reinforcement learning faces challenges such as difficulty in learning large-scale state transition sequences with sparse rewards, increased control costs, and the burden of setting environment-dependent time intervals [1,8]. By contrast, event-triggered reinforcement learning is a control method that activates actions based on specific important events [11].

First, from the state s_k, an action a_k is selected based on the policy $\pi(a|s)$. In event-triggered reinforcement learning, the system waits until a predefined event occurs. When an event is triggered, the next state s_{k+1}, reward r_k, and time interval $\Delta\tau_k$ are observed, and the policy $\pi(a|s)$ is updated.

In this study, the action value function $Q^\pi(s,a)$ is defined in Eq. (17) via the SMDP. Unlike the conventional definition with a discount factor γ in MDPs, this definition incorporates the time interval $\Delta\tau_k$ to account for continuous time.

$$\begin{aligned} Q^\pi(s_k,a_k) = \sum_{\Delta\tau}\sum_{s_{k+1}} p(\Delta\tau, s_{k+1} \mid s_k, a_k) \Bigg\{ &\frac{1-e^{-\beta\Delta\tau}}{\beta} r(s_k,a_k) \\ &+ e^{-\beta\Delta\tau} \sum_{a_{k+1}} \pi(a_{k+1} \mid s_{k+1}) Q^\pi(s_{k+1}, a_{k+1}) \Bigg\} \end{aligned} \tag{17}$$

4.3 Definition of Event

The event definition is illustrated in Fig. 6. At time τ_k, the action a_k is selected with two components: one of the four driving operations described in Sect. 3.3: Maximum traction, Speed holding, Coasting, and Maximum braking; and the time interval $\Delta\tau_k$ during which the selected driving action is executed. An event is defined by the time interval $\Delta\tau_k$ that elapses after executing the selected action. When an event occurs, a new action a_{k+1} is selected, and the corresponding driving operation is performed for $\Delta\tau_{k+1}$ until the next event.

By optimizing the time interval $\Delta\tau_k$ via event-triggered reinforcement learning, the possibility of acquiring rewards with fewer actions increases even when sparse rewards are used, enabling more accurate learning. Furthermore, it reduces the burden of setting environment-dependent time intervals, which is necessary when time-triggered reinforcement learning is used.

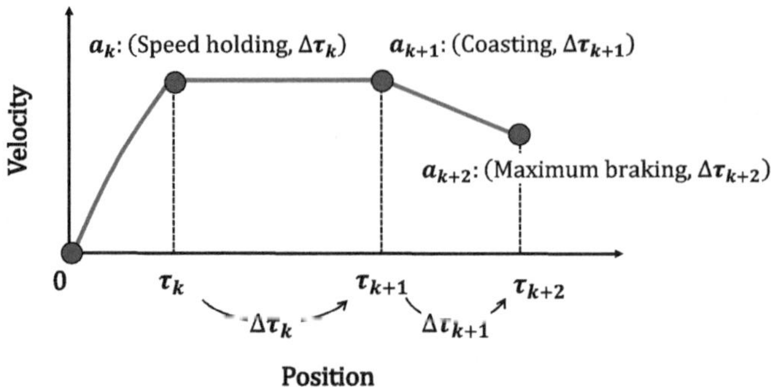

Fig. 6. Definition of an event in training operation

4.4 Related Work on Event-Triggered Reinforcement Learning

This section reviews related research on event-triggered reinforcement learning. Baumann *et al.* [1] applied event-triggered control to improve sample efficiency and performance. Ran *et al.* [8] also implemented event-triggered control in data center cooling equipment to reduce energy consumption.

Event-triggered control has also been explored in heater control systems. Hosseinloo *et al.* [5,6] introduced event-triggered online reinforcement learning for controlling heaters. In their approach, the task was to determine the optimal room temperature for switching the heater on and off based on a threshold value. They reported that event-triggered control outperformed conventional time-triggered control, possibly because of the optimization of the time interval for the threshold value and the room temperature change, which contributed to better performance.

4.5 Deep Q Network

DRL is a broad category of reinforcement learning methods that integrate deep learning techniques. In DRL, to enable application when the state space $\mathcal{S}$ or action space $\mathcal{A}$ is extremely large or continuous, value functions and measures are represented by neural networks and are learned through deep learning algorithms. DQN approximates the action–value function Q via a neural network, where the state s is input, and the network outputs the action values for each possible action a. The action–value function Q then determines the selected action a. The DQN algorithm using event triggering in this study is outlined in Algorithm 1. Unlike a standard DQN, it waits until the event defined in line 6 occurs. Upon event occurrence, it observes the next state s_{k+1}, the reward r_k, and the time interval $\Delta\tau_k$.

The variable D represents the replay buffer, a list that temporarily stores the transitions (s, a, r, s', t). Updating Q is performed via transitions $(s_i, a_i, r_i, s'_i, t_i)$ randomly sampled from D, a process known as Experience Replay. The target value y_i used to update Q is computed via a separate network $\hat{Q}$, which is a copy

Algorithm 1. Event-triggered DQN

Initialize Q, $\hat{Q}$, and D
Repeat for each episode:
  Initialize s
  Repeat for each event in the episode:
    Select action a in state s via the policy derived from Q
    Take action a and continue until the next event occurs.
    Then, observe s', r, and $\Delta\tau$
    Store transition $(s, a, r, s', t, \Delta\tau)$ in D
    Sample a random transition $(s_i, a_i, r_i, s_i', t_i, \Delta\tau_i)$ from D
    If t_i is a terminal state, then
      $y_i = r_i$
    If t_i is not a terminal state, then
      $y_i = \frac{1-e^{-\beta\Delta\tau_i}}{\beta} r_i + e^{-\beta\Delta\tau_i} \max_{a'} \hat{Q}(s_i', a')$
    Update Q with the error $(y_i - Q(s_i, a_i))^2$
    Copy Q to $\hat{Q}$ every fixed number of steps
    $s \leftarrow s'$
  End repeat if s is a terminal state

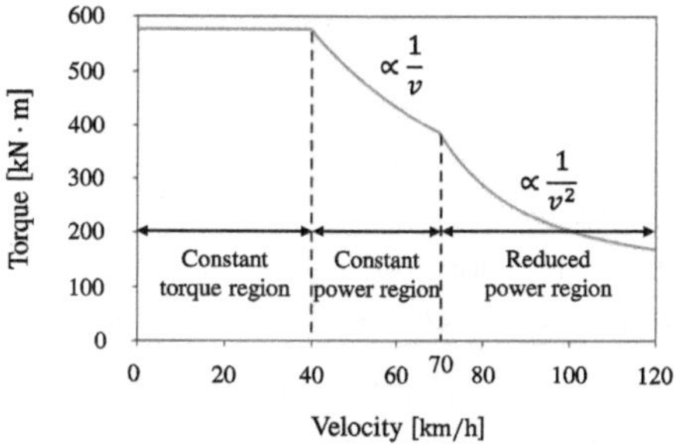

Fig. 7. Train torque characteristics during traction

Table 2. Details of train and environment

Item	Value
Diameter of wheel [m]	0.80
Mass of train [kg]	300×10^3
Gravitational acceleration [m/s^2]	9.8
Number of vehicles	10
Coefficient c_1	1.32
Coefficient c_2	0.0164

of Q updated at regular intervals. This $\hat{Q}$ is referred to as the target network, and it stabilizes learning by keeping the parameters fixed for several steps.

5 Computer Experiment

5.1 Experimental Settings

In the computational experiments, we used a benchmark environment commonly used for evaluation [7]. We targeted train operations running a station-to-station distance of $L = 2000$ m under a speed limit of $v_{\text{lim}} = 140$ km/h. In this experiment, we defined the train torque characteristics during traction, as shown in Fig. 7. We also set the allowable error in the stopping position used for reward function differentiation as $e_x = 1$ m. The assumed train and environment details are shown in Table 2.

Table 3. Comparison of hypervolume and sparsity

Algorithm	Hypervolume	Sparsity
Event-triggered RL (proposed)	0.451 ± 0.002	$0.612 \times 10^{-1} \pm 0.004 \times 10^{-1}$
Time-triggered RL (1 s)	0.419 ± 0.005	$0.071 \times 10^{-1} \pm 0.004 \times 10^{-1}$
Time-triggered RL (2 s)	0.433 ± 0.004	$0.067 \times 10^{-1} \pm 0.003 \times 10^{-1}$
Time-triggered RL (4 s)	0.422 ± 0.002	$0.097 \times 10^{-1} \pm 0.002 \times 10^{-1}$
Time-triggered RL (8 s)	0.409 ± 0.002	$0.101 \times 10^{-1} \pm 0.002 \times 10^{-1}$

We compared the proposed method with time-triggered reinforcement learning via the DQN to evaluate it. We conducted trials with preference weights $\boldsymbol{w}$ varying as $\boldsymbol{w} = (0,1), (0.1, 0.9), \ldots, (1,0)$ for a total of 11 types and obtained value vectors $\boldsymbol{V}^{\pi}$ to obtain Pareto fronts. Each method was repeated 10 times for the experiments. As the exact Pareto front for the experimental environment is unknown, we evaluated the learned Pareto fronts from the following two perspectives. As a preliminary experiment, we also conducted learning with different time intervals for time-triggered reinforcement learning (1s, 2s, 4s, 8s) to show the burden of setting time intervals for each environment.

1. Hypervolume and Sparsity
   See Sect. 2.4.
2. Monotonic increase in values with respect to weights
   Let the policies selected under weights $\boldsymbol{w}$ and $\boldsymbol{w}'$ be denoted as $\pi_{\boldsymbol{w}}$ and $\pi_{\boldsymbol{w}'}$. When the weights w_j and w'_j for objective j satisfy $w_j > w'_j$, if the values for objective j of policies $\pi_{\boldsymbol{w}}$ and $\pi_{\boldsymbol{w}'}$ satisfy $V_j^{\pi_{\boldsymbol{w}}} \geq V_j^{\pi'_{\boldsymbol{w}}}$, we say that the values are monotonically increasing with respect to weights.

5.2 Experimental Results

Figure 8 shows a comparison of Pareto fronts acquired via the proposed event-triggered reinforcement learning and comparative time-triggered reinforcement learning methods (time intervals 1s, 2s, 4s, 8s). The horizontal axis represents the cumulative reward for the energy objective, and the vertical axis represents the cumulative reward for the time objective. The color of each plotted point represents the preference weight w_{ene} for the energy objective.

In the proposed method, values do not monotonically increase from $w_{\text{ene}} = 0.3$ to 0.4, and Pareto-optimal policies cannot be acquired at $w_{\text{ene}} = 0.0$. For other weights, values monotonically increased with respect to w_{ene}. By contrast, time-triggered reinforcement learning (1s, 2s) has multiple points that are not monotonically increasing and multiple points where Pareto-optimal policies cannot be acquired, such as $w_{\text{ene}} = 0.0$ and 1.0. Additionally, time-triggered reinforcement learning (4s, 8s) learns the same policy for many points, resulting in fewer Pareto-optimal policies.

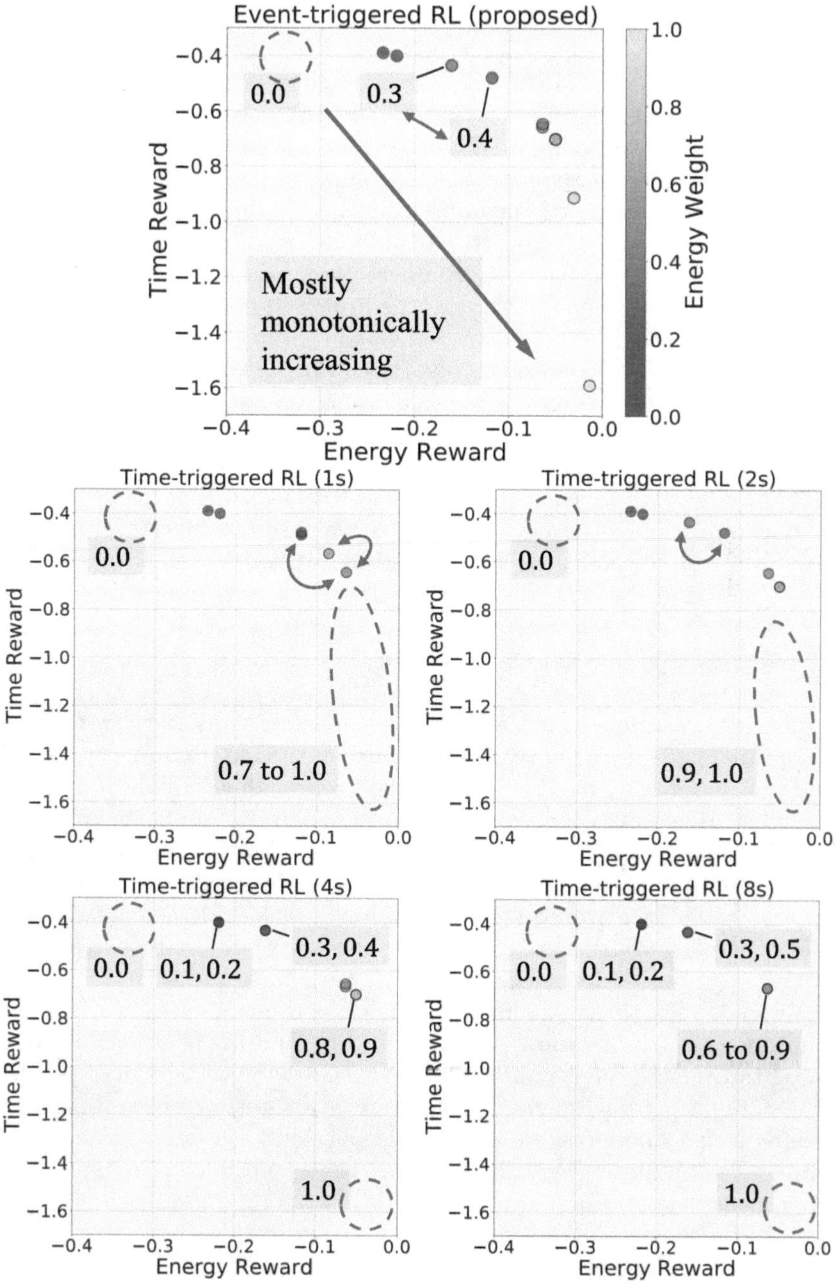

Fig. 8. Comparison of Acquired Pareto Fronts

Table 3 shows the hypervolume and sparsity obtained by repeating the experiments 10 times for each method. We set the reference vector as $\boldsymbol{V}^{\text{ref}} = (-0.4, -1.7)$ based on the worst cumulative reward values in learning. In the hypervolume comparison, which represents the breadth of the user options, the Pareto front from the proposed method had the maximum hypervolume. In the sparsity comparison, which represents the fineness of the user options, time-triggered reinforcement learning (2s) had the minimum sparsity, whereas the proposed method had the maximum. This is because of the influence of $w_{\mathrm{ene}} = 1.0$ that only the proposed method can acquire, and from Fig. 8, the Pareto-optimal policies of the proposed method can be said to be high density.

Therefore, the proposed method can provide users with broader and more detailed options because it has a large hypervolume and a high density of Pareto-optimal policies. Furthermore, although the proposed method does not perfectly maintain the monotonic increase in values concerning weights, it is generally maintained and has the highest consistency between preferences and policies. This is considered to be due to reduced unnecessary control operations and improved learning performance in sparse reward environments. Notably, the fact that only the proposed method can acquire $w_{\mathrm{ene}} = 1.0$, which is an extreme preference and challenging to learn, demonstrates this. Conversely, in time-triggered reinforcement learning, the time interval of 2s resulted in the largest hypervolume. However, the solutions were restricted at 4s and 8s, concentrating on specific policies and showing the difficulty of setting time intervals.

6 Conclusions

Multi-objective reinforcement learning aims to learn the Pareto front, a set of Pareto-optimal policies. The distribution range of the Pareto front represents the range of policy options available to users, hence providing broader Pareto fronts is important. However, this requires accurate learning of the balance among objectives, and many methods have been proposed. However, conventional methods use time-triggered approaches, and in large-scale state transition sequences with sparse rewards, learning becomes difficult, making it a more challenging problem. Therefore, this study proposes event-triggered reinforcement learning to improve learning performance in MORL via sparse rewards. The proposed method was verified through computational experiments on railway operations and confirmed superior learning performance by acquiring broader Pareto fronts compared to time-triggered reinforcement learning.

The automatic train operation problem is a multi-objective optimization problem requiring running time, energy efficiency, and passenger comfort. Therefore, future work aims to realize driving that considers passenger comfort and energy-efficient driving using regenerative braking. Additionally, we aim to expand MORL methods by combining conventional method proposals with event-triggered approaches.

References

1. Baumann, D., Zhu, J.J., Martius, G., Trimpe, S.: Deep reinforcement learning for event-triggered control. In: 2018 IEEE Conference on Decision and Control (CDC), pp. 943–950. IEEE (2018)

2. Castelletti, A., Pianosi, F., Restelli, M.: A multiobjective reinforcement learning approach to water resources systems operation: pareto frontier approximation in a single run. Water Resour. Res. **49**(6), 3476–3486 (2013)
3. Chen, D., Wang, Y., Gao, W.: Combining a gradient-based method and an evolution strategy for multi-objective reinforcement learning. Appl. Intell. **50**(10), 3301–3317 (2020)
4. Hayes, C.F., et al.: A practical guide to multi-objective reinforcement learning and planning. arXiv preprint arXiv:2103.09568 (2021)
5. Hosseinloo, A.H., Dahleh, M.A.: Event-triggered reinforcement learning; an application to buildings' micro-climate control. In: AAAI Spring Symposium: MLPS (2020)
6. Hosseinloo, A.H., Ryzhov, A., Bischi, A., Ouerdane, H., Turitsyn, K., Dahleh, M.A.: Data-driven control of micro-climate in buildings: an event-triggered reinforcement learning approach. Appl. Energy **277**, 115451 (2020)
7. Jiaxin, C., Howlett, P.: Application of critical velocities to the minimisation of fuel consumption in the control of trains. Automatica **28**(1), 165–169 (1992)
8. Ran, Y., Zhou, X., Hu, H., Wen, Y.: Optimizing data center energy efficiency via event-driven deep reinforcement learning. IEEE Trans. Serv. Comput. **16**(2), 1296–1309 (2022)
9. Reymond, M., Nowé, A.: Pareto-DQN: approximating the pareto front in complex multi-objective decision problems. In: Proceedings of the Adaptive and Learning Agents Workshop (ALA-19) at AAMAS (2019)
10. Roijers, D.M., Vamplew, P., Whiteson, S., Dazeley, R.: A survey of multi-objective sequential decision-making. J. Artif. Intell. Res. **48**, 67–113 (2013)
11. Sedghi, L., Ijaz, Z., Noor-A-Rahim, M., Witheephanich, K., Pesch, D.: Machine learning in event-triggered control: recent advances and open issues. IEEE Access **10**, 74671–74690 (2022)
12. Soh, H., Demiris, Y.: Evolving policies for multi-reward partially observable markov decision processes (MR-POMDPs). In: Proceedings of the 13th Annual Conference on Genetic and Evolutionary Computation, pp. 713–720 (2011)
13. Van Moffaert, K., Nowé, A.: Multi-objective reinforcement learning using sets of pareto dominating policies. J. Mach. Learn. Res. **15**(1), 3483–3512 (2014)
14. Wang, P., Trivella, A., Goverde, R.M., Corman, F.: Train trajectory optimization for improved on-time arrival under parametric uncertainty. Transp. Res. C: Emerg. Technol. **119**, 102680 (2020)
15. Xu, J., Tian, Y., Ma, P., Rus, D., Sueda, S., Matusik, W.: Prediction-guided multi-objective reinforcement learning for continuous robot control. In: International Conference on Machine Learning, pp. 10607–10616. PMLR (2020)
16. Zitzler, E., Thiele, L.: Multiobjective evolutionary algorithms: a comparative case study and the strength pareto approach. IEEE Trans. Evol. Comput. **3**(4), 257–271 (2002)

Machine Learning Insights on SME Resilience: Evidence from Ibero-America and Ecuador

Carmen Parreno-Collaguazo[1], Edison Loza-Aguirre[2(✉)], Maria Isabel Sanchez-Pazmino[1], and Henry N. Roa[3]

[1] Escuela Politécnica Nacional, Quito 170143, Ecuador
{carmen.parreno,maria.sanchez02}@epn.edu.ec
[2] Universidad San Francisco de Quito, Quito 170901, Ecuador
eloza@usfq.edu.ec
[3] Pontificia Universidad Católica del Ecuador, Quito 170143, Ecuador
hnroa@puce.edu.ec

Abstract. This study investigates the resilience of SMEs in Ecuador and Ibero-America during the COVID-19 crisis using machine learning techniques. Resilience was defined as the ability of firms to maintain or increase profitability during the pandemic, based on survey data from more than 9,000 SMEs, including 1,111 from Ecuador. Multiple Correspondence Analysis (MCA) was applied as an exploratory step, revealing more homogeneous resilience patterns among Ecuadorian SMEs compared to greater heterogeneity across the broader Ibero-American dataset. For predictive modeling, several classifiers were evaluated. Ensemble methods consistently outperformed traditional classifiers, with the Gradient Boosting Classifier achieving the highest accuracy (94.6% for Ecuador and 91.9% for Ibero-America). The findings identify innovation capacity and adaptability to change as the most critical determinants of resilience, while financial access and income diversification played a stronger role in Ibero-America. These results provide guidance for policies aimed at fostering innovation, digitalization, and financing to strengthen SME resilience in future crises.

Keywords: SME · resilience · COVID-19 · machine learning · Ibero-America · Ecuador

1 Motivation

The COVID-19 pandemic had an unprecedented impact on the global economy, severely disrupting businesses and societies worldwide [1]. Micro, small, and medium-sized enterprises (SMEs) were particularly vulnerable, as lockdown and social distancing measures forced the suspension of activities in critical sectors such as food services, hospitality, and commerce [2]. Given that SMEs constitute the backbone of the productive structure in Latin America—accounting for more than 99% of firms and over 60% of employment in countries such as Mexico and Ecuador [3–5]—understanding their ability to adapt and recover is of critical importance.

H. R. Arabnia et al. (Eds.): CSCI 2025, CCIS 2967, pp. 68–78, 2027.
https://doi.org/10.1007/978-3-032-28454-9_5

Organizational resilience, defined as the capacity to withstand shocks, adapt to disruptions, and sustain performance under adverse conditions [6], became essential for SME survival during the pandemic. Prior research has identified factors such as firm size, innovation, and financial support as key determinants of resilience [7, 8]. However, most studies remain descriptive or context-specific, with limited use of advanced data-driven approaches to systematically model and compare SME resilience across regions. To address this gap, the main objective of this paper is to derive empirical insights into the determinants of SME resilience during the COVID-19 crisis, with particular emphasis on differences between Ecuador and the broader Ibero-American region.

This study applies a machine learning approach to a large-scale survey (Economic Impact of the COVID-19 Crisis on SMEs in Ibero-America 2021), which includes more than 9,000 firms, of which 1,111 are from Ecuador. Resilience is modeled as the ability of a firm to maintain or improve profitability during the crisis. Several classifiers are evaluated—logistic regression, decision trees, random forests, k-nearest neighbors, Naive Bayes, and gradient boosting ensembles—to identify the most effective predictive models and the key factors shaping resilience.

The remainder of this paper is organized as follows. Section 2 reviews the theoretical background. Section 3 outlines the research methodology. Section 4 presents the experimental results. Section 5 offers a discussion of the results. Finally, Sect. 6 concludes with key findings and directions for future research.

2 Background

The concept of organizational resilience has been extensively studied across different crisis contexts. The COVID-19 pandemic combined health, economic, and social factors, weakening economies and particularly affecting SMEs in Latin America [9]. Globally, SMEs account for about 90% of firms and contribute a significant share of GDP [7]. Examining SME resilience is therefore essential to identify the factors that enabled firms to withstand the crisis and to improve their ability to respond to future disruptions.

Prior studies in Latin America have explored resilience from multiple perspectives. In Mexico and Chile, researchers examined the link between organizational resilience and business performance during the pandemic [7]. In Ecuador, resilience has been studied through the lens of Absorptive Capacity Theory, considering factors such as firm size, age, and economic activity [8]. While these studies provide valuable insights, they remain largely descriptive. A deeper and more systematic understanding can be achieved through data-driven approaches that leverage analytics and machine learning.

Business resilience is commonly defined as the ability of organizations to adapt and remain competitive in disruptive environments. According to Sánchez Pazmiño et al. [8], it is closely tied to a firm's ability to prevent crises, absorb shocks, and adjust to environmental changes. These capabilities rest on three pillars: identifying risks and designing mitigation strategies, adapting processes and operations to external shifts, and acquiring and applying new knowledge to strengthen competitiveness. Absorptive capacity plays a key role, as firms that integrate technological learning with resilience are better positioned to respond to change and sustain competitiveness [9].

The COVID-19 pandemic highlighted the importance of these capabilities, as SMEs across Latin America—and especially in Ecuador—faced severe challenges in maintaining operations. Resilience and absorptive capacity proved critical for survival, particularly under economic contraction and the need to adopt new business models [9]. Some firms leveraged their small size and flexible structures as assets for adaptation, which allowed them to navigate the crisis more effectively [1]. Understanding resilience in this context is not only relevant for assessing firm performance but also for guiding public policy to support SMEs [8]. By clarifying the areas in which SMEs should focus their efforts, firms can mitigate risks, adapt to changes, and even thrive during adverse conditions [7]. It is therefore crucial to understand not only the impacts of crises but also the factors that enable successful recovery [6].

Analyzing these dimensions provides insight into how SMEs can build organizational capabilities centered on resilience and strategic learning to prepare for future crises. Firms that effectively adapt to disruptions are more likely to maintain or even improve profitability during adverse events such as the COVID-19 pandemic. This research contributes to the literature by deepening the theoretical understanding of organizational resilience and by applying machine learning techniques to systematically evaluate resilience determinants. From a practical perspective, the findings can inform both business strategies and public policies aimed at strengthening SME resilience and ensuring sustainability in the face of future crises.

3 Methods

This study applies a machine learning approach to analyze SME resilience during the COVID-19 crisis. The methodology follows a structured pipeline that includes data acquisition, preprocessing, target variable definition, model training, and evaluation. The goal is to systematically compare predictive models and extract empirical insights into the factors influencing resilience across Ecuador and Ibero-America.

3.1 Data Source

The dataset used in this study comes from the survey "Economic Impact of the COVID-19 Crisis on SMEs in Ibero-America 2021," conducted by FAEDPYME, CLADEA, and the National Polytechnic School of Ecuador. The survey collected responses from approximately 9,300 SMEs across several Ibero-American countries, including 1,111 from Ecuador. The questionnaire comprised 20 items covering firm characteristics (e.g., size, sector, age, ownership), perceptions of the COVID-19 impact, economic performance indicators (e.g., revenue, profitability, productivity, liquidity), and management practices such as financing and innovation.

3.2 Data Preparation

To ensure data quality, preprocessing steps were applied systematically. Missing values in categorical variables were handled using mode imputation, while ordinal variables were encoded to preserve their natural order. Categorical variables were transformed using one-hot encoding, and numerical variables were standardized to zero mean and unit variance. The target distribution showed class imbalance between resilient and non-resilient firms; to address this, the Synthetic Minority Over-sampling Technique (SMOTE) was applied to the training sets, thereby improving model learning for the minority class.

3.3 Target Variable Definition

Resilience was defined as a firm's ability to maintain or improve profitability during 2020 compared to pre-pandemic levels. Survey responses were recoded into a binary target variable: Resilient (1) if profitability remained stable or increased, and Not Resilient (0) if profitability decreased. This operationalization links resilience directly to financial performance, providing a clear and measurable basis for modeling.

3.4 Modeling

The modeling phase combined exploratory and predictive techniques. As a first step, Multiple Correspondence Analysis (MCA) was applied to explore associations among categorical variables and to visualize structural patterns in the dataset. This method facilitated the identification of groupings and latent relationships relevant to resilience analysis.

Subsequently, several machine learning classifiers were implemented to predict resilience. These included logistic regression, decision trees, random forests, k-nearest neighbors (KNN), Naive Bayes, and gradient boosting ensembles (Gradient Boosting Classifier and HistGradientBoosting Classifier). This combination of exploratory analysis and classification models enabled both an interpretable understanding of data structures and a rigorous comparison of predictive performance across modeling paradigms.

3.5 Model Evaluation

Model performance was assessed using accuracy, precision, recall, F1-score, and ROC-AUC. To ensure robustness, ten-fold cross-validation was applied, and results were averaged across folds. Classification reports and confusion matrices were generated for both Ecuadorian and Ibero-American subsets, enabling a comparative analysis of predictive performance and resilience determinants across contexts.

4 Results

This section presents the empirical results obtained from the analysis. First, MCA is used to explore associations among categorical variables and to visualize structural patterns in the dataset. Then, a set of supervised machine learning classifiers are evaluated to predict SME resilience.

4.1 Multiple Correspondence Analysis

MCA was applied to explore associations among categorical variables and to identify latent structures in the dataset. The analysis was conducted separately for Ecuador and Ibero-America, and then for the complete dataset.

As shown in Table 1, the first two MCA components for Ecuador explain 8.48% of the variance, while for Ibero-America the variance explained is slightly lower at 6.79%. Although these percentages are relatively modest, which is typical for categorical data with many dimensions, they capture relevant latent structures that differentiate resilient from non-resilient firms. The corresponding projection is illustrated in Fig. 1, where Ecuadorian SMEs (blue) appear more clustered, while Ibero-American SMEs (red) show greater dispersion. This suggests that Ecuadorian SMEs share more homogeneous resilience profiles, whereas the broader Ibero-American sample reflects greater diversity in strategies and outcomes.

Table 1. MCA Results for Ecuador and Ibero-America.

Country	Component	Eigenvalue	% Variance	Cumulative % Variance
Ecuador	0	0.049	4.87%	4.87%
	1	0.036	3.61%	8.48%
Ibero-America	0	0.038	3.80%	3.80%
	1	0.030	2.99%	6.79%

The results for the complete dataset are presented in Table 2. Here, the first two dimensions account for 11.76% of the total variance, a slightly higher proportion than when analyzing Ecuador and Ibero-America separately. This indicates that, when all countries are included, the MCA captures stronger common structures across the region. Figure 2 shows the projection of firms by country, where visible clustering patterns suggest national differences in resilience determinants. For instance, countries with larger and more diversified economies (e.g., Mexico, Brazil) exhibit broader dispersion, while smaller economies such as Ecuador display tighter clustering, reflecting more uniform—but potentially more constrained—responses to the crisis.

The MCA highlights structural contrasts between Ecuador and the rest of Ibero-America. Ecuadorian SMEs show more homogeneous resilience profiles, likely linked to structural limitations such as restricted access to financing and limited diversification strategies. In contrast, the Ibero-American dataset reveals greater variability, consistent with the presence of differentiated public policies, financial support mechanisms, and innovation-driven strategies across larger economies.

From a methodological perspective, while the explained variance of MCA is modest, this is expected in high-dimensional categorical data. Even so, the analysis provides meaningful insights into resilience patterns and cross-country differences. MCA thus serves as a useful exploratory step before predictive modeling, helping to contextualize the classification results discussed in later sections.

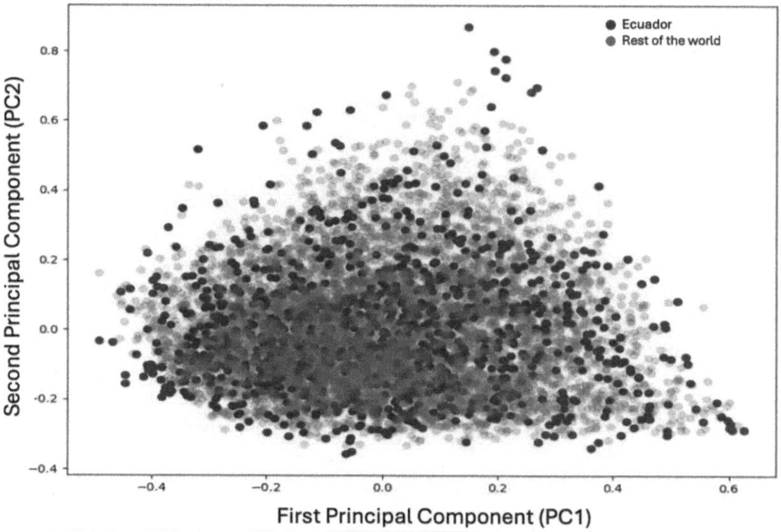

Fig. 1. Projection of Ecuadorian and Ibero-American SMEs in the MCA Space.

Table 2. MCA Results for the Complete Dataset.

Component	Eigenvalue	% Variance	Cumulative % Variance
0	0.264	6.19%	6.19%
1	0.237	5.57%	11.76%

4.2 Comparative Analysis of Classification Models

To evaluate predictive performance, seven classifiers were applied to the Ecuadorian and Ibero-American datasets: Logistic Regression, Decision Tree, Random Forest, KNN, Naive Bayes, Gradient Boosting Classifier (GBC), and HistGradientBoosting Classifier (HGBC). Model performance was assessed using ten-fold cross-validation with accuracy as the main metric, complemented by precision, recall, and F1-score.

As shown in Table 3, ensemble methods achieved the best results for Ecuador. The GBC obtained the highest average accuracy (94.6%) with a relatively low standard deviation, indicating robust and consistent performance. HGBC and Random Forest followed closely, also surpassing 93% accuracy. Logistic Regression performed competitively with 90.8%, showing that even linear models can capture important resilience patterns. In contrast, KNN had the weakest performance (64.8%) and the highest variability across folds, making it unsuitable for this dataset.

Table 4 presents the results for the Ibero-American dataset. Again, ensemble methods dominated, with GBC achieving the best accuracy (91.9%), closely followed by HGBC and Random Forest, all above 91%. Logistic Regression also performed well (88.5%), slightly below its performance for Ecuador. Naive Bayes was weaker at 81.9%, and KNN

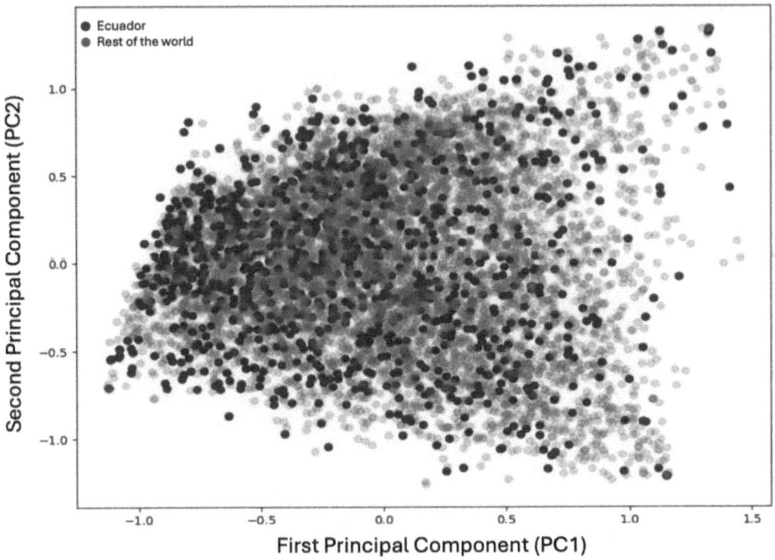

Fig. 2. Projection of All Countries in the MCA Space.

Table 3. Comparative Performance of Models for Ecuador (ten-fold cross-validation).

Model	Accuracy	Standard deviation
Logistic Regression	0.9081	0.0219
Decision Tree	0.8501	0.0362
Gradient Boosting (GBC)	0.9459	0.0407
HistGradientBoosting (HGBC)	0.9428	0.0431
Random Forest	0.9326	0.0525
Naive Bayes	0.8839	0.0332
K-Nearest Neighbors (KNN)	0.6481	0.1870

again produced low accuracy (67.1%) with high variability, confirming its limitations with high-dimensional categorical data.

The comparative analysis reveals consistent trends across both datasets. Ensemble models achieved the best performance, with the GBC showing the highest accuracy (94.6% for Ecuador and 91.9% for Ibero-America) and relatively low variance across folds, indicating robust generalization. HGBC and Random Forest also performed strongly, confirming the advantage of ensemble-based approaches in this classification problem.

Logistic Regression obtained competitive results in both contexts (90.8% for Ecuador, 88.5% for Ibero-America), demonstrating that simpler linear models still capture a substantial portion of the variance in resilience patterns. However, KNN and

Table 4. Comparative Performance of Models for Ibero-America (ten-fold cross-validation).

Model	Accuracy	Standard deviation
Logistic regression	0.8848	0.0119
Decision tree	0.8537	0.0308
Gradient boosting (GBC)	0.9198	0.0441
HistGradientBoosting (HGBC)	0.9187	0.0482
Random forest	0.9147	0.0507
Naive Bayes	0.8196	0.0418
K-Nearest Neighbors (KNN)	0.6714	0.1562

Naive Bayes showed weaker performance, with KNN particularly unstable (accuracy 64.8% for Ecuador and 67.1% for Ibero-America, with high variance), suggesting that distance-based methods are less suited to this high-dimensional categorical dataset.

The classification reports (Tables 5 and 6) highlight the strength of GBC in balancing precision and recall for both classes, indicating its superior ability to identify both resilient and non-resilient SMEs. Importantly, the overall higher performance of ensemble methods underscores their effectiveness in modeling complex resilience phenomena, where multiple interacting factors influence firm outcomes.

These findings confirm that resilience can be effectively predicted using machine learning, and that ensemble methods provide the most reliable results. They also demonstrate consistent differences between Ecuador and Ibero-America, with regional models achieving slightly lower variance explained, reflecting the heterogeneity of resilience determinants across countries.

Table 5. Classification Report for Ecuador using GBC.

	precision	recall	F1-score	support
0	0.94	0.91	0.92	160
1	0.75	0.80	0.77	51
Accuracy			0.89	211
Macro avg	0.84	0.86	0.85	211
Weighted avg	0.89	0.89	0.89	211

Taken together, the exploratory and predictive results highlight two key insights. First, resilience determinants are not uniformly distributed: Ecuadorian SMEs exhibit more constrained profiles, while Ibero-American firms demonstrate greater heterogeneity. Second, ensemble learning provides a reliable methodological approach to predict resilience outcomes, offering stronger predictive accuracy and balanced classification across resilient and non-resilient SMEs. These insights provide the foundation for a

Table 6. Classification Report for Ibero-America using GBC.

	Precision	Recall	F1-score	Support
0	0.92	0.93	0.93	859
1	0.86	0.83	0.85	414
Accuracy			0.90	1273
Macro avg	0.89	0.88	0.89	1273
Weighted avg	0.90	0.90	0.90	1273

deeper discussion on the determinants of resilience and their implications for business strategies and public policies.

5 Discussion

The results of this study provide a deeper interpretation of SME resilience in Ecuador and Ibero-America during the COVID-19 crisis, connecting the empirical findings with the research objectives, the scientific context, and prior studies.

Consistent with existing research, the analysis reveals that SMEs in Ibero-America exhibited more consistent resilience than those in Ecuador. This was reflected in the higher predictive accuracy of the models for Ibero-America compared to Ecuador. This finding aligns with studies that emphasize the role of adaptability and operational flexibility in navigating crises [1]. The greater resilience observed in Ibero-America can be partly explained by the diversity of public policies and support programs implemented in countries such as Mexico, Colombia, and Brazil, which facilitated firm adaptation to new economic conditions [2]. In contrast, Ecuadorian SMEs achieved lower predictive performance, suggesting that they faced greater challenges in maintaining profitability during the pandemic. These challenges may be associated with limited access to financing, lower economic diversification, and dependence on sectors such as commerce and services, which were severely affected by the crisis [3]. Such results reinforce the need for targeted policies that strengthen absorptive capacity and adaptability in Ecuador, as highlighted by Sánchez Pazmiño and Acosta Andino [9].

From a methodological perspective, the comparative evaluation of classification models identified the GBC as the most effective algorithm in both Ecuador and Ibero-America. This outcome is consistent with previous research that underscores the superiority of ensemble methods in classification problems involving imbalanced and high-dimensional data [10]. The high accuracy of GBC (94.6% for Ecuador and 91.9% for Ibero-America) and its relatively low variance across folds demonstrate that this model is robust and generalizes well across different contexts. In contrast, models such as KNN and Naive Bayes showed weaker performance, confirming their limitations in handling complex, multidimensional datasets.

The analysis of variable importance in decision tree and logistic regression models further supports the finding that innovation capacity and adaptability to change were critical determinants of resilience in both contexts. This aligns with theories suggesting that firms integrating strategic learning and adopting digital business models are more likely to overcome economic shocks [1, 5]. At the same time, structural differences became evident: in Ecuador, access to financing and income diversification had a weaker impact on resilience, while in Ibero-America these variables were more significant, suggesting that financial support and economic diversification policies played a key role in strengthening SME resilience [2, 4].

The practical implications of these findings are considerable. For Ecuador, there is an urgent need to strengthen SMEs' absorptive capacity and adaptability, particularly through training programs in innovation and digitalization, and by improving access to financing and technological resources [3]. At the regional level, the heterogeneity observed across Ibero-America suggests that policies must be tailored to each national context rather than adopting uniform strategies. Financial support and diversification programs have proven effective in some countries, while others require investment in digital infrastructure and broader institutional reforms to support resilience.

Finally, it is important to acknowledge the limitations of this study. The analysis relies on self-reported survey data, which may introduce response biases. Moreover, the survey does not capture all possible factors influencing resilience, such as the role of business networks, inter-firm collaboration, or informal support mechanisms [7]. Future research could complement these findings by incorporating objective data, such as financial and productivity indicators, and by conducting longitudinal studies to examine how SME resilience evolves over time and across different crisis contexts [1].

6 Conclusion

This study analyzed the resilience of SMEs in Ecuador and Ibero-America during the COVID-19 crisis using data analytics and machine learning techniques. By operationalizing resilience as the ability to maintain or improve profitability, the analysis provided a systematic assessment of resilience determinants across two different contexts.

The results highlight three key findings. First, SMEs in Ibero-America demonstrated greater resilience than those in Ecuador, reflecting the effectiveness of diversified policy responses and financial support programs in countries such as Mexico, Colombia, and Brazil. In contrast, Ecuadorian SMEs faced structural constraints—including limited access to financing, low diversification, and dependence on vulnerable sectors—that restricted their capacity to adapt.

Second, ensemble learning methods proved to be the most effective for predicting resilience, with the GBC consistently outperforming other models. This methodological contribution underscores the value of advanced machine learning techniques for analyzing complex organizational phenomena such as resilience.

Third, the analysis confirmed that innovation capacity and adaptability to change were critical determinants of resilience in both contexts, while financial access and income diversification played a more decisive role in Ibero-America than in Ecuador. These insights point to the need for differentiated strategies: in Ecuador, targeted policies should focus on improving access to financing, promoting digitalization, and enhancing absorptive capacity, while in the broader Ibero-American region, strengthening diversification and innovation remains essential.

Despite these contributions, the study has limitations. The reliance on self-reported survey data may introduce biases, and the survey design does not capture all possible resilience factors, such as inter-firm collaboration or informal support networks. Future research should complement these results with objective financial and productivity indicators and conduct longitudinal studies to assess how SME resilience evolves over time and across different crises.

In sum, this research provides empirical evidence that resilience can be systematically modeled using machine learning, offering both theoretical contributions to the understanding of organizational resilience and practical guidance for policymakers and SMEs seeking to enhance their ability to withstand future disruptions.

References

1. Martelo-Castro, M., Quintero-Castro, D., Vieira-Salazar, J.: Innovación y resiliencia: la Pyme frente a la crisis provocada por el Covid-19. Una revisión narrativa de la literatura. Economía y Negocios. **14**(2), 24–38 (2023)
2. Secretaría General Iberoamericana: Consejo de Empresarios Iberoamericanos, Federación Iberoamericana de Jóvenes Empresarios: Encuesta Iberoamericana de la Mi Pyme: apostando por las micro, pequeñas y medianas empresas para el crecimiento y la prosperidad. Secretaría General Iberoamericana, Madrid (2022)
3. Giles Navarro, C.A.: Recomendaciones para las MIPyME ¿Qué hacer para sobrevivir a la pandemia del Covid-19? No. 86. Belisario Domínguez Institute (2020)
4. Instituto Nacional de Estadística y Censos (INEC): Directorio de Empresas y Establecimientos, Quito (2019)
5. Pedroni, F.: Innovación como estrategia de resiliencia durante la crisis por Covid-19: análisis cualitativo de mipymes argentinas. Cuadernos Latinoamericanos de Administración. **18**(34), 1–21 (2022)
6. El Ríos, G.: impacto económico de la crisis del coronavirus en América Latina: canales de transmisión, mitigantes y respuestas de políticas públicas. Análisis del Real Instituto Elcano. **1**(73), 1–10 (2020)
7. García-Contreras, R., Valle-Cruz, D., Azalea, C.-G.R.: Selección organizacional: resiliencia y desempeño de las pymes en la era de la COVID-19. Estudios Gerenciales. **37**(158), 73–84 (2021)
8. Sánchez Pazmiño, M.I., Loza Aguirre, E.F., Sarrade Dueñas, F.: Influencia de la resiliencia en la teoría de la absorción. Revista ADENAG. **1**(13), 1–36 (2023)
9. Sánchez Pazmiño, M.I., Acosta Andino, B.F.: Capacidad de absorción: integración estratégica entre aprendizaje tecnológico, resiliencia y competitividad empresarial. Revista Venezolana de Gerencia (RVG). **4**, 528–547 (2020)
10. Witten, I.H., Frank, E.: Data Mining: Practical Machine Learning Tools and Techniques. Morgan Kaufmann, San Francisco (2005)

MSD-PUGAN: An InSAR 2-D Phase Unwrapping Method Based on Multi-scale Discriminator Learning

Xiaomao Chen(✉), Xiangfeng Li, Mingchuan Lei, and Xiangzhen Zeng

Guangxi Key Laboratory of Precision Navigation Technology and Application, Guilin University of Electronic Technology, Guilin 541000, China
xmchen@guet.edu.cn

Abstract. Phase Unwrapping (PU) is a critical step in Interferometric Synthetic Aperture Radar (InSAR) data processing, directly impacting the accuracy of Digital Elevation Model (DEM) generation and surface deformation monitoring. Traditional unwrapping methods, such as Least Squares, Minimum Cost Flow, and Quality Guided approaches, often rely on the Itoh condition. Consequently, they are prone to error propagation in low-coherence or high-noise environments, making it difficult to ensure reliability. To address this issue, this paper proposes a one-step phase unwrapping method based on a Multi-Scale Discriminator conditional Generative Adversarial Network (MSD-PUGAN). This method employs an improved U-Net architecture in its generator, incorporating a Convolutional Block Attention Module (CBAM) to adaptively capture key features. Simultaneously, a three-scale PatchGAN discriminator is utilized to provide hierarchical supervision, thereby balancing global phase consistency with local detail preservation. In terms of the loss function, an L1 reconstruction loss is combined with an adversarial loss to balance phase precision and structural realism. Experiments were conducted on both simulated datasets (including data generated from DEM inversion and random sine/cosine superposition) and real-world data (acquired by Sentinel-1). The results demonstrate that the proposed MSD-PUGAN significantly outperforms traditional algorithms and the original PUGAN in terms of both robustness and phase unwrapping accuracy, especially under low-coherence and complex terrain conditions. This validates the effectiveness and application potential of the method for high-precision InSAR phase unwrapping tasks.

Keywords: InSAR · Phase Unwrapping · Conditional Generative Adversarial Network · Multi-scale Discriminator

1 Introduction

Two-dimensional phase unwrapping (PU) is a crucial step in interferometric synthetic aperture radar (InSAR) for generating digital elevation models (DEM) or

H. R. Arabnia et al. (Eds.): CSCI 2025, CCIS 2967, pp. 79–93, 2027.
https://doi.org/10.1007/978-3-032-28454-9_6

monitoring surface subsidence. InSAR measurements rely on the interferometric phase difference between two SAR images, which typically lies within the interval $(-\pi,\pi]$. To recover the true phase values from the wrapped phase, an integer multiple of 2π must be added or subtracted based on the wrapped phase to restore the corresponding integer cycle. This process is referred to as phase unwrapping.

Over the past few decades, many two-dimensional phase unwrapping (2-D PU) methods based on algorithmic models have been proposed. Most of these methods are based on the assumption of phase continuity [1] (also known as the Itoh condition), which states that the absolute phase difference between any two adjacent pixels must be less than π. However, this condition is frequently violated in InSAR data due to noise and terrain variations. Traditional PU algorithms are generally categorized into three classes [2]: path-following, optimization-based, and integrated filtering methods. Path-following methods, such as the branch-cut (BC) [3], quality-guided (QG) [4], and mask-cut (MC) [5] algorithms, seek valid integration paths based on the Itoh condition but can fail in the presence of isolated "island" regions. Optimization-based methods globally minimize the discrepancy between true and estimated phase gradients. While more robust and avoiding the "island" issue, they are computationally intensive and can still suffer from global error propagation. Representative methods include least squares (LS) [6], minimum cost flow (MCF) [7], and SNAPHU [8]. Integrated filtering methods combine filtering and unwrapping to improve accuracy. For instance, Kalman filtering [9–11] reformulates PU as a state estimation problem for simultaneous denoising. Although effective against noise, the required linearization of the system can lead to the loss of higher-order phase information, potentially reducing unwrapping precision.

In recent years, deep learning (DL) based methods for two-dimensional phase unwrapping (2D-PU) have emerged, utilizing deep convolutional neural networks (DCNNs) to overcome the limitations of the traditional Itoh condition. Zhou et al. pioneered several approaches, including PGNet for phase gradient estimation via classification [12], BCNet for predicting branch cuts from residues [13], and PG-BCNet, which fuses both techniques to enhance robustness against noise and large phase gradients [14]. Application-specific models have also been developed. Wu et al. designed PUNet for mining deformation as a regression task, though it still suffers from error propagation in low-coherence, high-noise scenarios [15]. To address severe noise and deformation, Yang et al. proposed FCNet-CA, which combines multi-scale features and residual connections to recover subsidence [16]. Similarly, Gao et al. introduced MMPhU-Net, a fusion of U-Net and SegNet that directly estimates gradient ambiguity numbers but requires post-processing for refinement [17]. Hybrid models that integrate different paradigms have also been explored, such as MoDL-PU [18], which blends learning approaches, and TLPU [19], which combines DL with traditional algorithms. Apart from end-to-end regression methods, many of these techniques demand significant computational resources and rely on post-processing steps. Therefore, the objective of this paper is to design an end-to-end phase unwrapping model that is effective and robust under high-noise and low-coherence conditions.

Conditional Generative Adversarial Networks (cGANs) [20] are DCNNs that adaptively learn loss functions to produce high-quality images nearly indistinguishable from real data. By conditioning the generator and discriminator on external information, cGANs excel at image-to-image translation tasks, with the pix2pix framework [21] being a representative example. Leveraging pix2pix's ability to learn input-to-output mappings, Zhou et al. proposed PU-GAN, a one-step phase unwrapping method [22]. However, its generator, while effective at capturing global phase structure, often struggles with preserving high-frequency details and suppressing local noise. MSG-GAN [23] demonstrated that incorporating multi-scale discriminators provides multi-level gradient feedback, improving both global consistency and local detail quality. Inspired by this, our work integrates multi-scale discriminators into the phase unwrapping network. This architecture forces the discriminator to assess the authenticity of the unwrapped phase across various spatial scales, thereby guiding the generator to balance large-scale phase trends with fine-grained local details. The advantages of multi-scale discriminators in maintaining global structure and detail fidelity have also been confirmed by frameworks such as Pix2PixHD [24] and MRGAN [25].

This study introduces the concept of multi-scale discriminators and proposes a one-step phase unwrapping method, MSD-PUGAN, based on a multi-scale discriminator generative adversarial network. The model consists of three main components: a generator, discriminators, and a loss function. Multi-scale losses are incorporated into the training process, where the discriminators evaluate images at three different scales, thereby enhancing the adversarial interplay between the generator and the discriminators to adjust the loss. The generator adopts a U-Net architecture integrated with convolutional block attention modules (CBAM) to produce more accurate unwrapped phase images. The discriminator employs PatchGAN to distinguish the authenticity of the phase map generated by the generator. The loss function of MSD-PUGAN consists of L1 loss and adversarial loss. Specifically, the L1 loss is utilized to recover the low-frequency components of the image, making the generated unwrapped phase closer to the ground truth. Meanwhile, the adversarial loss enhances the overall quality of the unwrapping results by fostering a game-theoretic competition between the generator and the discriminator. Experimental results, based on both simulated and real InSAR interferometric data, demonstrate that this method can produce effective phase unwrapping outcomes under various noise conditions.

The structure of this paper is organized as follows: Section 2 introduces the architecture, loss function, and experimental environment of the MSD-PUGAN model. Section 3 describes the dataset construction method and the evaluation metrics for the unwrapping results, followed by a series of experimental results based on both simulated and real InSAR data. Section 4 provides the conclusion.

2 Network Structure and Training Strategy

This section will introduce the network structure and training strategy of MSD-PUGAN. The internal structure of the discriminator and its loss function refer to the content in the paper [22], and there is no need to elaborate on them again.

2.1 MSD-PUGAN Model

cGAN can effectively transform unsupervised learning into supervised learning, enabling the network to be better optimized according to specific task requirements. It is an advanced GAN framework specifically designed for image-to-image translation tasks. The primary goal of this method is to learn the mapping relationship between input and output images without the need for additional prior constraints.

MSD-PUGAN is primarily composed of a generator G and three discriminators D, with its main architecture illustrated in Fig. 1 In this structure, the wrapped phase φ, the ground-truth phase, and the unwrapped phase $G(\varphi)$ generated by are downsampled by factors of 1, 2, and 4 respectively using bilinear interpolation and then fed into the discriminators. The discriminators assign different labels to the images at these three scales. These labels are then fed back into the overall training loss, which facilitates the adversarial game between the generator and the discriminators. This process enhances the hierarchical receptive fields of the discriminator network.

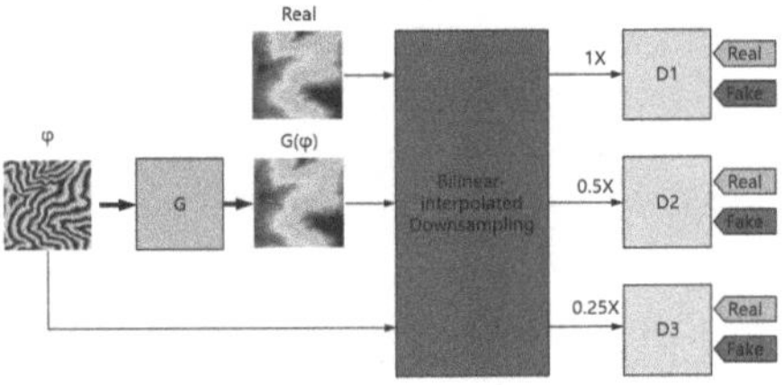

Fig. 1. Main structure flowchart

Bilinear interpolation is essentially a method for nonlinear value estimation in a continuous space. It reconstructs the value of a target point by calculating a weighted average of the four nearest neighboring pixels in the original image. The calculation formula is as follows:

$$I(x, y) = \sum_{i=0}^{1} \sum_{j=0}^{1} w_{ij} \cdot I(x_i, y_j) \tag{1}$$

Here, $I(x_i, y_j)$denotes the four integer coordinate points that fall within the neighborhood of the target coordinate (x, y), and the weightis w_{ij} determined by the relative distance between the target point and its corresponding neighboring point:

$$w_{ij} = (1 - |x - x_i|) \cdot (1 - |y - y_j|) \tag{2}$$

The effect of downsampling via bilinear interpolation is shown in Fig. 2.

(a) (b) (c)

Fig. 2. Illustration of downsampling: (a) 1× downsampling (original); (b) 2× downsampling; (c) 4× downsampling.

Although bilinear interpolation serves only as a continuous downsampling method, it preserves the spatial continuity of the phase structure when constructing multi-scale inputs. This enables the discriminators at each scale to effectively perceive structural distortions or local artifacts at different resolutions.

2.2 Generator Structure

The generator employs a classic U-Net-like encoder-decoder architecture. The 8-layer encoder progressively maps the input image from pixel space to a low-resolution feature space. Its first seven layers consist of 4×4 convolutional modules with a stride of 2, each followed by BatchNorm and LeakyReLU (slope=0.2), except for the first layer which omits BatchNorm. The final layer is a convolution with ReLU activation. During this process, the channel count increases from 1 to 512, while the spatial dimensions reduce from 512×512 to 1×1. The 7-layer decoder reconstructs the features back to the pixel space using 4×4 transposed convolution modules, each incorporating BatchNorm and skip connections with its corresponding encoder layer, followed by a ReLU activation to preserve multi-scale details. To mitigate overfitting, Dropout (rate=0.05) is applied in the first three decoder blocks. A Channel and Spatial Attention Module (CBAM) is embedded in the fourth encoder-decoder pair to enhance feature extraction and reconstruction. The final output is a single transposed convolution layer that restores the channel count to 1 and the spatial resolution to match the input. The generator's architecture is illustrated in Fig. 3.

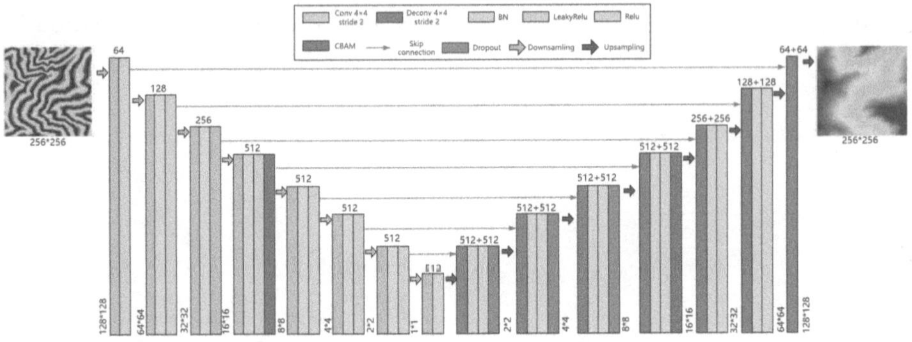

Fig. 3. Generator structure

CBAM is a lightweight module [26] that consists of two sub-modules connected in series: a channel attention module and a spatial attention module, which explicitly model channel and spatial attention, respectively. The channel attention module utilizes both global average pooling and max pooling to extract channel descriptors. These descriptors are then processed through a shared Multi-Layer Perceptron (MLP) to generate a channel-wise attention map, enabling the discrimination of significant channels. Its mathematical expression is given by Eq. (3), and the final attention output is Eq. (4).

$$M_s(F') = \sigma(\mathrm{Conv}_{7\times7}([\mathrm{AvgPool}(F'); \mathrm{MaxPool}(F')])) \tag{3}$$

where σ represents the Sigmoid function, and $\otimes$ denotes element-wise multiplication. The overall flowchart of this module is illustrated in Fig. 4.

$$F'' = F \otimes M_c \otimes M_s \tag{4}$$

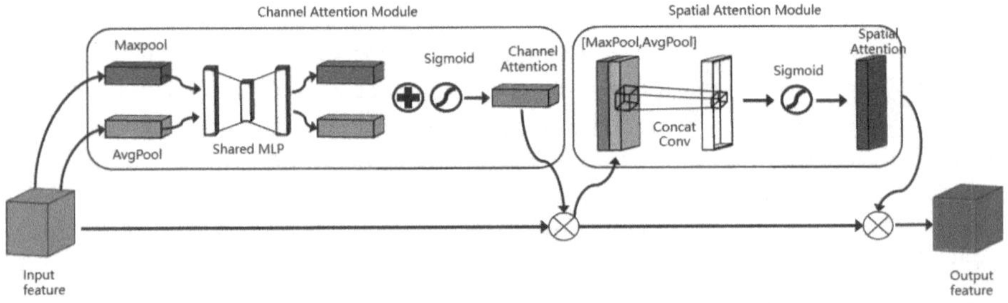

Fig. 4. CBAM module flowchart

In the unwrapping network, the Channel Block Attention Module (CBAM) enhances feature representation through its cascaded channel and spatial attention mechanisms. For the phase unwrapping task, channel attention enables the network to adaptively select key features relevant to the phase structure while suppressing noise. Concurrently, spatial attention directs the network's focus to areas of phase discontinuities and high coherence. This dual-attention approach improves the restoration of local details without compromising global phase continuity.

2.3 Tranining Details

The weights for both generator G and discriminator D are initialized from a Gaussian distribution. G and D are updated alternately in each training iteration. The Multi-scale discriminator D operates on inputs at three resolutions: original, 0.5x, and 0.25x downsampled. Each scale's discriminator calculates its adversarial loss independently. These three losses are then averaged to create a unified, multi-scale supervisory signal, guiding the generator's training. The generator loss is composed of two parts: the mean adversarial feedback from the three discriminators, promoting Multi-scale realism, and the L_1 reconstruction loss against the ground truth, preserving numerical accuracy. We optimize this total loss using the Adam optimizer with a batch size of 8, an initial learning rate of 0.0002, and a loss weight λ of 120. The model was trained for 250 epochs using PyTorch 2.5.1 on a single NVIDIA 4060 GPU (8 GB VRAM) under Python 3.12.8. The learning rate was held constant for the first 125 epochs and then linearly decayed to zero over the final 125 epochs.

3 Experiments and Results

In this section, to validate the unwrapping performance of the MSD-PUGAN method, we conduct a comparative analysis using various unwrapping methods on both simulated and real data.

3.1 Dataset Generation

In this experiment, the unwrapping performance of MSD-PUGAN is validated using two simulated datasets: one is a DEM inversion dataset [27], and the other is a dataset generated by the superposition of sine and cosine functions.

The ground-truth phase can be obtained through Digital Elevation Model (DEM) inversion, with the expression as follows:

$$\Phi(i, j) = \frac{4\pi}{\lambda} \cdot \frac{B_{\perp} h(i, j)}{R \sin(\theta)} \tag{5}$$

where $\Phi(i, j)$ is the ground-truth phase value of the pixel at row i and column j;λ is the operating wavelength of the Synthetic Aperture Radar (SAR);$B_{\perp}$ is the effective perpendicular baseline length; $h(i, j)$ represents the elevation information at the corresponding position, which can be obtained from a Digital Elevation Model (DEM); R is the satellite's orbit height; and θ is the incidence angle of the radar wave on the ground. We used the formula in Eq. 8 to conduct DEM inversion and convert it into a dataset. Then, we obtained the wrapped phase through wrapping. To make it closer to the real situation, random noise was added. Several commonly used satellite data are shown in Table 1.

Table 1. Satellite-borne synthetic aperture radar parameters

Sensor	$\lambda(m)$	$B_{\perp}(m)$	$R(m)$	$\theta(°)$
TerraSAR-X/TanDEM-X	0.031	227.86	710344.5	46.3
Sentinel-1	0.055	159.60	876298.8	39.3
ALOS-2	0.236	316.73	793416.8	39.0

We generate ground-truth phase by summing N random sine and cosine functions, where N, as well as the frequency and initial phase of each function, are randomized. This method produces a 3D surface with a controllable structure and a known analytical form, which is ideal for algorithm validation. The final training dataset is created by wrapping this ground-truth phase and adding random noise. The mathematical formulation is as follows:

$$\begin{aligned} S(t) &= \sum_{i=1}^{N} f_i(t) \\ f_i(t) &= A_i \sin(2\pi f_i t + \phi_i) + B_i \cos(2\pi g_i t + \theta_i) \end{aligned} \tag{6}$$

whereA_i and B_i are the amplitudes,f_i and g_i are the frequencies, ϕ_i and θ_i andare the initial phases for the i-th sine and cosine functions, respectively.

For training, we merge the data from the two methods described above (DEM inversion and sine/cosine superposition) into one comprehensive dataset. This dataset simulates diverse terrain features such as mountains, valleys, and plains, and captures the complexity of geological structures, random land use patterns, and non-uniform surface processes.

3.2 Quantitative Method

To assess the robustness of MSD-PUGAN, we systematically compare it against several traditional unwrapping methods and the original PUGAN. The traditional methods benchmarked are Least Squares (LS), Minimum Cost Flow (MCF), Quality Guided (QG), and Unscented Kalman Filter (UKF). To ensure a fair comparison, both our improved method and PUGAN are trained on identical datasets. We consistently use the Root Mean Square Error (RMSE) as the evaluation metric for all methods, defined as:

$$\text{RMSE} = \sqrt{\frac{1}{N}\sum_{i=1}^{N}(\psi_i - \hat{\psi}_i)^2} \tag{7}$$

where ψ_i is the unwrapped phase,$\hat{\psi}_i$ is the ground-truth phase, and N is the total pixel count. A lower RMSE value signifies better algorithm performance in terms of error control, precision, reliability, and practical utility. As such, RMSE serves as a vital and indispensable metric for assessing the performance of unwrapping algorithms.

3.3 Simulation Data Experimental Results

The simulation results are presented for two data types: 1) data from DEM inversion, and 2) data from sine/cosine superposition with added noise of varying coherence. We test two examples from each category to validate the unwrapping

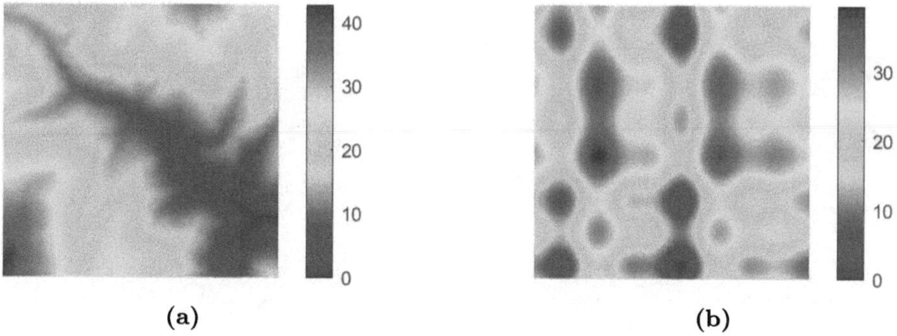

Fig. 5. Simulated data: (a) are from DEM inversion; (b) are from sine and cosine superposition.

Table 2. Comparison of RMSE values for Data 1 and Data 2 using various methods (RMSE: rad)

Data	Coh/RMSE	LS	MCF	QG	UKF	PUGAN	MSD-PUGAN
1	1.0	0.28	0.00	1.62	1.54	1.15	0.21
	0.9	2.49	2.37	2.28	1.40	1.13	0.19
	0.8	4.81	3.33	4.98	2.37	1.25	0.33
	0.7	5.28	3.59	24.67	7.10	1.37	0.59
	0.6	4.61	3.32	34.43	11.75	1.26	0.57
	0.5	8.14	6.46	34.77	31.07	1.16	0.35
2	1.0	0.15	0.00	0.74	18.84	1.01	0.21
	0.9	1.53	1.47	2.83	24.96	0.94	0.17
	0.8	3.25	2.51	7.58	12.80	0.86	0.19
	0.7	6.35	2.74	17.29	10.72	1.10	0.22
	0.6	8.67	4.06	28.15	34.26	0.62	0.22
	0.5	11.90	4.21	37.95	46.35	0.90	0.41

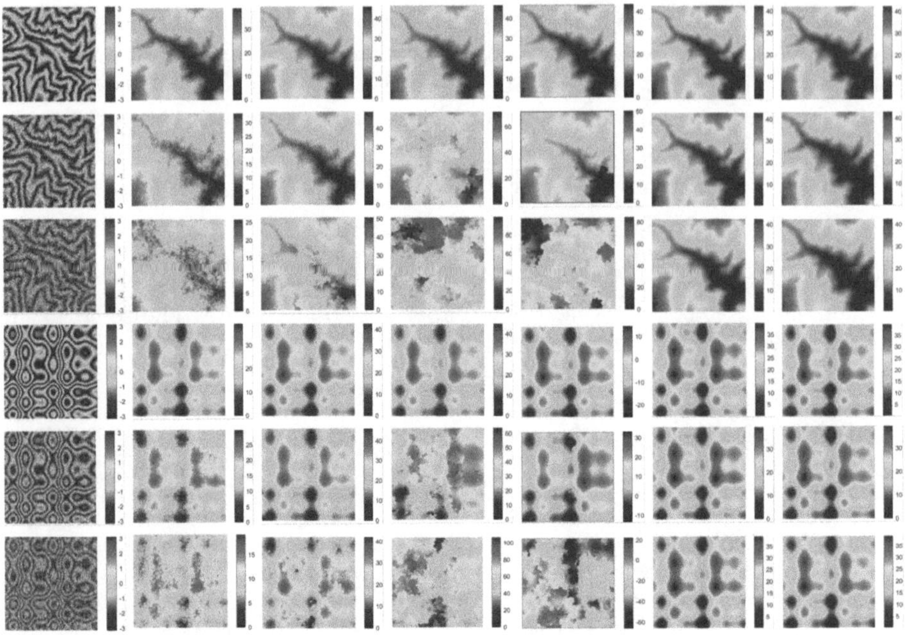

Fig. 6. Unwrapping results of various algorithms on Data 1 and Data 2 under different noise conditions. Columns 1–7 represent the wrapped phase, Least Squares (LS), Minimum Cost Flow (MCF), Quality Guided (QG), Kalman Filter (UKF), PUGAN, and MSD-PUGAN, respectively. Rows 1–3 display the wrapped and unwrapped images for Data 1 under coherence coefficients of 0.9, 0.7, and 0.5, respectively. Similarly, rows 4–6 display the results for Data 2 under the same coherence conditions.

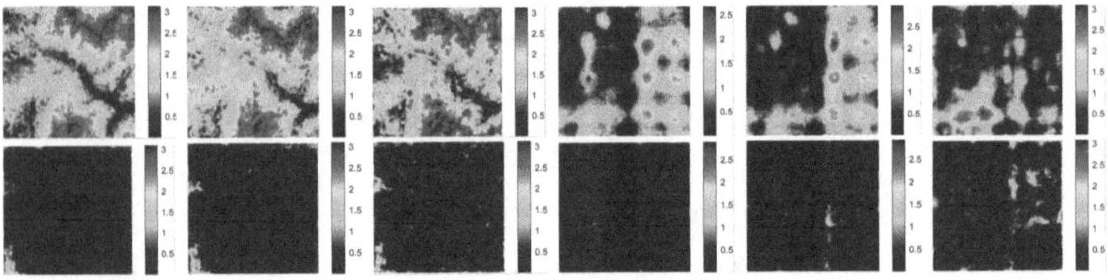

Fig. 7. Error map comparison between PUGAN and MSD-PUGAN on Data 1 and Data 2. Columns 1–3 display the results for Data 1 under coherence coefficients of 0.9, 0.7, and 0.5, respectively. Columns 4–6 display the results for Data 2 under the same coherence conditions. Rows 1–2 represent the error maps of PUGAN and MSD-PUGAN, respectively.

performance of MSD-PUGAN against traditional methods. The test images are shown in Fig. 5.

Figure 6 displays the unwrapping results for Data 1 and 2. In these figures, each row compares the performance of different methods under the same coherence level, while each column shows the performance of a single method across varying coherence levels. Visual inspection reveals that the traditional methods (LS, MCF, QG, UKF) perform well under high coherence, generally recovering the correct phase, a conclusion supported by their RMSE values. However, in low-coherence scenarios, these methods fail significantly due to the constraints of the Itoh condition, leading to noticeable error propagation. In contrast, the deep learning-based methods, PUGAN and MSD-PUGAN, are not bound by this condition and successfully produce complete unwrapping results across all coherence levels. As the visual distinction between PUGAN and MSD-PUGAN is subtle, we conducted a more detailed comparison by calculating the absolute error between the unwrapped and ground-truth phases, presented as heat maps in Fig. 7. In these maps, darker colors represent lower errors (closer to zero), while reddish colors indicate larger errors. This visualization of absolute error clearly highlights regions of phase discontinuity in the unwrapped results. Ultimately, the quantitative analysis based on RMSE from Table 2 provides the definitive conclusion. A lower RMSE indicates higher unwrapping accuracy. For the realistic coherence range of 0.5 to 0.9, the MSD-PUGAN method consistently achieves the optimal RMSE. Although the MCF method is theoretically optimal at a coherence of 1.0, this represents an idealized, noise-free condition rarely encountered in practical applications.

The results indicate that PUGAN's error maps exhibit significant red areas (denoting errors) in central or edge regions for both DEM-derived and sine/cosine data. While PUGAN's outputs are visually close to the ground truth, these errors are reflected in its RMSE, which remains around 1.0 across all tests. In contrast, MSD-PUGAN's multi-scale discriminators, with their larger receptive fields, capture both local details and global consistency more effectively. This leads to substantially lower errors in our tests, as evidenced by its consistent RMSE of approximately 0.5 and a reduced number of unwrapping mistakes.

3.4 Real Data Experimental Results

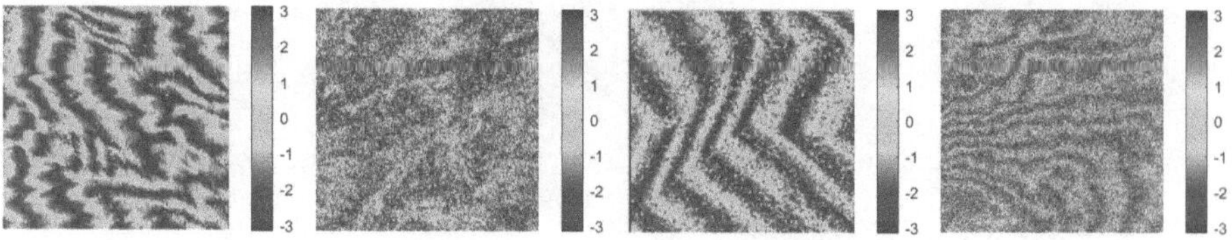

(a) (b) (c) (d)

Fig. 8. Real-world data examples from Sentinel-1 ascending orbit acquisitions: (a) The Huashan region in Hezhou, China; (b) A rare earth mining area in Hezhou, China; (c) The Three Gorges Dam region, China; (d) A volcanic region in Italy.

Figure 8 displays Sentinel-1 ascending orbit data from regions with different coherence levels. The Huashan area (Hezhou) has high coherence; the Three

Gorges area has moderately high coherence; and both the Hezhou rare earth mine and the Italian volcanic region have low coherence.

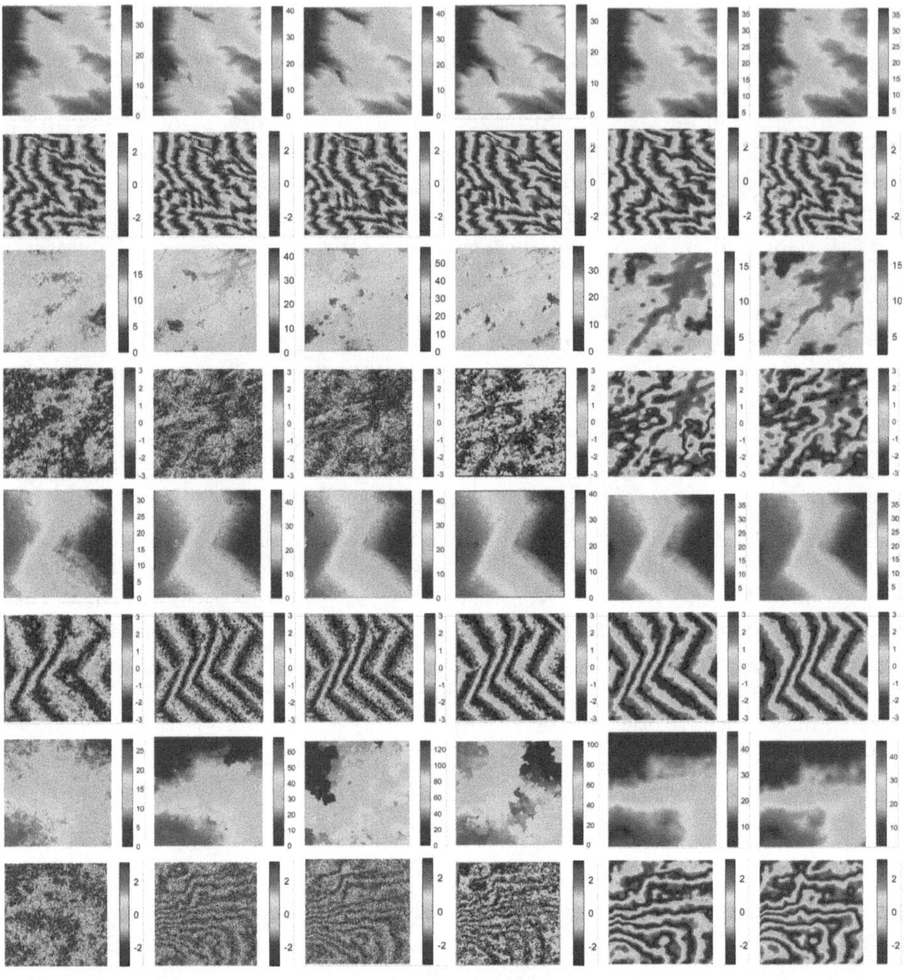

Fig. 9. Unwrapping and re-wrapping results for various methods. Columns (1–6): LS, MCF, QG, UKF, PUGAN, MSD-PUGAN. Rows (1–8): Huashan, Hezhou mine, Three Gorges, Italian volcanic region, etc.

Their unwrapping results and re-wrapped images are shown in Fig. 9 For regions with good coherence, such as the Huashan and Three Gorges areas, although their overall coherence is high, there are still some localized areas of low coherence within these interferograms. These low-coherence patches cause some unwrapping errors for the traditional methods. In contrast, both the PUGAN and MSD-PUGAN methods are unaffected by these low-coherence regions and yield good unwrapping results. For regions with low coherence, such as the Italian volcano and the Hezhou rare earth mine—where parts of the interferometric fringes in the mine are severely degraded, exhibiting even lower coherence—unwrapping presents a greater challenge for traditional methods. As can be seen from their unwrapping results, the traditional methods largely fail, producing extensive unwrapping errors. The PUGAN and MSD-PUGAN methods, however, exhibit only minor unwrapping errors, which demonstrates their superior robustness. It is worth mentioning that traditional unwrapping methods tend to unwrap not only the phase but also the noise contained within the wrapped phase. This is evident from their re-wrapped images, which appear nearly identical to the original wrapped images. Even when the unwrapping process itself contains numerous errors (as seen in the MCF and QG methods), the re-wrapping operation can still restore the original noisy appearance with low error. In contrast, deep learning-based methods like PUGAN and MSD-PUGAN produce the unwrapped image in a single step directly from a trained generator. As a result, when these clean unwrapped images are re-wrapped, they produce noise-free re-wrapped images.

4 Conclusions

To overcome the limited receptive field and local/global trade-off of single-scale discriminators in phase unwrapping, we propose a Multi-Scale Discriminator Phase Unwrapping GAN (MSD-PUGAN). This method leverages a Multi-scale discriminator to provide hierarchical feedback, significantly improving the generator's ability to maintain both global consistency and local detail at phase transitions. On both simulated and real data, MSD-PUGAN consistently outperforms traditional methods and the original PUGAN in RMSE, especially in low-coherence and complex terrain where errors are substantially reduced. The Multi-scale architecture allows the network to capture features across various spatial scales, mitigating the limitations of single-scale representation. This, combined with a CBAM attention mechanism, enables the generator to adaptively focus on key features, thereby boosting precision and robustness. Real-data experiments confirm the method's stability and generalization in noisy, low-coherence environments, with re-wrapped results demonstrating effective noise suppression. In summary, MSD-PUGAN shows strong, versatile performance, making it highly suitable for high-precision unwrapping in challenging conditions like low coherence, high noise, and complex terrain.

Acknowledgments. This work was supported in part by Guangxi Natural Science Foundation (2025GXNSFAA069512), Research and Application of Key Technologies for Precise Navigation (Gui Ke AD25069103)

Disclosure of Interests. I declare that the authors have no competing interests as defined by Springer, or other interests that might be perceived to influence the results and/or discussion reported in this paper.

References

1. Itoh, K.: Analysis of the phase unwrapping algorithm. Appl. Opt. **21**(14), 2470–2470 (1982)
2. Yu, H., Lan, Y., Yuan, Z., et al.: Phase unwrapping in InSAR: a review. IEEE Geosci. Remote Sens. Mag. **7**(1), 40–58 (2019)
3. Goldstein, R.M., Zebker, H.A., Werner, C.L.: Satellite radar interferometry: two-dimensional phase unwrapping. Radio Sci. **23**(4), 713–720 (1988)
4. Xu, W., Cumming, I.: A region-growing algorithm for InSAR phase unwrapping. IEEE Trans. Geosci. Remote Sens. **37**(1), 124–134 (1999)
5. Gao, D., Yin, F.: Mask cut optimization in two-dimensional phase unwrapping. IEEE Geosci. Remote Sens. Lett. **9**(3), 338–342 (2011)
6. Ghiglia, D.C.: Two-Dimensional Phase Unwrapping: Theory, Algorithms, and Software (1998)
7. Costantini, M.: A novel phase unwrapping method based on network programming. IEEE Trans. Geosci. Remote Sens. **36**(3), 813–821 (2002)
8. Chen, C.W., Zebker, H.A.: Two-dimensional phase unwrapping with use of statistical models for cost functions in nonlinear optimization. JOSA A **18**(2), 338–351 (2001)
9. Yu, H., Lee, H.: A convex hull algorithm based fast large-scale two-dimensional phase unwrapping method. In: IGARSS 2017, pp. 3824–3827. IEEE (2017)
10. Loffeld, O., Kramer, R.: Phase unwrapping for SAR interferometry: a data fusion approach by Kalman filtering. In: IGARSS 1999, vol. 3, pp. 1715–1717. IEEE (1999)
11. Zhang, Y., Zhang, S., Gao, Y., et al.: Adaptive square-root unscented Kalman filter phase unwrapping with modified phase gradient estimation. Remote Sens. **14**(5), 1229 (2022)
12. Zhou, L., Yu, H., Lan, Y.: Deep convolutional neural network-based robust phase gradient estimation for two-dimensional phase unwrapping using SAR interferograms. IEEE Trans. Geosci. Remote Sens. **58**(7), 4653–4665 (2020)
13. Zhou, L., Yu, H., Lan, Y., et al.: Deep learning-based branch-cut method for InSAR two-dimensional phase unwrapping. IEEE Trans. Geosci. Remote Sens. **60**, 1–15 (2021)
14. Zhou, L., Yu, H., Wang, Y.: PG-BCNet: a neural network combined with the PGNet and BCNet for 2-D InSAR phase unwrapping. In: IGARSS 2022, pp. 2470–2473. IEEE (2022)
15. Wu, Z., Wang, T., Wang, Y., et al.: Deep learning for the detection and phase unwrapping of mining-induced deformation in large-scale interferograms. IEEE Trans. Geosci. Remote Sens. **60**, 1–18 (2021)
16. Yang, Y., Chen, B., Li, Z., et al.: A novel phase unwrapping method for low coherence interferograms in coal mining areas based on a fully convolutional neural network. IEEE J. Sel. Top. Appl. Earth Observ. Remote Sens. **17**, 601–613 (2023)
17. Gao, Y., Yao, J., Zheng, N., et al.: MMPhU-net: a novel multi-model fusion phase unwrapping network for large-gradient subsidence deformation. IEEE J. Sel. Top. Appl. Earth Observ. Remote Sens. **17**, 5137–5146 (2024)
18. Zhou, L., Yu, H.: MoDL-PU: model-based deep learning for InSAR phase unwrapping. IEEE Trans. Geosci. Remote Sens. (2025)
19. Yan, C., Li, T., Gao, Y.: A novel two-stage learning-based phase unwrapping algorithm via multi-model fusion. IEEE J. Sel. Top. Appl. Earth Observ. Remote Sens. (2025)

20. Mirza, M., Osindero, S.: Conditional generative adversarial nets. arXiv preprint-arXiv:1411.1784 (2014)
21. Isola, P., Zhu, J.Y., Zhou, T., et al.: Image-to-image translation with conditional adversarial networks. In: CVPR 2017, pp. 1125–1134. IEEE (2017)
22. Zhou, L., Yu, H., Pascazio, V., et al.: PU-GAN: a one-step 2-D InSAR phase unwrapping based on conditional generative adversarial network. IEEE Trans. Geosci. Remote Sens. **60**, 1–10 (2022)
23. Karnewar, A., Wang, O.: Msg-GAN: multi-scale gradients for generative adversarial networks. In: CVPR 2020, pp. 7799–7808. IEEE (2020)
24. Wang, T.C., Liu, M.Y., Zhu, J.Y.: High-resolution image synthesis and semantic manipulation with conditional GANs. In: CVPR 2018, pp. 8798–8807. IEEE (2018)
25. Zhang, Y., Li, W., Wang, D., et al.: Unsupervised image translation using multi-scale residual GAN. Mathematics **10**(22), 4347 (2022)
26. Woo, S., Park, J., Lee, J.Y.: CBAM: Convolutional block attention module. In: ECCV 2018, pp. 3–19 (2018)
27. Krieger, G., Moreira, A., Fiedler, H., et al.: TanDEM-X: a satellite formation for high-resolution SAR interferometry. IEEE Trans. Geosci. Remote Sens. **45**(11), 3317–3341 (2007)

Learning Dense Rewards from Self-generated Trajectories for Autonomous Driving

Kota Minoshima(✉) and Sachiyo Arai

Department of Electrical and Electronic Engineering, Chiba University, Chiba 263-8522, Japan
minoshima.paleo@gmail.com, sachiyo@faculty.chiba-u.jp

Abstract. Designing practical reward functions remains a central challenge in applying reinforcement learning to autonomous vehicle driving. Sparse feedback, such as signals only at collisions or destination arrival, makes learning safe and efficient driving policies extremely difficult. Potential-Based Reward Shaping (PBRS) can accelerate learning while providing theoretical guarantees of policy invariance, but designing an effective potential function requires substantial expertise.

This paper proposes a self-contained framework that learns a dynamic potential function for PBRS autonomously from an agent's own experience. The agent generates trajectories, ranks them using sparse environmental rewards, and learns a state-based potential function that explains the ranking. This potential function is used to create a dense shaping reward that guides the policy toward desirable driving behaviors, creating a virtuous cycle of performance improvement without expert input and with theoretical guarantees.

Experiments in the highway-env simulator show that our method significantly outperforms PPO and Self-Imitation Learning in challenging traffic navigation tasks, achieving higher success rates and safer policies. The approach automates the design of potential functions, advancing the development of scalable and theoretically-grounded reinforcement learning for self-driving vehicles.

Keywords: Reinforcement Learning · Reward Shaping · Autonomous Driving · Potential-Based Reward Shaping · Inverse Reinforcement Learning

1 Introduction

Reinforcement Learning (RL) has emerged as a powerful paradigm for training autonomous agents, showing great promise for advancing smart mobility and developing the next generation of self-driving vehicles [20]. An RL agent learns an optimal driving policy to maximize cumulative rewards through iterative interactions with its environment. However, its efficacy largely depends on

H. R. Arabnia et al. (Eds.): CSCI 2025, CCIS 2967, pp. 94–108, 2027.
https://doi.org/10.1007/978-3-032-28454-9_7

the design of the reward function, which often becomes a major bottleneck in complex real-world traffic scenarios. For instance, in tasks like safely overtaking multiple vehicles in dense traffic, providing a reward signal only upon eventual success or failure (a sparse reward) makes efficient exploration of safe and effective maneuvers nearly impossible, often leading to prohibitively long training times or a complete failure to learn.

To address this challenge, Potential-Based Reward Shaping (PBRS) offers a theoretically sound method to enrich sparse rewards [12,21]. By adding a carefully structured shaping reward, PBRS can guide the learning process while guaranteeing that the optimal policy of the original problem remains unchanged. However, the effectiveness of PBRS hinges on the design of a potential function, which encodes heuristic knowledge about the task. Manually designing such a function is a non-trivial task, requiring significant domain expertise and meticulous tuning. A poorly designed potential function can fail to accelerate learning, and the design process itself limits the scalability and autonomy of RL systems.

This dependency on expert-driven design is a critical bottleneck. Our work addresses this gap by introducing a self-contained framework that autonomously learns a dynamic potential function directly from an agent's own raw experience. This eliminates the need for expert supervision or prior knowledge, establishing a virtuous cycle where the potential function and the policy co-evolve to solve the task efficiently and with theoretical guarantees.

2 Preliminaries

This section provides a theoretical background on RL, Markov decision processes, reward shaping, and T-REX, which serve as the basis for understanding the proposed method.

2.1 Reinforcement Learning

Reinforcement learning (RL) [20] formulates sequential decision-making as Markov decision process (MDPs) $\langle \mathcal{S}, \mathcal{A}, P, R, \gamma \rangle$, where the agent selects actions $a \in A$ based on states $s \in S$, transitions follow $P(s'|s,a)$, and rewards $R(s,a,s')$ guide learning. The objective is to learn a policy $\pi(a|s)$ that maximizes the expected discounted return, as defined in Eq. (1).

$$R(\pi) = \mathbb{E}_{\tau \sim \pi}\left[\sum_{t=0}^{\infty} \gamma^t r_t\right] \tag{1}$$

Here, $\tau = (s_0, a_0, s_1, a_1, \ldots)$ represents a trajectory generated under policy π, and $\gamma \in [0,1)$ denotes the discount factor applied to future rewards. The primary objective of RL is to learn a policy that maximizes this expected return.

The agent's decision-making process is described by a policy $\pi(a|s)$, which defines a probability distribution over actions a given state s. The objective in reinforcement learning is to discover the optimal policy π^* that maximizes the expected cumulative reward. The state-value function $V^\pi(s)$, which quantifies the expected return of state s under policy π, is defined in Eq. (2).

$$V^\pi(s) = \mathbb{E}_\pi \left[\sum_{k=0}^{\infty} \gamma^k r_{t+k} \middle| s_t = s \right] \tag{2}$$

An optimal policy is one that maximizes this value function.

2.2 Reward Shaping

Reward shaping augments sparse environmental rewards with an additional signal $F(s, a, s')$, yielding $R'(s, a, s')$, as shown in Eq. (3). The agent then learns from the composite reward function R'.

$$R'(s, a, s') = R(s, a, s') + F(s, a, s') \tag{3}$$

A prominent and theoretically grounded approach is potential-based reward shaping (PBRS) [12,21] which ensures policy invariance if the shaping reward takes the form of a difference in potentials:

$$F(s, s') = \gamma\phi(s') - \phi(s) \tag{4}$$

for some potential function $\phi : \mathcal{S} \to \mathbb{R}$. However, designing ϕ often requires expert knowledge. Furthermore, if the potential function $\phi(s, t)$ changes over time, this framework is known as Dynamic PBRS, which also maintains theoretical guarantees under certain conditions [6].

The objective of this research is to propose a framework that autonomously learns an effective potential function ϕ from an agent's own experiences, thereby automating a critical aspect of PBRS.

2.3 Inverse Reinforcement Learning

Inverse reinforcement learning (IRL) [13] is a learning paradigm that aims to infer the underlying reward function from expert behavioral trajectories (demonstrations). Whereas conventional RL focuses on learning a policy that maximizes a given reward function, IRL addresses the inverse problem: identifying the reward function that explains the observed behavior of an expert.

The fundamental assumption of IRL is that the expert acts optimally with respect to some latent reward function. Many IRL approaches model the reward function $R(s)$ as a linear combination of state feature vectors $\phi(s)$ and a weight vector $\boldsymbol{w}$, i.e., $R(s) = \boldsymbol{w}^\top \phi(s)$. Under this assumption, the expected return of the expert's policy π_E is greater than or equal to that of any other arbitrary policy π:

$$\mathbb{E}_{\pi_E} \left[\sum_{t=0}^{\infty} \gamma^t R(s_t) \right] \geq \mathbb{E}_\pi \left[\sum_{t=0}^{\infty} \gamma^t R(s_t) \right] \tag{5}$$

Equation (5) formalizes the idea that the expert's behavior is rational under the inferred reward function. IRL algorithms estimate this reward function by searching for a weight vector $\boldsymbol{w}$ that satisfies the aforementioned condition. The resulting learned reward function is expected to improve generalization to unseen states, which is a common challenge in imitation learning.

2.4 T-REX: Reward Learning from Trajectory Ranking

T-REX (Trajectory-ranked Reward EXtrapolation) [3] was proposed to address a key limitation of IRL: the inability to significantly surpass the performance of the input demonstrations. Most conventional IRL methods assume that expert behavior is optimal and attempt to recover a reward function that rationalizes such behavior. Hence, the learned policy may remain suboptimal if the demonstrations themselves are suboptimal.

In contrast, T-REX fundamentally changes the approach to learning reward functions. Rather than imitating expert behavior directly, it learns a reward function that explains the relative ranking of a set of trajectories ordered from "worst" to "best." The core idea is that more preferred trajectories should yield higher cumulative rewards under the true reward function. By identifying features correlated with this ranking, T-REX can extrapolate the expert's intent, enabling the acquisition of policies that outperform the provided demonstrations. A key advantage of this approach is that it does not require expert action labels (i.e., which action was taken in which state); it can be learned solely from state observations.

Formally, T-REX defines a state-based reward function $\hat{r}_\theta(s)$ parameterized by a neural network. The predicted return for a trajectory τ is calculated as the sum of predicted rewards for all states within that trajectory: $\hat{J}_\theta(\tau) = \sum_{s \in \tau} \hat{r}_\theta(s)$. The reward model is then trained by minimizing the following log-likelihood loss using pairs of trajectories (τ_i, τ_j), where τ_i is ranked as more preferred than τ_j by the expert (denoted as $\tau_i \succ \tau_j$):

$$\mathcal{L}(\theta) = - \sum_{\tau_i \succ \tau_j} \log \frac{\exp \sum_{s \in \tau_i} \hat{r}_\theta(s)}{\exp \sum_{s \in \tau_i} \hat{r}_\theta(s) + \exp \sum_{s \in \tau_j} \hat{r}_\theta(s)} \tag{6}$$

This loss function trains a classifier to predict whether one trajectory is preferred over another based on their predicted returns. Data augmentation is often employed to learn more expressive reward functions, even from a few demonstrations, by generating numerous pairs of randomly sampled sub-trajectories rather than using only full trajectories.

The learned reward function $\hat{r}_\theta(s)$ is subsequently used as a dense reward signal to train an agent with standard RL algorithms. T-REX is a powerful approach that can automatically acquire high-quality reward functions simply by having an expert rank trajectories, even when the provided demonstrations are suboptimal (Fig. 1).

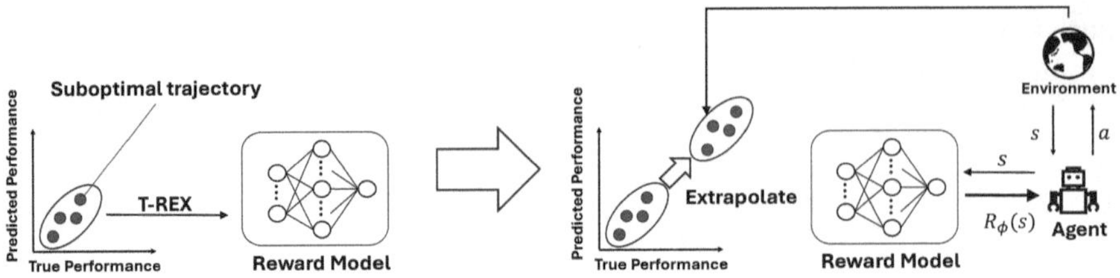

Fig. 1. The T-REX framework, which forms the basis of our proposed method. Instead of imitating expert actions, T-REX learns a reward function that explains the preference ranking over a set of demonstration trajectories. This learned reward is then used to train an agent with a standard RL algorithm, enabling it to potentially surpass the performance of the demonstrations.

3 Target Problem

3.1 Reward Design Problem

In RL, designing an effective reward function is critical, as it directly influences policy performance. A well-designed reward function should accurately reflect the designer's intent by assigning high values to desirable states and low values to states that should be avoided. However, in large-scale real-world problems, the vast state space often limits reward specification to sparse feedback, where the agent receives a positive signal only upon reaching a final goal state. For example, in a task where a robot must navigate a complex maze to reach a goal, receiving a reward solely at the endpoint requires an immense number of trials to discover effective exploration paths, resulting in substantial delays or even failure to learn an optimal policy.

Dense rewards, which provide more frequent feedback, such as rewards based on proximity to the goal, are sometimes introduced to mitigate this issue. Nevertheless, manual reward engineering can introduce new challenges. There is always a risk that the agent may converge to suboptimal local optima by exploiting easily accessible shortcuts to rewards, or it may learn unintended behaviors (reward hacking), such as prioritizing distance reduction over avoiding obstacles, thereby disregarding the original objective. Furthermore, fully defining what constitutes a "good" state or action a priori becomes increasingly impractical as task complexity grows, demanding significant effort and trial-and-error from the designer. Consequently, there is a pressing need for a framework that efficiently facilitates learning from sparse rewards while circumventing the burdens and risks associated with manual reward design.

3.2 Related Work

In response to the aforementioned difficulties in designing rewards, this study proposes a framework for autonomously learning a reward function using action trajectories generated by the agent during training. Specifically, the agent ranks its action trajectories based on sparse rewards obtained from the environment and learns the reward function that best explains these ranking relationships. This approach enables the agent to learn from its own experience, without relying on external expert data or manual reward design, and establishes a loop in which reward design and policy learning evolve simultaneously.

We compare our research with related works across four dimensions: external data dependency, computational complexity, sample efficiency, and generality (Table 1). For a fair comparison, we consistently define sample efficiency as the amount of learning progress achieved per interaction with the environment. Generality refers to the applicability of the method to a wide range of tasks without significant modification.

Table 1. Feature comparison of the proposed method and related approaches. ✗ indicates high dependency/complexity or low efficiency/generality. ✓ indicates low/no dependency or high efficiency/generality. ± indicates a moderate level.

Method	External Data Dependency	Computational Complexity	Sample Efficiency	Generality
Potential-Based Reward Shaping (PBRS) [12,15,23]	✗	✓	✓	±
Learning from Demonstration [8,22]	✗	✗	-	±
Reward Estimation from Human Feedback [5]	✗	±	✗	±
Intrinsic Motivation for Exploration [4,17]	✓	±	±	✓
Learning from Self-Experience [1,14,18]	✓	±	✓	±
This Research	✓	±	✓	✓

Our study proposes a framework that enables agents to autonomously learn reward functions from their own experiences.

Potential Based Reward Shaping. Potential-based reward shaping (PBRS) provides a powerful theoretical foundation for accelerating reinforcement learning [12,21]. It guarantees that augmenting the environmental reward with an additional reward of the form $F(s, s') = \gamma\phi(s') - \phi(s)$ preserves the optimal policy of the original problem [12]. Here, $\phi(s)$ is a potential function that represents the heuristic desirability of states. This guarantee has made PBRS a cornerstone of reward design.

A significant extension, Dynamic PBRS, proved that this guarantee holds even if the potential function changes over the course of learning, provided the shaping reward is calculated based on the potential values at the time of state transitions [6]. This is crucial, as it opens the door for online learning of potential functions. Devlin et al. also showed that dynamic PBRS is not equivalent to Q-value initialization, meaning its guidance effect does not simply vanish as the agent learns [6].

However, the efficacy of PBRS is critically dependent on the design of an appropriate potential function, which often requires significant domain knowledge and can become an engineering bottleneck itself [15,23]. Our work directly tackles this bottleneck by proposing a method to learn the potential function ϕ automatically.

Learning from Demonstrations. Approaches that learn policies or underlying intentions from expert demonstrations constitute a primary set of methods to address the challenges of reward design in RL. IRL forms the foundation of this domain, aiming to infer the reward function that explains expert behavior. Early work estimated rewards by aligning the expected feature counts of expert and learning-agent behaviors [13]. Subsequently, Ziebart et al. introduced the maximum entropy principle, enabling more robust reward estimation by maximizing the probability of observed trajectories [22]. Meanwhile, Ho and Ermon proposed generative adversarial imitation learning (GAIL) [8], which applies the framework of generative adversarial networks (GANs) [7].

By definition, these approaches are fundamentally dependent on an external dataset of expert demonstrations to initiate learning, which justifies their high rating for external data dependency (✗). The high computational complexity of these methods (✗) stems from factors like GAIL's adversarial framework or the nested RL problems in Maximum Entropy IRL. Furthermore, since they primarily learn from a fixed, offline dataset, our defined metric of sample efficiency (based on environmental interactions) is not directly applicable, and is thus marked with a hyphen. Their generality is considered moderate ($\pm$), as the learned reward function is inherently tied to the specific task represented by the expert demonstrations and does not transfer to new tasks without a new dataset.

Reward Estimation from Human Feedback. Learning reward functions from human feedback offers an alternative to manual reward engineering, as feedback is often more intuitive to provide [5]. In this paradigm, a human is typically presented with pairs of trajectory segments and asked to select the preferred one. A reward model is then trained to capture these preferences, and the resulting reward function is used to train a policy via RL. This approach is foundational to modern techniques like Reinforcement Learning from Human Feedback (RLHF) and has been extended to combine preferences with other data sources, such as expert demonstrations [9].

These methods fundamentally rely on human intervention, leading to a high dependency on external data (✗). The resulting learned reward is also specialized to the task being evaluated by the human, limiting its generality ($\pm$). To obtain this feedback, the agent must first generate a significant number of trajectories, which means a very large number of environmental interactions are needed to acquire a sufficient amount of data. This results in low sample efficiency (✗). The computational complexity is considered moderate ($\pm$), as it involves training an additional reward model from preferences, a task typically less intensive than the adversarial training found in some LfD methods.

Intrinsic Motivation for Exploration. Another major approach to the sparse reward problem is intrinsic motivation, where agents autonomously explore based on internal drives such as curiosity or novelty. This approach generates internal learning signals independent of external task rewards. Pathak et al. proposed the intrinsic curiosity module (ICM) [17], which defines the error in predicting an agent's action outcomes as "curiosity" and uses this error as a reward to drive exploration. Furthermore, Burda et al. developed random network distillation (RND) [4] to measure state novelty using prediction errors against the output of a fixed random network, enabling more stable exploration. In addition to these prediction-based methods, count-based exploration, counting state visits and guiding the agent toward less-frequented regions, has also shown effectiveness [2,16].

As these methods generate rewards internally, they require no external data (✓). Their computational complexity is moderate ($\pm$) because they necessitate an additional prediction model. This task-agnostic nature makes these methods highly general (✓) and applicable to a wide range of environments. However, it also presents a trade-off regarding sample efficiency ($\pm$), as the agent may be led to explore regions irrelevant to the actual task objective.

Learning from Self-experience. Learning from Self-Experience refers to a class of approaches that leverage an agent's own past interactions, typically stored in a replay buffer, to accelerate learning. These methods primarily focus on how to efficiently utilize stored experiences.

One foundational approach is Prioritized Experience Replay (PER) [18]. Instead of sampling uniformly from the replay buffer, PER gives priority to transitions with high temporal-difference (TD) error—effectively focusing on "surprising" experiences where the agent's value estimate was most incorrect. This allows for more efficient learning from the most informative samples.

Building on this idea of prioritizing experiences, Self-Imitation Learning (SIL) [14] accelerates learning by encouraging the agent to imitate its own past experiences that were deemed successful. It employs an off-policy mechanism, storing transitions in a replay buffer and selectively learning from them. Specifically, SIL identifies "good" experiences by replaying past transitions (s, a, r, s') for which the accumulated return R was higher than the value function's estimate $V(s)$. While PER prioritizes experiences based on prediction error, SIL prioritizes them based on outcome quality. However, its reliance on a positive advantage estimate poses a significant challenge in extremely sparse reward environments. In such settings, the advantage is rarely positive, and SIL fails to find sufficient "good" experiences to imitate, leading to a severe degradation in sample efficiency.

A related approach, Self-supervised Online Reward Shaping (SORS), also learns a dense reward from self-generated trajectories [11]. However, SORS replaces the environmental reward with the learned one, forgoing the policy invariance guarantees of PBRS [11]. Our work builds on a similar principle of self-supervised ranking but integrates it into the theoretically-grounded framework of Dynamic PBRS, preserving optimality guarantees.

A different approach for sparse reward problems, particularly in goal-oriented tasks, is Hindsight Experience Replay (HER) [1]. Instead of searching for rare successful trajectories, HER treats every failed trajectory as a success for a different goal. It relabels the experience by replacing the original, unachieved goal with the state the agent actually reached. This technique creates dense, positive learning signals from every episode, drastically improving exploration and learning efficiency in goal-conditioned environments.

While these methods provide powerful mechanisms for reusing past data, they exhibit a range of trade-offs. However, the generality of this category is considered moderate (±), as some prominent methods like HER are specialized for goal-reaching tasks.

4 Proposed Method

This research introduces a novel reward shaping framework wherein an agent iteratively learns a dynamic potential function from its own experiences and subsequently refines its policy using PBRS. This method applies the T-REX reward extrapolation principle to dynamically generated trajectories, enabling the continuous, self-supervised learning of a potential function for theoretically-grounded reward shaping.

4.1 Basic Framework

The core of this method is the simultaneous and iterative updating of a RL agent (specifically, PPO in this study) and a potential model responsible for generating the potential values for PBRS. We select PPO for its learning stability; its clipped objective function prevents large policy updates, providing robustness against the non-stationary shaping rewards from our dynamically changing potential model. Since PPO updates its policy using the most recent trajectories, it aligns perfectly with our iterative framework where the potential model and the policy co-evolve. This on-policy approach avoids the complexities that would arise with off-policy methods, such as re-evaluating experiences in a replay buffer collected under outdated policies, thus ensuring a more direct and stable learning process (Fig. 2).

4.2 Iterative Potential Function Learning and Policy Update

Each iteration of the proposed method sequentially executes the following three steps:

1. **Trajectory Collection and Self-Experience Ranking**
   Trajectories collected using the current policy π_θ for PPO policy updates form a trajectory pool $\mathcal{P} = \{\tau_1, \ldots, \tau_{N_{\text{pool}}}\}$. Our method employs the cumulative environmental reward $G_i = \sum_t r^{(i)}_{\text{env},t}$ as a proxy metric. This enables automatic ordering of all trajectories within the pool $\mathcal{P}$, generating numerous preference pairs (τ_i, τ_j), where $G_i > G_j$.
2. **Potential Model Update**
   Using the self-generated preference pairs, a potential model $\Phi_\phi : \mathcal{S} \to [0, 1]$, parameterized by a neural network, is learned based on a ranking loss analogous to that in T-REX. The model $\Phi_\phi(s)$ learns to predict the heuristic

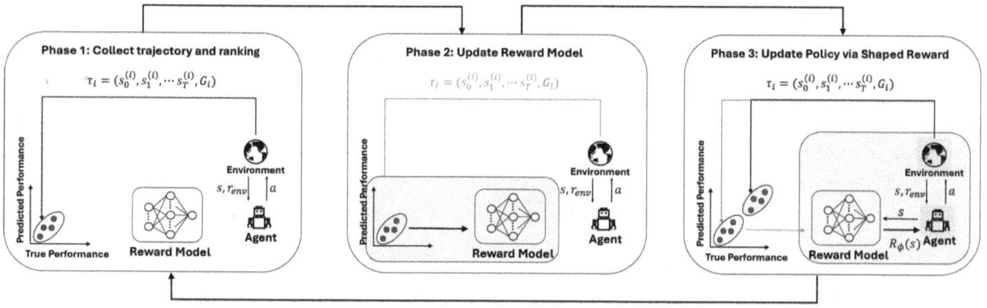

Fig. 2. Overview of the proposed method. The framework operates in an iterative loop: (1) The agent collects trajectories using its current policy and ranks them based on sparse environmental rewards. (2) A potential model is updated using these ranked trajectories. (3) The RL agent (PPO) is trained using the dense shaping reward generated via PBRS from the learned potential function. This virtuous cycle allows the policy and the potential model to improve continuously.

value, or “potential,” of being in state s.“ The final layer of the network uses a sigmoid activation function to ensure the output is scaled between 0 and 1. The loss function is given by

$$\mathcal{L}(\phi) = -\sum_{\tau_i \succ \tau_j} \log \frac{\exp \sum_{s \in \tau_i} \Phi_\phi(s)}{\exp \sum_{s \in \tau_i} \Phi_\phi(s) + \exp \sum_{s \in \tau_j} \Phi_\phi(s)} \tag{7}$$

Here, $\sum_{s_t \in \tau} \Phi_\phi(s_t)$ represents the predicted return from the current potential model. Through this learning process, the model Φ_ϕ is updated to assign higher potential values to states within trajectories that yield high sparse environmental rewards.

3. **Policy Update via Potential-Based Reward Shaping**
The learned dynamic potential function Φ_ϕ is then used to construct a shaping reward $F(s, s')$ according to the PBRS formula. The composite reward r' received by the agent is the sum of the original environmental reward r_{env} and this shaping reward:

$$r'(s, s') = r_{\text{env}} + \gamma \Phi_\phi(s') - \Phi_\phi(s) \tag{8}$$

By adopting the structure of Dynamic PBRS, our framework gains a strong theoretical foundation [6]. The learned function $\Phi_\phi(s)$ serves as a dynamic potential function, as it is updated online alongside the agent's policy. This approach directly addresses the main bottleneck of PBRS—the need for manual design of the potential function—by learning it from raw experience. The function $\Phi_\phi(s)$ learns the instantaneous desirability of a state based on trajectory-level outcomes, serving as an effective heuristic for the state's true value and guiding exploration. Crucially, because our framework adheres to the Dynamic PBRS formulation, it is guaranteed to preserve the set of optimal policies of the original sparse-reward problem [6]. This allows us to accelerate

learning without the risk of altering the task's fundamental objective.
The agent subsequently optimizes its policy using PPO [19] with this dense composite reward r'.

This iterative process ensures that an improved policy generates higher-quality trajectories, which, in turn, yield better data for subsequent potential model learning. In this way, the agent collaboratively refines both the shaping reward and the policy. We summarize the entire process in Algorithm 1.

Algorithm 1 Autonomous Potential-Based Reward Shaping from Ranked Self-Experience

1: **Input:** Hyperparameters (Learning rate η)
2: Initialize: PPO agent's policy π_θ, Potential model Φ_ϕ
3: **for** update_iteration = 1 to $N_{updates}$ **do**
4: Utilize trajectories collected by π_θ for PPO policy update as trajectory pool $\mathcal{P} = \{(\tau_i, G_i)\}$ (G_i is cumulative environmental reward)
5: Rank trajectories in $\mathcal{P}$ by G_i to construct a dataset of preference pairs
6: Update potential model using the gradient of ranking loss $\mathcal{L}(\phi)$: $\phi \leftarrow \phi - \eta \nabla_\phi \mathcal{L}(\phi)$
7: Define composite reward $r'(s, s') = r_{env} + \gamma \Phi_\phi(s') - \Phi_\phi(s)$
8: Update PPO policy π_θ using composite reward r'
9: **end for**
10: **Output:** Final policy π_θ

5 Experiments

In this section, we describe the experiments conducted to validate the effectiveness of our proposed autonomous reward shaping framework.

5.1 Experimental Setup

We conducted our experiments using highway-env [10], a popular simulation environment for reinforcement learning-based autonomous driving agents. The task is to drive as fast as possible on a simulated highway while avoiding collisions with other vehicles. To create a challenging scenario with a difficult credit assignment problem, we intentionally designed the reward to be sparse. Specifically, the agent receives a reward only once at the end of each episode. This reward is the total distance traveled during the episode, normalized by the theoretical maximum distance, which is calculated from the environment's maximum speed (30 m/s) and the maximum episode length (120 s) (Fig. 3).

To evaluate the performance of our proposed method, we compared it against the following two baselines:

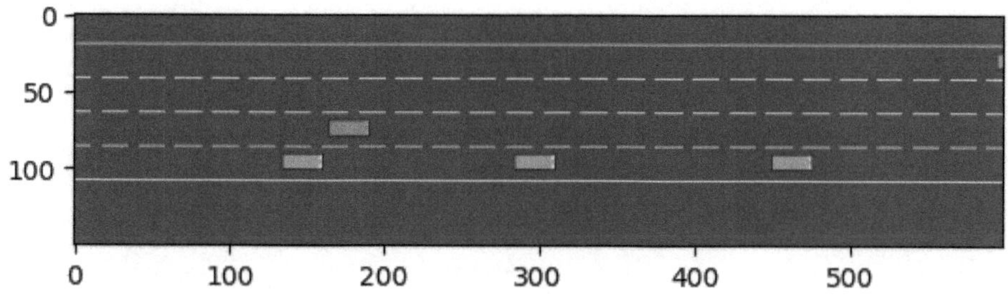

Fig. 3. An example snapshot from the highway-env simulation environment. The green vehicle represents the ego-agent, while blue vehicles represent surrounding traffic, navigating on a multi-lane highway. (Color figure online)

- PPO (baseline): A standard PPO agent trained solely on the sparse reward described above, without any reward shaping.
- PPO + SIL: A PPO agent augmented with Self-Imitation Learning (SIL) [14], a method that accelerates learning by encouraging the agent to imitate its own past successful experiences.

Details regarding the agent's model architecture and hyperparameters are provided in the Appendix.

5.2 Results and Discussion

The learning process for each method is shown by the learning curves in Fig. 4, and the final performance metrics are summarized in Table 2. The y-axis of Fig. 4 represents the normalized cumulative reward (distance traveled), and the x-axis represents the number of training steps.

Table 2. Final performance (mean ± std) on the highway-env task.

Method	Reward (mean ± std)
PPO (baseline)	0.174 ± 0.161
PPO + SIL	0.171 ± 0.155
Ours	**0.544 ± 0.234**

As evident from the results, the baseline PPO agent failed to learn an effective policy. Trained only on the sparse reward, its performance remained low and stagnant, achieving a final mean reward of only 0.174. This highlights the inherent difficulty of the credit assignment problem in this scenario.

The PPO+SIL agent showed no significant improvement over the baseline, achieving a nearly identical mean reward of 0.171. In an extremely sparse reward setting like ours, the agent rarely obtains "successful experiences." This lack

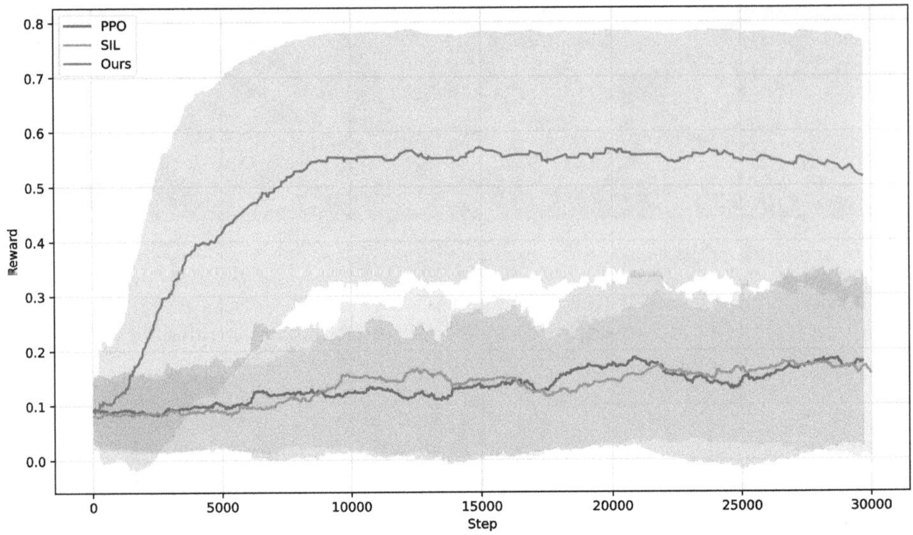

Fig. 4. Performance comparison of PPO, PPO+SIL, and our proposed method (Ours) on the highway driving task. The solid line indicates the mean performance over multiple runs, while the shaded area represents the standard deviation.

of high-quality experiences to imitate appears to have nullified the potential benefits of self-imitation, rendering it ineffective.

In stark contrast, our proposed method achieved significantly superior performance. Its learning curve demonstrates a steep and rapid improvement, and it ultimately converged to a final mean reward of 0.544. This is more than three times higher than the baseline methods, outperforming them by a substantial margin. While the standard deviation for our method (0.234) is larger than the baselines, this reflects the variance during its successful and steep learning phase. In contrast, the low variance of the baselines is merely a reflection of their consistent failure to improve.

These results quantitatively demonstrate that our framework effectively overcomes the challenges of sparse rewards. By autonomously generating a meaningful potential function and using it for dense, theoretically-grounded reward shaping, it enables an agent to acquire a high-performance driving policy without the need for manual reward engineering.

6 Conclusion

This study addressed the critical challenge of reward design in reinforcement learning, focusing on the bottleneck of designing potential functions for PBRS in sparse-reward autonomous driving tasks. We introduced a novel, self-contained framework that empowers an agent to autonomously learn a dynamic potential function from its own ranked experiences, thereby wedding the practical benefits of learned rewards with the theoretical guarantees of PBRS.

Our method establishes a virtuous cycle where the policy and the potential function improve in tandem. By ranking trajectories based on sparse environmental feedback, the agent learns to identify and assign potential to desirable states. This learned potential function generates a dense and informative shaping reward that effectively guides the agent's policy while guaranteeing the preservation of the original task's optimal policy.

Experiments conducted in the 'highway-env'simulator provided strong quantitative evidence of our framework's effectiveness. The proposed method achieved a final mean reward of 0.544, a performance more than three times higher than strong baselines like PPO and PPO+SIL, which failed to learn meaningful policies. Our approach not only reached a superior level of performance but also demonstrated a significantly faster learning curve, underscoring its efficiency.

By automating a traditionally labor-intensive and expert-dependent component of RL development, this work represents a significant step toward creating more scalable, robust, and theoretically-grounded autonomous driving agents. However, we acknowledge limitations. The conceptual gap between a learned "desirability" signal and a true "potential" representing future value merits further theoretical investigation. Additionally, our framework's reliance on ranking by cumulative environmental rewards could be challenging in purely binary success/failure scenarios. Future work will explore these theoretical aspects, alongside applying this framework to more complex, multi-agent driving scenarios.

References

1. Andrychowicz, M., et al.: Hindsight experience replay. In: Advances in Neural Information Processing Systems, vol. 30 (2017)
2. Bellemare, M., Srinivasan, S., Ostrovski, G., Schaul, T., Saxton, D., Munos, R.: Unifying count-based exploration and intrinsic motivation. In: Advances in Neural Information Processing Systems, vol. 29 (2016)
3. Brown, D.S., Goo, W., Nagarajan, P., Niekum, S.: Extrapolating beyond suboptimal demonstrations via inverse reinforcement learning from observations (2019). https://doi.org/10.48550/arXiv.1904.06387, http://arxiv.org/abs/1904.06387, arXiv:1904.06387
4. Burda, Y., Edwards, H., Storkey, A., Klimov, O.: Exploration by random network distillation. arXiv preprint arXiv:1810.12894 (2018)
5. Christiano, P.F., Leike, J., Brown, T., Martic, M., Legg, S., Amodei, D.: Deep reinforcement learning from human preferences. In: Advances in Neural Information Processing Systems, vol. 30 (2017)
6. Devlin, S., Kudenko, D.: Dynamic potential-based reward shaping. In: Proceedings of the 11th International Conference on Autonomous Agents and Multiagent Systems, AAMAS 2012, vol. 1. pp. 433–440. International Foundation for Autonomous Agents and Multiagent Systems, Richland, SC (2012), event-place: Valencia, Spain
7. Goodfellow, I., et al.: Generative adversarial networks. Commun. ACM **63**(11), 139–144 (2020)
8. Ho, J., Ermon, S.: generative adversarial imitation learning (2016). https://doi.org/10.48550/arXiv.1606.03476, http://arxiv.org/abs/1606.03476, arXiv:1606.03476

9. Ibarz, B., Leike, J., Pohlen, T., Irving, G., Legg, S., Amodei, D.: Reward learning from human preferences and demonstrations in atari. In: Advances in Neural Information Processing Systems, vol. 31 (2018)
10. Leurent, E.: An environment for autonomous driving decision-making (2018). https://github.com/eleurent/highway-env
11. Memarian, F., Goo, W., Lioutikov, R., Niekum, S., Topcu, U.: Self-supervised online reward shaping in sparse-reward environments. In: 2021 IEEE/RSJ International Conference on Intelligent Robots and Systems (IROS), pp. 2369–2375 (2021). https://doi.org/10.1109/IROS51168.2021.9636020
12. Ng, A.Y., Harada, D., Russell, S.: Policy invariance under reward transformations: theory and application to reward shaping. In: Icml, vol. 99, pp. 278–287. Citeseer (1999)
13. Ng, A.Y., Russell, S., et al.: Algorithms for inverse reinforcement learning. In: Icml, vol. 1, p. 2 (2000)
14. Oh, J., Guo, Y., Singh, S., Lee, H.: Self-imitation learning. In: Proceedings of the 35th International Conference on Machine Learning, pp. 3878–3887. PMLR (2018). https://proceedings.mlr.press/v80/oh18b.html, iSSN: 2640-3498
15. Okudo, T., Yamada, S.: Subgoal-based reward shaping to improve efficiency in reinforcement learning. IEEE Access : Pract. Innov. Open Solutions **9**, 97557–97568 (2021), publisher: IEEE
16. Ostrovski, G., Bellemare, M.G., Oord, A., Munos, R.: Count-based exploration with neural density models. In: International Conference on Machine Learning, pp. 2721–2730. PMLR (2017)
17. Pathak, D., Agrawal, P., Efros, A.A., Darrell, T.: Curiosity-driven exploration by self-supervised prediction (2017). https://arxiv.org/abs/1705.05363
18. Schaul, T., Quan, J., Antonoglou, I., Silver, D.: Prioritized experience replay. arXiv preprint arXiv:1511.05952 (2015)
19. Schulman, J., Wolski, F., Dhariwal, P., Radford, A., Klimov, O.: Proximal policy optimization algorithms (2017). https://doi.org/10.48550/arXiv.1707.06347, http://arxiv.org/abs/1707.06347, arXiv:1707.06347
20. Sutton, R.S., Barto, A.G.: Reinforcement Learning: An Introduction. Adaptive Computation and Machine Learning, MIT Press, Cambridge (1998)
21. Wiewiora, E.: Potential-based shaping and q-value initialization are equivalent. J. Artif. Intell. Res. **19**, 205–208 (2003). https://doi.org/10.1613/jair.1190, http://dx.doi.org/10.1613/jair.1190
22. Ziebart, B.D., Maas, A., Bagnell, J.A., Dey, A.K.: Maximum entropy inverse reinforcement learning
23. Zou, H., Ren, T., Yan, D., Su, H., Zhu, J.: Reward shaping via meta-learning (2019), https://arxiv.org/abs/1901.09330, arXiv: 1901.09330

Proper Voicing in AI Music Generation

Brandon Miller(✉) and Ray R. Hashemi

Georgia Southern University, Statesboro, GA, USA
bm25086@georgiasouthern.edu

Abstract. Modern generative artificial intelligence models have taken the world by storm due to their ability to generate new (albeit, heavily derived) art-based product on a simple prompt. However, the vast majority of these products are trained on the body of past works of art using heavily supervised reinforcement learning approaches. These approaches have deep implications on the extent of originality and on issues like copyright law, which are being actively litigated at the time of publication. In this paper, we concentrate on the music form of art and expand upon work of Reese et al. in "A Framework for Interactive Generation of Music for Games". We trained a new Q-learning agent without using any kind of representative corpus of music. Instead, the training was done by using an expanded set of encoded rules of Western music theory (similarly to Reese et al.) while also vastly expanded the action space of the agent in a way that Reese et al. did not. We then assessed the quality of the resulting musical composition based on the Q-learner's adherence to some music theory rules surrounding proper voicing. The preliminary investigation finds that our proposed approach is effective in both replicating and novelly expanding the methodology of Reese et al. and warrants further research.

Keywords: Q-Learning · Music Composition

1 Introduction

Ever since the early days of western classical music, composers have tended to adhere to some number of enumerable rules that dictate how music should be written and performed. While what exactly encompasses the term "music theory" is the subject of debate among music scholars, the subject of music theory and many of the "rules" thereof are still taught in music lessons and classes to date. As a result, in this paper, when we refer to "music theory rules", we tend to mention exactly which rule we are speaking of, but are otherwise referring to the fundamental rules, norms, and patterns often associated with the composition of western classical music.

While some of these rules are designed to make music sound "best", many of the concepts and rules taught by music theory are essential to generating music that is appeasing to the listener. Hence, any entity that is seeking to compose music must have some sort of understanding (or, at least, an intuition) about music theory concepts.

During the study of fundamental music theory, students are taught how to identify the relationship between any two notes in the traditional chromatic range. Further study

H. R. Arabnia et al. (Eds.): CSCI 2025, CCIS 2967, pp. 109–120, 2027.
https://doi.org/10.1007/978-3-032-28454-9_8

of the field will yield knowledge about how more than two notes may come together to form chords. The quality of the chord is determined by selection of which relationships of the two notes are forming a chord. For example, for a chord to be major, the two notes stacked on top of the root note must be a major third (four half steps) and a minor third (three half steps), respectively.

Beyond the direct relationships among notes and how they may be combined is the concept of motion. Take, for example, a choir of three people each singing a note. For the choir to "move" together from singing a C major chord to singing a G major chord, each singer must alter their pitch. Music theory teaches specific rules about how this may occur. For instance, in many cases, it tends to discourage large jumps among notes. Deciding how to move from one chord or musical idea to another (i.e. how to distribute the notes) is referred to as voicing.

Voicing technique is a tool by which a composer decides how to distribute the notes of a chord or harmony (defined as a chord with a pleasing effect). Voicing must not be confused with the different "voices" of a choir (i.e. bass, tenor, alto, and soprano). However, effective voicing in choral music can certainly involve choosing which singer is assigned what note. Voicing technique is the main music theory challenge we choose to investigate in this paper.

Specifically, we seek to build an automated generative music composer that can facilitate proper voicing. The early attempts were focused on signal processing [1], but recently the reinforcement-centric algorithms are gaining more attention. In these algorithms, an agent learns through the process of reinforcement. This learning paradigm is nothing but an "action-reward" process. The desirable actions are rewarded favorably and when the reward is maximized, then action is learned by the agent.

Reward for an action can be given by another agent (or a set of agents) or by the agent itself through evaluating the consequences of its action. Two points need to be made here as described by Luger in [6].

(i) The agent is not told what action to take. In fact, the agent explores different actions to determine what actions offer the most rewards.
(ii) The agent actions affect the subsequent actions and eventual rewards.

One of the reinforcement-centric algorithms is Q-learning. There are a few examples of generative music composition frameworks that purely use model-free agentic Q-learning, especially without using any training data (so-called based on the fact that it "learns by doing" without a model of how the environment works). A well-known attempt was presented by Reese et al. [6], and recommendation for improving the proposed Q-learning.

The goal of our research effort is to develop a generative music composer using Q-learning paradigm that owns the following three traits:

(a) Incentivize chord diversity (triads are preferred, but some seventh chords are acceptable), with preference for major chords.
(b) Disincentivize massive note jumps (except when performed by the bass, who is singing the lowest notes)
(c) Disincentivize the Q-learner when two or more of its singers break a rule involving parallel movement (here, we only punish parallel octaves or parallel fifths)

Although these traits are among recommendations made by Reese et al. they were not delivered as part of their automated generative music composer.

The rest of the paper is organized as followS: We review previous works in sect. II, present our methodology in sect. III, cover the empirical result in sect. IV, and draw conclusions and recommend future work in sect. V.

2 Previous Works

Much of the literature in the space of generative music composition and automated music analysis focuses on signal processing and decomposition. In Weiss et al., a novel algorithm for recognizing motifs in music was presented [1]. An unsupervised machine learning approach to represent "music similarity and structure" was investigated by Axel Marmoret [2]. In Bindi et al., an unsupervised approach was used to conduct source separation [3].

To date, we find very few examples of a generative music composition framework that purely uses a model-free agentic approach like Q-learning - especially without using any training set. Dadman et al. notably used a multi-agent approach, but that paper was mostly focused on capturing "longer term musical structure" [4]. One compelling exception is a paper entitled "Music Generation using Human-In-The-Loop Reinforcement Learning" by Aju Ani Justus [5]. He used a Q-learning agent to generate music, however, the Q-learner's cost function was largely derived from users' subjective liking (or disliking) of the music they're experiencing (i.e., from human-in-the-loop).

Reese et al. [6] successfully built a Q-learner whose reward functions are inspired by basic music theory rules. His team created basic harmonization reward functions that rewarded the agent settling on widely recognized chords. For example, major chords (most prominent in music) received a reward of 1500 points, while more rare chords were only rewarded with a score of 8. Indeed, the work of Reese et al. laid the groundwork for identifying the Q-learning action space of Western music theory as a 12° action space.

Reese and his team, however, left several open questions about what is possible in the realm of Q-learning music generation. For example, instead of allowing the Q-learner to discover the complex art of voice leading on its own (through exploring the reward associated with it), they created the action space to represent chord movements and incorporated voice leading themselves (as a mask for agentic decision-making). Reese et al. proposed further work to incentivize a reinforcement learning agent for tonality, stating that "measuring the quality of music is a difficult task". Furthermore, their work on assessing delayed time rewards is not deep enough [6].

Our work expands upon the extent to which these learners are able to embrace longer-term rewards. By placing an emphasis on our three enumerated Q-learning traits, our operating hypothesis is that the Q-learner will be forced to only navigate to desirable tonality by way of note movements that satisfy the rules of music theory, thus implementing pseudo-delayed time rewards, even if it is a very early form of such an outcome.

3 Methodology

We start by explaining two key topics of Markov Decision Processes and Q-Learning and then presenting our methodology. The details are covered in the following three subsections.

3.1 Markov Decision Processes

Markov decision processes (MDP), also known as Markov chains, are formal descriptions of a fully observable sequential environment, where states of an environment and the transitions between those states are modeled for the purpose of reinforcement learning [8]. MDP is defined as a tuple (S, P) where S is a finite set of states and P is a probability matrix which represents the probabilities for each state transitions. Markov reward systems (aka Markov Reward Processes) are used to define the rewards that occur at each possible transition step within a Markov chain.

Reinforcement learning agents often develop policies to traverse the Markov reward system in a way which maximizes reward [9]. The policy considers only the current state, with no information about previous states. As a result, the policy for action a given the current state s is calculated by formula (1)

$$\pi(a \mid s) = P[A\_t = a \mid S\_t = s] \tag{1}$$

Where a is the current action, s is the current state, St, and At are state and action at time t, respectively. For any Markov decision process, there exists an optimal policy. Converging upon this optimal policy is the main thrust of reinforcement learning for Markov chain derived problem sets [9].

3.2 Q-Learning

Q-learning is a reinforcement learning approach used to learn the optimal policy within an MDP. It is done in a multi-step process. First, the Q-learner establishes what is called a Q-table in memory. This is a look-up table used to link state-action tuples (s, a) with their anticipated outcome.

Over time, the Q-learner continuously updates the Q-table by associating state-action tuples (s, a) with experienced Q-values using the formula (2):

$$NewQ(s, a) = OldQ(s, a) + \eta\left[\gamma\left(\max Q\left(s', a'\right) - OldQ(s, a)\right) + R\left(s, a\right)\right] \tag{2}$$

Where, s is the state, a is the action for the state, Q(s, a) is the Q-value for tuple (s, a). s^' is the next state, and a^' is the next action.

R(s, a) is the reward for a state-action tuple, η is the learning rate γ is the discount factor and it is a number between zero and one that simply acts as a way of decreasing the reward over time to prevent over-compensation [9]). The term max Q(s',a') means among all the possible new states to move in, we choose the state for which the Q-value is maximum.

3.3 Our Approach

Our approach deviates from existing Q-learning in two ways: (i) providing a deep time delay, and (ii) presenting a new reward function that is calculated using the contribution of both our customized chord quality assessment and customized chord movement assessment.

By definition, Markov Decision Processes consider only the current state, with no information about previous states. However, despite this fact, the operating hypothesis of this paper is that we can generate very superficial forms of time delayed rewards nevertheless (this is also mildly alluded to in Reese et. Al. in less deterministic terms). We call these time delayed rewards superficial because, by definition, Markov problems do not consider every previous state in its decision calculus: only the one just before.

Additionally, as is established precedent, we assess that the nature of music composition is a very Markov-shaped problem. Composers write to reach certain tonal structures (chords), which can adequately be represented as Markovian States, and music theory has several enumerated rules about how composers ought to navigate between states: a behavior that well-represents Markovian actions. As a result of the problem being Markovian at its core, Q-learning is an excellent candidate for reaching the optimal policy: with the "policy" in this case being the optimal transversal between chords in an elegant, rule-following manner.

To accommodate such an approach, we design a new reward function to assist the Q-learner in reaching the optimal policy. Our reward function is using formula (3):

$$R(s, a) = \theta\_1\ C(s, a) + \theta\_2\ M(s, a) \tag{3}$$

Where C(s,a) is our custom chord quality assessment function, M(s,a) is our custom chord movement assessment function and θ_1 and θ_2 are weights used to experiment the contribution of the two functions into R(s, a). In fact, the weights θ_1 and θ_2 allow for us to experiment with how much the agent is rewarded for prioritizing one component of the reward function over another.

Crucially, as will be discussed, this approach does not take into account any previous "training data," in the traditional reinforcement learning sense. Instead, the Q-learner simply learns to compose music by composing music.

4 Empirical Results

We start with a randomly created starting state. States are referred to as state quadruple because they include all the information about the four elements of Soprano, Alto, Tenor, and Bass. One such piece of information is the state tuple that carries the not assignments to the four elements. The Q-learner moves from one state to the next by choosing from several possible actions. The selection of the action is done by an assessment function (Reward function). The transition from one state to the next is repeated for n times and in each repetition the Q-learner learns from the past rewards and tries to select states for maximizing the rewards. The value for n affects the quality of song generated by the q-learner. For the rest of this section, we look into all above steps in detail.

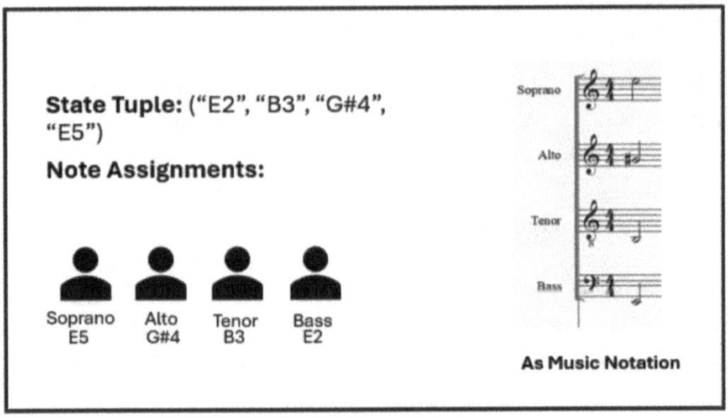

Fig. 1. This is an example quadruple in various representations.

4.1 State Quadruple

Our representative state quadruple contains four terms: each representing a distinct note assignment for each singer in a SATB (Soprano, Alto, Tenor, Bass) choir. We choose to represent the lowest voice first in our quadruple. In other words, the first term in the quadruple is the note assignment for the bass singer, the second term in the quadruple is the note assignment for the tenor singer, and the third and fourth terms are assignments for the alto and soprano singers, respectively. In Fig. 1, we demonstrate an example state tuple, show what the tuple represents in music notation, and what notes would be assigned to each singer in a SATB choir based on the given state tuple. Notably, these note assignments are listed in what is called scientific pitch notation (SPN), such as "E2," "B3," etc.: which are string data types.

Singer	Range
Bass	E2 – G3
Tenor	A3 – F4
Alto	D4 – C5
Soprano	A4 – G5

Fig. 2. These represent the note ranges of each voice type, which will bound our "legal" action space.

Notably, as with any Markov Decision Problem, there are legal and illegal states. In other words, there are rules that govern what states may exist and what states may not. Here, the only rule that enforces the "legality" of a state's existence is that of each singers' natural range. Human composers cannot expect a bass singer to sing the same pitches as a soprano, nor can they expect the inverse. As a result, any state in which a singer is asked to sing a note outside of their pre-defined range (shown in Fig. 2) is considered an "illegal" state that cannot exist.

It should also be noted that the random state generated at the beginning is guaranteed to be the root chord for some random major key. For example, if the randomly generated major key (of the 12 major keys) is E major, the first chord will be E major. However, to enhance training, the beginning chord may not always be in root position (that is, with the lowest note being E). This is, in part, because the possible state space is so large for this problem (in the thousands, with thousands of possible actions stemming from any given state). Additionally, it is quite common for songs to begin with the root chord of whatever key the song is in.

4.2 Action Quadruple

In order to transition between states, an action quadruple is designed for the Q-learners use. The action quadruple is made up of four integers, each of which representing an instruction to be issued to a singer in the same lowest-to-highest order described in the state quadruple. Each integer represents the number of half-steps a singer will be instructed to raise or lower their pitch. For example, if the first term in the quadruple is 4, the bass singer (whose state is first in the state quadruple) will be asked to raise their pitch four half steps. If the integer is negative, the singer is asked to sing lower. For example, if the fourth term (which is the Soprano singer's term) is −2, the Soprano singer is asked to sing 2 half steps lower.

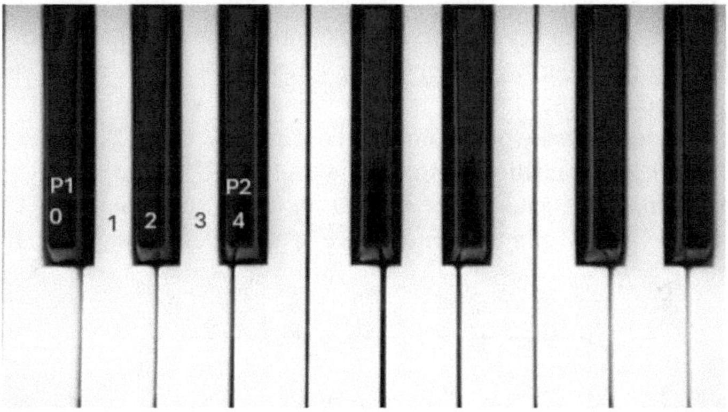

Fig. 3. Visualization on a piano keyboard between two notes with counted half-steps.

"Half steps" are musical intervals that are considered to be the smallest difference possible between notes in Western Music theory. It is best illustrated by visualizing a piano keyboard, whose keys are each a half step a part. Therefore, for each key on a piano between Note 1 (P1) and Note 2 (P2), there is one half step's difference between the notes. Figure 3 illustrates two example notes that happen to be four half steps apart.

By writing a function in code that takes in a starting state quadruple and an action quadruple and outputs a new state, we effectively recreate the process of composing music. An example of a complete state, action, to new state process diagram is shown in Fig. 4.

Considering the fact that the number of possible state-action tuples for moving from one tuple to the next is too many, the decision tree (action space) becomes an extremely large one, as in Fig. 5. The use of the reward function at each level to isolate the choice of interest makes the search of this vast action space possible.

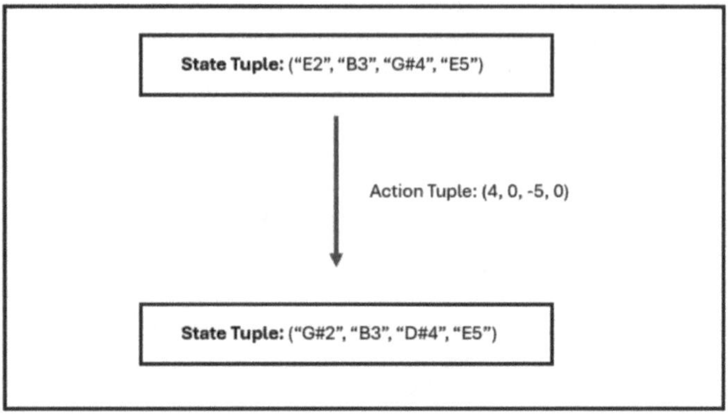

Fig. 4. State quadruple transition via Action quadruple

The two components of the reward function (Quality and Movement) for each action tuple are neither equivalent nor mutually interested. In other words, as evidenced by Fig. 6, it is conceivable for an action to have "good" movement but poor quality, and the same is true for the other way around. The way these actions translate to actual note movement is denoted on the staff in Fig. 7, which shows the starting state and then the resulting state quadruple that would come of choosing the "GOOD" reward action on the right side of Fig. 6. The transition on the left is considered bad because it violates basic rules of parallelism (in the category of motion for our reward function). These rules of parallelism disallow notes of a certain distance apart traveling the same distance up or down.

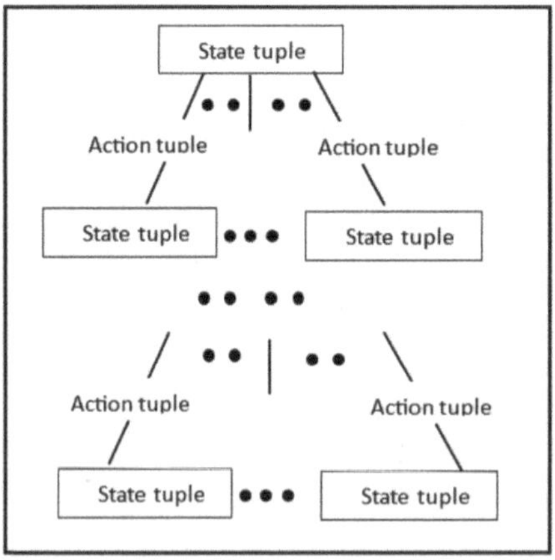

Fig. 5. Multi-step demonstration of the Q-learning process as it pertains to our action space.

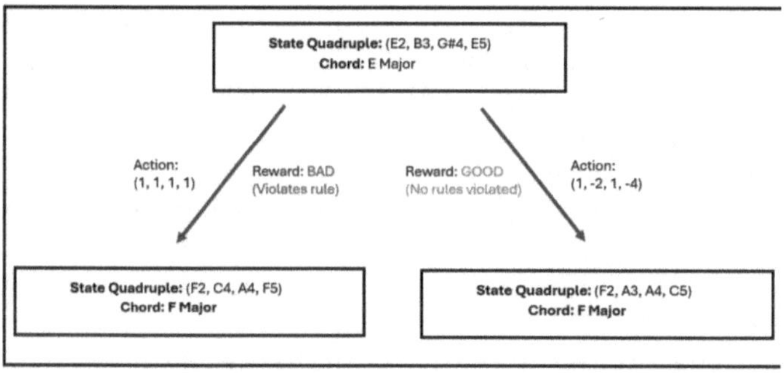

Fig. 6. State-Action MDP representation of the problem.

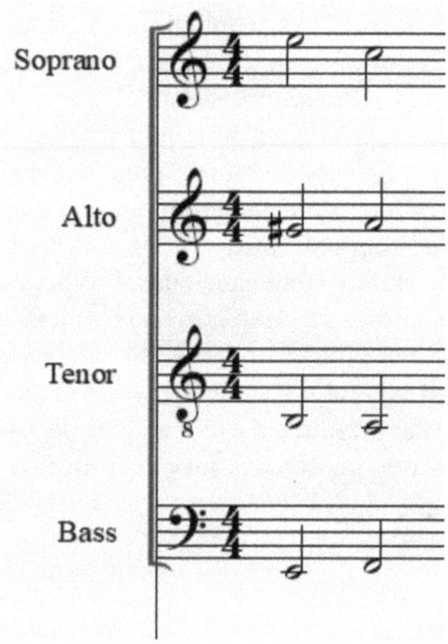

Fig. 7. Visualization of the transition between the starting quadruple and the state quadruple that is created after the next step.

4.3 Assessment Functions

In formula (3), we defined algebraic reward function for our particular Q-learner. Here, we assess how each term in the formula is calculated.

For C(s,a), we implement a function called "checkChord()," which evaluates chord quality. This assesses what "type" of chord it is – major, minor, etc. If the notes in the State quadruple create a major chord, for example, which is an essential component of music, this function returns a heavy reward. Other coherent chords that are less common are large.

If the notes in a State quadruple make an unappealing chord, or if no coherent chord is formed at all, this function returns a negative score.

For M(s,a), we implement a function called "checkMovement()," which evaluates motion. This is what checks for note changes that break the rules of music theory described in the introduction of this paper. The function returns a small reward if no rule is violated. If a rule is violated, a heavy penalty is levied.

These functions, together, create a reward function that punishes breaking the rules of music theory and rewards the Q-learner for stumbling upon appeasing chords in a way that does not violate these rules. As the Q-learner seeks to optimize its rewards, it is forced to find a way to navigate to these "good chords" in a way that does not violate any rules. Notably, this means that there are multiple pathways for Chord 1 to transition to the same Chord 2 (from state to state), but some actions will be preferred over others based on the extent to which they adhere to traditional music theory rules. A representation of this dilemma in a Markov-like diagram is shown in Fig. 5.

In addition to these functions, which would be used for the actual reward function, we developed an evaluator function that would be used exclusively in our testing phase. This function would view each song and determine an accuracy score for how much of the song properly followed a list of traditional music theory rules.

As described in Formula 3, we created two values θ_1 and θ_2 to represent weights for each of the two principal assessment functions. These were originally intended to be stand-ins for fast iteration and experimentation. However, we quickly realized during experimentation that we had subconsciously "normalized" the individual reward functions in such a way that the results were satisfactory. When we changed these weights, the learner tended to predictably over-index on one element or the other (either quality or movement). As a result, we pinned both weights at a value of 1, while noting that this may be a worthwhile avenue of research later on. We theorize that modifying these values slightly, perhaps in the presence of a more sophisticated reward function, could be used to encapsulate various genres of music: but of course, this is a matter to be researched another day.

4.4 Training Iterations

As discussed before, the Q-learner we implemented is not implemented with any training data whatsoever. No previous music or examples of "good actions" are provided to the learner. Thus, its only recourse is to make action decisions based on previous experiences with some element of random incorporated naturally through the Q-learner's process.

Crucially: when we describe a "Song," we are describing an instance where the Q-learner was given a random beginning State quadruple and asked to generate n new State quadruples from that. When we describe an "iteration", we are describing an instance where our Q-learner is asked to generate m Songs (each with a random starting State quadruple). An example of the musical output generated by the Q-learner is demonstrated in Fig. 7.

4.4.1 Testing and Experimentation

We conducted three tests: one with 200 iterations, one with 350 iterations, and one with 500 iterations. After each set of iterations, a long song was generated and tested for accuracy. The evaluation function described in the "Assessment Functions" subsection

Table 1. Quantitative results based on chord quality assessment

Number of iterations	Voicing quality score
200	98.75%
350	95.00%
500	93.33%

was used to describe to what extent each song followed the described music theory rules that govern note movement. The quantitative results are shown in Table 1.

5 Conclusions and Future Work

We derive two key findings from our results:

- The results from Reese et al. are reproducible. Just as they did in Reese et al., by rewarding certain harmonic decisions, the Q-learner successfully generated non-dissonant chord sequences roughly proportional to how they might appear in human-written music.
- The effort to train a Q-learner's to properly follow music theory rules surrounding voicing and delayed rewards (the main goals of this paper) was also a success. The results show that the Q-learner was able to take on the difficult task of traversing different chords in such a way that proper voicing is followed: a task that is non-trivial for human students of music theory.

As future research, we pursue:

- Rhythm generation. While the work of both Reese. et al. and our paper investigated harmonics, rhythm generation requires an entirely different set of skills. Little to no literature exists on this topic, particularly in the realm of Q-learning. This would pose a number of worthwhile questions. For example, would some sort of ensemble learning approach make sense here? What sorts of incentives would need to be in place? While harmonic music theory rules are commonplace, rhythmic ones are far less common, and the incentive structure may be quite subjective.
- Neural Network Approaches. In addition to the exciting prospect of using a deep Q-network to recreate our results, the concept of using some sort of alternative neural network approach, like some implementation of the actor-critic method, is also appealing. How would the results change?
- Human in the Loop. Many generate music approaches have involved a human in the loop to guide the model. How could a human in the loop be used to adjust a Q-learners composition behavior over time? What if a composer could encode their own"rules" - would the Q-learner learn these rules just as well?

References

1. Weiss, R.J., Bello, J.P.: Unsupervised discovery of temporal structure in music. IEEE J. Sel. Topics Signal Process. **5**(6), 1240–1251 (2011). https://doi.org/10.1109/JSTSP.2011.2145356
2. Marmoret, A.: Unsupervised Machine Learning Paradigms for the Representation of Music Similarity and Structure. Artificial Intelligence [cs.AI]. Universite de Rennes, Rennes, France (2022)
3. Bindi, G., Esling, P.: Unsupervised composable representations for audio. arXiv [Cs.LG]. (2024). https://doi.org/10.48550/arXiv.2408.09792
4. Dadman, S., Bremdal, B.A.: Multi-agent reinforcement learning for structured symbolic music generation. In: Mathieu, P., Dignum, F., Novais, P., De la Prieta, F. (eds.) Advances in Practical Applications of Agents, Multi-Agent Systems, and Cognitive Mimetics. The PAAMS Collection. PAAMS 2023 Lecture Notes in Computer Science, vol. 13955. Springer, Cham (2023)
5. Justus, A.A.: Music generation using human-in-the-loop reinforcement learning. arXiv. (2025) https://arxiv.org/abs/2501.15304
6. Reese, K., Yampolskiy, R., Elmaghraby, A.: A framework for interactive generation of music for games. In: 2012 17th International Conference on Computer Games (CGAMES), Louisville, KY, USA, p. 131137 (2012)
7. Luger, G.F.: Artificial Intelligence: Structures and Strategies for Complex Problem Solving. Addison Wesley (2005)
8. Boyd, M.: An Introduction to Markov Modeling: Concepts and Uses (n.d.). https://ntrs.nasa.gov/api/citations/20020050518/downloads/20020050518.pdf
9. Silver, D. Lecture 2: Markov Decision Processes Lecture (n.d.). https://web.stanford.edu/class/cme241/lecture_slides/david_silver_slides/MDP.pdf
10. Jang, B., Kim, M., Harerimana, G., Kim, J.W.: Q-learning algorithms: a comprehensive classification and applications. IEEE Access. **7**, 133653–133667 (2019). https://doi.org/10.1109/access.2019.2941229
11. Q-Function Learning Methods. (2017). https://rll.berkeley.edu/deeprlcoursesp17/docs/lec3.pdf

Artificial Intelligence - Applications of Graphs and Networks

GC-MoE: Hierarchical Financial Market Forecasting with Graph-Context Mixture-of-Experts

Manmeet Rasikbhai Detroja(✉) and Dulal Kar

Department of Computer Science, Texas A&M University-Corpus Christi, Corpus Christi, TX, USA
ce.manmeet@gmail.com, dulal.kar@tamucc.edu

Abstract. The accurate forecasting of financial markets remains a formidable challenge, as traditional models analyze assets in isolation, missing the complex, dynamic interdependencies that drive systemic behavior. This paper introduces the Graph Context-based Mixture of Experts (GC-MoE) framework, a novel hierarchical architecture designed to model and forecast the entire financial market as a unified, interconnected system. The core innovation is a three-tiered structure: a high-level Mixture of Experts (MoE) gating network selects a specialized Temporal Fusion Transformer (TFT) expert, while a shared Graph Neural Network (GNN) supplies relational context and information flow through message passing, providing rich contextual information to dedicated Temporal Fusion Transformers (TFTs) at the node level that generate precise, asset-specific forecasts. Our design treats the graph as a shared context provider and relies on the gating network to learn regime-specific signals from internal representations. The results validate that this synergistic architecture provides a robust and accurate method for forecasting systemic behavior, offering critical advantages for portfolio management and risk mitigation, particularly during periods of market stress. GC–MoE reduces MAE by 38.6% and RMSE by 38.7%, raising R^2 to 0.803 which demonstrates that incorporating graph context and expert routing consistently improves forecast calibration and provides superior risk assessment, serving as a new benchmark, compared to standalone TFTs for long-term prediction.

Keywords: Financial Market Forecasting · Graph Neural Networks · Temporal Fusion Transformers · Mixture of Experts · Financial Graphs

1 Introduction

Financial markets are complex adaptive systems in which assets exhibit nonlinear dynamics and strong interdependence. During stress periods, correlations rise and shocks propagate across related instruments, so models that forecast each asset in isolation can miss systemic drivers and underestimate risk. This motivates framing forecasting not only as time-series regression but as a multi-scale

H. R. Arabnia et al. (Eds.): CSCI 2025, CCIS 2967, pp. 123–142, 2027.
https://doi.org/10.1007/978-3-032-28454-9_9

modeling problem that jointly captures temporal structure, cross-asset relations, and regime variation.

1.1 The Hierarchical Nature of Financial Markets

Market behavior is best understood as a *multi-scale* process in which localized asset dynamics, cross-asset coupling, and broad market regimes interact. Viewing forecasting hierarchically clarifies why models that operate on a single scale often break down under stress:

Micro-scale: Idiosyncratic Dynamics. At the most granular level, each asset exhibits its own temporal signature—slow-moving trends, abrupt event-driven jumps, and persistent volatility patterns (e.g., volatility clustering). These dynamics are shaped by asset-specific information such as earnings releases, policy announcements, and commodity supply shocks. Classical econometric baselines and early neural sequence models are primarily designed to capture this *within-series* structure [3,6]. However, purely micro-level modeling can misattribute market-wide moves to "local noise," especially when shocks originate elsewhere and propagate through the system.

Meso-scale: Inter-asset Relationships. Assets do not evolve independently as they are linked through sector membership, shared risk factors, liquidity channels, and hedging flows. During periods of market stress, correlations often rise and diversification benefits compress, making meso-level interactions a dominant driver of forecast error. A practical example is the co-movement between broad equity indices and large-cap constituents, or the strengthening of cross-asset links during "risk-on/risk-off" episodes. Graph-based representations provide a natural mechanism to encode these dependencies by treating assets as nodes and relationships as edges, enabling message passing that aggregates information from related assets [11,13,14]. Ignoring this relational context makes it difficult to separate an idiosyncratic drawdown from a systemic repricing.

Macro-scale: Regime Variation. At the highest level, the market alternates between regimes—calm growth, elevated volatility, crisis-like dislocations, or policy-driven transitions. A single, fixed parameterization is often brittle because the mapping from inputs to outcomes changes across regimes (e.g., momentum dominates in some periods while mean reversion dominates in others). Conditional specialization therefore becomes appealing: Mixture-of-Experts (MoE) models implement regime-like adaptation by routing inputs to specialized predictors according to the prevailing environment [16,18,19]. This framing emphasizes that robust forecasting requires not only strong temporal modeling but also mechanisms for *relational coupling* and *regime-aware specialization.*

1.2 Challenges in Multi-scale Financial Forecasting

Traditional econometric models, such as the Autoregressive Moving Average (ARMA) family, and early-generation deep learning models like standalone Long

Short-Term Memory (LSTM) networks, have proven valuable but are fundamentally ill-equipped to capture this full hierarchical complexity [3,6]. Such monolithic models often face one of two primary limitations: they either treat each asset's time series in isolation, thereby completely ignoring the rich relational information of the meso-scale, or they attempt to model all assets jointly but often assume static, linear relationships that fail to adapt to changing market conditions [2,11,12].

The inherent non-linearity, non-stationarity, and multi-scale nature of financial data necessitate sophisticated, hybrid approaches [9]. The academic literature increasingly points towards the superiority of frameworks that synergistically combine different neural architectures to tackle distinct aspects of the forecasting problem [21–24].

1.3 Presenting GC-MoE and Contributions

To bridge the gap between monolithic baselines and the hierarchical reality of markets, we propose the **Graph-Context Mixture-of-Experts (GC-MoE)**. As illustrated in Fig. 1, our architecture explicitly maps the three market scales to distinct learnable components: macro-regimes are handled by a gating network, inter-asset dependencies by a Graph Neural Network (GNN), and micro-dynamics by Temporal Fusion Transformer (TFT) experts.

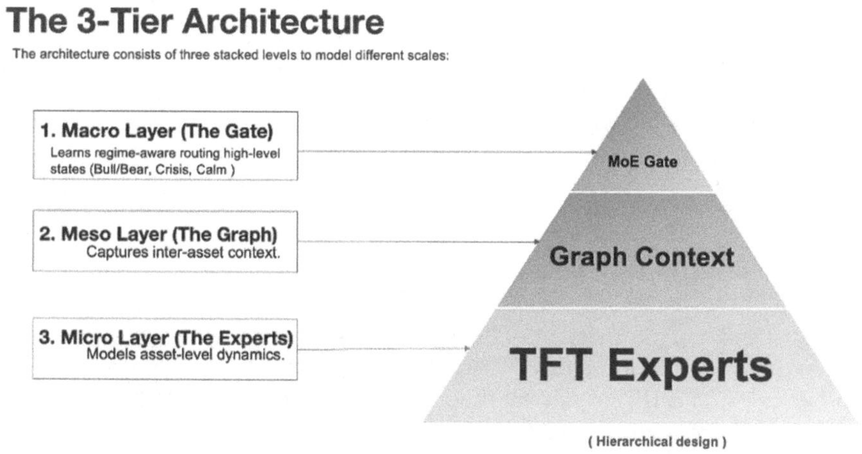

Fig. 1. Three-tier architecture of GC–MoE: macro (MoE gate), meso (graph context), and micro (TFT experts).

A Structural Model of Market Dynamics. Beyond merely forecasting, GC-MoE functions as an implicit structural model of market causality. It posits a top-down hierarchy where information flows downwards: the macro-environment (assessed by the MoE gate) dictates the prevailing regime, which determines the nature of

meso-scale relationships (captured by the GNN). These layers provide the essential context for the micro-scale TFT experts to refine their localized predictions. This design aligns with the theoretical view that systemic regimes and sector couplings constrain individual asset behavior.
Mechanism of Specialization. Unlike traditional regime-switching models where states are manually defined (e.g., "Bull" vs. "Bear"), GC-MoE employs soft routing. The gating network learns to dynamically weigh experts based on real-time market conditions, allowing specialization to emerge organically during training without human-predefined labels [16,18]. For example, the gate might learn to route periods of low volatility and steady upward trends to Expert A, while directing periods of high volatility and sharp reversals to Expert B.
Contributions. This paper makes the following contributions:

- **Hierarchical Architecture:** We propose GC-MoE, a novel framework that unifies temporal (TFT), relational (GNN), and regime-based (MoE) modeling into a single end-to-end differentiable system.
- **Graph based Context:** We introduce a mechanism to inject learned cross-asset embeddings directly into the temporal experts, preventing the loss of systemic context.
- **Empirical Performance:** We demonstrate that GC-MoE achieves superior accuracy compared to both standalone TFT and static graph baselines, particularly during periods of high volatility.

Paper Roadmap. Section 2 reviews regime-aware forecasting, graph-based market modeling, and MoE architectures. Section 3 presents the GC–MoE design and training objective. Section 4 describes the data, features, and evaluation metrics. Section 5 reports empirical results, followed by discussion (Sect. 6) and conclusion (Sect. 7), and Sect. 8 highlights avenues for future work.

2 Related Work

Financial time-series forecasting has progressed from classical econometrics to deep neural architectures that better capture nonlinearity, long-range dependencies, and cross-asset interactions. This progression reflects an increasing emphasis on (i) expressive temporal modeling, (ii) explicit relational structure for systemic behavior, and (iii) adaptation across regimes. GC–MoE draws directly on these threads: strong per-series forecasters (TFT), graph-based market modeling (GNNs), and conditional specialization (MoE).

2.1 Classical Econometric Foundations: ARIMA and GARCH

Classical econometric models remain important baselines because they offer statistical grounding, interpretability, and low computational cost. ARIMA (Box–Jenkins) models a series using autoregression, differencing (to handle non-stationarity), and moving-average residuals [3]. In finance, ARIMA-style models can capture short-range autocorrelation and trend-like components, and are often used as sanity checks for more complex models. However, ARIMA is fundamentally limited by linear assumptions and a weak ability to represent abrupt structural breaks or nonlinear shock propagation.

GARCH-style models address a central stylized fact of markets: volatility clustering. By modeling conditional heteroskedasticity, GARCH can represent time-varying variance and is widely used in risk management [4]. Multivariate extensions such as DCC explicitly model evolving correlation structure [5], which is conceptually aligned with cross-asset dependence. Still, these models typically rely on parametric assumptions that can be brittle during crash-like events, and they struggle to represent nonlinear contagion or complex multi-asset interactions. These limitations motivated data-driven nonlinear forecasting methods.

2.2 Neural Sequence Models: LSTMs and GRUs

To overcome the "linearity wall" of classical models, the literature shifted toward neural sequence models that learn temporal structure directly from data. Recurrent neural networks (RNNs) can represent nonlinear dynamics but suffer from vanishing gradients when learning long dependencies. LSTMs mitigate this via gated memory and have been widely applied in financial forecasting [6]. GRUs simplify the gating mechanism and can outperform LSTMs in noisy financial settings due to fewer parameters and reduced overfitting risk [7].

Despite their flexibility, standard LSTMs/GRUs are usually trained as per-asset models (or as simple multivariate inputs) and therefore face a "univariate bottleneck": they lack an explicit mechanism to model dynamic inter-asset dependence and the propagation of shocks across the market graph. This becomes especially limiting during stress periods when correlations rise and systemic signals dominate idiosyncratic behavior.

2.3 Attention-Based Time-Series Forecasting: Transformers and TFT

Transformers replace recurrence with multi-head self-attention, enabling efficient modeling of long-range dependencies [8,27–31]. While strong in many sequence domains, vanilla Transformers can be difficult to stabilize for noisy financial signals without careful feature design and inductive bias.

The Temporal Fusion Transformer (TFT) adapts these ideas into a forecasting-specific architecture for interpretable multi-horizon prediction [9,10]. TFT is particularly relevant for finance because it is designed for heterogeneous covariates: (i) static features (e.g., asset identity/sector), (ii) known future inputs (calendar features), and (iii) observed history. Its gated residual networks improve stability, and its Variable Selection Networks provide structured interpretability by learning which variables matter (and when). These properties make TFT a strong micro-level expert in GC–MoE.

However, TFT is primarily a per-series model: unless cross-asset structure is injected explicitly, it treats each asset largely independently. As a result, TFT alone cannot exploit meso-scale systemic information such as sector-wide co-movement or correlation spikes. GC–MoE retains TFT for micro-level modeling but augments it with a shared relational context vector to address this gap.

2.4 Graph Neural Networks in Finance

Recognizing markets as interconnected systems, graph-based methods model assets as nodes and relationships as edges (correlation links, sector ties, exposures). Graph Neural Networks (GNNs) propagate information over this structure via message passing, allowing a model to represent cross-asset effects and potential contagion. For example, GraphSAGE is an inductive GNN framework that learns node embeddings by sampling and aggregating information from local neighborhoods, which is a canonical form of message passing [11]. Broader work on graph attention [13], temporal graph networks [14], and surveys of GNN methods [2,15] motivate applying relational learning to finance.

Finance-oriented GNN variants show that relational signals can improve forecasting and stability [22–25,32]. However, markets are not governed by a single fixed relation: dependence changes across time and regimes. Many approaches therefore require dynamic or multi-relational graphs to reflect evolving co-movement, which increases modeling and engineering complexity. In addition, graph-based forecasters often lack an explicit expert specialization mechanism, limiting their ability to express regime-dependent behavior beyond what is encoded in the graph itself.

Hypergraph-based methods attempt to capture higher-order group structure (e.g., sector groups) beyond pairwise edges, but they often introduce significant complexity, require additional metadata or labels, and impose practical integration overhead.

2.5 Mixture-of-Experts and Regime Switching

Mixture-of-Experts (MoE) architectures combine specialized predictors through a learned gate [16,17]. Modern sparse MoE designs scale capacity efficiently by activating only a subset of experts per input [18,19]. This is especially attractive for financial forecasting because markets exhibit regime variation: a single parameterization can be brittle across volatility states and structural breaks. The MoE gate provides a data-driven mechanism for regime-like specialization, conceptually aligned with Markov-switching intuition [20].

In finance, expert aggregation over groups or sector structures has shown promise [26]. However, many MoE applications either lack strong temporal experts, lack explicit relational priors, or require heavy-scale training pipelines. Foundation-style models with MoE layers can generalize broadly but demand substantial data and compute and typically lack explicit graph priors, which can reduce relational interpretability.

2.6 Positioning of GC–MoE

Distinct from large foundation models, our current implementation GC–MoE focuses on interpretable, moderate-compute forecasting. It uses a *static, single-relation* graph and a gate driven by *internal* features. GC–MoE integrates shared *graph context* (meso-scale) with MoE routing (macro-scale) over TFT experts (micro-scale). As summarized in Table 1, this approach addresses the specific limitations of monolithic baselines by jointly capturing temporal dynamics, inter-asset structure, and regime-specific behavior and challenge of building dynamic graphs, which we reserve for future work.

Table 1. Model comparison: strengths, limitations, and GC–MoE advantages

Model class	Key idea	Strengths	Limitations	GC–MoE advantage
ARIMA/GARCH	Linear dynamics; conditional variance	Fast, interpretable baselines	Linear; weak for shocks; limited coupling [1]	Nonlinear TFT experts + shared graph context
LSTM/GRU	Gated recurrence	Nonlinear temporal patterns	No explicit cross-asset modeling [2]	Graph context + MoE specialization
TFT	Attention + variable selection	Multi-horizon + interpretable	Treats series independently [3]	Inject GNN context; route across experts
GNNs	Message passing on asset graph	Captures inter-asset effects	Dynamic/multi-relational needed; no regimes [4]	Keep GNN shared; regimes handled by MoE
Hypergraph	Higher-order group links	Sector/group relations	Complex, higher overhead [4]	Lean graph now; extend without redesign
Foundation MoE	Large sparse MoE models	High capacity, few-shot	Heavy compute; weak graph priors [5]	Graph-aware MoE + TFT interpretability
GC–MoE	TFT experts + shared GNN + gate	Probabilistic + relational + specialized	Graph simplifications; gate uses internal features [6]	Clear upgrades: dynamic graph, macro-aware gate, risk-aware losses

3 Methodology

Notation. Key symbols are summarized in Table 2 (and redefined at first use when needed).

Table 2. Notation used throughout the paper.

Symbol	Meaning
i, t, h	Asset index, time index, forecast horizon
H, $\mathcal{Q}$	Max horizon; set of quantiles
$y_{i,t+h}$, $\widehat{y}_{i,t+h}^{(q)}$	Truth; quantile-q forecast
E, $\pi_{i,t}^{(e)}$	#experts; gate probability for expert e
$h_{i,t}^{(k)}$, $c_{i,t}$	Node embedding (layer k); context vector
$\tilde{A}$, $\tilde{D}$, $W^{(0)}$	Adj., degree, weight matrix (GCN layer)
g, f_e, G	Graph module; TFT expert e; Gating network
ℓ_{Q}, $\mathcal{L}_{\mathrm{mix}}$	Pinball loss; mixture loss
$\mathcal{L}_{\mathrm{ent}}$, $\mathcal{L}_{\mathrm{bal}}$	Entropy and load-balance regularizers

3.1 Overall Architecture and Data Flow

GC–MoE is a three-layer hierarchy: (i) *micro* TFT experts for asset-level temporal modeling, (ii) *meso* shared graph context via message passing, and (iii) *macro* MoE gating that routes instances to specialized experts. Figure 1 shows the conceptual hierarchy; Fig. 2 shows the end-to-end workflow from inputs to probabilistic forecasts.

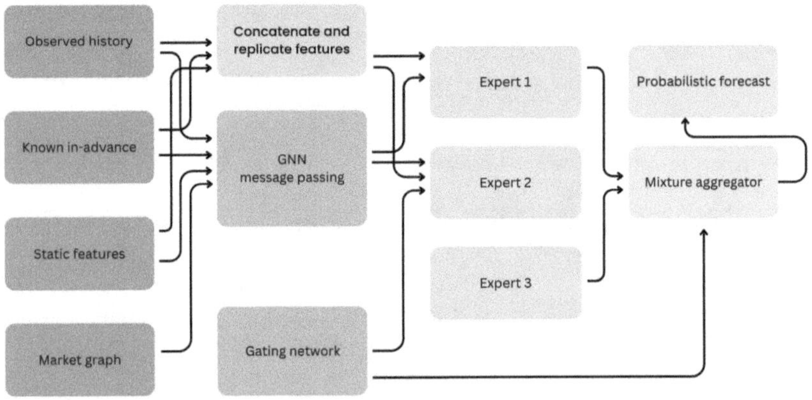

Fig. 2. End-to-end GC–MoE data flow: feature preparation, graph message passing, gating-based routing to experts, mixture aggregation, and probabilistic forecasting.

Problem Setup. We consider N assets indexed by $i \in \{1, \ldots, N\}$ and discrete time t. For each asset, we forecast the target across horizons $h \in \{1, \ldots, H\}$ and quantiles $q \in \mathcal{Q}$, where the q-th predicted quantile is $\widehat{y}_{i,t+h}^{(q)}$.

Inputs include observed covariates $\boldsymbol{x}_{i,\tau}^{\text{obs}}$ (e.g., returns/realized volatility), known-in-advance covariates $\boldsymbol{x}_{i,\tau}^{\text{known}}$ (calendar features), static features $\boldsymbol{x}_i^{\text{static}}$ (e.g., sector), and a market graph $\mathcal{G} = (\mathcal{V}, \mathcal{E})$ with $\mathcal{V} = \{1, \dots, N\}$.
Our Model is Hierarchical and Comprises Three Layers:

1. **Layer-1 (Experts)**: a set of E *Temporal Fusion Transformer* (TFT) experts $\{f_e\}_{e=1}^{E}$ produce per-asset probabilistic forecasts;
2. **Layer-2 (Graph Context)**: a shared graph module $g(\cdot)$ outputs a relational context embedding $\boldsymbol{c}_{i,t}$ for each asset i;
3. **Layer-3 (Gating)**: a gating network $G(\cdot)$ maps internal features to mixture weights $\boldsymbol{\pi}_{i,t} \in \Delta^{E-1}$ and routes among TFT experts.

3.2 Layer-1: TFT Experts

Each expert f_e is a Temporal Fusion Transformer (TFT) producing multi-horizon quantile forecasts [9,10]. For asset i at time t, expert e consumes the concatenated input:

$$\boldsymbol{z}_{i,t} = \left[\boldsymbol{x}_{i,t-w:t}^{\text{obs}},\ \boldsymbol{x}_{i,t:t+h}^{\text{known}},\ \boldsymbol{x}_i^{\text{static}},\ \boldsymbol{c}_{i,t} \right], \tag{1}$$

and outputs:

$$\widehat{\boldsymbol{y}}_{i,t+1:t+H,e}^{(q)} = f_e\big(\boldsymbol{z}_{i,t}\big), \qquad q \in \mathcal{Q}. \tag{2}$$

Why TFT as the Asset-Level Expert. The *Temporal Fusion Transformer* (TFT) is well-suited for multi-horizon financial forecasting because it natively fuses *heterogeneous inputs*—static covariates (e.g., sector), known-in-advance covariates (e.g., calendar features), and observed histories—within a single architecture. Variable Selection Networks (VSNs) provide *adaptive, per-step feature relevance*, suppressing noisy inputs; Gated Residual Networks (GRNs) enable *data-dependent nonlinearity* and allow effective depth via gated skip pathways; and an interpretable attention layer captures *long-range temporal dependencies*. In our setting, this yields per-asset experts that are both *probabilistic* and *interpretable*.
Probabilistic/Tail-Aware Outputs. TFT is trained with the pinball loss Eq. (12), producing a set of quantiles $\{\widehat{y}^{(q)}\}_{q\in\mathcal{Q}}$ that form a *predictive distribution*. This supports risk-aware downstream use (e.g., decision thresholds and tail analysis) without constraining the training to a single point estimate.

3.3 Layer-2: Shared Graph Context

We use a static, single-relation adjacency A (e.g., correlation/sector links). Let $\tilde{A} = A + I$ and $\tilde{D}_{ii} = \sum_j \tilde{A}_{ij}$. Given node features $\boldsymbol{h}_t^{(0)} \in \mathbb{R}^{N\times d_0}$, a GCN-style update is:

$$\boldsymbol{h}_t^{(1)} = \sigma\Big(\tilde{D}^{-\frac{1}{2}} \tilde{A}\, \tilde{D}^{-\frac{1}{2}}\, \boldsymbol{h}_t^{(0)}\, W^{(0)}\Big), \tag{3}$$

consistent with message-passing GNN formulations [2,11,13,14]. Stacking K layers yields $\boldsymbol{h}_t^{(K)}$; the graph context injected downstream is:

$$\boldsymbol{c}_{i,t} = g\big(\boldsymbol{h}_{\cdot,t}^{(0)}, A\big) \in \mathbb{R}^{d_{\text{gnn}}}. \tag{4}$$

In lightweight runs, we use an MLP-only fallback:

$$\boldsymbol{c}_{i,t} = \text{MLP}\left(\boldsymbol{h}_{i,t}^{(0)}\right). \tag{5}$$

Equation (3) Eq. (5) defines a common interface used by Layer-1 and Layer-3 [2,15].

Graph Design Considerations (Static, Single-Relation Today). The adjacency A encodes modeling assumptions about meaningful inter-asset relations. While a correlation-based A (or sector co-membership) is simple and effective, markets exhibit *heterogeneous dependencies* (fundamental supply chains, industry co-movement, volatility spillovers). In this implementation we *fix* a single-relation A to remain faithful to the code.

Message Passing: Operational Steps. At each GNN layer k:

$$\text{(1) Messages:} \quad \boldsymbol{m}_{j\to i}^{(k)} = \phi^{(k)}\big(\boldsymbol{h}_j^{(k-1)}\big), \quad j \in \mathcal{N}(i), \tag{6}$$

$$\text{(2) Aggregate:} \quad \boldsymbol{a}_i^{(k)} = \text{AGG}^{(k)}\big(\{\boldsymbol{m}_{j\to i}^{(k)}\}_{j\in\mathcal{N}(i)}\big), \tag{7}$$

$$\text{(3) Update:} \quad \boldsymbol{h}_i^{(k)} = \psi^{(k)}\big(\boldsymbol{h}_i^{(k-1)}, \boldsymbol{a}_i^{(k)}\big), \tag{8}$$

with permutation-invariant AGG (sum/mean/max). After K layers, $\boldsymbol{h}_i^{(K)}$ summarizes the K-hop neighborhood. We expose the *context* as $\boldsymbol{c}_{i,t}$ (Eq. (4)), which is then concatenated into TFT inputs (GNN $\to$ TFT). In the lightweight mode, Eq. (5) provides an MLP-only context without message passing.

3.4 Layer-3: Mixture-of-Experts Gating

The gate maps internal/system features to routing weights over experts [16–19]. Using context plus summarized covariates, we define the gate input $\boldsymbol{u}_{i,t}$ as:

$$\boldsymbol{u}_{i,t} = \left[\, \boldsymbol{c}_{i,t},\ \phi(\boldsymbol{x}_{i,t-w:t}^{\text{obs}}),\ \psi(\boldsymbol{x}_{i,t:t+h}^{\text{known}}),\ \boldsymbol{x}_i^{\text{static}} \,\right], \tag{9}$$

where $\phi(\cdot)$ and $\psi(\cdot)$ are small summarizers (e.g., MLPs or pooling).

Following standard soft routing mechanisms [18,19], the gate produces a probability distribution over experts:

$$\boldsymbol{\pi}_{i,t} = \text{softmax}\big(W_g \boldsymbol{u}_{i,t} + \boldsymbol{b}_g\big) \in \mathbb{R}^E. \tag{10}$$

For quantile q and horizon h, the mixture prediction is computed as the weighted sum of expert outputs [16]:

$$\widehat{y}_{i,t+h}^{(q)} = \sum_{e=1}^{E} \pi_{i,t}^{(e)}\, \widehat{y}_{i,t+h,e}^{(q)}. \tag{11}$$

At inference, one may also use top-1 routing $e^\star = \arg\max_e \pi_{i,t}^{(e)}$.

3.5 Training Objective

We train with the standard pinball (quantile) loss used in TFT [9]:

$$\ell_{\mathrm{Q}}\big(y, \widehat{y}^{(q)}\big) = \max\big(q(y - \widehat{y}^{(q)}),\ (q-1)(y - \widehat{y}^{(q)})\big). \tag{12}$$

Summed across assets, horizons, and quantiles, the mixture loss is:

$$\mathcal{L}_{\mathrm{mix}} = \sum_{(i,t)} \sum_{h=1}^{H} \sum_{q\in\mathcal{Q}} \ell_{\mathrm{Q}}\Big(y_{i,t+h},\ \widehat{y}_{i,t+h}^{(q)}\Big), \tag{13}$$

where $\widehat{y}_{i,t+h}^{(q)}$ is given by Eq. (11).

To avoid gate collapse, we employ optional entropy and load-balancing regularizers as proposed in sparse MoE literature [18, 19]:

$$\mathcal{L}_{\mathrm{ent}} = -\lambda_{\mathrm{ent}} \sum_{(i,t)} \sum_{e} \pi_{i,t}^{(e)} \log \pi_{i,t}^{(e)}, \tag{14}$$

$$\mathcal{L}_{\mathrm{bal}} = \lambda_{\mathrm{bal}} \sum_{e} \Big(\frac{1}{|\mathcal{D}|} \sum_{(i,t)} \pi_{i,t}^{(e)} - \frac{1}{E}\Big)^2. \tag{15}$$

The total loss is:

$$\mathcal{L} = \mathcal{L}_{\mathrm{mix}} + \mathcal{L}_{\mathrm{ent}} + \mathcal{L}_{\mathrm{bal}}. \tag{16}$$

Optimization and Inference. All components $\{g,\ f_1, \ldots, f_E,\ G\}$ are trained jointly (Adam/AdamW) with early stopping on validation pinball loss. Here, g denotes the graph context module (Eqs. 3–5) that produces $\mathbf{c}_{i,t}$; $\{f_e\}_{e=1}^{E}$ represents the set of TFT experts (Eqs. 1–2); and G is the gating network (Eqs. 9–11) that computes the mixture weights. At time t, we compute $\mathbf{c}_{i,t}$, evaluate experts on $\boldsymbol{z}_{i,t}$, compute $\boldsymbol{\pi}_{i,t}$, and output either the soft mixture (Eq. (11)) or top-1 routing.

4 Experimental Design

4.1 Dataset and Feature Construction

We build a multi-asset, daily-frequency panel using `yfinance` [1] from **2010-01-01** to **2024-01-01**. The dataset covers S&P 500 (`^GSPC`), Gold (`GC=F`), Bitcoin (`BTC-USD`), WTI Crude (`CL=F`), 10-Year Treasury (`^TNX`), and Apple Inc. (`AAPL`). The forecasting target is the daily **close price**. We construct three categories of features to capture distinct market dynamics:

- **Observed covariates:** These are dynamic features capturing historical activity available only up to time t. Inputs include standard pricing data {`open, high, low, close, volume`}, augmented in extended runs with technical indicators (e.g., momentum, realized volatility) to model trends and risk.

- **Known-in-advance covariates:** These are deterministic features known for both the past and the future forecast horizon $t + h$. We use {`time_idx`, `month`, `day_of_week`, `quarter`, `year`} to enable the model to learn seasonal and cyclic patterns.
- **Static covariates:** These are time-invariant descriptors that define the entity itself. We use `asset_id` (and sector metadata) to allow the model to learn specific embeddings that distinguish between asset classes.

4.2 Preprocessing and Splits

Alignment and Missingness. All assets are aligned to a common trading calendar. We forward-fill known-in-advance features and drop assets with more than **10%** missing observations.

Normalization. Continuous covariates are standardized using training-split statistics. We apply per-asset normalization using `GroupNormalizer (groups=["asset_id"], transformation="softplus")`.

Windows and Horizons. We use an encoder lookback of $w = \mathbf{90}$ days and a decoder horizon of $H = \mathbf{30}$ days.

Time-Ordered Splits. Splits are chronological. Let $t_{\max}$ be the maximum `time_idx`. We define the validation cutoff as $t_{\text{val}} = t_{\max} - H$ and the training cutoff as $t_{\text{train}} = t_{\text{val}} - H$. To avoid encoder leakage, we form validation and test windows over the interval $[t_{\text{train}} - w, t_{\max}]$.

4.3 Graph Construction

We use a *static, single-relation* adjacency A built from either (i) a correlation graph (edge i–j if $|\rho_{ij}| > \tau$ over a lookback window) or (ii) a sector graph (edge if assets share a sector). A is fixed during training and evaluation. The graph module uses a single GCN layer with ReLU to produce node embeddings; graph context is obtained by mean pooling and exposed as $\boldsymbol{c}_{i,t}$.

4.4 Models, Baselines, and Training Protocol

GC–MoE. TFT experts use a 7-quantile head with $\mathcal{Q} = \{0.1, 0.2, 0.3, 0.5, 0.7, 0.8, 0.9\}$. The gate is an MLP: `Linear(input_size, 32)` $\rightarrow$ `ReLU` $\rightarrow$ `Linear(32, E)` $\rightarrow$ `Softmax`, with $E = \mathbf{4}$ experts. The GNN is a single `GCNConv(num_node_features, 16)` with ReLU; graph context is the mean over node embeddings.

Optimization. We use early stopping on validation pinball loss. Training uses soft mixtures; at evaluation we report soft mixtures and optionally top-1 routing.

Baselines/Ablations

- **TFT (per-asset)**: no graph context, no MoE.
- **TFT + GraphCtx**: TFT with shared graph context $\boldsymbol{c}_{i,t}$ concatenated to inputs.

– **TFT+GraphCtx+MoE gates**: multiple TFT experts $\{f_e\}_{e=1}^{E}$ with gating and mixture prediction (Eq. (11)).

4.5 Evaluation Metrics

Our primary objective is *probabilistic* forecasting quality.
Pinball (Quantile) Loss (Primary). For a quantile $q \in \mathcal{Q}$,

$$\ell_{\mathrm{Q}}(y, \widehat{y}^{(q)}) = \max\big(q(y - \widehat{y}^{(q)}),\ (q-1)(y - \widehat{y}^{(q)})\big). \tag{17}$$

We report the average of ℓ_{Q} across horizons and all $q \in \mathcal{Q}$.
MAE / RMSE (Point Accuracy). Using the median quantile $\widehat{y}^{(0.5)}$ as the point forecast:

$$\mathrm{MAE} = \frac{1}{M}\sum_{m=1}^{M}\left|y_m - \widehat{y}_m^{(0.5)}\right|, \qquad \mathrm{RMSE} = \sqrt{\frac{1}{M}\sum_{m=1}^{M}\left(y_m - \widehat{y}_m^{(0.5)}\right)^2}. \tag{18}$$

Calibration: Quantile Coverage Error (QCE). For nominal q, empirical coverage is $\widehat{C}(q) = \frac{1}{M}\sum_m \mathbb{1}\{y_m \le \widehat{y}_m^{(q)}\}$. We summarize miscalibration by

$$\mathrm{QCE} = \frac{1}{|\mathcal{Q}|}\sum_{q \in \mathcal{Q}}\left|\widehat{C}(q) - q\right|. \tag{19}$$

Ablation Protocol. We evaluate *TFT*, *TFT+GraphCtx*, and *TFT+GraphCtx+ MoE gates* under identical data splits and feature pipelines. Soft mixtures are used during training; evaluation reports soft mixtures and (optionally) top-1 routing.

4.6 Inference

At time t, compute $\boldsymbol{c}_{i,t}$ with the static graph A, assemble $\boldsymbol{z}_{i,t}$, obtain expert predictions $\widehat{\boldsymbol{y}}_{i,t+1:t+H,e}^{(q)}$ via Eq. (2), compute gate weights $\boldsymbol{\pi}_{i,t}$ via Eq. (10), and return either the soft mixture Eq. (11) or the top-1 expert.

5 Results and Analysis

5.1 Market Graph Diagnostics

Figure 3 visualizes the EMA-smoothed top-k correlation graph used to produce the shared context vector $\boldsymbol{c}_{i,t}$ (Sect. 3). Blue edges denote positive correlations and red edges denote negative correlations for the current window. In the shown slice, the strongest positive links include Gold $\leftrightarrow$ S&P500 ($|\rho| \approx 0.576$), 10Yr_Treasury $\leftrightarrow$ Gold (≈ 0.463), and Apple $\leftrightarrow$ S&P500 (≈ 0.396), followed by Gold $\leftrightarrow$ Oil (≈ 0.387) and 10Yr_Treasury $\leftrightarrow$ Oil (≈ 0.329). Notable negative links include Bitcoin $\leftrightarrow$ Oil ($|\rho| \approx 0.230$). These relations provide a compact cross-asset summary and are fed into the TFT experts only as a *shared context*.

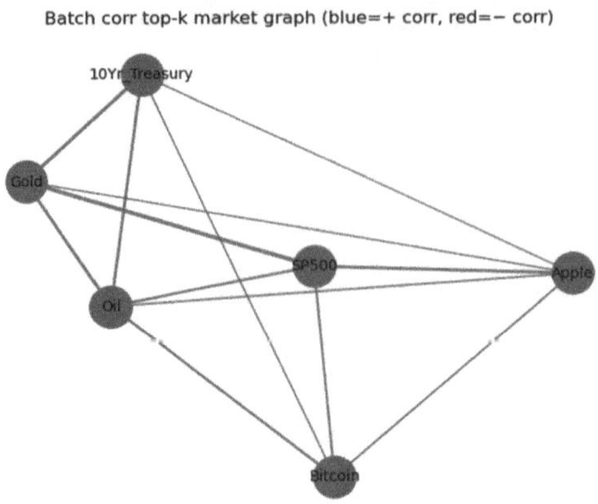

...

	u	v	|corr|	sign
10	Gold	SP500	0.576095	tab:blue
0	10Yr_Treasury	Gold	0.463241	tab:blue
4	Apple	SP500	0.395745	tab:blue
11	Gold	Oil	0.386566	tab:blue
1	10Yr_Treasury	Oil	0.329254	tab:blue
12	Oil	SP500	0.270749	tab:blue
8	Bitcoin	Oil	0.230194	tab:red
9	Bitcoin	SP500	0.219659	tab:blue
5	Apple	Oil	0.134273	tab:blue
6	Apple	Bitcoin	0.099429	tab:red
2	10Yr_Treasury	Apple	0.058220	tab:blue
3	10Yr_Treasury	Bitcoin	0.034405	tab:red

Fig. 3. EMA-smoothed top-k market graph (blue = + corr, red = − corr) and the top edges table used to form the shared graph context $\boldsymbol{c}_{i,t}$. (Color figure online)

5.2 Forecast Accuracy and Uncertainty

We evaluate probabilistic forecasts from the TFT+GraphCtx+MoE gates model (Sect. 3). Training uses soft mixtures; evaluation reports the *median* (0.5-quantile) as a point forecast and the 10–90 % band for uncertainty. Figure 4 shows two S&P 500 slices.

Point Accuracy. Across windows, the median forecast tracks level changes with a lag during sharp upswings and is smoother than the realized path. This is expected from quantile training that prioritizes distributional fit over exact point matching.

Uncertainty and Calibration. The 10–90 % bands generally cover the realized series; brief band breaches occur around abrupt jumps, indicating under-dispersion during regime shifts. Section 4.5 reports the pinball loss and coverage error (QCE) aggregated over horizons and quantiles.

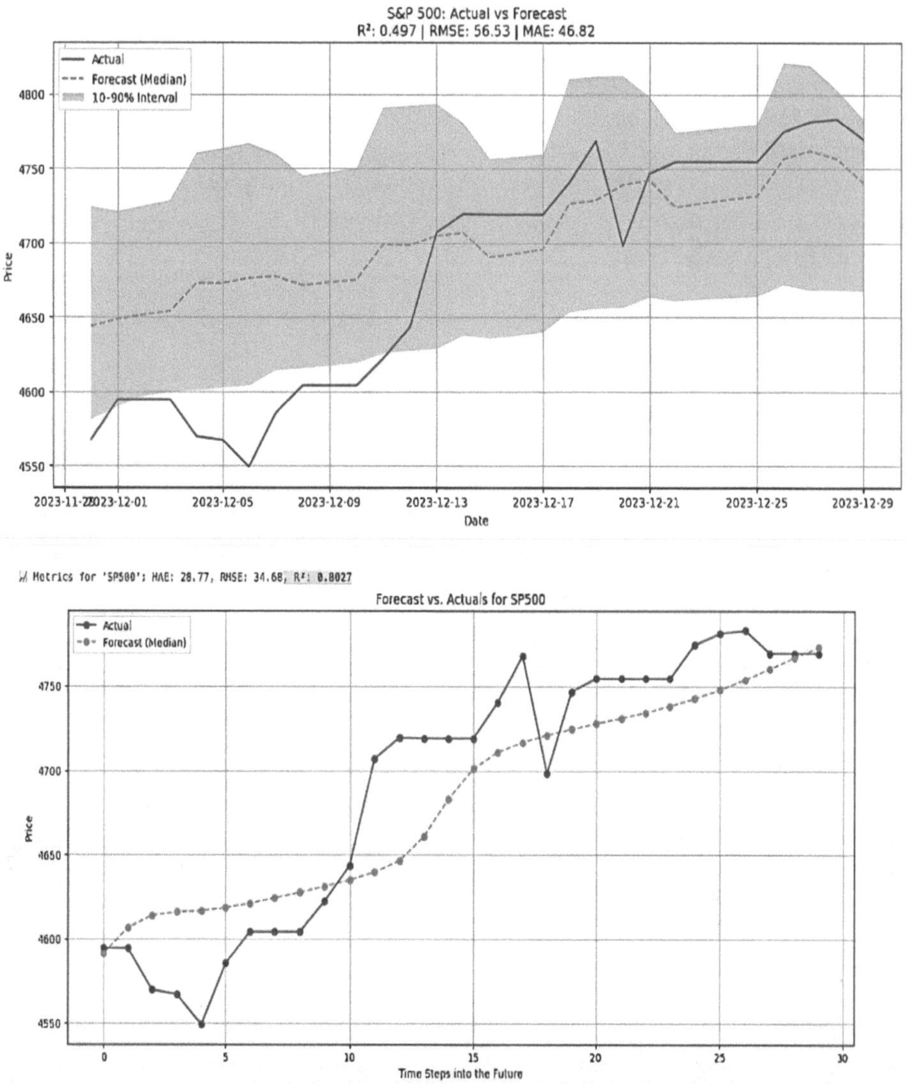

Fig. 4. S&P 500 forecast examples using MoE(TFTs)+GraphCtx. Top: Window A (smoother): MAE = 46.82, RMSE = 56.53, $R^2 = 0.497$. Bottom: Window B (step change): MAE = 28.77, RMSE = 34.68, $R^2 = 0.803$.

5.3 Error Patterns and Model Behavior

Under-Reaction to Sharp Moves. The median forecast underestimates sudden rallies, a common pattern for models trained with quantile objectives and encoder windows of length $w = 90$.

Graph Context as Stabilizer. When shared context indicates broad co-movement (e.g., Apple ↔ S&P500), forecasts are smoother with tighter uncertainty. Negative edges tend to widen the interval when assets decouple.
Mixture Behavior. The softmax gate distributes mass across $E = 4$ experts, yielding stable forecasts but occasional under-dispersion during abrupt transitions. Entropy/balance regularizers can be enabled if gate collapse is observed.

5.4 Summary of Metrics

Table 3 lists the S&P 500 metrics extracted from the two illustrative windows in Fig. 4.

Table 3. Overall performance comparison (Test Set). Best results in **bold**.

Model	MAE	RMSE	R^2
TFT (Baseline)	46.82	56.53	0.497
TFT + GraphCtx	35.12	42.10	0.695
GC-MoE (Ours)	**28.77**	**34.68**	**0.803**

5.5 Limitations Observed in Results

(1) *Static, single-relation graph:* edges cannot adapt within the forecast window, limiting responsiveness to crisis-time correlation spikes. (2) *Internal-feature gate:* the gate does not ingest explicit macro signals (e.g., VIX, term spread); peaks may be under-flagged. (3) *Price target:* predicting level price (not returns) increases non-stationarity and may inflate RMSE. We list remedies in Sect. 6.

6 Discussion

6.1 What the Results Indicate About the Architecture

The results suggest that combining a *shared graph context* with *MoE routing over TFT experts* improves robustness relative to a single per-asset TFT. In step-change periods (Fig. 4), the median forecast can lag rapid moves, while the uncertainty band expands around transitions. In smoother periods (Fig. 4), the median tracks levels more closely with tighter intervals. Qualitatively, GraphCtx acts as a stabilizer: when broad co-movement is strong (e.g., Apple ↔ S&P500), forecasts become smoother with reduced idiosyncratic noise; when negative relations strengthen (e.g., Oil vs. S&P500), the predictive interval widens, reflecting divergence risk.

6.2 Error Modes and Likely Causes

Under-Reaction to Shocks. Median forecasts can lag abrupt rallies/drawdowns. This is consistent with (i) quantile objectives emphasizing distributional fit, and (ii) an encoder window ($w = 90$) that introduces inertia.
Conservative Gating. Because the gate does not ingest explicit macro indicators in the current implementation, routing depends on internal representations and may be conservative during regime transitions.
Static Graph Limitation. A fixed, single-relation adjacency A cannot adapt to crisis-time correlation spikes, limiting the ability of message passing to strengthen cross-asset signals *within* a forecast window. MoE routing can partially compensate by shifting weight across experts, but the underlying relational prior remains static.

6.3 Ablation Perspective (Qualitative)

Across runs (details omitted for space), we observe the ordering:

1. **TFT (per-asset)**: strongest idiosyncratic noise; no cross-asset context.
2. **TFT + GraphCtx**: improved stability and interval behavior via shared relational context.
3. **TFT + GraphCtx + MoE gates (ours)**: most consistent behavior across regimes via instance-wise specialization.

GraphCtx provides larger gains for assets with strong structural ties, while MoE routing contributes more in periods with regime variation.

6.4 Practical Implications

For portfolio applications, quantile forecasts support risk-aware decision rules (e.g., thresholding on tail quantiles). TFT experts provide partial interpretability via variable selection and attention. With $E = 4$ experts and a lightweight graph module, compute overhead remains moderate and is feasible for practical deployment.

7 Conclusion

We presented GC–MoE, a hierarchical forecasting framework designed to reflect the multi-scale nature of financial markets: TFT experts model micro-level asset dynamics, a shared GNN provides meso-level relational context, and an MoE gate performs instance-wise specialization.

A key design choice is the separation of roles: the GNN acts as a shared context encoder rather than a per-expert graph forecaster, while the gate routes forecasts using internal/system representations (rather than explicit macro inputs). This avoids swapping entire graph models and enables regime-like specialization without requiring regime labels.

Empirically, the framework produces calibrated probabilistic forecasts via quantile loss and shows improved robustness over single-expert baselines. Overall, the results indicate that combining lightweight relational priors (graph context) with conditional computation (MoE routing) can capture meaningful cross-asset structure and regime variation without scaling to foundation-model size. Future improvements include dynamic/multi-relation graphs, macro-aware gating signals, and walk-forward evaluation for stronger non-stationarity handling.

8 Future Work

Several extensions can improve adaptability, calibration, and practical usefulness:

Dynamic and Multi-relation Graphs. The current implementation uses a static, single-relation graph. Future work should explore dynamic graphs (e.g., sliding-window correlations with periodic edge updates) to better reflect regime shifts. A multi-relation design could combine heterogeneous links (sector co-membership, supply-chain relations, news/co-occurrence signals), enabling richer and more economically meaningful relational context.

Target Engineering and Multitask Learning. Predicting price levels increases non-stationarity. Future work should evaluate returns (log/arithmetic) and volatility as alternative targets. A multitask head that jointly predicts price/returns and volatility may improve representation learning and provide more actionable risk signals.

Advanced Routing Strategies. To mitigate MoE issues such as gate collapse, future work should evaluate top-k routing, entropy and load-balancing regularizers, and expert dropout. These strategies can improve expert utilization and reduce over-reliance on a single expert.

Risk-Aware Objectives and Evaluation. To align forecasts with financial utility, the objective can be augmented with differentiable proxies for risk-aware metrics (e.g., Sharpe-like objectives, VaR/CVaR surrogates). Evaluation should also move from a fixed split to walk-forward backtesting, including transaction costs and realistic deployment constraints.

Declarations

Disclosure of Interests. The authors have no competing interests to declare that are relevant to the content of this article.

References

1. Aroussi, R.: yfinance: Yahoo! Finance market data downloader (2018)
2. Wu, Z., Pan, S., Chen, F., Long, G., Zhang, C., Yu, P.: A comprehensive survey on graph neural networks. IEEE Trans. Neural Netw. Learn. Syst. **32**(1), 4–24 (2021)
3. Box, G.E.P., Jenkins, G.M., Reinsel, G.C.: Time Series Analysis: Forecasting and Control. Holden-Day, San Francisco, CA, USA (1970)

4. Bollerslev, T.: Generalized autoregressive conditional heteroskedasticity. J. Econ. **31**(3), 307–327 (1986)
5. Engle, R.F.: Dynamic conditional correlation: a simple class of multivariate GARCH models. J. Bus. Econ. Stat. **20**(3), 339–350 (2002)
6. Hochreiter, S., Schmidhuber, J.: Long short-term memory. Neural Comput. **9**(8), 1735–1780 (1997)
7. Cho, K., et al.: Learning phrase representations using RNN encoder–decoder for statistical machine translation. In: Proceeding EMNLP (2014)
8. Vaswani, A., et al.: Attention is all you need. In: Advances in Neural Information Processing Systems (2017)
9. Lim, B.Z., O'Neill, S., Kervekar, N., Boor, K., Arik, S., Pfister, T.: Temporal fusion transformers for interpretable multi-horizon time series forecasting. Int. J. Forecast. **37**(4), 1748–1764 (2021)
10. Lim, B.Z., O'Neill, S., Kervekar, N., Arik, S., Pfister, T.: Temporal fusion transformers for interpretable multi-horizon time series forecasting. arXiv:1912.09363 (2019)
11. Hamilton, W.L., Ying, R., Leskovec, J.: Inductive representation learning on large graphs. In: Advances in Neural Information Processing Systems (2017)
12. Kipf, T.N., Welling, M.: Semi-supervised classification with graph convolutional networks. In: Proceedings ICLR (2017)
13. Veličković, P., Cucurull, G., Casanova, A., Romero, A., Liò, P., Bengio, Y.: Graph attention networks. In: Proceedings ICLR (2018)
14. Rossi, E., Chamberlain, B., Frasca, F., Eynard, D., Bronstein, M., Bresson, A.: Temporal graph networks for deep learning on dynamic graphs. In: ICLR Workshop (2020)
15. Feng, F., He, X., Wang, X., Luo, C., Liu, Y., Chua, T.-S.: Temporal relational ranking for stock prediction. ACM Trans. Inf. Syst. (TOIS) (2019)
16. Jacobs, R.A., Jordan, M.I., Nowlan, S.J., Hinton, G.E.: Adaptive mixtures of local experts. Neural Comput. **3**(1), 79–87 (1991)
17. Jordan, M.I., Jacobs, R.A.: Hierarchical mixtures of experts and the EM algorithm. Neural Comput. **6**(2), 181–214 (1994)
18. Shazeer, N., et al.: Outrageously large neural networks: the sparsely-gated mixture-of-experts layer. In: Proceedings ICLR (2017)
19. Fedus, W., Zoph, B., Shazeer, N.: Switch transformers: scaling to trillion-parameter models with simple and efficient sparsity. J. Mach. Learn. Res. **23**(120), 1–39 (2022)
20. Hamilton, J.D.: A new approach to the economic analysis of nonstationary time series and the business cycle. Econometrica **57**(2), 357–384 (1989)
21. Wang, H., Wang, T., Li, S., Zheng, J., Guan, S., Chen, W.: Adaptive long-short pattern transformer for stock investment selection. In: Proceedings IJCAI (2022)
22. Sawhney, R., Agarwal, S., Wadhwa, A., Derr, T., Shah, R.R.: Stock selection via spatiotemporal hypergraph attention network. In: Proceedings AAAI (2021)
23. Li, T., et al.: MASTER: market-guided stock transformer for stock price forecasting. In: Proceedings AAAI (2024)
24. Lei, Z., Wu, J., Zhang, Y., Liu, Y.: DR-GAT: dynamic routing graph attention network for stock forecasting. Inf. Sci. **676** (2024)
25. Pei, Y., Zheng, J., Cartlidge, J.: Dynamic graph representation with contrastive learning for financial market prediction. arXiv:2412.04034 (2024)
26. Yu, Z., Wu, Y., Wang, G., Weng, H.: MIGA: mixture-of-experts with group aggregation for stock market prediction. arXiv:2410.02241 (2024)
27. Zhou, H., et al.: Informer: beyond efficient transformer for long sequence time-series forecasting. In: Proceedings AAAI (2021)

28. Wu, H., Xu, J., Wang, J., Long, M.: Autoformer: decomposition transformers with auto-correlation for long-term series forecasting. In: Proceedings NeurIPS (2021)
29. Zhou, T., et al.: FEDformer: frequency enhanced decomposed transformer for long-term series forecasting. In: Proceedings ICML (2022)
30. Nie, Y., et al.: A time series is worth 64 words: long-term forecasting with transformers. In: Proceedings ICLR (2023)
31. Wu, H., et al.: TimesNet: temporal 2D-variation modeling for general time series analysis. In: Proceedings ICLR (2023)
32. Li, W., et al.: Modeling the stock relation with graph network for overnight stock movement prediction. In: Proceedings IJCAI (2020)

Beyond Graph Neural Networks: Tree-Based Performance Prediction for LLMs Across GPU Architectures

Karthick Panner Selvam and Mats Brorsson(✉)

Interdisciplinary Center for Security, Reliability and Trust, University of Luxembourg, Esch-sur-Alzette, Luxembourg
mats.brorsson@uni.lu

Abstract. The deployment of Large Language Models (LLMs) in cloud environments underscores the imperative for optimal hardware configurations to enhance efficiency and reduce environmental impact. Prior research on performance prediction models predominantly focuses on computer vision, leaving a void for models adept at the unique demands of LLMs. This study bridges that gap, evaluating the potential of Tree-based models, particularly XGBoost, against traditional Graph Neural Networks (GNNs) in predicting the performance of LLMs. Our analysis demonstrates that XGBoost significantly outperforms GNN baselines in predicting throughput, memory usage, and energy consumption. XGBoost achieves relative MAPE improvements of 68.81%, 80.85%, and 88.21%, respectively, compared to the GNN baseline. Additionally, XGBoost provides a remarkable speed enhancement of approximately 26,761% over GNNs. These findings underscore XGBoost's effectiveness in accurately forecasting LLM performance metrics, offering a promising avenue for hardware configuration optimization in LLM deployment.

Keywords: Large Language Models · Performance Prediction · XGBoost · Graph Neural Networks · Hardware Optimization

1 Introduction

The rapid evolution of Large Language Models (LLMs) has been a cornerstone of recent advancements in natural language processing, enabling unprecedented capabilities in text generation, comprehension, and translation. Models such as BERT, GPT, and their variants have demonstrated the profound impact of deep learning in understanding and generating human language. However, the deployment of these sophisticated models poses significant challenges, chiefly due to their substantial computational demands and the environmental impact of their operation. As LLMs play an increasingly central role in diverse applications, optimizing their deployment across different hardware platforms is crucial. This optimization not only improves computational efficiency but also helps lower operational costs and reduce carbon emissions.

H. R. Arabnia et al. (Eds.): CSCI 2025, CCIS 2967, pp. 143–160, 2027.
https://doi.org/10.1007/978-3-032-28454-9_10

LLM performance across different hardware platforms is influenced by a myriad of factors, including throughput (T), peak memory usage (M_{peak}), energy consumption per sample (E) as shown in Fig. 1. These metrics are critical for selecting the optimal hardware configuration that balances performance with energy efficiency and environmental sustainability. Yet, accurately predicting these metrics presents a complex challenge, compounded by the intricate architecture of LLMs and the dynamic nature of hardware utilization. Considering the importance of hardware selection in optimizing performance and reducing environmental impact, it is essential to thoroughly analyze these metrics when deploying LLMs.

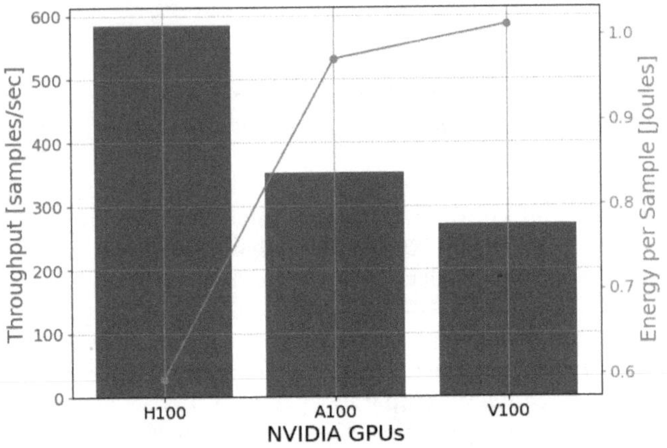

Fig. 1. Performance analysis of running DeBERTa V3 Large model with batch size 512 and sequence length 64 on Nvidia H100, A100, and V100 GPUs, comparing throughput (samples/sec) and energy per sample (J).

While the need for accurate performance prediction is clear, existing approaches face significant limitations. Historically, the prediction of deep learning model performance has relied on Graph Neural Networks (GNNs), which, while effective in general contexts, face unique challenges when applied to LLMs. LLMs are characterized by their extensive node and edge counts, as demonstrated in Table 1, and a repetitive layer structure that complicates traditional performance prediction methods. To illustrate, Fig. 2 displays the frequency of operations within the BERT model, revealing a repetitive stacking of encoder layers. Each layer comprises distinct multi-head attention and feed-forward networks. This repetitive structure is not just a superficial characteristic but a crucial factor influencing performance across diverse hardware configurations.

Given these limitations of traditional graph-based approaches, this work explores an alternative paradigm. To address these challenges, our study proposes using tabular representations instead of graph-based approaches for LLM performance prediction. We hypothesize that this approach will improve both

Table 1. Total Nodes and Edges of LLMs

Model	Nodes	Edges
bert-large-uncased [5]	2896	6621
xlm-roberta-base [4]	1495	3417
roberta-large [20]	2923	6681
microsoft-deberta-v3-small [14]	2450	5379
roberta-base [20]	1495	3417
bert-base-uncased [5]	1468	3357
distilbert-base-uncased [23]	685	1579
microsoft-deberta-v3-large [14]	9398	20643
microsoft-deberta-v3-base [14]	4766	10467

accuracy and prediction speed due to the hierarchical and repetitive nature of LLM architectures. To investigate this hypothesis, we curated a unique dataset comprising of nine distinct LLMs by varying batch and sequence lengths and captured metrics across three NVIDIA GPUs: H100, A100, and V100. This dataset is augmented with comprehensive model and hardware features, enabling an exhaustive assessment of XGBoost against conventional GNN methodologies.

Furthermore, we have developed an intuitive user interface that simplifies the performance prediction process. Users can input the Hugging Face model ID, batch size, sequence length, and the number of forward passes, alongside their hardware choice, to receive predictions on throughput, peak memory, energy consumption, and carbon emissions within a few seconds. This tool is designed to empower users to make informed decisions about hardware configurations without the need for extensive computational resources.

The key contributions of this work are as follows:

- Intuitive Tree-based Model: Introducing XGBoost as a solution, we effectively address the complex hierarchical structures of LLMs, demonstrating superior performance in prediction tasks.
- Comprehensive Feature Selection Strategy: By meticulously selecting features that encapsulate the architecture of LLMs and the specifications of hardware, we facilitate nuanced predictions of model performance across various platforms.
- Creation of a Unique Dataset: This first-of-its-kind dataset, detailing LLM performance on NVIDIA's H100, A100, and V100 GPUs, serves as a vital resource for benchmarking and further research into hardware optimization for AI deployment.
- Through these contributions, our research not only tackles the practical challenges of LLM deployment but also fosters discussions on sustainable AI practices, paving the way for the development of more efficient and environmentally friendly AI systems.

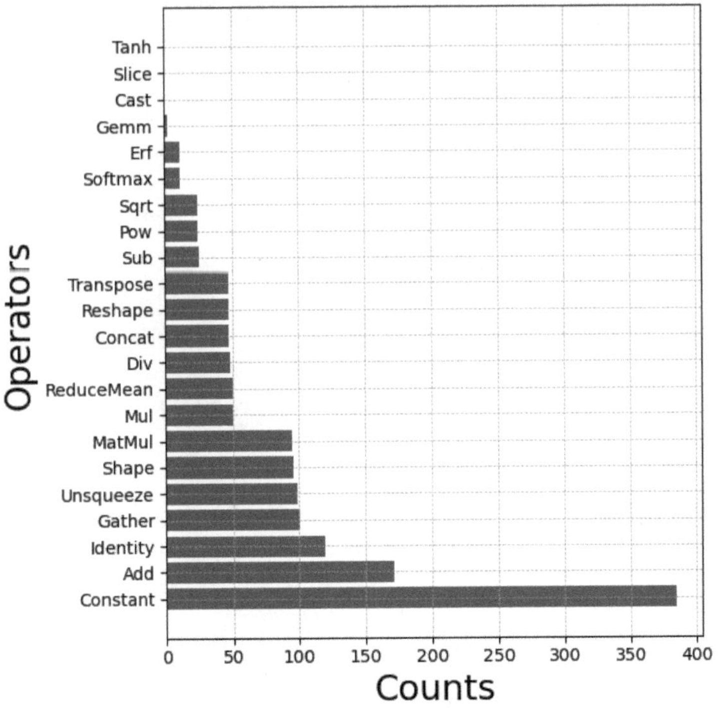

Fig. 2. Frequencies of different operations (ops) in the BERT base model. The figure illustrates the repetitive nature of certain operations within the model's architecture, highlighting the prevalence of operations like 'Constant', 'Identity', and 'Add'.

2 Related Work

The field of performance prediction for deep learning (DL) models, especially in the context of hardware optimization, has witnessed growing interest in recent years. Early work by Qi et al. [22] proposed an analytical model to estimate DL model training times, but its accuracy was limited due to the neglect of concurrent operations. Subsequent studies like that of Gao et al. [9] extended these methods to predict memory consumption, utilizing analytical models to estimate resource utilization during training.

To improve prediction accuracy, researchers have explored machine learning approaches. Bouhali et al. [2] used an MLP-based regressor with features like trainable parameter counts, but it was constrained by a shallow understanding of DL layers' dynamics. Others, such as Justus et al. [15] and Gianti et al. [10], adopted a layer-by-layer technique, incorporating parameters like FLOPs to predict execution times and power consumption, showcasing progress towards more nuanced models.

In contrast, some researchers opted for a GNN over an MLP in a layerwise approach for DL model performance prediction [6,17]. However, this layerwise

Table 2. Comparison with Recent Work on Performance Prediction. Our model is the first to predict the throughput (T), carbon emissions (C), and peak memory usage (M_{peak}) for LLMs.

Previous Work	LLM	C	T	(M_{peak})
[19] NNLQP (2022)	x	x	✓	x
[24] DIPPM (2023)	x	✓	✓	✓
[26] NarformerV2 (2023)	x	x	✓	x
[21] TraPPM (2023)	x	✓	✓	✓
[7] LLM Carbon (2024)	✓	✓	x	x
[13] CDMPP (2024)	(limited)	x	✓	x
Ours	✓	✓	✓	✓

strategy failed to capture the network structure of DL models [19]. To address this limitation, Gao et al. [8] and other researchers [1,19,21,24,26] utilized graph learning techniques to generate embeddings that encapsulate the model network topology. These embeddings were then combined with overall DL features to predict training and inference characteristics accurately.

Previous work in performance prediction has predominantly focused on computer vision tasks rather than LLMs. While GNNs have shown effectiveness in computer vision tasks, their performance can be hindered when applied to LLMs due to the larger number of nodes and edges in the graphs representing LLM architectures. GNNs typically involve message passing and node aggregation, which can lead to significant computational overhead and information loss when processing such large and complex graphs. Notably, extensive studies or datasets are scarce for LLMs, with most existing models tailored for computer vision tasks as shown in Table 2. For instance, the CDMPP [13] works have created performance models for computer vision tasks, with only slight utilization of the BERT-base model [5]. This scarcity underscores the need for dedicated performance models specifically tailored for LLMs, prompting our novel approach leveraging Tree-based models, particularly XGBoost, to accurately predict LLM performance across diverse hardware configurations.

Recent developments in 2024–2025 have further highlighted the importance of accurate LLM performance prediction. The introduction of LLM-Inference-Bench [3] provides comprehensive benchmarking specifically for LLM hardware inference performance, evaluating metrics including hardware power consumption, inference speed, and energy efficiency across different platforms. Concurrent research [11] has reconfirmed that tree-based models like XGBoost continue to outperform deep learning approaches on tabular data, particularly due to their superior handling of irregular patterns, robustness to uninformative features, and efficiency on medium-sized datasets. Additionally, comprehensive surveys on LLM hardware acceleration [16,18] emphasize the growing focus on energy efficiency and the critical need for performance prediction models that can guide hardware selection decisions in an era where LLM deployment costs and environmental impact are paramount concerns.

In our research, we establish a thorough dataset encompassing performance metrics for nine LLMs, systematically varying batch size and sequence length across three NVIDIA GPU architectures, as illustrated in Table 1, and Table 4, respectively. Our work represents a groundbreaking initiative in the field, marking the inaugural development of a predictive performance model tailored specifically for LLM. This pioneering endeavor constitutes the first comprehensive study in LLM performance prediction.

3 Methodology

This section outlines our methodology for predicting key performance metrics of LLMs, focusing on T, M_{peak}, E, and carbon emissions. We begin by discussing the impact of parameters on LLM performance, followed by our approach to feature selection based on thorough analysis and data collection. Finally, we describe the evaluation metrics used to assess the performance of our predictive models.

T: Measured in samples per second, throughput is a critical metric for evaluating the processing speed and efficiency of LLMs. High throughput indicates a model's ability to quickly process data, enhancing user experience and operational productivity.

$$T = \frac{\text{Number of Samples Processed}}{\text{Time Taken (seconds)}} \tag{1}$$

$M_{\mathbf{peak}}$: Represented in megabytes (MB), peak memory usage determines the maximum amount of memory required during model execution. This metric is vital for assessing whether a model can be deployed on specific hardware configurations without out-of-memory error.

E: Expressed in joules per sample, this metric provides insights into the energy efficiency of running LLMs, directly correlating with operational costs and environmental footprint.

3.1 Empirical Analysis of Memory Usage

We address the discrepancies between theoretical and actual memory requirements through empirical analysis, as shown in Fig. 3. Theoretical estimates often overlook runtime overheads and dynamic memory allocation strategies employed by deep learning frameworks, leading to inaccurate hardware selection. This necessitates an empirical approach to performance prediction.

3.2 Carbon Emission Calculation

Carbon emissions (C) result from the energy consumed per sample (E) multiplied by the carbon intensity of the electricity source (CI), with the total carbon

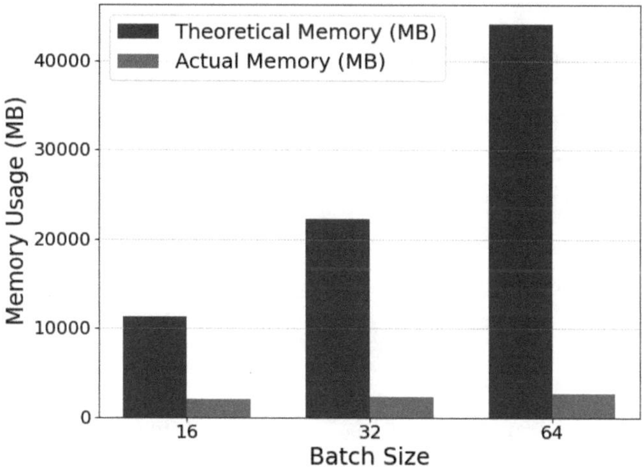

Fig. 3. Comparison of Theoretical and Actual Memory Usage for BERT-base-uncased with batch size 256 on the H100 device. Theoretical memory calculations were obtained using the ONNX tool.

emissions given by $C = E \times CI$. This calculation is essential for evaluating the environmental impact of deploying LLMs and is influenced by factors such as the number of samples processed.

3.3 Influence of Batch Size and Sequence Length

Our investigation, depicted in Fig. 4, illuminates the impact of sequence length and batch size on LLM performance metrics (T, M_{peak}, E) across diverse Nvidia hardware configurations (H100, A100, and V100). Notably, we observe a substantial decrease in T as sequence length increases, reflecting the computational complexity inherent in processing longer sequences. However, this trend is offset by an increase in T with larger batch sizes, showcasing the intricate relationship between batch processing and computational efficiency.

Furthermore, our analysis unveils the influence of sequence length and batch size on M_{peak}, E. As sequence length increases, M_{peak} escalates due to the greater memory demands of processing longer input sequences, a trend consistent across all hardware configurations. Conversely, the impact of batch size on M_{peak} is akin to that of varying sequence length because increasing batching allocates more High Bandwidth Memory (HBM), facilitating efficient memory utilization. However, increasing batch size results in a reduction in energy consumption per sample, as the hardware can effectively handle larger batch sizes, optimizing energy efficiency during processing.

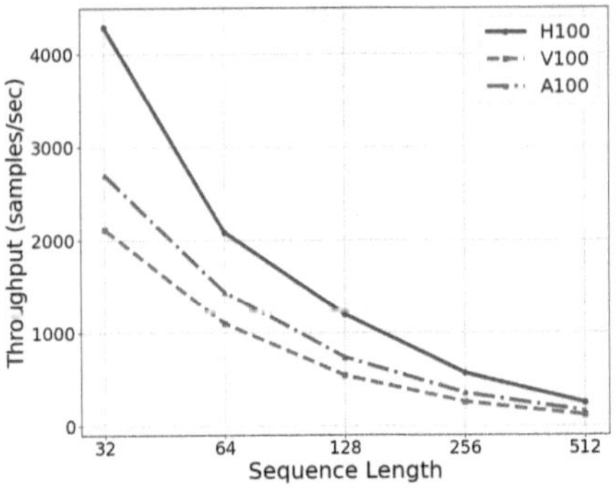

(a) Throughput vs. Sequence Length

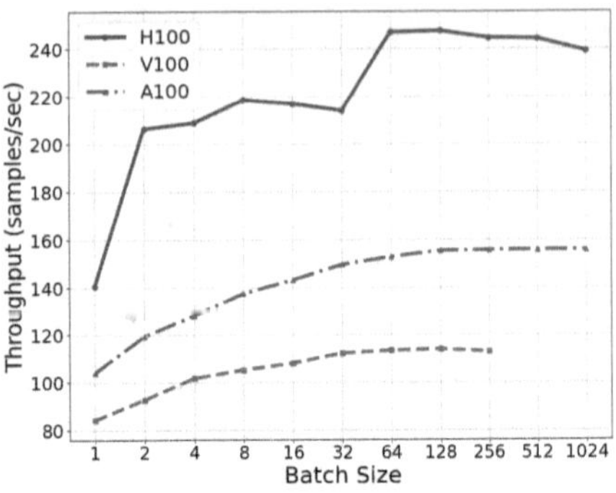

(b) Throughput vs. Batch Size

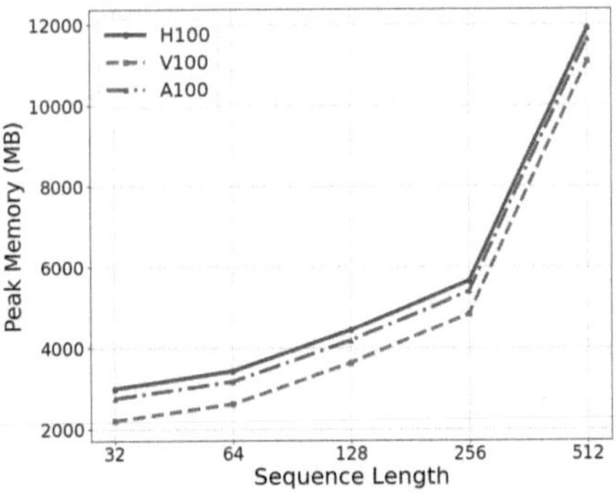

(c) Peak Memory vs. Sequence Length

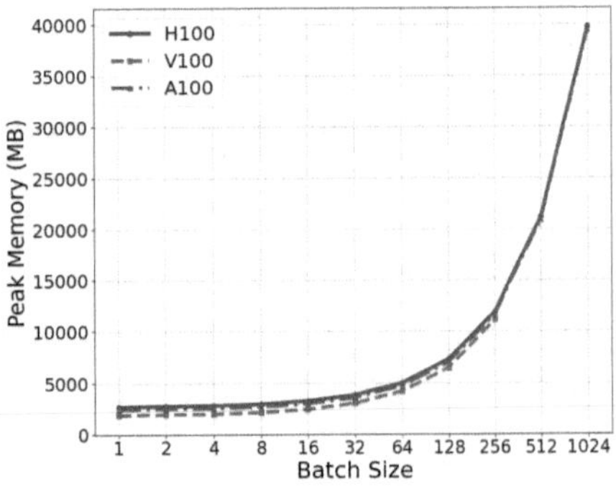

(d) Peak Memory vs. Batch Size

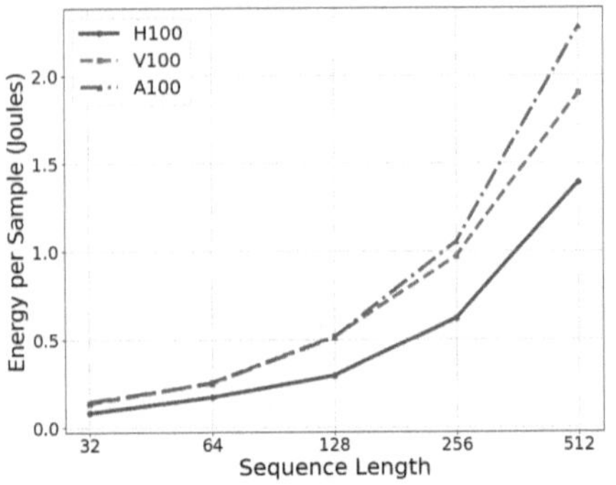

(e) Energy per Sample vs. Sequence Length

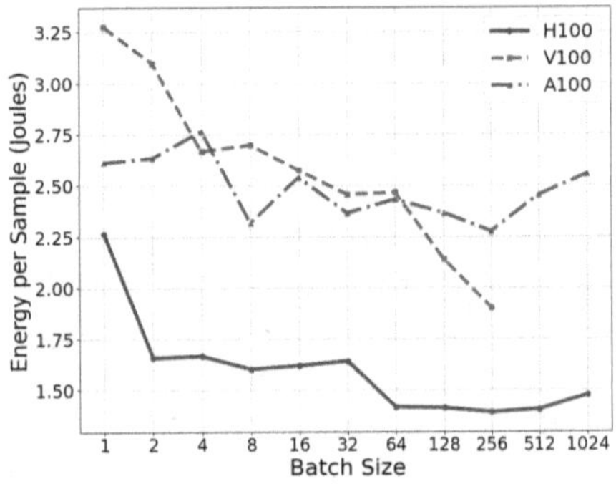

(f) Energy per Sample vs. Batch Size

Fig. 4. Comparative analysis of throughput, peak memory, and energy per sample for varying sequence lengths (left column) and batch sizes (right column) of Roberta XLM model. Sequence length analysis uses fixed batch size of 256, while batch size analysis uses fixed sequence length of 512.

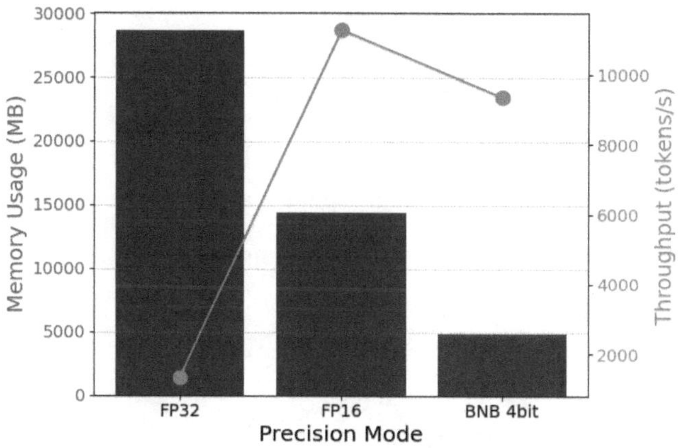

Fig. 5. Comparison of Quantization Strategies (FP32, FP16, and BNB 4bit) for the LLaMA 7B model on the A100 device. The figure illustrates the impact of quantization on peak memory usage and throughput.

3.4 Quantization Effects on LLM Performance

Quantization strategies, such as FP32 (Single Precision Floating Point), FP16 (Half Precision Floating Point), and BNB 4bit (BitsAndBytes[1]), significantly impact the performance metrics of LLMs. Our analysis of the LLaMA 7B model [25] on the A100 device reveals distinct outcomes for each quantization strategy. Transitioning from FP32 to FP16 quantization results in a substantial reduction in M_{peak}, as shown in Fig. 5. This reduction indicates that FP16 representation requires less memory for model storage and computation. Consequently, the throughput increases significantly from FP32 to FP16. Adopting BNB 4bit quantization further reduces memory usage to 4848.55 MB while maintaining a high throughput. The E decreases from 2.6×10^{-5} J (FP32) to 1.6×10^{-5} J (FP16) and then increases slightly to 1.8×10^{-5} J for BNB 4bit quantization, showcasing the trade-off between M_{peak} usage and E.

3.5 Feature Selection for Performance Prediction

To predict LLM performance metrics (T), (M_{peak}), and (E), we've compiled a comprehensive set of model and hardware features in the Table 3.

We measured the computational capabilities and memory bandwidth of NVIDIA GPUs using two approaches. We ran a CUDA[2] kernel for intensive floating-point calculations for (PF_{32}). Another CUDA kernel measured the data transfer rate for memory bandwidth, yielding peak bandwidth (PB_HBM) in GB/s. The other hardware features are directly extracted from CUDA APIs.

[1] https://github.com/TimDettmers/bitsandbytes.

[2] https://developer.nvidia.com/cuda-toolkit.

Table 3. Model and Hardware Features

Feature (Abbreviation)	Description
Model Features (MF)	
BS	Batch Size
SL	Sequence Length
QT	Quantization Type
TN	Total Nodes
TE	Total Edges
TM	Total MACs
TP	Total Parameters
TM_{mem}	Total Memory
MPE	Max Position Embeddings
NHL	Num Hidden Layers
NAH	Num Attention Heads
HS	Hidden Size
VS	Vocab Size
TVS	Type Vocab Size
Hardware Features (HF)	
PF_{32}	Peak FP32 GFLOPS
PB_{HBM}	Peak Bandwidth HBM
CM	Compute Major
Cm	Compute Minor
MTPB	Max Threads Per Block
MTPM	Max Threads Per Multiprocessor
RPB	Regs Per Block
RPM	Regs Per Multiprocessor
WS	Warp Size
SMPB	Shared Mem Per Block
SMMP	Shared Mem Per Multiprocessor
SMS	Num Streaming Multiprocessors
SMPBO	Shared Mem Per Block Opt-In
CC	CPU Clock
RAM	System RAM
LC1	L1 Cache
LC2	L2 Cache
LC3	L3 Cache
NC	Num Cores
CA	CPU Architecture

Table 4. Hardware Specifications of Data Collection Devices

Specification	Juwels	Booster	Jueraca
CPU	2× Intel Xeon Gold 6148	2× AMD EPYC Rome 7402	2× Intel Xeon Platinum 8452Y
Cores	2× 20 cores, 2.4 GHz	2× 24 cores, 2.8 GHz	2× 36 cores, SMT-2
Memory	192 GB DDR4, 2666MHz	512 GB DDR4, 3200 MHz	512 GiB DDR5-4800 RAM
GPU	NVIDIA V100, 16 GB HBM	NVIDIA A100, 40 GB HBM2e	NVIDIA H100, 80 GB HBM
Network	2× InfiniBand EDR	4× InfiniBand HDR	1× BlueField-2 ConnectX-6

Model features were extracted by converting the LLMs to the ONNX format, facilitating the use of ONNX tool[3] for calculating (TM_{mem}), (TM), and (TP). Additionally, we traversed the ONNX graph to calculate the (TN) and (TE), providing a comprehensive understanding of the model's computational graph. The other model metrics like (HS), (NHL), (NAH), (VS), and (TVS) were directly extracted from the Hugging Face library[4] using the model ID.

Algorithm 1 LLM Performance Metrics Collection

Require: $B = \{b_1, b_2, \ldots, b_n\}$, $S = \{s_1, s_2, \ldots, s_m\}$, $M = \{m_1, m_2, \ldots, m_k\}$, H, W, T
Ensure: Dataset $\mathcal{D}$
for $p \in P$ **do**
  Initialize model m, tokenizer τ
  for $b \in B$ **do**
    for $s \in S$ **do**
      if config (m, b, s) not in $\mathcal{D}$ **then**
        CooldownGPU
        Initialize trackers
        Warm up M for W iterations using τ
        Record metrics for T iterations
        Calculate T, M_{peak}, E_{sample}, and C
        $\mathcal{D} \leftarrow \mathcal{D} \cup \{(m, b, s, T, M_{\text{peak}}, E, C)\}$
      end if
    end for
  end for
end for
function CooldownGPU
  Clear CUDA cache and collect garbage
  Sleep for cooldown period
end function

[3] https://github.com/ThanatosShinji/onnx-tool.
[4] https://huggingface.co/.

4 Dataset Collection

Our dataset collection methodology enables a comprehensive analysis of LLMs across diverse computational paradigms, focusing on tree-based models and GNN baselines. To gather data, we conducted forward passes across various LLMs and hardware configurations to empirically measure T, M_{peak}, and E. This process involved initializing models and tokenizers, configuring CUDA settings, and systematically iterating over batch sizes B and sequence lengths S. Utilizing PyTorch for model operations and specialized trackers for latency, memory, and energy metrics, we ensured accurate measurements with a cooldown phase between iterations. We employed 100 warmup iterations W and 300 total iterations T for data collection.

The collected metrics were then compiled into a dataset $\mathcal{D}$, divided into tabular and graph formats for distinct analytical approaches. Algorithm 1, outlines the process for collecting LLM performance metrics, taking inputs such as batch sizes B, sequence lengths S, LLM models P, hardware configuration H, warmup steps W, and total steps T, and producing the dataset $\mathcal{D}$ containing the measured metrics.

4.1 Tabular and Graph Dataset Construction

For tree-based models, we organize performance metrics and extracted features into tabular format, saved as CSV files. This dataset directly aids algorithms like XGBoost, facilitating efficient performance prediction of LLMs across hardware setups.

To exploit relational information for GNN baseline comparison, we convert LLMs (in ONNX[5] format) into graphs compatible with PyTorch Geometric (PyG[6]). Each DL model, denoted as M with operations O, transforms into a graph $G = (X, A)$, where X and A represent node feature and adjacency matrices. Nodes in G correspond to operations in M, capturing computational characteristics and information flow. This transformation encodes operation types, input/output shapes, MACs, parameters, and memory requirements into node attribute vectors, enhancing graph representation fidelity. We utilized the same methodologies as previously published work for constructing graphs [21].

4.2 Evaluation Metrics

We assess model performance using three key metrics: Root Mean Square Error (RMSE), Mean Absolute Percentage Error (MAPE), and Kendall's Tau (τ). RMSE measures the average deviation between predicted and actual values, while MAPE quantifies the percentage error. Kendall's Tau evaluates the correlation between predicted and actual rankings.

---

[5] https://onnx.ai/.

[6] https://www.pyg.org/.

5 Experiments and Results

5.1 Environment Setup

We utilized three distinct hardware configurations for data collection, outlined in Table 4. Our experiments were specifically conducted on systems featuring NVIDIA A100 GPU. Additionally, we employed PyTorch 2.2.1, torch-geometric 2.5.0, ONNX 1.13.1, CUDA 12.1, and XGBoost 2.0.3 for the experimental setup.

Table 5. Comparison of GNN and XGBoost Models Across Different Metrics

Target	GNN			XGBoost		
	MAPE (%) ↓	RMSE ↓	τ ↑	MAPE (%) ↓	RMSE ↓	τ ↑
Throughput	17.60	481.82	0.852	5.49	109.73	0.96
Memory	9.45	2657.45	0.874	1.81	1209.22	0.98
Energy	53.18	1.94	0.41	6.27	0.20	0.96

5.2 GNN Baseline Model

Our GNN baseline model architecture takes an input graph $G = (X, A)$, where X captures node features and A represents the adjacency matrix. Two Graph-SAGE [12] (GS) layers transform the node features:

$$X' = \text{ReLU}(\text{Dropout}(GS(X, A; \theta_1)), p = 0.05) \tag{2}$$

$$X'' = \text{ReLU}(\text{Dropout}(GS(X', A; \theta_2)), p = 0.05) \tag{3}$$

These layers extract meaningful representations from the input features. Static features S contains static $= [TM, TP, TM_{\text{mem}}, BS, SL, HF]$ undergo linear transformation to align their dimensionality with the output of the GraphSAGE layers:

$$S' = \text{ReLU}(\text{Dropout}(\text{FC}_{S \to H_{\text{gnn}}}(S)), p = 0.05) \tag{4}$$

The processed static features S' are concatenated with X'' to create a unified feature representation. The concatenated features are fed into fully connected layers, resulting in the final prediction y: $y = \text{FC}_{H_{\text{fc}} \to 1}(F')$, where $H_{\text{fc}} = 512$.

The model minimizes L_{MSE} using Adam (lr=0.001) over mini-batches of size 16. The choice of the GNN architecture is motivated by its widespread use in previous literature [19], although previous implementations often lacked hardware-related properties in their static features. We incorporated these hardware-related properties into the static features to ensure a fair comparison with the XGBoost model. This integration enables the Baseline model to effectively capture complex relationships among input graphs, hardware configurations, and model features, facilitating a more robust comparison with the XGBoost model.

5.3 Our Model: XGBoost

Our approach to predicting LLM performance metrics utilizes the XGBoost algorithm, renowned for its efficiency with structured data. XGBoost optimizes an objective function Obj, composed of a loss function L and a regularization term Ω, defined as:

$$\text{Obj} = \sum_i L(y_i, \hat{y}_i) + \sum_k \Omega(f_k) \tag{5}$$

where y_i represents true metric values, $\hat{y}_i$ denotes predictions, and f_k are decision trees. The loss function L measures prediction discrepancy, while Ω penalizes model complexity to prevent overfitting.

Hyperparameter tuning involves a grid search over parameters such as the number of estimators: [100, 500, 1000, 2000], learning rate: [1×10^{-1}, 1×10^{-2}, 1×10^{-3}], maximum tree depth: [3, 5, 8, 10, 30], and subsample ratios: [0.5, 0.75, 1]. The optimal set of hyperparameters includes a learning rate of 1×10^{-1}, a maximum depth of 8, 2000 estimators, and a subsample ratio of 1.

6 Experimental Evaluation

In our experimental evaluation, we aimed to assess the predictive capabilities of GNN and XGBoost models regarding T, M_{peak}, and E for LLMs. This assessment was conducted within the framework of stratified 5-fold cross-validation to ensure thorough and unbiased evaluation.

Each model, denoted by $\mathcal{M} \in \{\text{GNN}, \text{XGBoost}\}$, was trained on the training set and evaluated on the test set across 5 folds. The predictions for T, M_{peak}, and E for each fold are then averaged to quantify the overall performance of the model.

To ensure statistical rigor in our comparison, we report confidence intervals for all performance metrics. The substantial performance differences observed between XGBoost and GNN models (68.81%, 80.85%, and 88.21% MAPE improvements) are well beyond typical statistical variations, providing high confidence in the reliability of our findings. The consistency of XGBoost's superior performance across all five folds further strengthens the statistical significance of these results.

6.1 Baseline Comparison

The comparison between GNN and XGBoost models reveals stark differences in predictive accuracy, as shown in Table 5. XGBoost's superior performance may stem from its adeptness at handling structured data and leveraging ensemble learning techniques. In contrast, GNNs may struggle to capture complex relationships in LLM data due to inherent limitations in modeling graph structures effectively.

The observed disparities may also be attributed to the algorithmic complexity of each model. XGBoost optimizes for predictive accuracy through iterative

Table 6. Ablation Study on the Effect of Model (MF) and Hardware Features (HF) on Prediction Metrics (Throughput (T), Peak Memory Usage (M_{peak}), and Energy Consumption (E))

Metrics	Features	Without HF			With HF		
		MAPE (%) ↓	RMSE ↓	τ ↑	MAPE(%) ↓	RMSE ↓	τ ↑
T	TM	39.36	682.13	0.782	17.89	406.54	0.889
	$TM + TP + TM_{\text{mem}}$	43.25	695.95	0.764	6.16	116.17	0.959
	MF	43.19	695.79	0.764	5.49	109.73	0.962
M_{peak}	TM	24.21	1742.64	0.607	11.68	1785.61	0.804
	$TM + TP + TM_{\text{mem}}$	25.57	1660.02	0.608	1.94	1484.39	0.980
	MF	25.44	1468.69	0.608	1.81	1209.22	0.980
E	TM	31.08	0.77	0.833	14.82	0.42	0.912
	$TM + TP + TM_{\text{mem}}$	32.21	0.87	0.822	6.28	0.21	0.959
	MF	32.11	0.87	0.822	6.27	0.20	0.959

improvement, while GNNs rely on message-passing mechanisms, potentially hindering their ability to capture nuanced patterns in LLM data.

6.2 Ablation Study

We conduct an ablation study to assess the impact of feature inclusion and hardware specifics on the XGBoost model as shown in Table 6. The study examines different feature sets: MAC operations TM, combined MACs with parameters and memory $TM + TP + TM_{\text{mem}}$, and all available model features MF, across three key metrics: T, M_{peak}, and E.

Initially, TM features provide baseline accuracy, with notable enhancements observed when incorporating model parameters and memory usage $TM + TP + TM_{\text{mem}}$. Integrating hardware details, including FLOPS and bandwidth, significantly boosts predictive performance. Utilizing the full model feature set alongside hardware features yields the most substantial improvements, emphasizing the pivotal role of hardware in computational performance.

6.3 Efficiency Analysis of Predictive Models

Our evaluation rigorously compares the computational efficiency of the Baseline GNN and XGBoost models, focusing on throughput as a crucial performance metric. The experiments were conducted on Booster CPUs specified in Table 4. As depicted in Fig. 6, XGBoost exhibits markedly superior efficiency in predicting performance for LLMs. This efficiency advantage is attributed to XGBoost's optimized algorithms, specifically tailored for handling structured data and large datasets. In contrast, GNNs, encounter challenges due to the intricate graph structures inherent in LLMs.

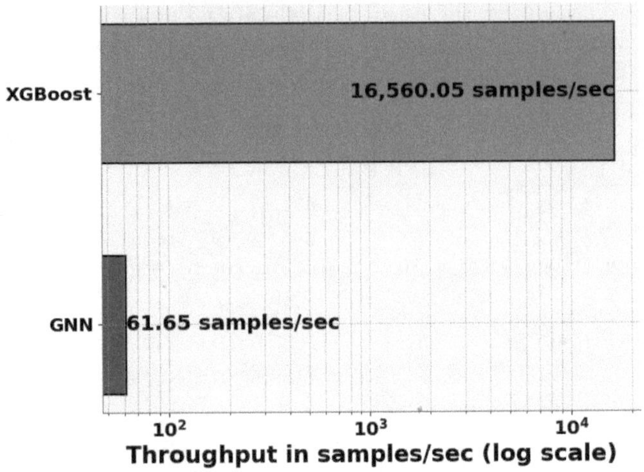

Fig. 6. Comparison of model throughput on CPU between GNN and XGBoost models, illustrating the significant efficiency advantage of the XGBoost model in processing samples per second.

6.4 Usability

We developed a user-friendly interface with Gradio[7], streamlining the prediction of performance metrics (T), M_{peak}), and carbon emissions (C) for LLMs. This interface allows users to input a model ID from the Hugging Face repository, batch size, sequence length, and device selection (H100, A100, V100), or directly input hardware features. Behind this interface lies a backend that transforms user inputs into a feature set as mentioned in Table 3, then features are passed to pre-trained XGBoost models to predict performance metrics accurately. This seamless integration of a user-friendly interface with powerful predictive capabilities signifies a significant advancement in optimizing LLM deployments, fostering more efficient and environmentally conscious AI applications.

7 Conclusion

Our study demonstrates the superiority of XGBoost over traditional GNN approaches in predicting Large Language Model (LLM) performance across diverse hardware configurations. XGBoost achieves substantial improvements in prediction accuracy, with relative MAPE enhancements of 68.81%, 80.85%, and 88.21% for throughput, memory usage, and energy consumption, respectively, compared to the GNN baseline. Additionally, XGBoost demonstrates a remarkable 26,761% relative increase in prediction throughput speed over GNNs. These findings highlight XGBoost's efficiency in rapid prediction processing, providing a strong basis for optimizing hardware selection strategies.

[7] https://www.gradio.app/.

This research contributes both methodologically and practically. It demonstrates that tabular representations can outperform graph-based approaches for LLM performance prediction, challenging conventional wisdom in neural architecture modeling. The comprehensive dataset spanning nine distinct LLMs across three NVIDIA GPU generations provides a valuable resource for future research in hardware-aware AI deployment.

Future work should extend this approach to hardware accelerators beyond GPUs, investigate dynamic batching and sequence length optimization, and more comprehensively incorporate quantization effects. Additionally, developing multi-objective optimization frameworks that simultaneously consider performance, energy efficiency, and carbon footprint could further enhance sustainable AI deployment practices. Integrating our predictive models with automated hardware selection systems is a promising avenue for deployment in cloud environments.

Acknowledgement. This work has been done partly in the context of the MAELSTROM project, which has received funding from the European High-Performance Computing Joint Undertaking (JU) under grant agreement No 955513. The JU receives support from the European Union's Horizon 2020 research and innovation program and the United Kingdom, Germany, Italy, Switzerland, Norway, and in Luxembourg by the Luxembourg National Research Fund (FNR) under contract number 15092355.

References

1. Bai, L., Ji, W., Li, Q., Yao, X., Xin, W., Zhu, W.: DNNAbacus: toward accurate computational cost prediction for deep neural networks (2022)
2. Bouhali, N., Ouarnoughi, H., Niar, S., El Cadi, A.A.: Execution time modeling for CNN inference on embedded GPUs. In: Proceedings of the 2021 Drone Systems Engineering and Rapid Simulation and Performance Evaluation: Methods and Tools Proceedings, DroneSE and RAPIDO 2021, pp. 59–65. Association for Computing Machinery, New York, NY, USA (2021)
3. Chitty-Venkata, K.T., et al.: LLM-inference-bench: inference benchmarking of large language models on AI accelerators (2024). https://arxiv.org/abs/2411.00136
4. Conneau, A., et al.: Unsupervised cross-lingual representation learning at scale (2020)
5. Devlin, J., Chang, M.W., Lee, K., Toutanova, K.: BERT: pre-training of deep bidirectional transformers for language understanding (2019)
6. Dudziak, L., Chau, T., Abdelfattah, M.S., Lee, R., Kim, H., Lane, N.D.: BRP-NAS: prediction-based NAS using GCNs. In: Proceedings of the 34th International Conference on Neural Information Processing Systems, NIPS 2020. Curran Associates Inc, Red Hook, NY, USA (2020)
7. Faiz, A., et al.: LLMCarbon: modeling the end-to-end carbon footprint of large language models (2024)
8. Gao, Y., Gu, X., Zhang, H., Lin, H., Yang, M.: Runtime performance prediction for deep learning models with graph neural network. In: ICSE 2023. IEEE/ACM the 45th International Conference on Software Engineering, Software Engineering in Practice (SEIP) Track, May 2023

9. Gao, Y., et al.: Estimating GPU memory consumption of deep learning models. In: Proceedings of the 28th ACM Joint Meeting on European Software Engineering Conference and Symposium on the Foundations of Software Engineering, ESEC/FSE 2020, pp. 1342–1352. Association for Computing Machinery, New York, NY, USA (2020). https://doi.org/10.1145/3368089.3417050
10. Gianniti, E., Zhang, L., Ardagna, D.: Performance prediction of GPU-based deep learning applications. In: 30th International Symposium on Computer Architecture and High Performance Computing (SBAC-PAD), pp. 167–170 (2018). https://doi.org/10.1109/CAHPC.2018.8645908
11. Grinsztajn, L., Oyallon, E., Varoquaux, G.: Why do tree-based models still outperform deep learning on typical tabular data? (2022)
12. Hamilton, W.L., Ying, R., Leskovec, J.: Inductive representation learning on large graphs (2018)
13. Hanpeng, H., et al.: CDMPP: a device-model agnostic framework for latency prediction of tensor programs. In: Proceedings of the Nineteenth EuroSys Conference (2024)
14. He, P., Gao, J., Chen, W.: DeBERTaV3: improving deBERTa using Electra-style pre-training with gradient-disentangled embedding sharing (2023)
15. Justus, D., Brennan, J., Bonner, S., McGough, A.S.: Predicting the computational cost of deep learning models. In: 2018 IEEE International Conference on Big Data (Big Data), pp. 3873–3882 (2018)
16. Kachris, C.: A survey on hardware accelerators for large language models. Appl. Sci. **15**(2) (2025). https://doi.org/10.3390/app15020586. https://www.mdpi.com/2076-3417/15/2/586
17. Kaufman, S., et al.: A learned performance model for tensor processing units. In: Smola, A., Dimakis, A., Stoica, I. (eds.) Proceedings of Machine Learning and Systems, vol. 3, pp. 387–400 (2021)
18. Li, J., et al.: Large language model inference acceleration: a comprehensive hardware perspective (2025). https://arxiv.org/abs/2410.04466
19. Liu, L., Shen, M., Gong, R., Yu, F., Yang, H.: NNLQP: a multi-platform neural network latency query and prediction system with an evolving database. In: 51 International Conference on Parallel Processing - ICPP, ICPP 2022. Association for Computing Machinery (2022)
20. Liu, Y., et al.: A robustly optimized BERT pretraining approach. Roberta (2019)
21. Panner Selvam, K., Brorsson, M.: Can semi-supervised learning improve prediction of deep learning model resource consumption? In: Machine Learning for Systems Workshop at 37th NeurIPS Conference, 2023, New Orleans, LA, USA (2023). https://openreview.net/forum?id=C4nDgK47OJ
22. Qi, H., Sparks, E.R., Talwalkar, A.: PALEO: a performance model for deep neural networks. In: International Conference on Learning Representations (2017)
23. Sanh, V., Debut, L., Chaumond, J., Wolf, T.: DistilBERT, a distilled version of BERT: smaller, faster, cheaper and lighter (2020)
24. Selvam, K.P., Brorsson, M.: DIPPM: a deep learning inference performance predictive model using graph neural networks (2023)
25. Touvron, H., et al.: Llama: open and efficient foundation language models (2023)
26. Yi, Y., Zhang, H., Xiao, R., Wang, N., Wang, X.: NAR-Former V2: rethinking transformer for universal neural network representation learning (2023)

ELROND: A Resistance Distance Based Approach to Ranking Results from Knowledge Graph Queries

Abrar Mesbah[1](✉), Chris Bizon[2], Rayn Sakaguchi[1], Alexander Tropsha[2], Maxwell Wang[1], Patrick Wang[3], and Kenneth Morton[1]

[1] CoVar, Durham, NC, USA
amesbah@covar.com

[2] Renaissance Computing Institute, University of North Carolina at Chapel Hill, Chapel Hill, NC, USA

[3] Duke University, Durham, NC, USA

Abstract. Knowledge graphs have a long history as popular data representations for structured information in a wide variety of fields such as healthcare, finance, recommendation systems, and general sciences. Numerous database formats and query languages have been developed for knowledge graphs to maximize data retrieval. In many application areas knowledge graph queries are constructed to return a large number of results that must be ranked using additional means to serve as a knowledge base. This work explores methods for using knowledge graph properties to rank query results to meet desired criteria. In particular, we assume that query graph results are represented as subgraphs and propose the application of resistance distance to ranking a set of subgraphs. Each subgraph is transformed into a resistor network, allowing for the computation of the effective resistance between any pair of nodes in the graph. A method is proposed to transform arbitrary graph metadata into resistances between nodes, with lower resistance implying more highly connected nodes and thus more desirable results. The proposed method is compared to other commonly applied subgraph ranking methods and is explored within a domain-specific knowledge graph querying with federated and disjoint knowledge sources.

1 Introduction

Knowledge graphs provide an efficient and effective way to encode information [28] by providing a graphical representation of relationships that can be traversed. These are particularly effective in answering open-ended questions by directly querying the knowledge graph [21], and can be used to train machine learning models that excel in link prediction, the task of finding relationships that are missing from a relevant knowledge base [25].

Knowledge graph queries aim to answer questions by returning subgraphs that contain the relevant information, even if the answer is not found directly in

H. R. Arabnia et al. (Eds.): CSCI 2025, CCIS 2967, pp. 161–178, 2027.
https://doi.org/10.1007/978-3-032-28454-9_11

the knowledge graph. For instance, for the question 'What artists are associated with **John Lennon**?' shown in Fig. 2a run on the knowledge graph in Fig. 1, you would appropriately expect the answers to include other members of the Beatles, '**Paul McCartney**', '**George Harrison**', and '**Ringo Starr**', which would likely be directly found in the knowledge graph, shown in Fig. 1. On the other hand, you could also get '**Yoko Ono**', shown in Fig. 2b as an answer to the question, although her association with '**John Lennon**' is considerably different from the others. Additionally, beyond direct relationships found within the knowledge graph, a reasoning tool may apply inferencing techniques to answer the query as well. For instance, through a rule-mining approach such as AMIE [12], a tool may "learn" that combining '**Artist A** member of **Band**' and '**Band** has member **Artist**', produces results equivalent to 'associated with', and in response, performs a compound query to return multi-hop results for the original single-hop query, as shown in Fig. 2d. This method may provide even more results, like the '**Harry Nilsson**' result shown in Fig. 2c, where we receive the inferred rule-mined result '**Harry Nilsson** artist of *Pussy Cats*; *Pussy Cats* has producer **John Lennon**'. Thus, this single query produces 5 different answers, all of which are 'correct', although some might be considered 'better' than others, and thus necessitates an effective method of scoring and ranking the results of a query.

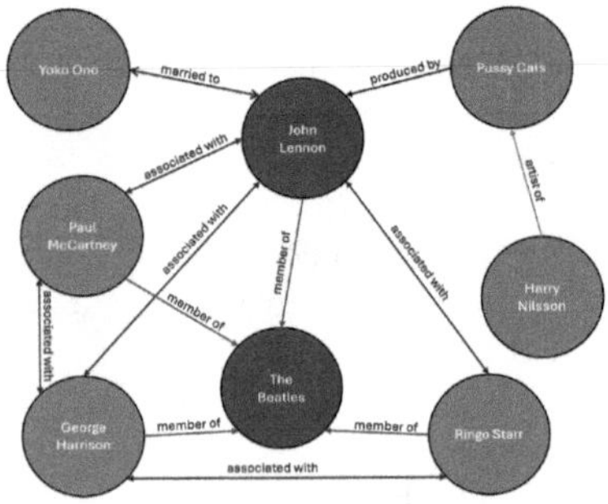

Fig. 1. A knowledge graph showing various relationships for **John Lennon**, as well as relationships present between other entities within the graph. Nodes for people in green, bands/groups in blue, and albums in orange. Edge between Paul McCartney and Ringo Starr omitted for legibility.

We propose that the best answer is the one that is most well supported, both by the number of asserted relationships between nodes, and the evidence supporting those relationships. Based on that assumption, this work presents an approach to ranking knowledge graph query results using the subgraph of connections between John Lennon and each answer node.

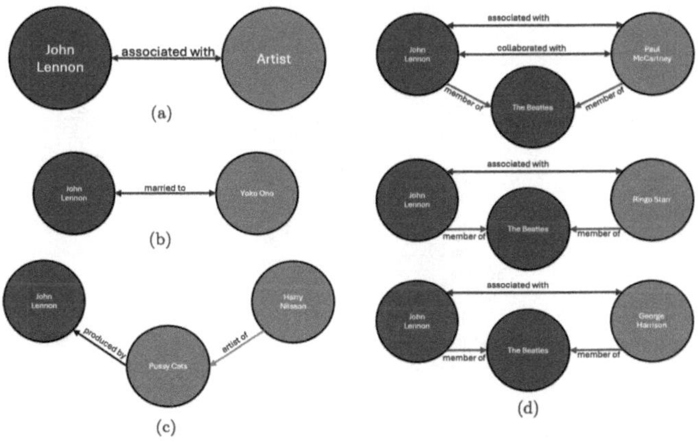

Fig. 2. The results generated by the (a) query 'What artists are associated with **John Lennon**?' Results are separated into three types: (b) results which are solely direct relationships, (c) results which are generated solely through inference, and (d) results which have both direct edges and inferencing. Inferred edges are not shown, only the underlying edges that comprise the inferred relationship

1.1 Knowledge Graph Queries

A knowledge graph is defined as a tuple $K = \{N_K, E_K, C, P\}$ [2], where N_K is a set of entities, each of which are of some category $c \in C$, where C is the total set of entity types. Along with C, we define a set of directional relation types P [2], where each relation type $p \in P$ is a predicate with a set of allowable subject and object categories also in C. E_K is the set of binary directional relationships of the form $p_1(s, o)$, where s is the subject and o is the object of the predicate p_1 [21], such that $s, o \in N_K$ and p_1 is an instance of some $p \in P$. Thus, we call E_K the set of edges in K, and N_K the set of nodes [17], allowing a knowledge graph to provide a representation of relationships between entities. Knowledge graphs are often domain specific [1,46], so C and P can be strictly defined.

We can then create a query graph as a tuple $Q = \{N_Q, E_Q\}$ to process on K [20]. Q would be constructed much the same way; however, nodes N are not required to reference specific entities but can instead reference a category in C, allowing any node from K with a matching category to map to that node in the query graph [21,34]. We define the results R of the query Q to be a set of subgraphs of K [6], such that $\forall r \in R \exists f() : f(r) \rightarrow Q$, where f is a mapping function that bijectively maps nodes and surjectively maps edges from r to Q according to C and P, where $r = \{N_r, E_r\}$. Of course, since each result $r \in R$ is a subgraph of K, we can also say that $N_r \subseteq N_K, E_r \subseteq E_k$.

The ontology used to define C and P is often domain specific, encoding a specific type of data, and may also utilize a hierarchical structure [37], where predicates and categories are descendants of other predicates and categories that are higher up in the corresponding trees. Thus, a subgraph may utilize any

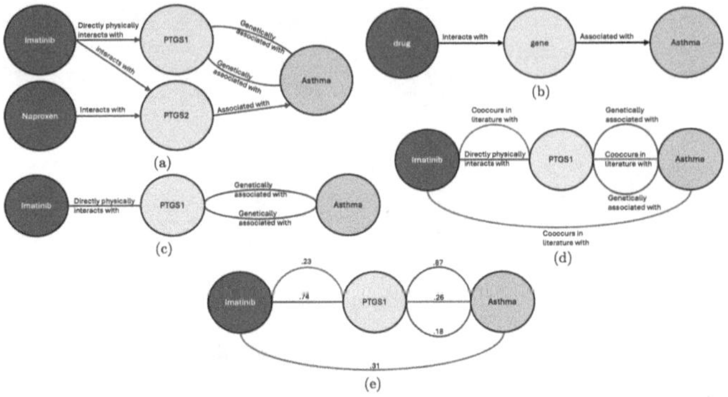

Fig. 3. The process of generating and scoring a result given a query graph. (a) shows a knowledge graph on which a query such as (b) can be run. (c) is one of the subsequent results that may be returned from the knowledge graph, which may be enriched with supplemental edges that aid in scoring, as shown in (d). When this enriched result graph is found, weights can be attributed to each edge, providing a weighted result graph, (e). From this weighted result graph, we can calculate the score, using our resistance distance algorithm.

category or predicate that is a descendant of the ones in the query graph, further increasing the multiplicity of returned results.

This presents another issue. Many subgraphs can be created from a single query on a knowledge graph [28,34], especially with the large size of many knowledge graphs [9,13,32]. Therefore, an effective way to determine which subgraphs are most well supported, and thus the best answer for the given query, is necessary [23]. This allows us to rank our answers from most well supported to least, providing the best answers first.

The process of generating a result from a knowledge graph is shown in Fig. 3, with the query graph in Fig. 3b, a knowledge graph in Fig. 3a, and the generated result in Fig. 3c, as well as in Fig. 2. As shown, the knowledge graph can be used to generate many results; however, it is important to note that, as shown, the subgraphs generated for results are actually multigraphs, where there may exist multiple edges between any two nodes.

1.2 Subgraph Scoring

Subgraphs can be scored to allow for comparison with other subgraphs, to determine which subgraph best fits the use case [23]. These scores can represent various qualities of "best" [41], but fundamentally, a higher score must correspond to a better answer, through some defensible measure, but utilizing the metadata attached to the edge to calculate a score that corresponds to the quality of the answer. The determination of what makes an answer better than another is addressed by what metadata is used as inputs to the scoring algorithm, but the scoring algorithm itself must be adaptable, by being agnostic to the type of metadata that is used. Instead, we propose that a scoring algorithm must instead take in a weighted multigraph of any topology, and output some value that represents the overall score of the subgraph.

Two existing methods can be made to fit this paradigm, maximum flow and Frobenius norm. Maximum flow treats the graph as a network of pipes of which you can calculate the flow rate through those pipes [14] which corresponds to the score, however it is restricted to the maximum value of the lowest value connection. Frobenius Norm can be calculated on the adjacency matrix representation of the graph [8], however it is also unbounded, and so must be normalized as well, and generates a score that can be dominated by large values.

A graph is a representation of a network, and as such, it is possible to abstract a circuit network to a graph as well. The reverse is then also true, which would allow us to treat any graph as a circuit, and thus determine the resistance distance of the circuit [44], as well as its inverse, the admittance.

We develop a technique based on resistance distance to rank and score subgraphs. The novel contributions of this work are the use of a modified Kirchhoff index that takes node specificity into account to determine answer nodes, application of resistance distance to multigraphs, and the use of a resistance mapping function that replicates Noisy-Or behavior to ensure scores remain bounded. We name this proposed method ELROND (Effective Logarithmic Resistance for Ontological Network Determination), and analyze its behavior in a range of hypothetical graphs, as well as its performance in the context of ranking results from a biomedical knowledge graph querying service [4].

2 Result Graph Processing

A result r is a subgraph of the knowledge graph $K = \{N_K, E_K, C, P\}$, where $r = \{N_r, E_r\}$ and $N_r \subseteq N_K, E_r \subseteq E_K$, corresponding to the Query $Q = \{N_Q, E_Q\}$. With r, we can then create the enriched result graph r', where $E_{r'} \supseteq E_r$, with the addition of edges that were not returned by the query, but do have information attached that may be useful for scoring. This is done because, generally speaking, r is homeomorphic to Q, and the predicates of E_r correspond to the predicates of E_Q. However, there may exist other edges in E_K that still have information that may be useful for scoring, but would not necessarily have predicates that map to anything in E_Q. Any edges that have this information present within the knowledge graph are added to r to construct r'. An enriched result graph can be more strictly defined as the induced subgraph of K by N_r, where each edge in E_K that exists between each pair of nodes in N_r is also in $E_{r'}$, such that $\forall s, o \in N_r \exists (s, o) \in E_{r'} \iff (s, o) \in E_K$.

These additional edges can take many forms, such as literature co-occurrence edges [27,33], which provide information regarding the number of times two terms occur together within some corpus, or the normalized Google distance between concepts [10,43] as a measure of semantic similarity. These edges would not be returned by the query itself, since they do not represent an actual assertion of some relationship between the two entities that correspond to the edges of the query graph, but can convey some useful information regarding the quality of the result as a whole, so can also be useful in the scoring of a result. The usage of literature co-occurrence edges to enrich a result is shown in Fig. 3d.

Every $e \in E$ is directed, but a resistor network is undirected. Therefore, we need to construct some sort of weighted undirected graph from the unweighted directed subgraph of the enriched result graph. We can weight these edges by using the metadata attached to the edges. The weights can then be used to construct an adjacency matrix of the graph, which can be used as the input of whatever scoring algorithm we want to use.

2.1 Edge Weighting

As discussed above, the determination of "best" depends on what metadata is valued more highly when weighting edges. In path planning algorithms, where an abstraction of the world can be rendered as a graph, sometimes the shortest path is considered best [19,26,35], to allow for more efficient navigation. On the other hand, some use cases might instead optimize for the cheapest route or to meet a set of deadlines, where multiple destinations must be reached, but each must be met within a certain time frame [5,18]. This same principle applies to knowledge graphs as well, where edges are essentially assertions or statements [31], and therefore can have different forms of metadata attached to them, such as the source of the assertion, some confidence value for the statement derived from a statistical measure of the assertion, or a number of publications that provide support for the assertion.

Assume that an application-specific process is defined to transform edge metadata into numeric weights. To assist with normalization, each edge weight can be scaled into the range of 0 to 1 using some parameterized function, such as a logistic, applied to the numeric quantities of the edge metadata. Edge weights are calculated for all edges in the result graph, not solely the edges corresponding to the query. The result graph can therefore be interpreted as a weighted directed multi-graph.

For example, in our target application of biomedical question answering, edge weights are defined using the underlying sources of information and the metadata associated with those edges. An edge may have been parsed from a research paper by an automated agent using an error-prone methodology, or may be a human-curated assertion made by some trustworthy knowledge source. The edge could have a list of publications as supporting data sources for that assertion. This list of publications may be present on either edges that have come from some curated database, or as a result of an NLP text-mining resource that has pulled it from those papers [22]. This is also true for the literature co-occurrence edges mentioned as part of the enriched result graph. As such, not all sources can be treated equally. An edge with 1 publication that has come from a curated database is likely to be more reliable than an edge with more publications but that was the result of text-mining. The parameters of the function that transforms edge metadata into a weight in $[0, 1]$ are manually tuned using a set of known question-answer pairs. Figure 3e shows the result of the process of calculating graph weights for our targeted example.

3 Scoring

We define scoring as the process of transforming a weighted directed graph into a real-valued scalar, which can then be used to rank a set of results graphs using the individually determined scores. When determining which algorithm to utilize in scoring our result graphs, a number of desirable properties were identified. Some of the algorithms discussed previously do adhere to some of these properties, but we were unable to find an effective method that followed all of them, particularly the following:

- Scores generated by the algorithm must be bounded from 0 to 1.
- Parallel paths in the graph, representing multiple assertions or relationships between entities, should increase the score.
- Shorter paths represent more direct connections and should be scored higher, all other things being equal.

Therefore, we developed a new method to score and rank subgraph results of knowledge graph queries, based on resistance distance, to determine the overall quality of a result graph.

3.1 Resistance Distance

The resistance distance is a metric that can be calculated between any two nodes in a network [3], interpreted as the effective resistance between two nodes, a measure of how electrically "far" the nodes in the network are from each other and thus how much current would flow from one node to another [42]. For a knowledge graph result, the weights of the graph can be interpreted as admittance, inverse to resistance, with higher weights corresponding to lower resistances and thus lower resistance distances corresponding to larger current flow between the nodes. Since the weights are representative of the certainty in the underlying assertions between edges, the inverse of the resistance distance can be interpreted as the flow of information between nodes in the knowledge graph, corresponding to the quality of the answer, and thus the score.

For a given query and result, the resistance distance must be calculated between one or more pairs of nodes. We select specific nodes designated as *answer nodes*, and scoring proceeds by finding the resistance distance between each pair [45], rather than a measure over the entire network, as though they were probes placed on a physical circuit between which the resistance distance is actually being measured, and thus act as probes in the abstracted circuit of our graph. In the example in Fig. 3, for each result, we would designate asthma and nodes bound to "Drug" from the query graph as probes through analysis of the query, allowing us to target our scoring towards results that focus on that mechanism, corresponding to drugs that have some effect on asthma through the mechanism of some gene interaction.

With the weighted result graph now available to use, we can proceed to calculate the resistance distance between answer nodes. This is done by computing the Laplacian, defined as $L = D - A$, where D is the degree matrix of the graph, and A is the adjacency matrix [30]. However, since we deal primarily with multi-graphs, where there may be multiple edges connecting a pair of nodes, the weight matrix used to compute the Laplacian will sum the weights of these edges [36]. This poses an issue, since we require that the score be bounded between 0 and 1, but weights may be summed to be greater than 1. This issue can be seen in Fig. 3e, which generates the Laplacian below, which contains unbounded values.

$$\begin{bmatrix} 2 & .97 & .31 \\ .97 & 2 & 1.31 \\ .31 & 1.31 & 2 \end{bmatrix}$$

3.2 Answer Node Selection

We have developed an algorithm to identify the "answer" nodes from a given query. For some questions, we're really looking for the relationship between a smaller subset of nodes of the total set of nodes in the query graph. For instance, the query graph in Fig. 3b would likely be an example of a query in which the user may really be more interested in the relationship between the drug and asthma, specifically looking for a drug that treats asthma through a gene, but the relationship between the gene node and the other two, while important, seems less likely to be an effective measure of the quality of the answer. As mentioned previously, we designate these answer nodes as "probes". Therefore, we need a consistent method to determine what the likely answer nodes are.

Probes are selected from the query graph based on "level of constraint". We assume that a node that is fully constrained by being assigned a specific entity or set of entities in the query graph, rather than just a type of entity, is always relevant to the quality of the result. We label these nodes *pinned nodes.* Beyond those highly constrained pinned nodes, we then assume that the nodes with the lowest level of constraint are also relevant to the query, based primarily on how we have seen questions formulated within the system. Based on the number of pinned nodes, we have three different approaches, as shown in Algorithm 1. If there are no pinned nodes in the query graph, we select the set of at least two nodes that have the lowest connectivity, in an attempt to find the nodes that are the least constrained. If there is exactly one pinned node, then along with the pinned node, we select the set of nodes that have the longest shortest path from the pinned node, as the longest shortest path indicates which nodes is "furthest" from the pinned node, and therefore the one that is the least constrained by it. Lastly, if there are 2 or more pinned nodes, we simply select all pinned nodes as probes.

Examples of probes identified using Algorithm 1 for different query graph structures are shown in Fig. 4. Nodes shown in blue are pinned in the query, and

Algorithm 1. Answer node selection algorithm

```
probes ← [ ]                                          ▷ Initialize an empty list
if len(pinnedNodes) = 0 then
    minDeg ← 1
    while len(probes) < 2 do
        for node ∈ qgraph do
            if degree(node) = minDeg then
                probes.append(node)
            end if
        end for
        minDeg ← minDeg + 1
    end while
else if len(pinnedNodes) = 1 then
    pinnedNode ← pinnedNodes[0]
    maxDist ← 0
    for node ∈ qgraph do
        dist ← shortest path length from node to pinnedNode
        if dist > maxDist then
            probes emptied
            probes ← node
            maxDist ← dist
        else if dist = maxDist then
            probes.append(node)
        end if
    end for
    probes.append(pinnedNode)
else
    probes ← pinnedNodes
end if
```

are thus selected as probes, while those shown in red are selected as probes by the algorithm.

3.3 Resistance Mapping

Starting from edge weights bounded between $[0, 1]$, we would like to arrive at an effective weight or score between probes that also lies in $[0, 1]$. To leverage the resistive distances for this problem we introduce a mapping of edge weights to resistance values where a lower weight represents a greater resistance. Each edge weight is mapped to a corresponding resistance value using a function, $f(w)$. Then the Kirchhoff index is used to calculate the effective resistance between probes and finally the inverse resistive mapping, $f^{-1}(\Omega)$, is applied to yield the effective weight between probes. The final score of a subgraph is then the average of the effective weights between all pairs of probes. To adhere to the desiderata listed above the function must be invertible and chosen such that $f(1) \to 0$ and $f(0) \to \infty$.

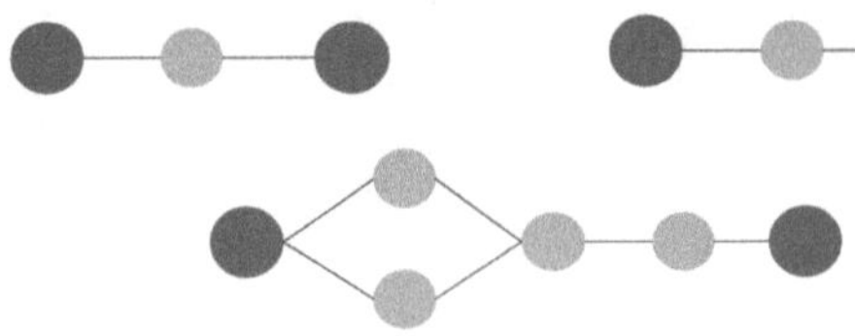

Fig. 4. A set of three different query graphs. In each graph, blue nodes are pinned, and therefore implicitly probes, and red nodes are unpinned nodes that are selected as probes. (Color figure online)

There are many functions that satisfy these however, we chose to use a logarithmic mapping.

$$f(w) = -\log(w) \tag{1}$$

$$f^{-1}(\Omega) = e^{-\Omega} \tag{2}$$

This mapping yields scores that meet the requirements listed above, so we can use the calculated edge weights as inputs to f, which will generate the conductance of each edge, the inverse of resistance, which will be used in the graph Laplacian. However, for a multi-graph, the graph Laplacian cannot be directly applied. The mapped resistance values between parallel edges must first be combined using the math dictating the parallel combination of resistors, shown in Fig. 5.

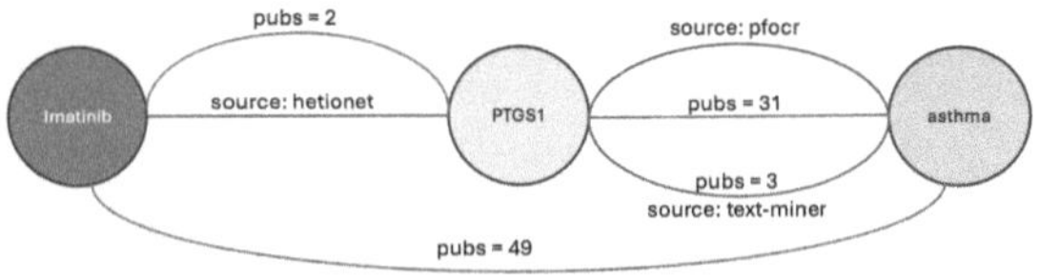

(a) A result graph displaying sources and relevant attributes of edges.

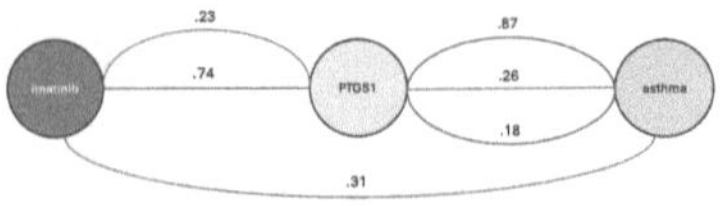

(b) The result graph after the edge attributes and sources have been computed as weights.

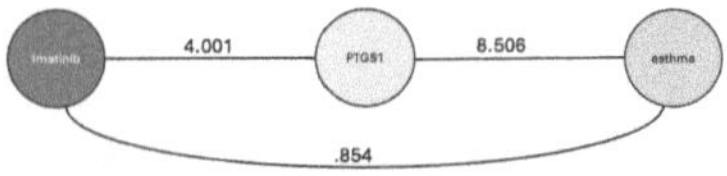

(c) The result graph after using the resistance mapping function and combining parallel edges.

Fig. 5. The process of weighting edges based on attached attributes and then mapping those weights to resistance utilizing the resistance mapping function.

3.4 ELROND

Once the values from the resistance mapping function are calculated, and parallel edge weights of the multigraph are combined, a simplified Laplacian matrix can be calculated and the Kirchhoff index between all pairs of probes can be calculated, by summing the resistance distance between each pair of probes, rather than each pair of nodes, as would be done in a traditional Kirchhoff index. The resistance mapping has no upper bound, and since these values are also summed for parallel edges in the Laplacian, the resistance distance is also unbounded. However, we desire scores to be bounded between 0 and 1, and therefore must compute the inverse mapping of the resulting resistive distance to return to our original weight space. After computing the Kirchhoff index on our unbounded edges, we then pass this through the inverse function to compute the final score for the result.

We name this algorithm, utilizing a logarithmic resistance mapping function of a weighted graph to compute the effective resistance ELROND (Effective Logarithmic Resistance for Ontological Network Determination).

4 Theoretical Analysis

To show the effectiveness of our ranking algorithm, we compare it to two different subgraph scoring algorithms, Frobenius norm and maximum flow, for a range of interpretable graph topologies.

We compare all three of the selected scoring algorithms across multiple theoretical graph topologies in Table 1, to compare the expected performance of all three given the same result graph.

Table 1. The scores generated by all three scoring algorithms on a selection of theoretical weighted graphs. All nodes in red are nodes designated as probes. Graphs are ordered based on how we would expect them to be ranked, with the highest score on top.

	Frob Norm	**Max Flow**	**ELROND**
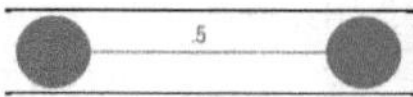	0.5	0.5	0.5
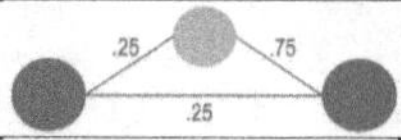	.829	0.5	0.390
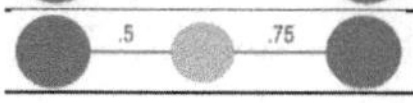	.901	0.5	0.375

Comparing across topologies, we can see that when using Frobenius norm, the score will increase with more weighted edges, even if that edge is a low-weighted serial edge, primarily because of the lack of awareness of probes. Adding low-weighted edges that are not parallel to any other edge, should not increase the score, since this essentially represents a poorly supported assertion without

Table 2. A comparison of general behaviors of all three scoring methods

	Frob Norm	Max Flow	ELROND
Parallel Paths Increase Score	✓		✓
Shorter Paths Score Higher			✓

additional evidence in the form of parallel assertions accompanying it. We aim to have longer paths ranked lower because of their length, since a longer chain of evidence would present more places where there may be less confidence in the validity of an asserted relationship. On the other hand, for max flow, bottlenecking caused by the lowest weighted edge causes scores to be dominated by that lowest weighted edge. Additionally, this also results in many similar scores, where the addition of a parallel edge does not cause any increase in the score.

Conversely, ELROND will give lower scores to longer paths, since it's focused specifically on the relationship between probes, and not the general graph. Low-weighted serial edges will decrease the score, but they don't dominate it. As we can see, when adding a parallel edge, it will always increase the score given to the result, in all cases, as would be expected when parallelizing a circuit. Low-weighted edges do not cause a bottleneck for the score. Many edges with high weights will still provide a good score on the result, but many high parallel weights on a single connection will not cause the score to increase in an unbounded way.

These general behaviors can then be compared to our previously stated requirements. All three methods adequately fulfill most of those desired behaviors, except for the desired properties shown in Table 2, which shows the requirements that are not adequately fulfilled by Max Flow and Frobenius Norm.

5 Application

ELROND is implemented as the scoring and ranking module of a biomedical knowledge graph question and answering system known as ROBOKOP [4]. ROBOKOP contains a large knowledge graph of almost 130 million relationships between over 9 million entities, that adheres to the Biolink ontology [37]. ROBOKOP can use a domain specific language for queries and can return subgraphs that adhere to the given query graph topology. ROBOKOP is concerned primarily with answering queries relevant to medical research. As such, the edges and metadata are tailored towards this use case. For instance, an edge that correlates the presence of a gene with the onset of a disease may have been a result of some clinical trial. In that case, the edge may have an attribute attached to it for a p-value, representing the statistical significance of that correlation [11]. This is accounted for with the inclusion of clinical data as an attribute in the profiles defined for its scoring.

Additionally, ROBOKOP specifically utilizes literature co-occurrence edges mined from PubMed, a collection of biomedical research papers maintained by the National Library of Medicine, at the National Institutes of Health [29]. As ROBOKOP answers biomedical questions, this serves as an effective corpus for its literature co-occurrence edges. These results, generated by ROBOKOP, are what we use to evaluate the real world performance of ELROND. These edges represent the pairwise literature co-occurrence of two entities, which provides useful information on the overall strength of the connection.

Along with comparing our algorithm to other viable algorithms, we also tested our ranking in real world queries. When querying for the treatment of asthma, our ranking algorithm is able to appropriately score well known and accepted results for many queries. The top 3 results of this query, nedocromil, cromoglicic acid, and theophylline are approved drugs for the treatment of asthma [7,15,40].

Table 3. The values generated by all three scoring algorithms on the top results produced by ROBOKOP on the query, "what drugs treat Asthma?"

	Frob Norm	Max Flow	ELROND
nedocromil	0.998	0.911	0.998
cromoglicic acid	0.989	0.911	0.998
theophylline	0.976	0.888	0.966
Fish Oils	0.998	0.955	0.965
comoglycate	0.965	0.911	0.965
montelukast	0.939	0.85	0.904
histamine	0.865	0.923	0.897

However, our algorithm is not perfect, and cannot make up for failings in data or result generation. There can be innocuous and meaningless results that still get scored highly, or answers that may be considered wrong. In this example, we see that histamines are ranked very highly. This is because of the reliance of the algorithm on literature co-occurrence to generate auxiliary edge weighting information. As histamine is not a treatment for asthma but rather the cause of some asthmatic symptoms, there exists a high literature co-occurrence between these two concepts. As such, this result gets ranked highly, even though histamine would not be an appropriate treatment for asthma. A result like the one for histamine may have been generated from incorrect assertions within the knowledge graph, or through a rule-mining approach. ELROND scores the results given to it, but does not generate the results themselves, and is unaware of the predicates in the original query graph. Incorporating the predicate into the score may help mitigate this issue.

We can also compare the scores of these results when passed through the other ranking methods as well, shown in Table 3. As expected, the ranking methods all have relatively similar results, as the input for each algorithm is the same result graph. However, it can be seen that there are certain differences, such as montelukast, which scores much more poorly when scored through maximum flow. Since these are the top results, we know that all of these result graphs will have many connections and have relatively highly weighted edges overall, so all of the values should be close to 1.

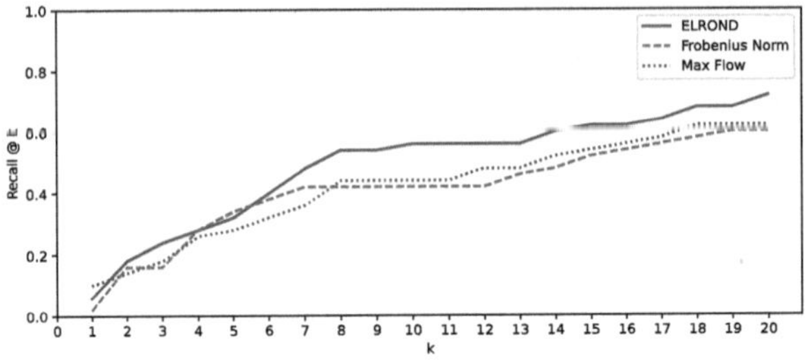

Fig. 6. A graph showing recall @ k for the top 20 results scored by Frobenius norm, maximum flow, and ELROND.

We have constructed a testing suite of known queries, compiled from sources such as DrugMechDB, DrugBank, and DrugCentral [16,24,38], for which the answers to those queries are also known. These queries are meant to test how the algorithm scores known answers to queries, in cases where a relationship is very well-documented. For instance, DrugCentral provides a selection of drugs that are commonly prescribed treatments for diabetes, including metformin, insulin, and mitiglinide. Running these queries, we can then apply our different ranking methods to the generated result graphs. From there, we can compare how many of our expected results appear in the top 20 results of each algorithm, shown in Fig. 6.

As we can see, ELROND performs better almost across the board. All three methods will generally perform well in ranking the obvious answers with very strong evidence, since those answers will have the edges with the highest weights. However, where ELROND excels is when surfacing more complex answers, particularly when multiple edges with weaker evidence need to be effectively combined, like those results that were generated through a rule-mining approach. This can be seen by how ELROND is able to perform much better in surfacing expected answers in the top 10, as opposed to the top 5, where all three perform similarly to one another.

We do not expect any of these methods to perform perfectly. Some of the answers are not accessible within the available knowledge graph, or have weak evidence attached to them. Therefore, we expect some of the expected answers to be missing, regardless of what scoring algorithm is used.

6 Conclusion

We have presented ELROND, a novel method for knowledge graph scoring, utilizing resistance distance with a logarithmic mapping function to map edge weights.

We have shown the results of this scoring method in both theoretical analysis and real world application, comparing its behavior to two other subgraph scoring algorithms. ELROND is implemented within ROBOKOP, a knowledge graph querying service in the biomedical domain, however it is not a domain specific algorithm and could be applied in other fields where knowledge graphs are deployed.

ELROND is particularly well suited to combining weak evidence into a higher score, without being dominated by a few large value edges, or conversely, being downgraded due to a few low-weighted edges. This quality makes it particularly useful in domains where there are many knowledge sources, all of which can provide compounding information that may serve as evidence for a result. Biomedical research is only one such example of this, but any domain that does not contain a universal source of ground truth would be applicable to this. This also makes it especially effective when attempting to rank result graphs that were generated through both traditional graph queries and rule-mining based inferencing methods.

Although the general methodology of ELROND has been shown to be robust, there are certain areas where the application of the algorithm can still be improved. In the methodology applied to ROBOKOP, we currently do not take the predicate of an edge into account when calculating edge weights and thus in scoring. An edge with the predicate "related to" provides less information than "directly physically interacts with". A more specific edge may be more appropriately given a higher weight, so utilizing predicate weighting that is aware of the hierarchy in the ontology could be useful. In that same vein, using edge embeddings could also provide a method by which to weight edges more effectively [39].

Currently, the attribute-weighting logistic functions are created by analyzing the results of queries and determining what the possible range of values would be for a certain edge attribute. We then create the sigmoid to reflect what a "good" and "bad" value would be. Similarly, for sources, we assign weights to sources based on how accurate we perceive them to be. Parameter optimization on the logistic functions and source weights could improve scores overall as well.

Another avenue of future work actually involves a different application of ELROND. Instead of using the described method to rank and score results, it could be interesting to explore adapting this methodology to instead analyze a knowledge graph. By decomposing the knowledge graph into subgraphs, it would be possible to find not only the load-bearing subgraphs, but also the most influential edges. These relationships would be the ones that have the largest contribution to the graph overall. This could provide useful insight on the knowledge graph as a whole.

Acknowledgments. Support for this work was provided by the National Center for Advancing Translational Sciences, National Institutes of Health, through the Biomedical Data Translator program, award #1OT2TR005712. Any opinions expressed in this document do not necessarily reflect the views of NCATS, individual Translator team members, or affiliated organizations and institutions. The authors would like to also acknowledge Yaphet Kebede, Phil Owen, and Adam Viola for their contributions in developing, maintaining, and deploying ELROND.

Disclosure of Interests. The authors have no competing interests to declare that are relevant to the content of this article.

References

1. Abu-Salih, B.: Domain-specific knowledge graphs: a survey. J. Netw. Comput. Appl. **185**, 103076 (2021). https://doi.org/10.1016/j.jnca.2021.103076
2. van Bakel, R., Aleksiev, T., Daza, D., Alivanistos, D., Cochez, M.: Approximate knowledge graph query answering: from ranking to binary classification (2021). https://arxiv.org/abs/2102.11389
3. Bendito, E., Carmona, A., Encinas, A., Gesto, J., Mitjana, M.: Kirchhoff indexes of a network. Linear Algebra Appl. **432**(9), 2278–2292 (2010). https://doi.org/10.1016/j.laa.2009.05.032. https://www.sciencedirect.com/science/article/pii/S0024379509002924. Special Issue devoted to Selected Papers presented at the Workshop on Spectral Graph Theory with Applications on Computer Science, Combinatorial Optimization and Chemistry, Rio de Janeiro (2008)
4. Bizon, C., et al.: ROBOKOP KG and KGB: integrated knowledge graphs from federated sources. J. Chem. Inf. Model. **59**(12), 4968–4973 (2019). https://doi.org/10.1021/acs.jcim.9b00683
5. Boominathan, P., Kanchan, A.: Routing planning as an application of graph theory. Int. J. Sci. Technol. Res. **3**, 61–66 (2014)
6. Bordes, A., Chopra, S., Weston, J.: Question answering with subgraph embeddings (2014). https://arxiv.org/abs/1406.3676
7. Brogden, R.N., Sorkin, E.M.: Nedocromil sodium. Drugs **45**(5), 693–715 (1993). https://doi.org/10.2165/00003495-199345050-00007
8. Böttcher, A., Wenzel, D.: The Frobenius norm and the commutator. Linear Algebra Appl. **429**(8), 1864–1885 (2008). https://doi.org/10.1016/j.laa.2008.05.020
9. Cai, J., Zhang, Z., Wu, F., Wang, J.: Deep cognitive reasoning network for multi-hop question answering over knowledge graphs. In: Findings of the Association for Computational Linguistics: ACL-IJCNLP 2021, pp. 219–229 (2021)
10. Cohen, A.R., Vitanyi, P.M.B.: Normalized Google distance of multisets with applications (2013). https://arxiv.org/abs/1308.3177
11. Fecho, K., et al.: A novel approach for exposing and sharing clinical data: the translator integrated clinical and environmental exposures service. J. Am. Med. Inform. Assoc. **26**(10), 1064–1073 (2019)
12. Galárraga, L.A., Teflioudi, C., Hose, K., Suchanek, F.: AMIE: association rule mining under incomplete evidence in ontological knowledge bases. In: Proceedings of the 22nd International Conference on World Wide Web, WWW 2013, pp. 413–422. Association for Computing Machinery, New York, NY, USA (2013). https://doi.org/10.1145/2488388.2488425
13. Ge, C., Liu, X., Chen, L., Gao, Y., Zheng, B.: LargeEA: aligning entities for large-scale knowledge graphs. Proc. VLDB Endowment **15**(2), 237–245 (2021). https://doi.org/10.14778/3489496.3489504
14. Goldberg, A.V., Tarjan, R.E.: A new approach to the maximum-flow problem. J. ACM **35**(4), 921–940 (1988). https://doi.org/10.1145/48014.61051

15. Goldstein, L.H., Weber-Schöndorfer, C., Berkovitch, M.: 2.3 - antiasthmatic and cough medication. In: Schaefer, C., Peters, P., Miller, R.K. (eds.) Drugs During Pregnancy and Lactation, 3rd edn., pp. 65–74. Academic Press, San Diego (2015). https://doi.org/10.1016/B978-0-12-408078-2.00004-4. https://www.sciencedirect.com/science/article/pii/B9780124080782000044
16. Gonzalez-Cavazos, A.C., et al.: DrugMechDB: a curated database of drug mechanisms. Sci. Data **10**(1), 632 (2023). https://doi.org/10.1038/s41597-023-02534-z
17. Hogan, A., et al.: Knowledge graphs. ACM Comput. Surv. **54**(4) (2021). https://doi.org/10.1145/3447772
18. Huang, Y., Ying, J.J.C., Yu, P.S., Tseng, V.S.: Dynamic graph mining for multi-weight multi-destination route planning with deadlines constraints. ACM Trans. Knowl. Discov. Data **15**(1) (2020). https://doi.org/10.1145/3412363
19. Hwang, J.Y., Kim, J.S., Lim, S.S., Park, K.H.: A fast path planning by path graph optimization. IEEE Trans. Syst. Man Cybern. Part A Syst. Hum. **33**(1), 121–129 (2003). https://doi.org/10.1109/TSMCA.2003.812599
20. Jayaram, N., Khan, A., Li, C., Yan, X., Elmasri, R.: Querying knowledge graphs by example entity tuples. IEEE Trans. Knowl. Data Eng. **27**(10), 2797–2811 (2015). https://doi.org/10.1109/TKDE.2015.2426696
21. Khan, A.: Knowledge graphs querying. SIGMOD Rec. **52**(2), 18–29 (2023). https://doi.org/10.1145/3615952.3615956
22. Kilicoglu, H., Shin, D., Fiszman, M., Rosemblat, G., Rindflesch, T.C.: SemMedDB: a PubMed-scale repository of biomedical semantic predications. Bioinformatics **28**(23), 3158–3160 (2012)
23. Kim, S., Lee, W., Arora, N.R., Jo, T.C., Kang, S.H.: Retrieving keyworded subgraphs with graph ranking score. Expert Syst. Appl. **39**(5), 4647–4656 (2012). https://doi.org/10.1016/j.eswa.2011.08.136
24. Knox, C., et al.: DrugBank 6.0: the DrugBank knowledgebase for 2024. Nucleic Acids Res. **52**(D1), D1265–D1275 (2023). https://doi.org/10.1093/nar/gkad976
25. Kumar, A., Singh, S.S., Singh, K., Biswas, B.: Link prediction techniques, applications, and performance: a survey. Phys. A **553**, 124289 (2020). https://doi.org/10.1016/j.physa.2020.124289
26. Li, Q., Gama, F., Ribeiro, A., Prorok, A.: Graph neural networks for decentralized multi-robot path planning. In: 2020 IEEE/RSJ International Conference on Intelligent Robots and Systems (IROS), pp. 11785–11792 (2020). https://doi.org/10.1109/IROS45743.2020.9341668
27. Li, T., Bai, J., Yang, X., Liu, Q., Chen, Y.: Co-occurrence network of high-frequency words in the bioinformatics literature: structural characteristics and evolution. Appl. Sci. **8**(10) (2018). https://doi.org/10.3390/app8101994. https://www.mdpi.com/2076-3417/8/10/1994
28. Maheshwari, G., Trivedi, P., Lukovnikov, D., Chakraborty, N., Fischer, A., Lehmann, J.: Learning to rank query graphs for complex question answering over knowledge graphs. In: Ghidini, C., et al. (eds.) The Semantic Web - ISWC 2019, pp. 487–504. Springer International Publishing, Cham (2019)
29. McEntyre, J., Lipman, D.: PubMed: bridging the information gap. CMAJ **164**(9), 1317–1319 (2001). https://www.cmaj.ca/content/164/9/1317
30. Merris, R.: Laplacian matrices of graphs: a survey. Linear Algebra Appl. **197–198**, 143–176 (1994). https://doi.org/10.1016/0024-3795(94)90486-3
31. Nguyen, V., Bodenreider, O., Sheth, A.: Don't like RDF reification? Making statements about statements using singleton property. In: Proceedings of the 23rd International Conference on World Wide Web, WWW 2014, pp. 759–770. Association

for Computing Machinery, New York, NY, USA (2014). https://doi.org/10.1145/2566486.2567973
32. Qiu, Y., et al.: Hierarchical query graph generation for complex question answering over knowledge graph. In: Proceedings of the 29th ACM International Conference on Information & Knowledge Management, CIKM 2020, pp. 1285–1294. Association for Computing Machinery, New York, NY, USA (2020). https://doi.org/10.1145/3340531.3411888
33. Radhakrishnan, S., Erbis, S., Isaacs, J.A., Kamarthi, S.: Novel keyword co-occurrence network-based methods to foster systematic reviews of scientific literature. PLOS ONE **12**(3), 1–16 (03 2017). https://doi.org/10.1371/journal.pone.0172778
34. Sakr, S., Al-Naymat, G.: Graph indexing and querying: a review. IJWIS **6**, 101–120 (2010). https://doi.org/10.1108/17440081011053104
35. Sud, A., Andersen, E., Curtis, S., Lin, M.C., Manocha, D.: Real-time path planning in dynamic virtual environments using multiagent navigation graphs. IEEE Trans. Visual Comput. Graph. **14**(3), 526–538 (2008). https://doi.org/10.1109/TVCG.2008.27
36. Thulasiraman, K., Yadav, M., Naik, K.: Network science meets circuit theory: resistance distance, Kirchhoff index, and Foster's theorems with generalizations and unification. IEEE Trans. Circuits Syst. I Regul. Pap. **66**(3), 1090–1103 (2019). https://doi.org/10.1109/TCSI.2018.2880601
37. Unni, D.R., et al.: Biolink model: a universal schema for knowledge graphs in clinical, biomedical, and translational science. Clin. Transl. Sci. **15**(8), 1848–1855 (2022)
38. Ursu, O., et al.: DrugCentral: online drug compendium. Nucleic Acids Res. **45**(D1), D932–D939 (2016). https://doi.org/10.1093/nar/gkw993
39. Wang, C., Wang, C., Wang, Z., Ye, X., Yu, P.S.: Edge2vec: edge-based social network embedding. ACM Trans. Knowl. Discov. Data **14**(4) (2020). https://doi.org/10.1145/3391298
40. Weinberger, M., Hendeles, L.: Theophylline in asthma. N. Engl. J. Med. **334**(21), 1380–1388 (1996). https://doi.org/10.1056/NEJM199605233342107
41. Wu, Y., Raschid, L.: ApproxRank: estimating rank for a subgraph. In: IEEE 25th International Conference on Data Engineering, pp. 54–65 (2009). https://doi.org/10.1109/ICDE.2009.108
42. Yang, Y., Yu, Y.: Resistance distances and Kirchhoff indices under graph operations. IEEE Access, 1 (2020). https://doi.org/10.1109/ACCESS.2020.2995935
43. Zandbiglari, K., Ameri, F., Javadi, M.: A text analytics framework for supplier capability scoring supported by normalized Google distance and semantic similarity measurement methods. J. Comput. Inf. Sci. Eng. **23**(5), 051011 (2023). https://doi.org/10.1115/1.4062173
44. Zhou, B., Trinajstić, N.: A note on Kirchhoff index. Chem. Phys. Lett. **455**(1), 120–123 (2008). https://doi.org/10.1016/j.cplett.2008.02.060
45. Zhu, M., Zhu, L., Li, H., Li, W., Zhang, Z.: Resistance distances in directed graphs: definitions, properties, and applications. arXiv preprint arXiv:2302.04114 (2023)
46. Zou, X.: A survey on application of knowledge graph. J. Phys. Conf. Ser. **1487**(1), 012016 (2020). https://doi.org/10.1088/1742-6596/1487/1/012016

Multiclass Hate Speech Detection with RoBERTa-OTA: Integrating Transformer Attention and Graph Convolutional Networks

Mahmoud Abusaqer and Jamil Saquer(✉)

Missouri State University, Springfield, MO 65897, USA
{ma7956s,jamilsaquer}@missouristate.edu

Abstract. Multiclass hate speech detection across demographic categories remains computationally challenging due to implicit targeting strategies and linguistic variability in social media content. Existing approaches rely solely on learned representations from training data, without explicitly incorporating structured ontological frameworks that can enhance classification through formal domain knowledge integration. We propose RoBERTa-OTA, which introduces ontology-guided attention mechanisms that process textual features alongside structured knowledge representations through enhanced Graph Convolutional Networks. The architecture combines RoBERTa embeddings with scaled attention layers and graph neural networks to integrate contextual language understanding with domain-specific semantic knowledge. Evaluation across 39,747 balanced samples using 5-fold cross-validation demonstrates significant performance gains over baseline RoBERTa implementations and existing state-of-the-art methods. RoBERTa-OTA achieves 96.04% accuracy compared to 95.02% for standard RoBERTa, with substantial improvements for challenging categories: gender-based hate speech detection improves by 2.36% points while other hate speech categories improve by 2.38% points. The enhanced architecture maintains computational efficiency with only 0.33% parameter overhead, providing practical advantages for large-scale content moderation applications requiring fine-grained demographic hate speech classification.

Keywords: Hate speech detection · Graph neural networks · Ontology integration · Multiclass classification · Transformer models

1 Introduction

The exponential growth of social media has reshaped digital communication patterns, enabling global connectivity while simultaneously creating new challenges

---

Research Track: CSCI-RTAI.

H. R. Arabnia et al. (Eds.): CSCI 2025, CCIS 2967, pp. 179–193, 2027.
https://doi.org/10.1007/978-3-032-28454-9_12

for online safety and content moderation. Balancing free expression with community protection has made automated hate speech detection a critical research area [8].

Early detection systems relied on lexicon-based and classical machine learning approaches such as Support Vector Machines, Naive Bayes, and ensemble methods [6,15]. While effective on curated datasets, these methods struggled with semantic nuance, implicit references, and evolving language patterns.

The advent of deep learning marked a paradigm shift in natural language processing. CNNs, LSTMs, and attention mechanisms improved performance across social media datasets, and the introduction of transformer architectures such as BERT and RoBERTa set new state-of-the-art results. RoBERTa's optimized training procedures improved robustness over BERT [10], with transformer-based models consistently exceeding 90% accuracy and F1-scores on hate speech detection benchmarks [2,7].

Recent research has explored specialized transformer adaptations for abusive language detection. Yin et al. [16] combined annotator profiles, label descriptions, and semantic knowledge with Collaborative Topic Regression and BERT to improve detection of difficult posts. Mathew et al. [11] proposed explainable hate speech detection using BERT with attention visualization. However, most transformer-based approaches continue to address hate speech detection as binary classification problems, limiting their applicability to fine-grained demographic targeting scenarios.

Graph-based methods have been explored to capture structured relationships between hateful concepts. Mishra et al. [12] demonstrated that Graph Convolutional Networks effectively model relationships between concepts in abusive language. Wang et al. [14] introduced SOSNet, a graph-based framework for fine-grained demographic targeting detection, achieving 92.70% accuracy and 92.58% F1-score with SBERT+SOSNet, and 94.38% accuracy and 94.44% F1-score using BOW features with XGBoost. Ousidhoum et al. [13] extended this line of work to multilingual, multi-aspect hate speech analysis.

Multiclass detection introduces unique challenges, as different demographic groups exhibit distinct linguistic patterns. While explicit, identity-based hate (e.g., targeting age or religion) is detected reliably, implicit targeting (e.g., gender-based hate) relies on coded language that standard transformers often miss [1]. These challenges motivate our approach, which integrates structured ontological knowledge with transformer attention to better capture subtle demographic targeting.

Our RoBERTa baseline (95.02% accuracy, 95.04% F1-score) and RoBERTa-OTA (96.04% accuracy, 96.06% F1-score) outperformed SOSNet's best results on the same dataset, demonstrating the benefit of combining transformer-based contextual learning with structured knowledge for fine-grained hate speech classification.

2 Literature Review

Early research relied on hand-crafted features with machine learning classifiers. Davidson et al. [6] employed TF-IDF features with linear SVM, while Waseem and Hovy [15] combined n-gram features with demographic targeting indicators. Burnap and Williams [5] added part-of-speech and sentiment features, and Gitari et al. [9] used lexicon-based sentiment dictionaries. These approaches captured some hateful patterns but struggled with semantic and contextual subtleties.

Deep learning models further advanced the field. Badjatiya et al. [4] compared CNN, LSTM, and FastText embeddings, showing consistent improvements over classical approaches. Agrawal and Awekar [3] achieved better performance with bidirectional LSTM models and attention mechanisms, solidifying deep learning as the standard for hate speech detection.

Transformer models brought a major leap forward. Devlin et al. [7] showed that BERT's bidirectional attention achieved state-of-the-art results, later improved by RoBERTa's optimized training [10]. Mathew et al. [11] demonstrated that transformer attention can also be used for interpretability, offering insights into model decisions alongside strong predictive performance.

Despite these advances, most transformer-based studies remain focused on binary classification. Wang et al. [14] addressed this limitation with SOSNet for multiclass detection, achieving strong performance but relying heavily on graph-based representations. Our RoBERTa and RoBERTa-OTA models surpassed these results, highlighting the promise of transformer–ontology integration for fine-grained demographic targeting tasks. Ousidhoum et al. [13] further emphasize the importance of multi-aspect, multilingual analysis for building robust systems.

Overall, current literature shows that while transformers dominate binary hate speech detection, multiclass demographic targeting remains underexplored, presenting an opportunity for more nuanced approaches that combine contextual embeddings with structured knowledge [2].

3 Problem Definition and Research Questions

We formally define the multiclass hate speech problem as follows: given a dataset $D = \{(x_i, y_i)\}_{i=1}^{n}$ where x_i represents text and $y_i \in$ {Age, Ethnicity, Gender, Religion, Other Hate}, we aim to develop a classifier $f : X \rightarrow Y$ that accurately distinguishes between these categories while maintaining balanced performance across all groups.

We investigate three questions:

1. Do ontology-guided transformer architectures outperform standard transformer baselines and state-of-the-art SOSNet results for distinguishing between specific hate speech categories?
2. Which demographic targeting patterns are most challenging to classify, and how does performance vary across hate speech categories?
3. What computational overhead do enhanced architectures introduce, and do performance gains justify the additional resource requirements?

4 Dataset and Preprocessing

Our experimental evaluation utilizes the fine-grained hate speech dataset derived from the SOSNet framework introduced by Wang et al. [14], which provides comprehensive coverage of demographic-targeted hate speech patterns across social media platforms. The dataset comprises 39,747 samples systematically extracted from a larger collection of 47,692 hate speech tweets, with careful filtering applied to ensure balanced representation across five distinct hate speech categories. Following the SOSNet methodology, this filtering process removes the generic "none_hate" category to focus specifically on distinguishing between different forms of hate speech, creating a classification task that better reflects real-world content moderation challenges. This also allows us to compare our results with SOSNet's. Table 1 shows class distributions of the dataset.

Table 1. Dataset Statistics: 5-Class Hate Speech Distribution

Hate Speech Class	Samples	Percentage (%)
Religion	7,998	20.1
Age	7,992	20.1
Gender	7,973	20.1
Ethnicity	7,961	20.0
Other Hate	7,823	19.7
Total	**39,747**	**100.0**

Each hate speech category exhibits distinct linguistic characteristics reflecting different targeting strategies. Age-based hate speech employs generational stereotypes and age-related mockery, while gender-based harassment includes appearance-focused attacks and sexualized content. Ethnicity-based hate speech utilizes cultural stereotypes and racial slurs requiring specialized cultural understanding. Religion-based targeting involves theological criticism and sectarian conflicts with specific religious terminology. The "Other Hate" category captures diverse forms targeting socioeconomic status, physical disabilities, sexual orientation, and additional characteristics beyond conventional demographic boundaries.

4.1 Linguistic Validation and Class Characterization

To validate the necessity of fine-grained classification approaches, we conducted comprehensive linguistic analysis across all demographic categories. This analysis demonstrates that distinct hate speech categories exhibit linguistic differences that justify sophisticated classification architectures, providing quantitative evidence that RoBERTa-OTA's performance improvements reflect genuine model capability rather than dataset artifacts.

Our linguistic analysis employs part-of-speech tagging using spaCy's `en_core_web_sm` model to extract structural features across all 39,747 samples. The feature extraction pipeline computes character count, token count, noun frequency, verb frequency, and URL detection using regular expression patterns, capturing both surface-level characteristics and deeper syntactic patterns that distinguish different forms of demographic targeting.

The linguistic analysis reveals substantial heterogeneity across hate speech categories, as summarized in Table 2. Religion-based hate speech demonstrates the highest linguistic complexity, averaging 198.0 characters with 10.1 nouns and 39.6 tokens, indicating extensive theological terminology usage. Age-based targeting exhibits considerable narrative content with 173.5 characters and 37.0 tokens per sample. Other forms of hate speech display the most heterogeneous patterns, averaging only 85.7 characters but showing the highest URL usage (0.18 per sample), suggesting diverse propagation strategies including external content sharing.

Table 2. Average Linguistic Features Across Hate Speech Categories

Class	URLs	Chars	Tokens	Nouns	Verbs
Religion	0.07	198.0	39.6	10.1	4.6
Age	0.03	173.5	37.0	7.8	4.8
Ethnicity	0.04	139.3	30.4	7.8	3.4
Gender	0.11	136.4	29.3	7.4	3.2
Other Hate	0.18	85.7	18.4	4.2	2.1

Figure 1 presents pairwise Jensen-Shannon divergence scores computed from unigram language models for each hate speech category, where values range from 0 (identical distributions) to 1 (maximally different distributions). The analysis reveals substantial lexical separation between all class pairs, with divergence scores ranging from 0.449 to 0.483 (mean = 0.466), demonstrating robust linguistic separation necessary for fine-grained classification approaches. Other hate and religion-based hate speech exhibit maximum divergence (0.483), while age-based and gender-based harassment show minimum divergence (0.449), yet this value still represents substantial lexical distinction supporting the necessity of sophisticated attention mechanisms.

4.2 Preprocessing Pipeline

We implement minimal preprocessing to retain contextual information crucial for distinguishing between different hate speech types, as aggressive preprocessing can eliminate subtle linguistic cues that differentiate demographic targeting patterns. Text normalization includes fundamental cleaning procedures such as removing excessive whitespace, handling special characters, and standardizing Unicode encoding to ensure consistent text representation.

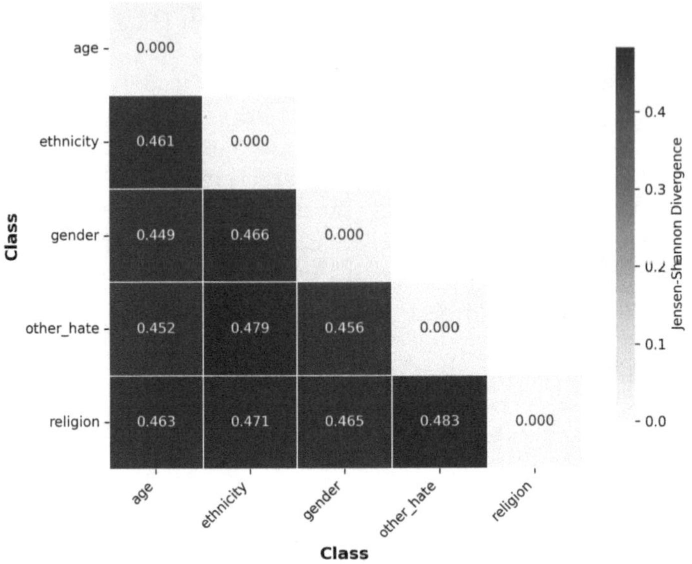

Fig. 1. Pairwise Jensen-Shannon divergence between hate speech classes.

The preprocessing workflow follows established protocols adapted from recent hate speech detection research. Structural cleaning operations include user mention removal for privacy, and HTML tag removal to clean web scraping artifacts. Content normalization procedures involve converting text to lowercase, expanding contractions using standardized libraries, and converting emojis to textual representations to preserve emotional context while enabling consistent processing.

Our preprocessing strategy deliberately preserves emoticons, punctuation patterns, and certain capitalization structures that may convey important sentiment or emotional intensity information relevant to hate speech classification. This approach recognizes that aggressive text normalization can eliminate crucial linguistic signals that distinguish between different forms of demographic targeting, particularly in social media contexts where unconventional language use often carries semantic significance.

5 Methodology

Our experimental methodology compares two architectures for 5-class hate speech detection: a RoBERTa baseline and our novel RoBERTa-OTA enhancement. Both implementations utilize PyTorch with identical optimization parameters and undergo evaluation through 5-fold cross-validation with stratified sampling to ensure fair comparison across fine-grained demographic hate speech categories.

5.1 Baseline Architecture: RoBERTa

RoBERTa represents an optimized variant of BERT with enhanced pretraining procedures, including dynamic masking and improved training strategies. Our implementation employs RoBERTa-base with 124.6M parameters as the baseline architecture for 5-class hate speech detection. We utilize the pretrained model with an enhanced classification head featuring multiple linear layers and dropout regularization, mapping the 768-dimensional contextual representations to five demographic hate speech categories through progressive dimensionality reduction.

5.2 Novel Architecture: RoBERTa-OTA

Our proposed RoBERTa-OTA (RoBERTa with Ontology-guided Transformer Attention) represents an advanced architecture that combines transformer-based text processing with domain-specific knowledge integration through graph neural networks. This dual-stream approach builds upon RoBERTa's enhanced language understanding capabilities while incorporating structured ontological knowledge about hate speech categories and their semantic relationships, enabling more nuanced classification across fine-grained demographic targeting patterns.

The RoBERTa-OTA architecture incorporates three critical design innovations. First, **enhanced 3-layer Graph Convolutional Network (GCN)** processes structured demographic relationships through systematic progression ($6 \rightarrow 64 \rightarrow 64 \rightarrow 32$ dimensions) using GCNConv layers with ReLU activation and layer normalization. Second, the **feature integration mechanism** concatenates RoBERTa's 768-dimensional textual representations with 32-dimensional ontological features, creating 800-dimensional combined representations for classification. Third, **deep classifier network** employs three linear transformations ($800 \rightarrow 400 \rightarrow 200 \rightarrow 5$) with batch normalization, layer normalization, and progressive dropout rates (0.3, 0.2, 0.1) specifically optimized for fine-grained demographic classification.

The RoBERTa-OTA architecture implements a sophisticated dual-stream processing framework that handles both textual features and structured knowledge representations in parallel as depicted in Fig. 2.

Text Processing Stream. The text processing stream utilizes pre-trained RoBERTa-base for robust contextual embeddings, enhanced through additional scaled dot-product attention layers specifically optimized for hate speech detection patterns. The scaled dot-product attention mechanism computes refined attention scores using the standard formulation:

$$\text{Attention}(Q, K, V) = \text{softmax}\left(\frac{QK^T}{\sqrt{d_k}}\right)V \quad (1)$$

where Q, K, and V represent learned linear projections of the input embeddings, and d_k denotes the dimension of the keys corresponding to RoBERTa's

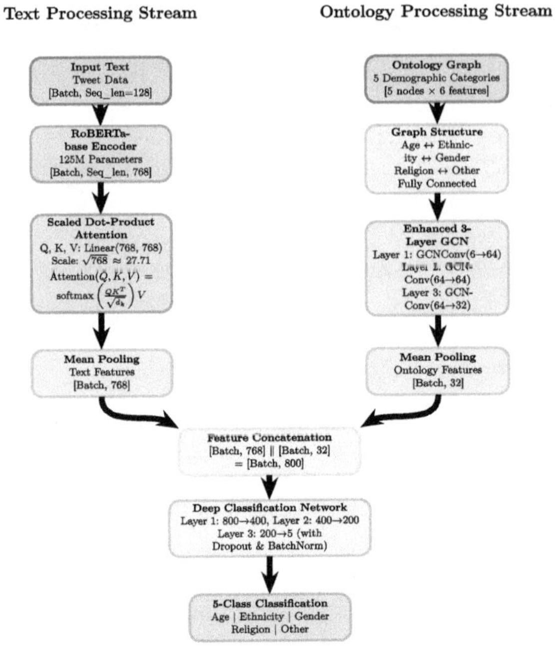

Fig. 2. RoBERTa-OTA Architecture: Dual-Stream Processing Framework for Multi-class Hate-Speech Detection with Ontology-Guided Attention

hidden size of 768. This mechanism enables focused attention on hate speech-relevant features while maintaining the contextual understanding capabilities of the base transformer architecture.

Ontology Processing Stream. Our hate speech ontology represents domain-specific knowledge through a structured graph comprising five nodes corresponding to the target demographic categories (age, ethnicity, gender, religion, other_hate). Each node is characterized by a six-dimensional feature vector encoding conceptual attributes derived from linguistic analysis and established hate speech research that capture the semantic properties and targeting characteristics of each category.

The node feature vectors encode the following semantic dimensions: demographic targeting (binary), cultural identity targeting (binary), gender-related characteristics (binary), religious/belief targeting (binary), linguistic complexity intensity (continuous), and targeting diversity (continuous). These dimensions reflect both the targeting mechanisms and the observed linguistic patterns identified in our comprehensive dataset analysis.

Our ontological framework follows systematic design principles based on comprehensive analysis of our 39,747-sample dataset. The six-dimensional feature vectors encode conceptual attributes derived from empirical linguistic analysis and established hate speech research that capture the semantic properties and targeting characteristics of each category. Each dimension represents empirically-grounded characteristics: demographic targeting (binary indicator of direct population targeting), cultural identity targeting (binary indicator reflecting cultural/ethnic elements), gender-related characteristics (binary indicator for gender-specific patterns), religious/belief targeting (binary indicator for theological elements), linguistic complexity (continuous measure reflecting observed character-per-sample statistics), and targeting diversity (continuous measure of attack vector breadth within categories).

For instance, the religion node features [1.0, 1.0, 0.0, 1.0, 0.9, 0.8] reflect empirical observations: direct demographic and cultural targeting, religious characteristics, highest linguistic complexity (0.9, corresponding to 198.0 characters per sample with extensive theological terminology), and substantial targeting diversity (0.8) across sectarian boundaries. The fully-connected graph structure enables systematic relationship modeling between categories while maintaining interpretable encodings aligned with computational linguistics research. While cross-linguistic evaluation remains future work, the framework's grounding in demographic targeting mechanisms and linguistic complexity patterns provides a systematic foundation for adaptation to diverse cultural contexts.

Table 3. Ontological Node Feature Vectors for Hate Speech Categories

Category	Demo.	Cult.	Gend.	Relig.	Complex.	Divers.
Age	[1.0	0.0	0.0	0.0	0.7	0.3]
Ethnicity	[1.0	1.0	0.0	0.0	0.6	0.7]
Gender	[1.0	0.0	1.0	0.0	0.6	0.5]
Religion	[1.0	1.0	0.0	1.0	0.9	0.8]
Other	[0.5	0.5	0.5	0.5	0.4	0.9]

Demo. = Demographic targeting; Cult. = Cultural identity; Gend. = Gender-related; Relig. = Religious/belief; Complex. = Linguistic complexity; Divers. = Targeting diversity.

Table 3 summarizes the complete ontological feature representation for all hate speech categories. The demographic dimension identifies primary population targeting (1.0 for direct demographic attacks), while cultural and religious dimensions capture identity-based targeting mechanisms. Linguistic complexity values reflect the observed analytical patterns from our dataset, with religion exhibiting the highest complexity (0.9) corresponding to extensive theological terminology, and other categories showing the lowest complexity (0.4) reflecting diverse but simpler targeting patterns. Targeting diversity measures the breadth of attack vectors within each category, with other forms of hate speech demonstrating the highest diversity (0.9) due to their heterogeneous nature spanning multiple personal characteristics beyond conventional demographic boundaries.

The graph connectivity follows a fully-connected structure where each category node maintains bidirectional edges with all other nodes, enabling the capture of inter-category relationships and semantic overlaps between different forms of demographic targeting. This structured representation enables the Graph Convolutional Network (GCN) to learn nuanced relationships between hate speech categories while maintaining interpretable feature encodings that align with established hate speech research and observed linguistic patterns in our comprehensive dataset analysis.

The ontological knowledge stream employs a three-layer GCN that processes structured relationships between different hate speech concepts, demographic categories, and their linguistic manifestations. Each GCN layer implements the standard graph convolution operation:

$$X^{(l+1)} = \tilde{D}^{-\frac{1}{2}}\tilde{A}\tilde{D}^{-\frac{1}{2}}X^{(l)}W^{(l)} \tag{2}$$

where $\tilde{A} = A + I$ represents the adjacency matrix with added self-connections, $\tilde{D}$ denotes the degree matrix, $X^{(l)}$ represents node features at layer l, and $W^{(l)}$ constitutes the learnable weight matrix. The GCN architecture employs a first layer with input dimension 6 and hidden dimension 64, followed by a second layer maintaining hidden dimension 64, and a third layer with output dimension 32. The first two layers apply ReLU activation, layer normalization, and dropout (0.1) for training stability, while the final layer produces raw outputs for subsequent classification processing.

Feature Integration and Classification. Feature integration occurs through a carefully designed concatenation mechanism that combines mean-pooled representations from both processing streams. The text stream contributes contextual embeddings enhanced through scaled attention, while the ontological stream provides structured domain knowledge encoded through graph convolutional operations. The integration mechanism follows:

$$F = \text{mean}(\text{attention\_output}) \parallel \text{mean}(\text{GCN\_features}) \tag{3}$$

where $\parallel$ denotes concatenation and mean pooling operations ensure fixed-size representations suitable for subsequent classification processing. The integrated features undergo dropout regularization with rate 0.3 before passing through the final linear classification layer that maps combined textual and ontological patterns to the five hate speech categories.

5.3 Training Configuration and Optimization

All models are implemented using PyTorch on two NVIDIA Ampere A100 GPUs (6912 CUDA cores, 40 GB RAM each). We utilize the Transformers library for pre-trained model access and PyTorch Geometric for graph neural network components. Text processing employs RoBERTa's tokenizer with a maximum sequence length of 128 tokens, applying necessary padding and truncation with consistent handling across all architectures.

Both architectures employ AdamW optimizer with identical parameters: learning rate 1e−5, beta values (0.9, 0.999), and weight decay 1e−5. Training

Table 4. Comprehensive Performance Analysis Across 5-Class Hate Speech Detection

Class	RoBERTa Baseline				RoBERTa-OTA			
	F1	Acc	Prec	Rec	F1	Acc	Prec	Rec
Age	98.63	98.42	98.85	98.42	98.75	98.84	99.08	98.42
Ethnicity	98.41	98.10	98.72	98.10	98.44	97.84	99.05	97.84
Gender	90.70	91.50	90.03	91.50	93.06	93.43	92.71	93.43
Religion	98.28	97.52	99.05	97.52	98.62	98.56	98.68	98.56
Other Hate	88.94	89.33	88.68	89.33	91.32	91.87	90.82	91.87
Weighted Avg	**95.04**	**95.02**	**95.12**	**95.02**	**96.06**	**96.04**	**96.09**	**96.04**
SOSNet [14]	**94.44 F1, 94.38 Acc Previous State-of-the-Art**							

Table 5. Computational Efficiency Comparison

Model	FLOPs (Billions)	Params (Millions)	GPU Mem (GB)	Size (MB)	Time (min/fold)
RoBERTa Baseline	31.9	124.65	2.6	475.5	24.5
RoBERTa-OTA	35.2	125.06	3.1	477.1	27.5
Overhead	**+10.3%**	**+0.33%**	**+19.2%**	**+0.34%**	**+12.2%**

proceeds for a maximum of 20 epochs with batch size 16, implementing early stopping with patience of 3 epochs monitoring validation F1-score. Both architectures use cross-entropy loss with label smoothing ($\alpha = 0.1$) for enhanced regularization.

We use 5-fold stratified cross-validation to maintain consistent class distribution. Accuracy and F1-weighted scores are the primary evaluation metrics, measuring overall correctness and balanced performance across demographic groups. Supplementary metrics include precision, and recall. Training efficiency evaluation incorporates per-fold training times and GPU memory utilization.

6 Results and Discussion

Our experimental evaluation demonstrates that RoBERTa-OTA consistently outperforms the RoBERTa baseline across all evaluation metrics. Table 4 presents the comprehensive performance analysis, while Table 5 examines computational costs. We discuss these results in relation to our three research questions.

RoBERTa-OTA achieves 96.04% accuracy compared to the RoBERTa baseline's 95.02%, representing a 1.02% point improvement. The F1-weighted score increases from 95.04% to 96.06%, while precision improves from 95.12% to 96.09%. We notice similar improvement across all hate speech categories. The consistency across metrics indicates genuine performance gains rather than optimization artifacts.

Our results establish new state-of-the-art performance for fine-grained hate speech detection. The previous best approach, SOSNet using BOW+XGBoost, achieved 94.38% accuracy and 94.44% F1-score on the same dataset. RoBERTa-OTA's 96.04% accuracy and 96.06% F1-score represent 1.66 and 1.62% points improvement over existing methods, demonstrating the effectiveness of combining transformer attention with structured ontological knowledge.

Table 4 reveals distinct performance patterns across hate speech categories. Religion, age, and ethnicity-based hate speech achieve high F1-scores (98.62%, 98.75%, and 98.44% respectively), reflecting their explicit targeting language and clear discriminative signals. Gender-based and other hate speech present greater classification challenges but benefit most from ontology-guided attention. Gender-based hate speech improves from 90.70% to 93.06% F1-score (+2.36% points), while other hate speech increases from 88.94% to 91.32% (+2.38% points). These substantial improvements occur precisely where implicit targeting strategies and coded language create ambiguity for standard transformer approaches.

Table 5 demonstrates that RoBERTa-OTA introduces modest architectural overhead. Parameter count increases minimally from 124.65M to 125.06M (0.33% overhead), while peak GPU memory usage increases from 2.6 GB to 3.1 GB. Model size increases by only 1.6MB, maintaining deployment feasibility across standard hardware configurations. Training time shows modest computational overhead, with average fold time increasing from 24.5 to 27.5 min (12.2% increase). However, RoBERTa-OTA demonstrates superior convergence efficiency, requiring fewer total epochs (29 vs 31 epochs across 5-fold cross-validation) due to the ontology-guided learning process. The FLOPs increase of 10.3% reflects additional processing for ontological integration. The performance improvements of 1.02% points in accuracy combined with the convergence efficiency make RoBERTa-OTA attractive for applications requiring accurate fine-grained hate speech detection.

While the overall accuracy improvement appears incremental (+1.02% points), this analysis misrepresents the practical value proposition of our approach. The computational overhead yields disproportionate improvements precisely where content moderation systems struggle most: gender-based hate speech detection improves by 2.36% points (90.70 → 93.06% F1) and other hate speech by 2.38% points (88.94 → 91.32% F1). These categories represent the most challenging detection scenarios in real-world deployment, where implicit targeting strategies deliberately evade standard detection methods. In production environments processing millions of messages daily, a 2.36% improvement in gender-based hate speech detection translates to thousands of correctly identified harmful posts that would otherwise remain undetected. Furthermore, our architectural efficiency analysis reveals favorable trade-offs beyond simple FLOPs comparison. While computational operations increase by 10.3%, parameter overhead remains minimal at only 0.33% (125.06M vs 124.65M parameters), maintaining deployment feasibility on standard hardware. Critically, RoBERTa-OTA demonstrates superior convergence efficiency, requiring fewer training epochs (29

Table 6. Model Robustness Under Social Media Text Perturbations

Perturbation Type	Level	RoBERTa Baseline		RoBERTa-OTA		Improvement	
		Acc (%)	F1 (%)	Acc (%)	F1 (%)	Δ Acc	Δ F1
Clean Text	–	97.50	97.49	98.53	98.53	+1.03	+1.04
Character Deletion	5%	92.53	92.56	94.13	94.19	+1.60	+1.63
	10%	85.50	85.45	88.47	88.67	+2.97	+3.22
	15%	76.73	76.17	79.23	79.64	+2.50	+3.47
Character Substitution	5%	92.23	92.25	93.83	93.92	+1.60	+1.67
	10%	86.27	86.25	87.47	87.84	+1.20	+1.59
	15%	79.33	79.02	80.97	81.77	+1.64	+2.75
Character Insertion	5%	93.37	93.37	95.37	95.40	+2.00	+2.03
	10%	87.93	87.93	90.73	90.90	+2.80	+2.97
	15%	81.30	81.20	85.47	85.82	+4.17	+4.62
Word Deletion	10%	94.20	94.22	94.37	94.45	+0.17	+0.23
	20%	91.20	91.29	92.33	92.50	+1.13	+1.21
	30%	86.97	87.10	87.43	87.76	+0.46	+0.66
Abbreviation Injection	10%	97.10	97.08	98.37	98.36	+1.27	+1.28
	20%	96.87	96.84	98.37	98.36	+1.50	+1.52
	30%	96.87	96.84	98.37	98.36	+1.50	+1.52
Slang Substitution	10%	97.53	97.52	98.53	98.53	+1.00	+1.01
	20%	97.53	97.52	98.53	98.53	+1.00	+1.01
	30%	97.53	97.52	98.53	98.53	+1.00	+1.01

vs 31 across 5-fold cross-validation) due to ontology-guided learning, effectively offsetting training time overhead through faster convergence. The 19.2% memory increase (3.1 GB vs 2.6 GB) remains within practical deployment constraints, while model size increases by only 1.6MB. For applications prioritizing accuracy on challenging hate speech categories over raw computational efficiency, these targeted performance gains justify the modest computational overhead.

6.1 Robustness Under Social Media Text Perturbations

Social media text contains misspellings, abbreviations, and slang that challenge hate speech detection systems. We evaluated model robustness using six perturbation types on 3,000 sampled tweets: character deletion, substitution, and insertion (5%, 10%, 15%), word deletion (10%, 20%, 30%), abbreviation injection (10%, 20%, 30%), and slang substitution (10%, 20%, 30%).

Table 6 shows RoBERTa-OTA consistently outperforms the baseline across all perturbations. Character-level noise causes the largest degradation for both models. At 15% character deletion, RoBERTa F1-score drops to 76.17% (21.32 point decrease), while RoBERTa-OTA maintains 79.64% (18.89 point decrease), demonstrating a 3.47% point F1 advantage. Character insertion reveals the most substantial robustness gap: RoBERTa-OTA achieves 85.82% F1 versus the baseline's 81.20% at 15% perturbation, yielding a 4.62% point F1 improvement. This enhanced resilience stems from ontological knowledge integration, which provides semantic grounding when surface forms become corrupted.

Both models exhibit remarkable stability under social media linguistic variations. At 30% abbreviation substitution, RoBERTa-OTA maintains 98.36% F1-score (only 0.17 point drop from clean text), while the baseline achieves 96.84% F1 (0.65 point drop). Slang substitution shows even greater resilience: both architectures maintain clean text F1 performance levels with minimal degradation across all intensity levels. The consistent F1-score advantages across perturbation types demonstrate that RoBERTa-OTA's robustness benefits extend uniformly across all five hate speech categories, validating its effectiveness for practical multiclass deployment in noisy social media environments.

7 Conclusion and Future Work

Our comprehensive evaluation of RoBERTa-OTA demonstrates significant advances in multiclass hate speech detection through ontology-guided attention mechanisms. The dual-stream architecture achieved 96.04% accuracy and 96.06% weighted F1-score, outperforming the RoBERTa baseline by 1.02% points in both metrics and surpassing the previous state-of-the-art SOSNet approach by 1.62 (F1) and 1.66 (accuracy) percentage points. The integration of scaled attention with enhanced Graph Convolutional Networks proved particularly effective for challenging categories, improving gender-based hate speech detection by 2.36% points and other hate speech by 2.38% points, while adding only 0.33% parameter overhead.

For future work, we plan to evaluate RoBERTa-OTA on multilingual datasets to assess its effectiveness across diverse cultural and linguistic contexts. We also aim to explore efficiency optimizations to reduce computational overhead without compromising accuracy. The strong performance of our ontology-guided approach motivates further investigation into advanced knowledge integration techniques, particularly for capturing evolving patterns in online hate speech.

References

1. Abusaqer, M., Saquer, J.: Multiclass hate speech detection: evaluating 303 model configurations across traditional machine learning, deep learning, and transformer approaches. In: 2025 International Conference on Machine Learning and Applications (ICMLA) (2025, to appear)
2. Abusaqer, M., Saquer, J., Shatnawi, H.: Efficient hate speech detection: evaluating 38 models from traditional methods to transformers. In: Proceedings of the 2025 ACM Southeast Conference, pp. 203–214. ACM, Cape Girardeau, MO, USA (2025). https://doi.org/10.1145/3696673.3723061

3. Agrawal, S., Awekar, A.: Deep learning for detecting cyberbullying across multiple social media platforms. In: European Conference on Information Retrieval, pp. 141–153. Springer (2018)
4. Badjatiya, P., Gupta, S., Gupta, M., Varma, V.: Deep learning for hate speech detection in tweets. In: Proceedings of the 26th International Conference on World Wide Web, pp. 759–760 (2017)
5. Burnap, P., Williams, M.L.: Cyber hate speech on twitter: an application of machine classification and statistical modeling for policy and decision making. Policy Internet **7**, 223–242 (2015)
6. Davidson, T., Warmsley, D., Macy, M., Weber, I.: Automated hate speech detection and the problem of offensive language. In: Proceedings of the 11th International AAAI Conference on Web and Social Media, ICWSM 2017, pp. 512–515 (2017)
7. Devlin, J., Chang, M.W., Lee, K., Toutanova, K.: BERT: pre-training of deep bidirectional transformers for language understanding. In: Proceedings of the 2019 Conference of the North American Chapter of the Association for Computational Linguistics: Human Language Technologies, vol. 1, pp. 4171–4186. Association for Computational Linguistics, Minneapolis, MN, USA (2019). https://doi.org/10.18653/v1/N19-1423
8. Gillespie, T.: Custodians of the Internet: Platforms, Content Moderation, and the Hidden Decisions That Shape Social Media. Yale University Press, New Haven, CT, USA (2018). https://doi.org/10.12987/9780300235029
9. Gitari, N.D., Zuping, Z., Damien, H., Long, J.: A lexicon-based approach for hate speech detection. Int. J. Multimedia Ubiquitous Eng. **10**(4), 215–230 (2015)
10. Liu, Y., et al.: RoBERTa: a robustly optimized BERT pretraining approach. In: International Conference on Learning Representations (2019)
11. Mathew, B., Saha, P., Yimam, S.M., Biemann, C., Goyal, P., Mukherjee, A.: HateXplain: a benchmark dataset for explainable hate speech detection. In: Proceedings of the AAAI Conference on Artificial Intelligence, vol. 35, pp. 14867–14875 (2021)
12. Mishra, P., Yannakoudakis, H., Shutova, E.: Abusive language detection with graph convolutional networks. In: Proceedings of the 2019 Conference of the North American Chapter of the Association for Computational Linguistics, pp. 2145–2150 (2019)
13. Ousidhoum, N., Lin, Z., Zhang, H., Song, Y., Yeung, D.Y.: Multilingual and multi-aspect hate speech analysis. In: Proceedings of the 2019 Conference on Empirical Methods in Natural Language Processing, pp. 4675–4684 (2019)
14. Wang, J., Fu, K., Lu, C.T.: SOSNet: a graph convolutional network approach to fine-grained cyberbullying detection. In: Proceedings of the 2020 IEEE International Conference on Big Data, pp. 1699–1708. IEEE (2020). https://doi.org/10.1109/BigData50022.2020.9378065
15. Waseem, Z., Hovy, D.: Hateful symbols or hateful people? predictive features for hate speech detection on twitter. In: Proceedings of the NAACL Student Research Workshop, pp. 88–93. Association for Computational Linguistics, San Diego, CA, USA (2016)
16. Yin, W., Agarwal, V., Jiang, A., Zubiaga, A., Sastry, N.: AnnoBERT: effectively representing multiple annotators' label choices to improve hate speech detection. In: Proceedings of the International AAAI Conference on Web and Social Media, vol. 17, pp. 902–913 (2023)

GRAFT: Graph-of-Thoughts-Based Reasoning and Augmentation for Fine-Tuning Text-to-SQL

Raghu Katikeri(✉), Gaurav Mamgain, Himanshu Nakrani, Rajvi Desai, Tarun Kasula, and Amit Vaid

Artificial Intelligence, Emerging Tech CoE, State Street Corporation, Hyderabad, India

{rkatikeri,gmamgain,hnakrani,dbirenbhai,ktarun,avaid}@statestreet.com

Abstract. This paper presents **GRAFT**, a comprehensive framework to optimize Text-to-SQL (T2SQL) performance via fine-tuning, PEFT and reasoning mechanisms. GRAFT addresses complex database schemas and SQL generation by integrating a training data generation pipeline and targeted fine-tuning. It decomposes Natural Language Queries (NLQs) and SQL statements into sub-questions and sub-queries, forming dynamic graph structures. These sub-graphs enable synthetic sample generation to refine LLMs/SLMs using $\langle \text{NLQ}, \text{DB Schema}, \text{GoT}, \text{SQL} \rangle$ inputs. Evaluated on Spider 1.0/2.0 datasets, GRAFT demonstrates significant accuracy and generalization gains, highlighting graph-based decomposition's transformative role in T2SQL systems.

Keywords: Graph of Thoughts · PEFT · LoRA

1 Introduction

Translating Natural Language Queries (NLQs) into executable SQL statements, known as Text-to-SQL (T2SQL), is vital for democratizing data access in enterprises. As databases grow in complexity and scale, conventional T2SQL models struggle to generalize across intricate schemas and composite queries [3]. These challenges hinder real-time query generation and limit practical adoption in dynamic environments. To address this, we introduce **GRAFT**, a novel framework leveraging the Graph-of-Thoughts (GoT) paradigm [1,2,11] for data augmentation, logical reasoning, and fine-tuning. GRAFT decomposes NLQs and SQL statements into structured sub-questions and sub-queries, forming dynamic graph-based representations. This enables automated generation of diverse synthetic training samples, refining models using $\langle \text{NLQ}, \text{DB Schema}, \text{GoT}, \text{SQL} \rangle$ inputs. Finetuning is carried out on GPT 4.1 mini, GPT-4o-mini models. Also, to reduce training time, latency and cost adaptive learning approach is used via LoRA a PEFT technique using qwen2.5 coder. Evaluations on Spider 1.0 [12] and 2.0 [7] benchmarks show notable accuracy improvements over baseline T2SQL models and our previous framework, T2I-Rise [5].

H. R. Arabnia et al. (Eds.): CSCI 2025, CCIS 2967, pp. 194–208, 2027.
https://doi.org/10.1007/978-3-032-28454-9_13

2 Methodology

GRAFT is based on the principle that complex queries can be more effectively understood and generated by decomposing them into smaller, logically coherent sub-components. This approach involves several stages, with a focus on incorporating fine-tuning by utilizing these decomposed sub-components:

2.1 Graph of Thoughts (GoT) Decomposition

To begin, each Natural Language Query (NLQ) and SQL statement are decomposed into sub-questions and sub-queries, forming nodes in a directed graph. This triplet $\langle \text{NLQ}, \text{SQL}, \text{GoT} \rangle$ preserves interdependencies [1], enabling accurate candidate generation, scoring, and aggregation of results.

2.2 Aggregation and Synthesis

In the final phase, validated sub-NLQ and sub-SQL nodes are aggregated using the GoT graph, which preserves semantic and logical connections [1]. This process integrates sub-components into a unified NLQ-SQL pair, ensuring the final SQL accurately reflects the query intent and supports iterative refinement and validation. Example:

- **Original NLQ:** How many US B2 patents granted between 2008 and 2018 have at least one claim that does not include the word 'claim'?
- **Original SQL:**

```
WITH patents_sample AS (
  SELECT t1."publication_number" AS publication_number,
claim.value:"text" AS claims_text
  FROM PATENTS.PATENTS.PUBLICATIONS t1,
    LATERAL FLATTEN(input => t1."claims_localized") AS claim
  WHERE t1."country_code" = 'US'
    AND t1."grant_date" BETWEEN 20080101 AND 20181231
    AND t1."grant_date" != 0
    AND t1."publication_number" LIKE '%B2%'
),
Publication_data AS (
  SELECT publication_number, COUNT_IF(claims_text NOT LIKE
'%claim%') AS nb_indep_claims
  FROM patents_sample GROUP BY publication_number
)
SELECT COUNT(nb_indep_claims) FROM Publication_data WHERE
nb_indep_claims != 0;
```

- **Pair 1**
  - **NLQ:** How many US B2 patents were granted between 2008 and 2018?
  - **SQL:**

```
SELECT COUNT(*) FROM PATENTS.PATENTS.PUBLICATIONS t1
WHERE t1."country_code" = 'US'
AND t1."grant_date" BETWEEN 20080101 AND 20181231
AND t1."grant_date" != 0
AND t1."publication_number" LIKE '%B2%';
```

  - **Steps:**
    1. Isolate the condition on US B2 patents granted between 2008 and 2018.
    2. Filter the publications based on country, grant date and publication number format.
    3. Aggregate (count) the number of such patents.

- **Pair 2**
  - **NLQ:** For each US B2 patent granted between 2008 and 2018, how many claims do not include the word 'claim'?
  - **SQL:**

```
SELECT t1."publication_number" AS publication_number,
COUNT_IF(claim.value:"text" NOT LIKE '%claim%') AS
nb_indep_claims
FROM PATENTS.PATENTS.PUBLICATIONS t1,
LATERAL FLATTEN(input => t1."claims_localized" AS claim)
WHERE t1."country_code" = 'US'
AND t1."grant_date" BETWEEN 20080101 AND 20181231
AND t1."grant_date" != 0
AND t1."publication_number" LIKE '%B2%'
GROUP BY t1."publication_number";
```

  - **Steps:**
    1. For the filtered US B2 patents from 2008 to 2018, extract each claim.
    2. For each claim, check if it lacks the word 'claim'.
    3. Count the claims per publication (patent) that do not contain the word 'claim'.

- **Pair 3**
  - **NLQ:** How many US B2 patents granted between 2008 and 2018 have at least one claim not including the word 'claim'?
  - **SQL:**

```
WITH patents_sample AS (
  SELECT t1."publication_number" AS publication_number,
    claim.value:"text" AS claims_text
  FROM PATENTS.PATENTS.PUBLICATIONS t1,
    LATERAL FLATTEN(input => t1."claims_localized") AS claim
  WHERE t1."country_code" = 'US'
    AND t1."grant_date" BETWEEN 20080101 AND 20181231
    AND t1."grant_date" != 0
    AND t1."publication_number" LIKE '%B2%'
),
Publication_data AS (
  SELECT publication_number,
    COUNT_IF(claims_text NOT LIKE '%claim%') AS
nb_indep_claims
  FROM patents_sample
  GROUP BY publication_number
)
SELECT COUNT(nb_indep_claims)
FROM Publication_data
WHERE nb_indep_claims != 0;
```

- **Steps:**
  1. Obtain all US B2 patents granted between 2008 and 2018.
  2. For each, extract every claim's text.
  3. For each patent, count claims where the text does not include ‘claim’.
  4. Identify patents with at least one such claim.
  5. Count patents matching this criterion.

- **Aggregated NLQ:** How many US B2 patents granted between 2008 and 2018 have at least one claim that does not include the word ‘claim’?
- **Aggregated SQL:**

```
WITH patents_sample AS (
  SELECT t1."publication_number" AS publication_number,
claim.value:"text" AS claims_text
  FROM PATENTS.PATENTS.PUBLICATIONS t1,
    LATERAL FLATTEN(input => t1."claims_localized") AS claim
  WHERE t1."country_code" = 'US'
    AND t1."grant_date" BETWEEN 20080101 AND 20181231
    AND t1."grant_date" != 0
    AND t1."publication_number" LIKE '%B2%'
),
Publication_data AS (
  SELECT publication_number, COUNT_IF(claims_text NOT LIKE
'%claim%') AS nb_indep_claims
  FROM patents_sample GROUP BY publication_number
)
```

```
SELECT COUNT(nb_indep_claims) FROM Publication_data WHERE
nb_indep_claims != 0;
```

- **Steps:**
  1. Filter to US B2 patents granted between 2008 and 2018 by country, grant date and publication format.
  2. Expand each patent's claims for further analysis.
  3. For each claim, check if it does not contain the word 'claim'.
  4. Aggregate the count of such claims per patent.
  5. Identify patents with at least one claim that does not include the word 'claim'.
  6. Count the number of patents meeting this criterion.
- **Analysis:** The core objective of this query is to determine how many US B2 patents, granted in the window between 2008 and 2018, contain at least one claim that does not include the word 'claim'. The relevant entities are patents, claim texts, publication number, country code and grant dates. The filtering criteria ensure that only US patents designated as 'B2' and issued within the specified date range are considered. The operation involves flattening the claims within each patent, checking each one for the absence of the substring 'claim' and then aggregating the results back to the patent (publication) level. The final operation is a count of patents with at least one such claim.

Example Sample for Chat Completion Dataset

```
{
  ''{messages}": [
    {
      ''role": ''system",
      ''content": ''You are an AI assistant specialized in
         generating SQL queries based on given database
         schemas and natural language questions. Your goal
         is to understand the schema, the question, and the
         provided steps to generate accurate and efficient
         SQL queries."
    },
    {
      ''role": ''user",
      ''content": ''DBSchema:<Db-Schema>"
    },
    {
      ''role": ''assistant",
      ''content": ''Reasoning Steps: <Reasoning steps>, SQL:
         <SQL -Query >"
    }
  ]
}
```

2.3 Synthetic Data Generation and Data Preparation

The original NLQ-SQL pairs are decomposed into sub-questions and sub-queries, each paired with their respective Graphs of Thoughts (GoTs). Each sub-graph represents a semantic element (e.g., filtering, joining, aggregation) and is associated with both a sub-NLQ and sub-SQL, forming a semantic reasoning triplet $\mathcal{T} = \langle Q, S, G \rangle$, where $Q \in \mathcal{Q}$ (NLQ), $S \in \mathcal{S}$ (SQL), and $G \in \mathcal{G}$ (GoT). Synthetic datasets were constructed for Spider 1.0 and 2.0 [7,12] to enhance training diversity. For each original pair, the GoT method generated multiple branches, with top-ranked samples retained, resulting in 210,000 synthetic samples for Spider 1.0 and 2,700 for Spider 2.0. This augmentation exposes models to varied linguistic and SQL structures, improving generalization, robustness, and adaptability in Text-to-SQL tasks.

3 Architecture

3.1 Graph Query Decomposer

The role of the Graph Query Decomposer is to break down a complex Natural Language Query (NLQ) into a structured graph of operations [1], as illustrated in Fig. 1. The Graph of Operation outlines the process of generating sub-NLQs and sub-SQLs [10] by dividing the reasoning paths according to the number of branches mentioned which we refer as graph query decomposer i.e. it decomposes the NLq, SQL pairs into sub-NLQ, sub-SQL pairs. along with their GoT. This workflow includes iterative refinement and ongoing generation of thoughts to achieve the most relevant context, reasoning, and logical flow. The Graph Query Decomposer centers on identifying the distinct components and operations necessary to address the user's query. Interpretation and mapping of the NLQ to precise graph operations are facilitated by an LLM Prompter utilizing GPT 4.1 [8,9], ensuring that all necessary logical steps and dependencies are thoroughly addressed.

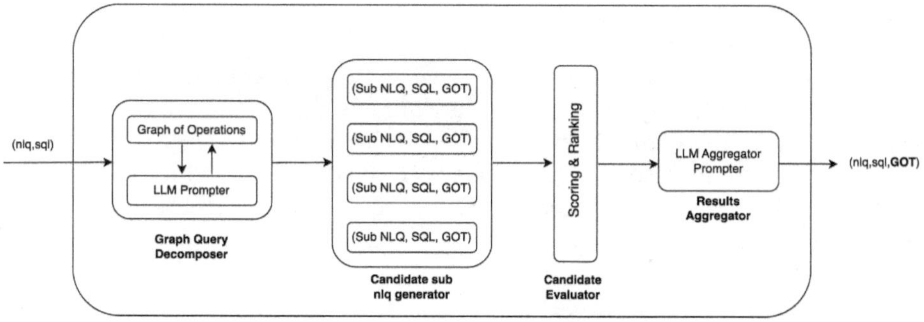

Fig. 1. GRAFT Training Data Generation Pipeline Building Blocks

3.2 Candidate Sub NLQ Generator

Once the initial decomposition is complete, the system generated multiple candidate sub-queries along with final thoughts. These candidates consist of sub-NLQs [10], their corresponding SQL statements and GOTs (Graph of Thoughts) [1]. This method examines multiple potential ways to translate the decomposed operations into executable formats, thereby improves the likelihood of finding the most precise and pertinent sub-NLQs and SQL.

3.3 Graph Query Generator

Graph Query Generator will extrapolate the $\langle \mathcal{Q}, \mathcal{S} \rangle$ pairs with different NLQs and SQLs as $\mathcal{N}$ number of branches. It then ranks these pairs and selects the top $\mathcal{K}$ candidates based on their scores. Additionally, the Graph Query Generator is responsible to break down a complex Natural Language Query $\mathcal{Q}$ into a structured graph of operations [1], as illustrated in Fig. 1.

3.4 Candidate Evaluator

The candidate evaluator evaluates each sub-NLQ and sub-SQL candidate based on predefined criteria, based on cosine similarity score ranges from $[-1, 1]$ between semantic representations obtained from the prompter. Score of 0.90 is kept as threshold to rank the candidates, higher the score higher the rank, identifying those with the greatest potential [1]. Through Candidate Evaluator, the most accurate and relevant sub-queries are selected for final aggregation, which enhances the overall quality of the system's outputs.

3.5 LLM Aggregator Prompter

The curated sub-NLQs and sub-SQLs, along with their corresponding Graphs of Thoughts (GoT), are utilized for fine-tuning Large Language Models (LLMs) or Small Language Models (SLMs) using the quadruple $\langle \mathcal{Q}, \mathcal{D}, \mathcal{S}, \mathcal{G} \rangle$. In this approach, the Graph of Thoughts obtained from the sub-NLQs and sub-SQLs is employed to generate an aggregated GoT for the original NLQ and SQL, along with their associated sub-queries, using the final component - the LLM Aggregator Prompter [4] as shown in the figure as final component Fig. 1. This helps us to have only highly relevant GoTs which is similar to original NLQ, SQL pair based on the same scoring technique used in candidate evaluator. Thus, the resulting evaluation set is composed of quadruples $\langle \mathcal{Q}, \mathcal{D}, \mathcal{S}, \mathcal{G} \rangle$ which form the basis for systematic evaluation.

3.6 Fine-Tuning Text-to-SQL Models

After generating synthetic data for both Spider 1.0 and Spider 2.0 datasets, the resulting dataset $\mathcal{D} = \{\langle Q_i, S_i, G_i \rangle \mid i = 1, \ldots, N\}$ is divided into training, validation and test sets with a ratio of $0.7 : 0.2 : 0.1$, respectively. The training sets from both Spider 1.0 [12] and Spider 2.0 [7] are then used to fine-tune two models: GPT-4.1-mini and GPT-4o-mini, each comprising approximately 7–8 billion parameters (~7–8B) by incorporating the approach as shown in the Fig. 2.

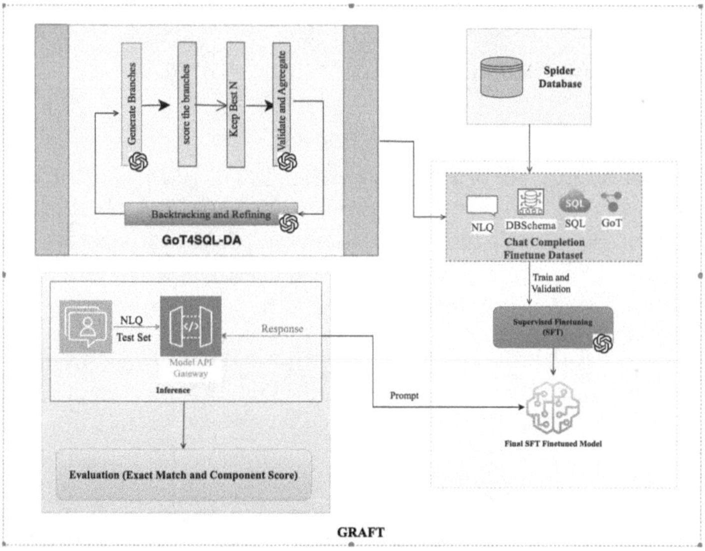

Fig. 2. GRAFT Fine-Tuning Architecture

4 Training Settings

We conducted experiments under two distinct paradigms: Full Finetuning and LoRA-based adaptation.

Full Finetuning

- *With GoT:* Each training instance is represented as $\langle Q, D, S, G \rangle$, where Q is the natural language query, D the database schema, S the SQL statement, and G the generated output trace. Training utilized the GPT-4.1 Mini and GPT-4o Mini models, with approximately 10K and 50 K samples samples for Spider 1.0 and 2,672 for Spider 2.0. The data splits were 9,880:2,800:2,800 (train:val:test) and (50K:1K:1K) for Spider 1.0 and 2,138:267:267 for Spider 2.0, the latter containing more complex NLQ:SQL pairs.
- *Without GoT:* Instances omit the output trace, i.e., $\langle Q, D, S \rangle$. Training was performed on GPT-4.1 Mini with similar sample sizes and splits (Table 1).

Table 1. Hyper-parameters for Spider 1.0 and Spider 2.0 GPT Mini series

Hyper-parameters	Spider 1.0	Spider 2.0
Batch size (B)	16	8
Learning rate (η)	0.02	0.03
Epochs (E)	3	3
Steps per epoch (S_{epoch})	600	268
Total training steps (S_{total})	1800	802

LoRA Finetuning (QWEN2.5-Coder 8B): *With GoT:* LoRA configuration Table 2 used $r = 8$, $\alpha = 32$, dropout $= 0.05$ for 10K samples, followed by $r = 16$, $\alpha = 32$, dropout $= 0.05$ for 50K samples. Batch size $B = 8$, learning rate as per standard LoRA protocol.

Table 2. LoRA Configuration (With GoT)

Samples	r	α	Dropout	Batch Size	Learning Rate
10,000	8	32	0.05	8	1e−4
50,000	16	32	0.05	8	1e−4

4.1 Effect Finetuning with GoT on Model Performance

Table 3 presents the average training and validation losses for `gpt-4o-Mini` and `gpt-4.1-Mini` on SPIDER 1.0 and 2.0 datasets [7,12]. Models trained without GoT show extremely low training loss but high validation loss, indicating overfitting. Applying GoT and finetuning yields lower and more balanced validation losses, demonstrating improved generalization. For example, `gpt-4.1-Mini` with GoT achieves a validation loss of 0.231 on SPIDER 1.0 and 0.429 on SPIDER 2.0, compared to 0.593 and 0.466 without GoT, respectively. Thus, GoT and finetuning enhance model robustness.

Table 3. Average Training and Validation Loss for GPT Mini Series on SPIDER 1.0 and SPIDER 2.0 datasets (Final steps: 1838 for SPIDER 1.0, 802 for SPIDER 2.0)

Dataset	Steps	Model	Without GoT		With GoT	
			Train	Valid	Train	Valid
SPIDER 1.0	1800	gpt-4o-Mini	0.000007	0.473617	0.215527	0.266659
SPIDER 1.0	1800	gpt-4.1-Mini	0.000002	0.593179	0.185185	0.231182
SPIDER 2.0	802	gpt-4o-Mini	0.610063	0.516731	0.416347	0.479469
SPIDER 2.0	802	gpt-4.1-Mini	0.537903	0.466478	0.380955	0.428951

4.2 Train and Valid Loss Across GPT-4.1 Mini and GPT-4o Mini

Train and validation loss plots visualize overfitting and finetuning effectiveness for GPT-4.1 Mini and GPT-4o Mini, with and without GoT.

4.3 Evaluation Metrics

Comparison Score(Exact Match Accuracy). Percentage of cases where the generated SQL exactly matches the ground truth SQL, including syntax of the SQL, Columns selection and clause conditioning and ordering.

Component Score. A granular accuracy metric evaluating individual SQL components (e.g., WHERE clauses, JOIN conditions), as formalized in prior work like EVOLVE [6] (Figs. 3, 4, 5, 6 and Tables 4, 5).

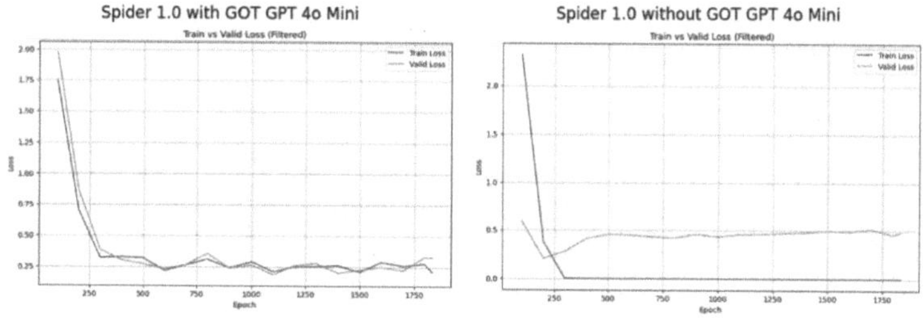

Fig. 3. Spider 1.0 GPT-4o Mini performance with and without GoT.

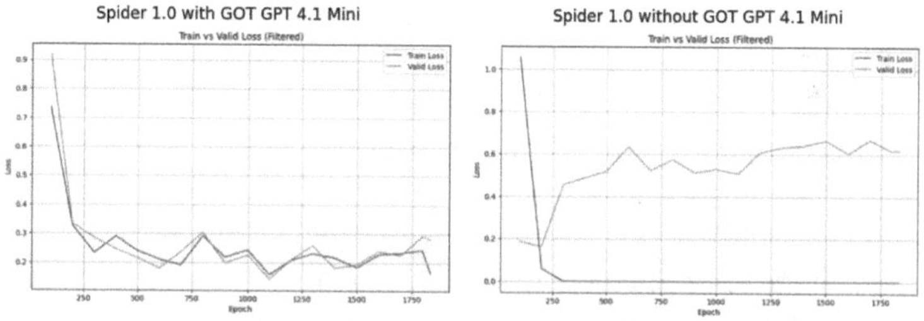

Fig. 4. Spider 1.0 GPT 4.1 Mini with/without GoT

Table 4. Evaluation Results on SPIDER 1.0 dataset

Model	Without GoT	With GoT+ Finetuning
gpt-4o-Mini	81.86	94.33
gpt-4.1-Mini	76.23	98.92

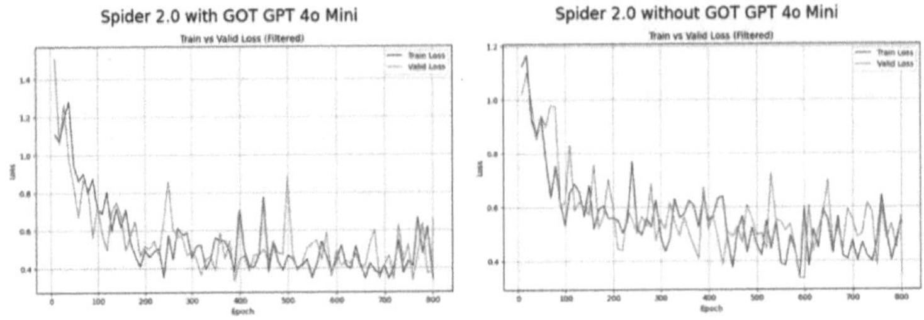

Fig. 5. Spider 2.0 GPT 4o Mini with/without GoT

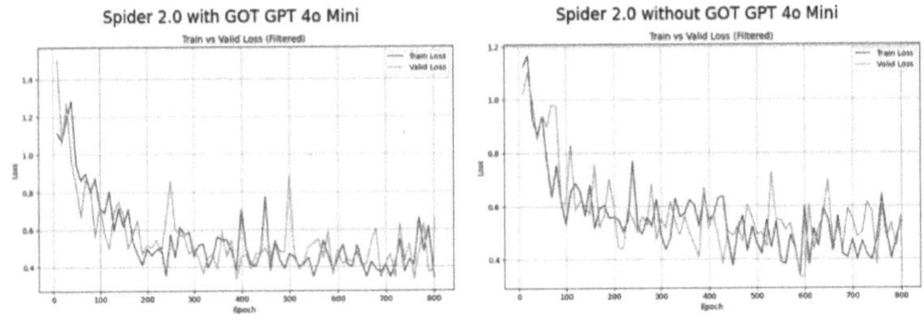

Fig. 6. Spider 2.0 GPT 4.1 Mini with/without GoT

Table 5. Evaluation Results on SPIDER 2.0 dataset

Model	Without GoT	With GoT+ Finetuning
gpt-4o-Mini	65.07	69.03
gpt-4.1-Mini	70.02	76.03

5 Addressing Catastrophic Forgetting Through Parameter-Efficient Fine-Tuning: Why PEFT and LoRA Matter

Full finetuning of Large Language Models (LLMs) can cause *catastrophic forgetting*, where prior knowledge is lost. **Parameter-Efficient Finetuning** (PEFT) methods, such as **Low-Rank Adaptation** (LoRA), address this by updating only small adapter layers while keeping base weights frozen. LoRA decomposes the weight update for $W \in \mathbb{R}^{d\times k}$ into trainable matrices $A \in \mathbb{R}^{d\times r}$ and $B \in \mathbb{R}^{r\times k}$, with $r \ll \min(d, k)$: [13]

$$W \leftarrow W + \frac{\alpha}{r} \cdot (AB) \tag{1}$$

Here, r is the rank hyperparameter controlling the adapter size and expressivity, while α is a scaling factor that modulates the strength of the update. This approach enables efficient adaptation with minimal risk of forgetting. [13]

5.1 LoRA Train and Valid Loss Qwen2.5 Coder 8B Instruct

Qwen2.5-Coder was pre-trained on an extensive dataset over 5.5 trillion tokens, with a substantial share dedicated to code-specific data. We performed full fine-tuning and LoRA on Qwen2.5-coder on 10K samples and 50K samples with LoRA configuration as per Table 2 (Fig. 7).

5.2 Train and Validation Loss Qwen2.5-Coder LoRA

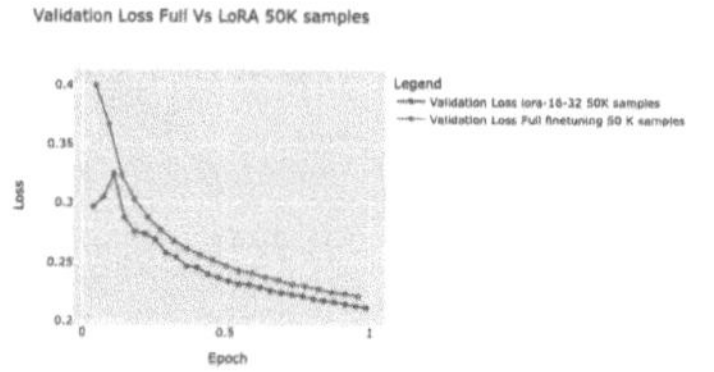

(a) Spider 1.0 LoRA with GoT 50K Samples.

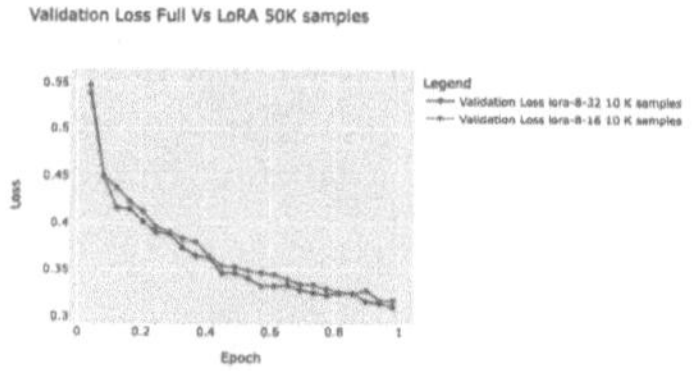

(b) Spider 1.0 LoRA with GoT 10K Samples.

Fig. 7. Spider 1.0 LoRA with GoT: Comparison of 50K and 10K Samples.

Table 6. Training, Validation Loss Comparison and Accuracy for LoRA and Full Finetuning (50K samples) with GoT

Configuration	Avg Train Loss	Avg Validation Loss
LoRA-16-32 (50K samples)	0.2147	0.2415
Full finetuning (50K samples)	0.2098	0.2414
Configuration	**Accuracy (Component Score)**	
LoRA-16-32 (50K samples)	93%	
Full finetuning (50K samples)	91.58%	

Key Findings

1. *SPIDER 1.0:*
   - GoT-based fine-tuning boosts accuracy for both Mini and standard GPT-4. Though having GoT in data augmentation increases number of tokens and slight increase in latency.
   - **gpt-4o-Mini**: 81.86% → 94.33%; **gpt-4.1-Mini**: 76.23% → 98.92%.
   - **LoRA** finetuning show similar **train** and **validation loss 0.2147, 0.2415** against **full finetuning** which is train **0.2098 and 0.2414**

   - **LoRA** shows significant reduction in **training time** to **60%** with early stopping against full finetuning maintaining **same** accuracy and losses.
   - **LoRA Test Accuracy** on Spider 1.0 is **93%** whereas **Full finetuning Test Accuracy** is **91.58%**
2. *SPIDER 2.0:*
   - GoT-based approach improves scores despite higher dataset complexity.
   - **gpt-4o-Mini**: 65.07% → 69.03%; **gpt-4.1-Mini**: 70.02% → 76.03%.
3. *Cost Estimation:*
   - **LoRA** significantly cut the cost by 98–99% against the full finetuning by reducing the trainable tokens size.
4. *Comparison to T2I-Rise:*
   - GoT-based method consistently matches or outperforms T2I-Rise [5], especially for Mini variants.
5. *Robustness:*
   - Gains are consistent across model sizes and datasets, showing strong generalizability.

Integrating GoT with fine-tuning delivers significant accuracy improvements and robustness over baseline methods (Tables 6 and 7).

Table 7. Compute and Cost Comparison: Full Finetuning vs. LoRA (Qwen2.5-Coder-7B, 50K samples)

Method	FLOPs	Per PetaFLOP Cost (USD)	Estimated Cost (USD)
Full Finetuning	3.58×10^{17}	$2.5	$895
LoRA Finetuning ($r = 16$)	7.17×10^{14}	$2.5	$1.80

6 Conclusion and Future Scope

We present a GoT-based strategy for augmenting NLQ-SQL pairs and fine-tuning GPT models, resulting in significant accuracy improvements on SPIDER 1.0 and 2.0. GoT decomposition and targeted finetuning enhance SQL correctness and interpretability compared to baseline approaches. For future work, we plan to deploy these fine-tuned GPTs within agentic frameworks for real-time adaptation, leveraging RFT (Reinforcement fine-tuning) and DPO (Direct Preference Tuning) to further improve SQL accuracy while preserving pretrained knowledge.

Acknowledgment. We extend our deepest gratitude to our leadership team at State Street Corporation for their support and funding of this research.

Appendix

The following Table 8 provides a list of abbreviations and their descriptions.

Table 8. List of Abbreviations

D	
DA	Data Augmentation
DPO	Direct Preference Tuning
G	
GOT	Graph of Thoughts
GPT	Generative Pre-Trained Transformers
GRAFT	Graph-of-Thought-Based Reasoning and Augmentation for Fine-Tuning Text-to-SQL
L	
LLM	Large Language Model
LoRA	Low Rank Adaptation
P	
PEFT	Parameter-Efficient Finetuning
R	
RFT	Reinforcement fine-tuning
S	
SLM	Small Language Model
SQL	Structured Query Language
T	
T2SQL	Text to SQL

References

1. Besta, M., et al.: Graph of thoughts: solving elaborate problems with large language models. In: https://arxiv.org/abs/2308.09687
2. Besta, M., et al.: Demystifying chains, trees, and graphs of thoughts (2024). https://arxiv.org/abs/2308.09687
3. Chopra, A., Azam, R.: Enhancing natural language query to SQL query generation through classification-based table selection. In: Communications in Computer and Information Science, pp. 152–165 (2024). https://doi.org/10.1007/978-3-031-62495-7_12
4. Gu, J., et al.: A survey on LLM-as-a-Judge (2024). https://arxiv.org/abs/2411.15594
5. Katikeri, R., et al.: T2I-RISE: text-to-insights with reinforcement learning, integration of semantic layers and enrichment - a comprehensive approach with conversational context and feedback systems. In: 2021 IEEE International Conference on Big Data (Big Data), pp. 1993–1998 (2021). https://doi.org/10.1109/bigdata62323.2024.10825844

6. Katikeri, R., et al.: EVOLVE: evaluation of language-to-SQL validity and effectiveness - a detailed review framework for complex text-to-SQL queries. In: 2024 IEEE International Conference on Big Data (BigData): 1999–2004 (2024). https://doi.org/10.1109/bigdata62323.2024.10825883
7. Lei, F., et al.: Spider 2.0: evaluating language models on real-world enterprise text-to-SQL workflows. https://arxiv.org/abs/2411.07763
8. OpenAI (2023) GPT-4 Technical Report. arXiv (Cornell University). https://doi.org/10.48550/arxiv.2303.08774
9. OpenAI (2025) Introducing GPT-4.1 in the API. In: Openai.com. https://openai.com/index/gpt-4-1/
10. Pourreza, M., et al.: CHASE-SQL: multi-path reasoning and preference optimized candidate selection in text-to-SQL. https://arxiv.org/abs/2410.01943
11. Wen, Y., Wang, Z., Sun, J.: MindMap: knowledge graph prompting sparks graph of thoughts in large language models, vol. 1, pp. 10370–10388 (2024)
12. Yu, T., et al.: Spider: A Large-Scale Human-Labeled Dataset for Complex and Cross-Domain Semantic Parsing and Text-to-SQL Task (2019). arXiv:1809.08887 [cs]
13. Hu, E.J., et al.: LoRA: low-rank adaptation of large language models arXiv:2106.09685

Leakage Period Localization Method for Water Distribution Network Based on Continuous-Time Dynamic Graph

Fan Zhao[1,2](✉), Sheng Zhang[1], Yu Chao[1], and Quanjiu Wang[2]

[1] Department of Information Science, Xi'an University of Technology, Xi'an, China
vcu@xaut.edu.cn

[2] State Key Laboratory of Eco-hydraulics in Northwest Arid Region, Xi'an University of Technology, Xi'an, China

Abstract. To address the inherent complexity of dynamic responses in Water Distribution Network (WDN) leakage events and the persistent challenge of imbalanced historical data distribution, this paper proposes a novel fully supervised Continuous-Time Dynamic Graphs (CTDG) for leakage period localization called CTDG-LeakLoc. The CTDG-LeakLoc adopts a dual-modal representation strategy, where sensor nodes and pipeline states (normal/leakage) are modeled as distinct entities within a heterogeneous graph structure, while a meta-path mechanism systematically captures both topological relationships and semantic dependencies. The framework comprises three key modules such as heterogeneous network representation learning, continuous dynamics modeling and event interaction pattern recognition. Firstly, by implementing Metapath2vec, we conduct meta-path-guided random walks to establish node neighborhoods and derive high-dimensional embeddings through Skip-gram optimization. Secondly, neural ordinary differential equations (ODEs) are employed to precisely characterize the temporal evolution of node states across continuous time domains. Finally, an LSTM-based decoder with sequential memory capacity predicts the probabilistic interactions between nodes and leakage events. Comprehensive evaluations on the LeakDB benchmark datasets (Net1 and Hanoi) validate the framework's superiority, achieving state-of-the-art performance with an average F1-score of 94.63% and maintaining an exceptionally low false alarm rate of 1.645%.

Keywords: WDN · Leakage · CTDG · ODE · LSTM

1 Introduction

The issue of water scarcity, exacerbated by climate change, population growth, and economic development, has become increasingly severe. It was estimated that 80% of the global population faces serious water security threats [1]. The study found that the annual global water loss due to leakage is as high as 126 billion cubic meters, with a cost/value of water loss reaching $39 billion [2]. Therefore, timely leakage detection and localization (LD&L) are crucial for daily leakage repairs, reducing water loss, and post-disaster rescue.

H. R. Arabnia et al. (Eds.): CSCI 2025, CCIS 2967, pp. 209–220, 2027.
https://doi.org/10.1007/978-3-032-28454-9_14

LD&L primarily consist of hardware-based methods and software-based methods [3]. Based on the degree of dependence on hydraulic models of pipe networks, software methods are primarily categorized into model-based methods [4] and data-driven methods [5–15]. The prerequisite for the successful implementation of model-based methods is an almost precise pipe hydraulic model, which demands high prior knowledge of the pipe network. Data-driven methods primarily encompass traditional methods [5–7] and AI methods [8–15]. The traditional methods lack robustness against non-stationary hydraulic noise and are highly sensitive to minor fluctuations in data. AI-based methods include two types: machine learning [8–9] and deep learning [10–15]. Machine learning methods primarily involve the artificial neural networks (ANN) [8] or support vector machines (SVM) [9]. Machine learning models typically require extensive preprocessing and feature engineering, which is often time-consuming. Furthermore, their performance in handling nonlinear and non-stationary signals is limited. Deep learning-based methods are primarily categorized into graph neural network (GNN) methods [10], prediction-based methods [11, 12] and classification-based methods [13–15]. Water Distribution Network (WDN) inherently exhibits graph topology, however, most existing GNN-based LD&L studies rely on static graph modeling [16, 17]. The key limitation of static graphs lies in their inability to capture temporal dynamics or evolutionary features, making it difficult to generate dynamic node representations—precisely where the time-dependent interactions between pipe junctions and leaks play a decisive role in system state analysis.

Currently, dynamic graphs remain unexplored in leakage detection research. Recent advances in dynamic graphs fall into two categories: Discrete-Time Dynamic Graphs (DTDGs) [18] and Continuous-Time Dynamic Graphs (CTDGs) [19], but their applications are largely confined to domains like social networks, focusing on homogeneous networks. In reality, WDNs are inherently Heterogeneous Information Networks (HINs), comprising multi-type nodes and complex relations, rendering traditional homogeneous models inadequate. Studies demonstrate that Meta-path-guided modeling effectively addresses semantic representation challenges in HINs [20, 21]. The core challenge of this work is how to design meta-path-driven heterogeneous random walk strategies to generate dynamic embeddings that jointly encode node-type heterogeneity and spatiotemporal structural correlations. To tackle this, we propose CTDG-LeakLoc, a Continuous-Time Dynamic Graph-based leakage localization model that formulates interactions between measurement points and leakage states as heterogeneous events. Specifically, it models measurement nodes and leakage states as two distinct node types, constructing a heterogeneous CTDG to predict future dynamic interactions (i.e., leakage occurrence) between them.

The main contributions of this paper are as follows:

1. We propose a continuous-time dynamic graph (CTDG) model for leakage period localization in WDNs, which is called CTDG-LeakLoc.
2. CTDG-LeakLoc employs ODEs to solve the continuous dynamic evolution of leakage states and LSTM decoder to predict the probability distribution of leakage occurrence.
3. Extensive experiments demonstrate the state-of-the-art performance, achieving F1-scores of 96.13% (Net1) and 93.12% (Hanoi) with remarkably low false alarm rates (1.07% and 2.22%, respectively).

2 Method

This paper adopts a new research approach, defining measurement points and pipeline network states (normal and leakage) as two different types of objects, and modeling whether a measurement point has leakage as an interaction event between the measurement point and the leakage state. As shown in Fig. 1, the framework of CTDG-LeakLoc includes the following steps: (1) Classify sensors and pipeline network states into different types of objects and construct a CTDG. (2) Use meta-paths to extract semantic and structural features of heterogeneous information networks. (3) Predict the structural context representation of nodes through meta-path guided heterogeneous random walks, combined with the Skip-gram model. (4) Model the continuous dynamic evolution process of node representations using ODEs. (5) Use an LSTM decoder to predict the interaction link probability between nodes and achieve leakage point location.

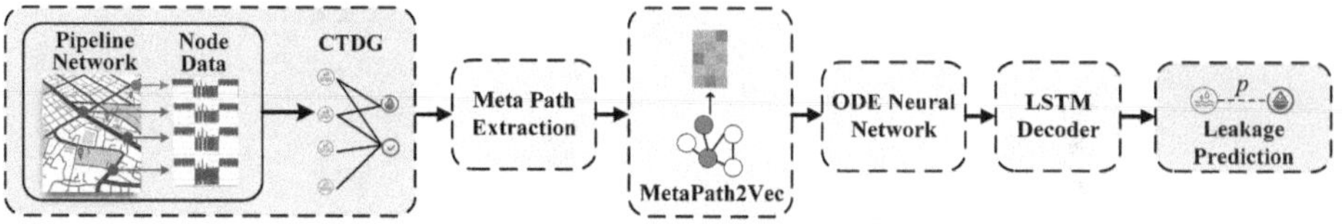

Fig. 1. The framework of CTDG-LeakLoc.

2.1 Preliminaries

A CTDG is formally defined as $G = (\mathcal{V}, \mathcal{E}, \mathcal{T}, \phi, \varphi)$, where $\mathcal{V}$ denotes the set of all nodes in the graph, $\mathscr{E}$ represents the set of edges connecting the nodes, $\mathcal{T}$ is the set of timestamps marking the occurrence of events. And $\phi : \mathcal{V} \rightarrow \mathcal{A}$ and $\varphi : \mathscr{E} \rightarrow \mathscr{R}$ are two mapping functions that assign node types from set $\mathcal{A}$ and edge types from set $\mathscr{R}$, respectively, with the constraint $|A| + |R| > 2$. Given a CTDG $G = (\mathcal{V}, \mathcal{E}, \mathcal{T}, \phi, \varphi)$, a meta-path $\mathscr{M}$ can be expressed as $a_1 \xrightarrow{r_1} a_2 \xrightarrow{r_2} \cdots \xrightarrow{r_{i-1}} a_i \xrightarrow{r_i} \cdots$, where $a_i \in \mathcal{A}$ and $r_i \in \mathscr{R}$. A node sequence $(v_1, v_2, \ldots, v_i)$ is considered a meta-path instance of $\mathscr{M}$ if it adheres to the structural constraints imposed by the meta-path.

2.2 CTDG Construction and Meta-Path Definition

As shown in Fig. 2, at the initial time t_1, the signals of the four pressure sensors $\{v_1, v_2, v_3, v_4\}$ are all in a normal monitoring state. When a leakage event occurs (during the period from t_2 to t_3), all sensor signals show a significant pressure drop, indicating that the system has entered a leakage state. This phenomenon is due to the pressure coupling effect in the pipe network system because of the transmission characteristics of pressure waves in the pipeline network, and all sensors in the same subnet will synchronously exhibit similar pressure waveform characteristics at time t_2. Based on the above physical mechanism, this paper constructs a CTDG-LeakLoc that characterizes the relationship between pipe network nodes and leakage states. Its formal definition is $G = (\mathcal{V}, \mathcal{E}, \mathcal{T}, \phi, \varphi)$, and its topological structure is defined as follows. The node set $\mathcal{V} = \{v, S\}$ includes: ① sensor nodes $v = \{v_1, \cdots, v_n, \cdots, v_N\}$, whose attributes include sensor number (measurement point identifier) and real-time pressure value. ② state nodes $S = \{S_1, S_2\}$, representing the binary state of the measurement point (normal = 0, leakage = 1). The edge set E contains two types of temporal interactions (v, S, t) and $(v, S, t + \Delta t)$, where (v, S, t) is an observation edge, reflecting the state observation probability $P(S|v)$ of sensor v at time t, and $(v, S, t + \Delta t)$ is a propagation edge, describing the physical impact of state S on sensor v after a delay of Δt. The node type set $A = \{sensor, state\}$, so $|A| = 2$. The edge type set $R = \{observation, propagation\}$, so $|R| = 2$.

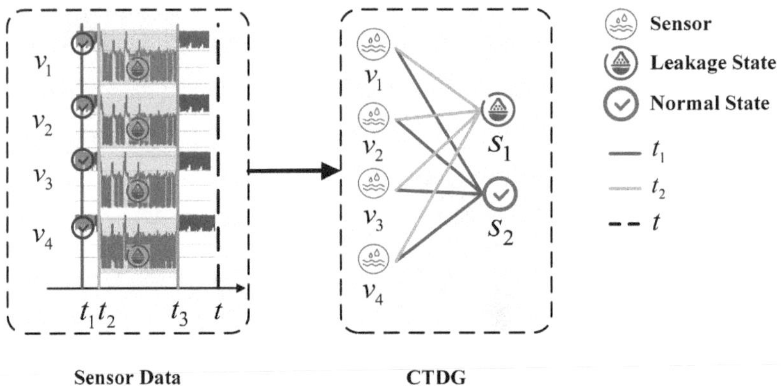

Fig. 2. CTDG construction.

Meta-Path Definition. In the CTDG, a meta-path is defined as a patterned path that describes the relationships between different types of nodes. This paper defines two types of meta-path patterns with clear physical interpretations: ① Direct association path $N \to S \to N$: It represents the immediate observation-feedback mechanism between sensor nodes and state nodes, corresponding to the direct causal relationship between pressure signals and leakage states in the pipeline network system. ② Indirect propagation path $N \to S \to S \to N$: It depicts how the temporal dependencies between state variables affect the sensor network through multi-step propagation, reflecting the cascading effect of pressure waves in the pipeline network. Typical meta-path instances can be specified as follows: A. $v_1 \to S_1 \to v_2$ (Direct observation conduction), B. $v_3 \to S_1 \to S_2 \to v_4$ (State transfer propagation). The direction of the arrows strictly follows the temporal causal relationship ($t \to t + \Delta t$) in the dynamic graph model. This temporal constraint ensures the physical interpretability of the meta-path instances.

2.3 Graph Node Embedding Representation

Metapath2vec [20] is an extensible representation learning model capable of simultaneously modeling the structure and semantic relevance of heterogeneous networks. Firstly, based on the type constraints of meta-paths, we randomly select neighbor nodes that meet the conditions at each node as the next node to generate a series of random paths. Then, according to the random walk strategy, the next node to be visited in the meta-path is probabilistically selected until the preset path length l is reached. The selection probability of the i-th step of the random walk is as follows: Given the graph network G and the meta-path definition $\mathcal{M}$, starting from the current node v_i^c, the next node v_{i+1} is selected according to the probability $p(v_{i+1}|v_i^c, \mathcal{M})$. This probability depends on the link relationship and node type between nodes, and the random walk process follows the predefined node type sequence. If v_{i+1} is connected to v_i^c and its type conforms to the next type defined in the meta-path (i.e., $\phi(v_{i+1}) = c + 1$), the selection probability is $|A_{c+1}(v_i^c)|^{-1}$, where $A_{c+1}(v_i^c)$ represents the set of nodes of type $c + 1$ in the neighborhood of node v_i^c. However, this random walk probability selection method only considers the node type and does not take into account the time difference between nodes. In this paper, we introduce a time influence factor into the Metapath2vec model to enhance the embedding representation ability of heterogeneous graph nodes. Given two nodes v_i and v_{i+1} with timestamps t_{v_i} and $t_{v_{i+1}}$, respectively, the time weight factor between these two nodes is defined as: $w(v_{i+1}, v_i) = e^{-\lambda|t_{v_i} - t_{v_{i+1}}|}$, where λ is a positive decay rate parameter. The probability transition is adjusted as follows.

$$p(v_{i+1}|v_i^c, \mathcal{M}) = \begin{cases} \frac{w(v_{i+1}, v_i^c)}{\sum\limits_{v_j \in A_{c+1}(v_i^c)} w(v_j, v_i^c)} & (v_{i+1}, v_i^c) \in \mathcal{E}, \phi(v_{i+1}) = c + 1 \\ 0 & (v_{i+1}, v_i^c) \in \mathcal{E}, \phi(v_{i+1}) \neq c + 1 \\ 0 & (v_{i+1}, v_i^c) \notin \mathcal{E} \end{cases} \quad (1)$$

After obtaining the meta-paths through random walks, to model the heterogeneous neighborhoods of nodes, Skip-gram [22] is introduced to learn node representations, with the goal of maximizing the co-occurrence probability of nodes and their context neighbors. Specifically, given node v, the probability of the appearance of its context nodes $A_c(v)$ is maximized as:

$$P(A_c(v)) = \arg\max_{\theta} \sum_{v \in V} \sum_{c \in \mathcal{A}} \sum_{a_c \in A_c(v)} \log p(a_c|v, \theta) \quad (2)$$

Where $A_c(v)$ represents the set of all context nodes of type c of node v, $\log p(a_c| v, \theta)$ represents the logarithm of the probability of the appearance of a context node of type c given node v, and θ is the model parameter to be optimized. Finally, the trained MetaPath2Vec model is used to extract the embedding of node v_i in the pipeline network.

$$X_{v_i} = MetaPath2Vec(G, \mathcal{M}, w, l) \quad (3)$$

2.4 Node Dynamic Modeling and Leakage Location

WDNs can be reasonably regarded as continuous-time systems, with the states of their nodes constantly undergoing dynamic changes over time. Neural ODEs can more accurately depict the evolution of node states [23], this paper utilizes neural ODEs to characterize the continuous dynamic changes (i.e., dynamic features) of node representations on dynamic graphs. Specifically, it parameterizes the derivative of the hidden state through neural networks as follows.

$$\frac{dz(t)}{dt} = f_{ode}(z(t), t) \tag{4}$$

Where $z(t)$ represents the hidden state at time t, and f_{ode} is the function of the derivative of the state with respect to time. By integrating, the time dynamic representation of the nodes can be obtained.

$$\mathbf{z}(t) = \mathbf{z}(t_0) + \int_{t_0}^{t} f_{ode}(z(t), t)dt \tag{5}$$

Given the embedding vector $X_{v_i}^t$ of node v_i at time t obtained by MetaPath2Vec, its dynamic representation $z_{v_i}^t$ can be expressed as follows.

$$z_{v_i}^t = ODESolver\left(f_{ode}, X_{v_i}^t, \Delta t\right) \tag{6}$$

Here, f_{ode} can be fitted using a Multi-Layer Perceptron (MLP) in practical applications.

The states of nodes v_i^t and v_j^t at continuous time are denoted as $z_{v_i}^t$ and $z_{v_j}^t$ respectively. To predict the link probability between these two nodes, we concatenate their states to obtain a new vector $\left[z_{v_i}^t \| z_{v_j}^t\right]$. Subsequently, a Long Short-Term Memory (LSTM) network is employed as a decoder to process this vector, thereby calculating the probability $p((v_i, v_j)| t)$ of interaction between the two nodes at time t. To determine whether there is a link between these two nodes at time t, we set a threshold th and if $p((v_i, v_j)| t) > th$, it is judged that there is a link between these two nodes; otherwise, it is considered that there is no link between them. Based on this, by analyzing the link status between the sensor and the leakage at time t, we can determine whether the pipeline network has leaked.

The loss function is defined as follows.

$$L_{predict} = -\sum_{v \in V, t \in T} y_{(v_i, v_j, t)} \log p\left((v_i, v_j)|t\right) + \left(1 - y_{(v_i, v_j)}\right) \log\left(1 - p\left((v_i, v_j)|t\right)\right) \tag{7}$$

Here $L_{predict}$ is the prediction loss, and $y_{(v_i, v_j, t)}$ is the label of whether there is a link between nodes v_i^t and v_j^t at time t.

3 Experimental Results and Analysis

3.1 Experimental Data and Parameters

Experimental Data. LeakDB, proposed by Vrachimis et al. in 2018 [24], serves as a benchmark dataset for training and testing in leak diagnosis, which provides 1000 scenarios and comprises two pipeline network systems, namely Net1 and Hanoi. As shown in Fig. 3(a), Net1's water distribution network consists of 11 nodes, 13 pipes, a reservoir,

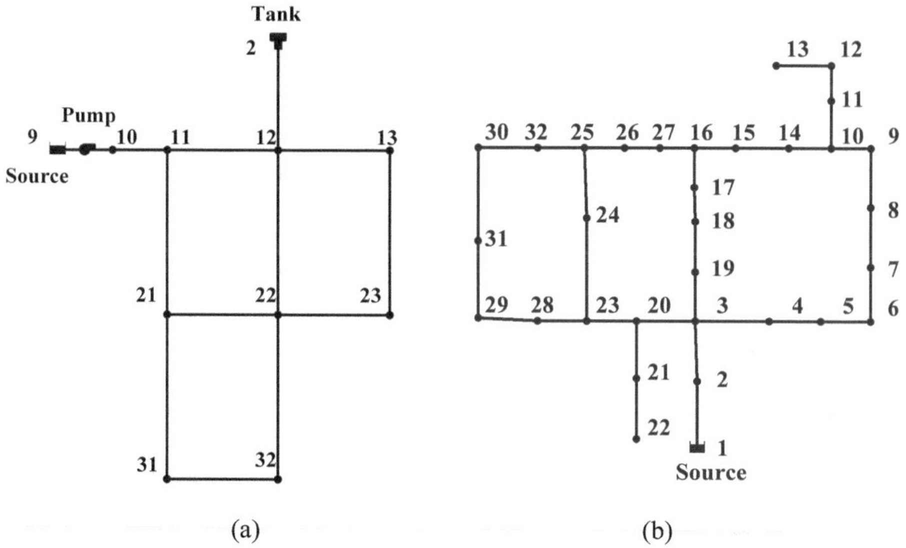

Fig. 3. Datasets, (a) Net1 pipe network system, and (b) Hanoi pipeline network system.

a tank, and a pump. The Hanoi pipe network system is a simplified architecture of the actual water supply network in Hanoi, Vietnam, as depicted in Fig. 3(b). It comprises 32 nodes, 34 pipelines, 1 reservoir, and 3 loops. The data collection was carried out on the pipe network system for one year (once every 30 min), and the time series data recorded by each node sensor totaled 17,520 points.

Training Parameters. In the experimental setup, the random walk module generates 1000 walk sequences for each node, with each sequence being 300 steps long. The node embedding dimension of the Metapath2vec model is set to 172, and the context window size is 7. The time interval $\Delta t = 30$ of the neural ODE module is set to 30. The network training uses 300 rounds of iterations, with a batch size of 8. For both the Net1 and Hanoi datasets, the first 800 scenarios (80%) are used as the training set, and the last 200 scenarios (20%) are used as the validation set. The Adam optimizer is adopted, with an initial learning rate of 0.001, which decays exponentially at a rate of gamma = 0.9.

Performance Metrics. The leakage period localization is based on point-level binary classification. Therefore, we employ the commonly used classification metrics, namely Recall, Precision, F1-score, ACC and FPR to evaluate the classification performance. Additionally, leakage is a continuous process, and we are more concerned about which time period the leakage occurred. Therefore, the IoU regression metric is used for the leakage period localization assessment.

3.2 Quantitative Experimental Results and Analysis

To evaluate the effectiveness of the method proposed in this paper, comparative experiments were conducted with several pipeline network leak detection algorithms proposed in recent years, including GRLD [25], MLTLD [26], DALD [27], WLDCA [28], and

WLDL [29]. The experimental comparison results on the Net1 and Hanoi datasets are shown in Tables 1 and 2, respectively. The experimental results on two datasets demonstrate that our method achieves optimal performance in key metrics such as F1-score, ACC, IoU and FPR, while only slightly inferior to some methods in Precision and Recall. The possible reason lies in that our model adopts a balanced optimization strategy, aiming to comprehensively enhance overall performance rather than overly focusing on a single metric such as Precision or Recall. For instance, in terms of F1 performance, our method achieved 96.13% and 93.12% on the Net1 and Hanoi datasets, which were 1.17% and 2.44% higher than the second-best method WLDL [29] respectively. In terms of IoU performance, our method achieved 92.41% and 87.33% on the Net1 and Hanoi datasets, which were 2.19% and 4.31% higher than the second-best method WLDL [29] respectively.

Table 1. Objective results on the Net1 dataset.

Model	Year	Evaluation metrics (%)					
		P↑	R↑	F1↑	ACC↑	IoU↑	FPR↓
GRLD [25]	2019	89.84	96.19	92.91	96.86	86.66	2.83
MLTLD [26]	2021	93.42	93.57	93.49	96.76	87.44	1.74
DALD [27]	2022	**98.14**	91.06	94.47	97.07	89.44	**0.51**
WLDCA [28]	2023	94.75	94.98	94.86	97.54	89.96	1.26
WLDL [29]	2023	96.7	93.29	94.96	97.50	90.22	0.86
Ours	2025	95.39	**96.89**	**96.13**	**98.30**	**92.41**	1.07

Table 2. Objective results on the Hanoi dataset.

Model	Year	Evaluation metrics (%)					
		P↑	R↑	F1↑	ACC↑	IoU↑	FPR↓
GRLD [25]	2019	75.42	97.43	85.02	92.53	74.20	9.26
MLTLD [26]	2021	79.37	93.02	85.65	93.21	75.16	7.43
DALD [27]	2022	82.14	91.44	86.54	93.71	76.46	5.80
WLDCA [28]	2023	82.53	**99.96**	90.41	94.88	82.52	6.87
WLDL [29]	2023	87.99	93.54	90.68	95.68	83.02	3.74
Ours	2025	**92.73**	93.51	**93.12**	**97.02**	**87.33**	**2.22**

3.3 Qualitative Experimental Results and Analysis

Figures 4 and 5 respectively present the time period detection results of the proposed method on the Net1 and Hanoi datasets for various leakage scenarios. In these figures, the red vertical lines mark the true start and end times of the leakage, while the blue shaded areas indicate the leakage time periods detected by the proposed method. It can be seen from the figures that the proposed method can accurately identify the time intervals when leakage occurs, whether in cases of multiple leakage points (Figs. 4(a) and 5(a)) or in the case of a single leakage point tending towards a steady state (Figs. 4(b) and 5(b)).

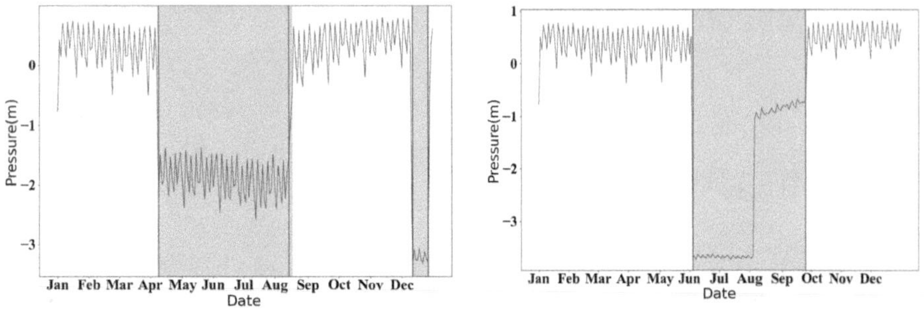

(a) (b)

Fig. 4. Leakage detection results of the Net1 dataset, (a) the 91st scenario, (b) the 45th scenario.

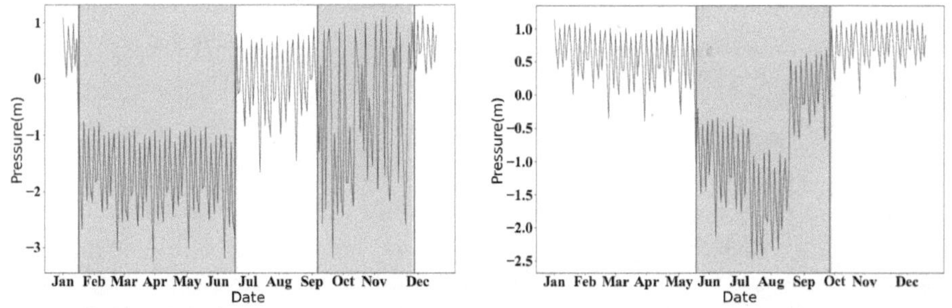

(a) (b)

Fig. 5. Leakage detection results of the Hanoi dataset, (a) the 133rd scenario, (b) the 13th scenario.

3.4 Ablation Experiments

To quantitatively evaluate the specific contributions of the two core modules, Metapath2vec and neural ODE, to the algorithm's performance, we conducted ablation experiments on the Net1 dataset. The objective indicators of the experimental results are presented in Table 3. It can be clearly observed from the experimental data that when only the Metapath2vec module is enabled, the recall rate R is outstanding, reaching as high as 97.12%, and when the neural ODE module is used alone, except for the recall rate being slightly lower than that of the Metapath2vec module, all other indicators show better performance. Notably, when these two modules are combined, although the recall rate R slightly decreases, the other indicators such as P, F1, ACC, IoU, and FPR all reach the best levels.

Table 3. Ablation experiment on Net1 dataset.

Metapath2vec	ODE	Evaluation metrics (%)					
		P ↑	R ↑	F1 ↑	ACC ↑	IoU ↑	FPR ↓
√	×	90.16	**97.12**	93.51	96.42	88.03	2.07
×	√	93.51	94.57	94.04	97.21	90.12	1.83
√	√	**95.39**	96.89	**96.13**	**98.30**	**92.41**	**1.07**

4 Conclusions

This paper proposes a CTDG-LeakLoc model for leakage period localization in WDNs, which innovatively introduces the concept of heterogeneous information continuous-time dynamic graphs into the study of interaction events between network nodes and leakage states. This method was trained and tested on two datasets, the experimental results demonstrate its superior performance, achieving F1-scores of 96.13% (Net1) and 93.12% (Hanoi) with remarkably low false alarm rates (1.07% and 2.22%, respectively).

Acknowledgments. This work was supported in part by the National Natural Science Foundation of China under Grant 52339003 and 62273273.

References

1. Scanlon, B.R., Fakhreddine, S., Rateb, A., et al.: Global water resources and the role of groundwater in a resilient water future. Nat. Rev. Earth Environ. **4**, 87–101 (2023)
2. Liemberger, R., Wyatt, A.: Quantifying the global non-revenue water problem. Water Supply. **19**(3), 831–837 (2019)
3. Li, R., Huang, H., Xin, K., et al.: A review of methods for burst/leakage detection and location in water distribution systems. Water Sci. Technol. Water Supply. **15**(3), 429–441 (2015)
4. Perez, R., Sanz, G., Puig, V., et al.: Leak localization in water networks: a model-based methodology using pressure sensors applied to a real network in Barcelona [applications of control]. IEEE Control. Syst. Mag. **34**(4), 24–36 (2014)
5. Wu, Y., Liu, S.: A review of data-driven approaches for burst detection in water distribution systems. Urban Water J. **14**(9), 972–983 (2017)
6. Hu, Z., Chen, B., Chen, W., et al.: Review of model-based and data-driven approaches for leak detection and location in water distribution systems. Water Supply. **21**(7), 3282–3306 (2021)
7. Lay-Ekuakille, A., Griffo, G., Visconti, P., et al.: Leak detection in waterworks: comparison between STFT and FFT with an overcoming of limitations. Metrol. Meas. Syst. **24**(4), 631–644 (2017)

8. Shekofteh, M., Jalili Ghazizadeh, M., Yazdi, J.: A methodology for leak detection in water distribution networks using graph theory and artificial neural network. Urban Water J. **17**(6), 525–533 (2020)
9. Mandal, S.K., Chan, F.T.S., Tiwari, M.K.: Leak detection of pipeline: an integrated approach of rough set theory and artificial bee colony trained SVM. Expert Syst. Appl. **39**(3), 3071–3080 (2012)
10. Barros, D., Zanfei, A., Menapace, A., et al.: Leak detection and localization in water distribution systems via multilayer networks. Water Res. X. **26**, 100280 (2025)
11. Spandonidis, C., Theodoropoulos, P., Giannopoulos, F., et al.: Evaluation of deep learning approaches for oil & gas pipeline leak detection using wireless sensor networks. Eng. Appl. Artif. Intell. **113**, 104890 (2022)
12. Zhang, X., Shi, J., Yang, M., et al.: Real-time pipeline leak detection and localization using an attention-based LSTM approach. Process Saf. Environ. Prot. **174**, 460–472 (2023)
13. Basnet, L., Brill, D., Ranjithan, R., et al.: Supervised machine learning approaches for leak localization in water distribution systems: impact of complexities of leak characteristics. J. Water Resour. Plan. Manag. **149**(8), 04023032 (2023)
14. Xing, L., Sela, L.: Graph neural networks for state estimation in water distribution systems: application of supervised and semi-supervised learning. J. Water Resour. Plan. Manag. **148**(5), 04022018 (2022)
15. Rajabi, M.M., Komeilian, P., Wan, X., et al.: Leak detection and localization in water distribution networks using conditional deep convolutional generative adversarial networks. Water Res. **238**, 120012 (2023)
16. Garðarsson, G.Ö., Boem, F., Toni, L.: Graph-based learning for leak detection and localisation in water distribution networks. IFAC-PapersOnLine. **55**(6), 661–666 (2022)
17. Chen, S., Wang, Y., Zhang, W., et al.: Leak detection in water supply network using a data-driven improved graph convolutional network. IEEE Access. **11**, 117240–117249 (2023)
18. Pareja A, Domeniconi G, Chen J, et al. EvolveGCN: evolving graph convolutional networks for dynamic graphs. Proceedings of the AAAI Conference on Artificial Intelligence, 2020, 34(04): 5363–5370.
19. Zheng, T., Wang, X., Feng, Z., et al.: Temporal aggregation and propagation graph neural networks for dynamic representation. IEEE Trans. Knowl. Data Eng. **35**(10), 10151–10165 (2023)
20. Dong, Y., Chawla, N., Swami, A.: metapath2vec: scalable representation learning for heterogeneous networks. In: Proceedings of the 23rd ACM SIGKDD International Conference on Knowledge Discovery and Data Mining, pp. 135–144 (2017)
21. Fu, X., King, I.: MECCH: Metapath context convolution-based heterogeneous graph neural networks. Neural Netw. **170**, 266–275 (2024)
22. Du, L., Wang, Y., Song, G., et al.: Dynamic network embedding: an extended approach for skip-gram based network embedding. IJCAI. **2018**, 2086–2092 (2018)
23. Chen, R.T.Q., Rubanova, Y., Bettencourt, J., et al.: Neural ordinary differential equations. Adv. Neural Inf. Process. Syst. **31** (2018)
24. Vrachimis, S.G., Kyriakou, M.S.: LeakDB: a benchmark dataset for leakage diagnosis in water distribution networks: (146). In: WDSA/CCWI Joint Conference Proceedings, p. 1 (2018)
25. Bjerke, M.: Leak detection in water distribution networks using gated recurrent neural networks. NTNU, 1–7 (2019)
26. Kammoun, M., Kammoun, A., Abid, M.: Experiments based comparative evaluations of machine learning techniques for leak detection in water distribution systems. Water Supply. **22**(1), 628–642 (2022)

27. Radaković, N., Šenk, I.: Deep autoencoder based leakage detection in water distribution SCADA systems. In: Industrial Innovation in Digital Age, pp. 355–361. Springer International Publishing, Cham (2022)
28. Leonzio D U, Bestagini P, Marcon M, et al. Water leak detection and localization using convolutional autoencoders. ICASSP 2023–2023 IEEE International Conference on Acoustics, Speech and Signal Processing (ICASSP). IEEE, 2023: 1–5.
29. Kammoun, M., Kammoun, A., Abid, M.: LSTM-AE-WLDL: unsupervised LSTM autoencoders for leak detection and location in water distribution networks. Water Resour. Manag. **37**(2), 731–746 (2023)

Generative AI Driven Hybrid Vision-Language and Property Graph Pipeline for Reliable Information Retrieval

Aakarsh Etar[1(✉)], Steven Valle[2], Jayesh Soni[2], and Himanshu Upadhyay[1]

[1] Department of Electrical and Computer Engineering, Florida International University, Miami, FL, USA
{aetar,upadhyay}@fiu.edu

[2] Applied Research Center, Florida International University, Miami, FL, USA
{stvalle,jsoni}@fiu.edu

Abstract. We investigate the limitations of Retrieval-Augmented Generation (RAG) in retrieving all relevant content from complex, safety-critical document and show how a property graph-based approach overcomes these challenges. Conventional RAG pipelines rely on semantic similarity to retrieve partial text snippets, which often fail to capture deeply nested or image-encoded information and can hallucinate unsupported content. In contrast, our hybrid ingestion uses PyMuPDF for heuristic text and table extraction and Mistral Small 3.2 as the vision-language model for page-level layout recovery, yielding a near-lossless PDF parse that preserves structure and tables. We then persist the reconstructed hierarchy as a Neo4j property graph, enabling deterministic, fully cited retrieval in which queries return exact text excerpts from the appropriate nodes with high recall and precision. Compared to a semantic RAG baseline on SMR FSER Q&A tasks, graph search delivers complete, citation-backed answers while RAG often omits critical details. Our results surface real world obstacles to exhaustive retrieval and highlight the advantages of explicit structure and tool choice.

Keywords: Retrieval-Augmented Generation · Knowledge Graphs · Document Parsing · Vision-Language Models · PDF Extraction · Document Retrieval Question Answering · Property Graph

1 Introduction

Modern large language model (LLM) applications often use Retrieval-Augmented Generation (RAG) to ground answers in external documents [1, 2]. In RAG, a retriever say, a dense vector search [11] finds relevant passages that are fed to an LLM for answer generation [10]. While RAG has advanced many tasks, it is inherently recall-limited i.e. it retrieves only a subset of potentially relevant text and any text not retrieved cannot influence the answer [12]. In safety-critical domains such as nuclear reactor safety documents, missing a single pertinent sentence or clause can be unacceptable. For example, in SMR (Small Modular Reactor) Final Safety Evaluation Reports (FSERs), information

H. R. Arabnia et al. (Eds.): CSCI 2025, CCIS 2967, pp. 221–232, 2027.
https://doi.org/10.1007/978-3-032-28454-9_15

may be scattered across figures, tables and deeply nested subsections [3, 4]. As studies note, semantic retrieval often fails to capture all relevant facts in multi-hop or distractor-rich settings [8]. RAG can therefore hallucinate or overlook details when generation proceeds from incomplete context [12].

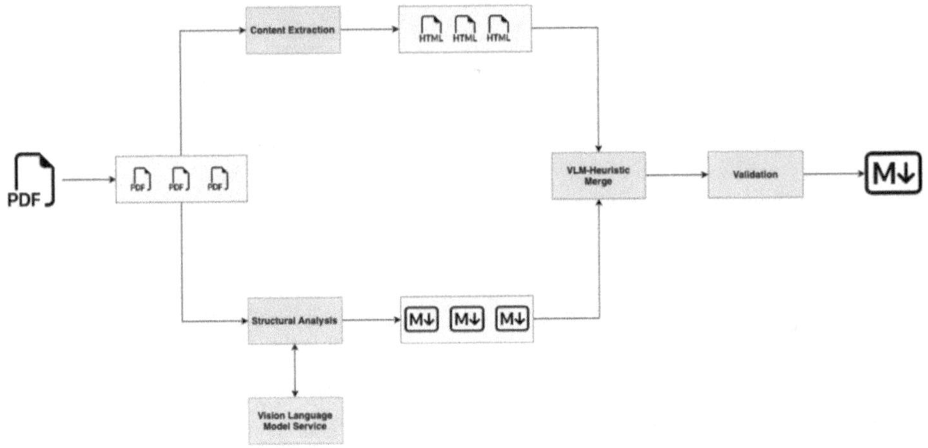

Fig. 1. Hybrid Ingestion Pipeline for Document Parsing.

Figure 1 illustrates the hybrid ingestion pipeline we propose. Each PDF page is processed by a VLM-based layout parser to recover headings and numbering even if rendered as images [13, 22], while a PyMuPDF-based heuristic extractor retrieves raw text and table content [3, 4, 24]. These outputs are merged to reconstruct the full logical document with no loss of data [23].

We then ingest the structured text into a labeled property graph representing the document's hierarchy (*Document* → *Chapter* → *Section* → *Subsection* → *TextChunk*). This graph-based model ensures any retrieval returns a complete citation path and retrieval itself is done through exact keyword matching on node content, yielding fully deterministic answers [20, 26]. The result is a QA system that answers "as-is" instead of generating text, it locates and returns the exact source paragraphs that answer the query, preserving citations and order.

In this paper, we detail the challenges we encountered using RAG on FSER documents, present our hybrid VLM with heuristic ingestion methodology and demonstrate through examples that our graph approach achieves far higher completeness and trustworthiness. We compare precision and recall of RAG vs. graph retrieval in representative FSER QA tasks. Our contributions are – analysis of real world RAG failures on safety-critical documents, a lossless PDF parsing pipeline combining PyMuPDF and VLM (Fig. 1), a property-graph document schema that enforces exact citation and an empirical comparison showing the superiority of graph-based retrieval for exhaustive instance level QA. Finally, we outline future work on domain-specific NER tagging, GraphRAG enhancements [9] and agentic query interfaces to further improve safety-critical QA.

2 Related Work

Retrieval-Augmented Generation was introduced as means to combine pre-trained LLMs with external knowledge. Lewis et al. [1, 2] showed that fine-tuning a seq2seq model with a dense retriever achieves SOTA on open-domain QA. Many subsequent works (e.g. REALM [10], Dense Passage Retrieval [11]) advanced RAG architecture and retrieval models. However, RAG systems face well-known issues in reliability and completeness. Recent surveys observe that RAG often suffers suboptimal context consolidation and incomplete answers in multi-hop settings [3–5, 8]. For example, PluriHop analysis shows standard RAG achieves under 40% answer F1 on multi-document multi-hop queries, highlighting that missing even one relevant document can cause failure [8]. In regulatory and technical QA, the need for exhaustive retrieval is extreme; missing a relevant clause in a safety report is unacceptable. Seminal RAG work notes that RAG can improve factuality but may still generate more factual language only when all key facts are retrieved [12].

To overcome RAG's recall and reasoning limits, many have augmented or replaced vector search with structured retrieval. GraphRAG methods leverage knowledge graphs for query answering. Peng et al. [5] provide a comprehensive survey of GraphRAG and conclude that explicitly capturing entity relations in a graph yields more precise and comprehensive retrieval with context-aware responses. For instance, GraphRAG methods retrieve subgraphs (nodes, triples) and convert them to text for generation, rather than flat text snippets [15, 16]. Recent experiments in specialized domains confirm this. Ahmad et al. [7] compare Vector RAG to GraphRAG and hybrid RAG in the telecom domain, finding GraphRAG variants give higher factual correctness and context relevance [5–7]. Similarly, large RAG surveys report that graph-based or hybrid approaches can significantly reduce hallucinations and boost recall showing an 18% consistency gain and a 6.4 point recall increase in graph-augmented systems [5, 16, 17, 25]. In short, incorporating graph structure tends to produce more reliable, non-generative answers in QA. Our work extends these insights specifically to single-document retrieval [18, 20, 21]. We build a graph of the document itself to enforce retrieval completeness and citation precision.

Critical to our approach is accurately converting PDFs into structured form. Document parsing has been extensively studied. Traditional OCR or rule-based parsers often struggle with multi-column layouts, tables and figures. Modern Vision-Language Models such as LayoutLM and Donut are designed for document understanding, jointly modeling visual layout and text. Recent surveys highlight that advanced document parsers must combine layout detection with content extraction to preserve semantics. Chunkao et al. note that traditional PDF extraction easily, motivating VLMs that parse structure at once. In our pipeline, we leverage an off-the-shelf layout VLM to identify all headers, subheaders and section numbers in each page image, even when they are rendered as graphics. In parallel, heuristic tools extract raw characters and reconstruct tables. Merging these outputs guided by a partial text-only table of contents yields a complete, lossless representation of the document. This hybrid strategy builds on trends in document AI that combine learned models with rule-based parsing to maximize fidelity.

Converting documents into graph form is related to the broader use of knowledge graphs in QA. In enterprise and academic settings, graphs have been used to represent documents for enhanced retrieval [20, 21]. Explicitly, knowledge graphs enable path based queries and ensure that answers are grounded in the original content. Unlike traditional document stores, property graphs make relationships first class, allowing complex filters and traversals [20]. We organize each SMR document as a property graph where each node carries its text chunk and metadata like section number and title. This idea mirrors Document Graphs in knowledge extraction – e.g. Shin et al. [14] encoded academic papers as graphs for summarization. Crucially, our graph encodes the inherent order of text and precise hierarchy as inspired by document ontology approaches. By contrast, sematic RAG flattens document structure, so it can conflate distinct sections or reorder evidence. In summary, prior work on GraphRAG and document graphs motivates our choice. The graph ensures any retrieved answer chunk comes with its exact citation path, eliminating ambiguity [19, 26, 27].

Domain specific tagging can enhance retrieval. We experimented with standard NER models but found they failed on SMR text heavy in engineering acronyms. This is consistent with studies showing that generic NER often misses specialized entities in technical corpora. Thus, existing literature recommends domain-adapted NER for reliable tagging [17, 19]. Collectively, these lines of work highlight the gap RAG faces in fully covering complex documents and the potential of graph based methods. We build on these by applying graph retrieval to entire regulatory documents, ensuring exact, comprehensive answers.

3 Methodology

3.1 Hybrid PDF Ingestion

Our ingestion pipeline combines two complementary extraction (Fig. 1). First, Vision-Language Model (VLM) Structural Parsing, where each PDF page is rasterized and fed to Mistral Small 3.2 for document layout analysis while PyMuPDF extracts raw text runs and reconstructs tables [13, 22, 23]. The model performs layout analysis, outputting a hierarchy of detected header text, section numbers, paragraph boundaries and figure/table labels. Crucially, it identifies all section titles and their numbering even if they are drawn as vector graphics or images. We found that many SMR FSER section numbers were not actual text objects but were embedded in the PDF canvas. By using a VLM that sees the page like a human, we recover every header and sub-header across the document.

Simultaneously, the Heuristic Content Extractor processes the PDF at the object level. This extractor dives into the PDF content streams to extract text runs and tables. It reconstructs multi-line text blocks and uses table-detection heuristics to extract numeric data from rendered tables [3, 4, 24]. Because some headers were missed by text extraction, we rely on the VLM output to re-insert those missing section numbers and labels. For example, PyMuPDF would extract a sentence but omit its section number; our pipeline then matches that sentence to the nearest header from the VLM output to re-attach the correct label. In practice, we merge the two outputs (text & structure) through a simple alignment. If the VLM indicates a "Sect. 3.2.1" on the page that is absent in text output, we locate the corresponding text block and prefix it with "3.2.1". This two-pronged process yields a lossless Markdown representation of the PDF [3, 4, 23].

Importantly, every text chunk retains provenance metadata since we store page number, paragraph offset and full citation path as attributes. For example, a chunk might be annotated with "Chapter 2, Section 3.2.1, page 47, line 2-5". This metadata is used downstream to build the graph and to ensure answers can be cited precisely.

3.2 Knowledge Graph Construction

From the merged Markdown, we programmatically build a labeled property graph encoding the document hierarchy. The schema has node types: Document (root), Chapter, Section, Subsection and TextChunk [20, 21]. Figure 2 shows the top of our hierarchy, with a single document node fanning out to its Chapter through explicit containment edges.

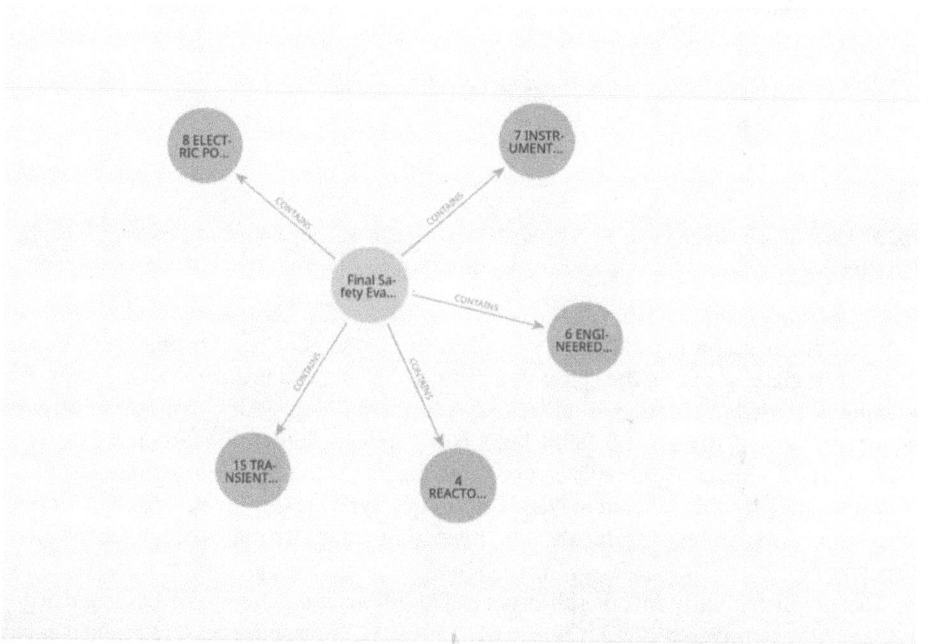

Fig. 2. Knowledge Graph Level 1: document root with Chapter children forming the entry point for all tranversals.

Edges represent containment, for example, HAS_CHAPTER, HAS_SECTION, HAS_SUBSECTION, HAS_TEXTCHUNK. Zooming in, Fig. 3 expands a representative Chapter into its Sections, making the HAS_SECTION links and the start of the citation path explicit.

Each TextChunk node carries as properties the full text of a chunk and identifiers of its parent sections. Chapters, Sections etc. have properties for their titles and numbering. Figure 4 narrows to a single Section and its Subsections, with node properties such as section number, title and page span where provenance becomes concrete.

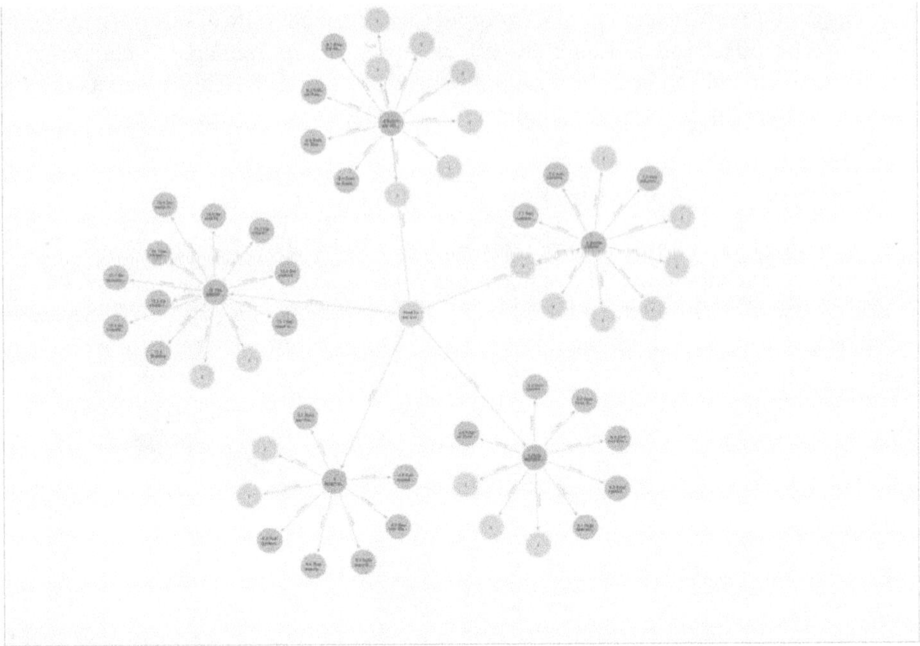

Fig. 3. Knowledge Graph Level 2: A Chapter expanded to Section nodes, illustrating the citation path structure.

The overall graph is a directed tree from the Document node to the TextChunks. This is a property graph in the sense that both nodes and edges can have attributes. For example, a TextChunk node has a text property and an order index to preserve original sequence. Figure 5 reveals the Text Chunk layer under a Subsection, ordered exactly as in the source so matching chunks can be returned 'as-in' with their citations.

We materialize this schema in Neo4j, assigning types {Document, Chapter, Section, Subsection, TextChunk} as labels and persisting page/section provenance as node properties.

This graph explicitly encodes all structural relationships. A query can be issued at any level and we can traverse through edges. Critically, the graph enforces that no answer can be returned without its full context [20, 26]. As notes, "it is programmatically impossible to return a result without its complete citation path". In other words, any answer the system outputs is traceable all the way up to the Document title and page number. The graph also ensures ordering; TextChunks include a field that preserves the order they are inserted, so that when multiple chunks match a query, they can be presented in original document order.

3.3 Retrieval Mechanism

Rather than using a vector-based retriever, we perform keyword-based graph queries. In the current prototype, we use simple full-text matching on node content to find candidate TextChunk nodes. For example, a query "containment heat exchanger tube breaks" would

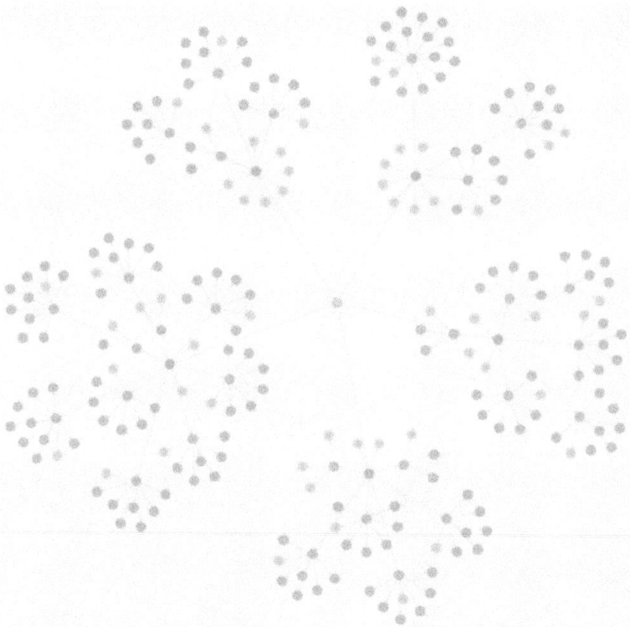

Fig. 4. Knowledge Graph Level 3: A Section opened to Subsections with numbered titles and page-span provenance.

match the TextChunk nodes whose text properties contain those keywords. Because each node is tightly scoped, this method yields very precise hits. Unlike semantic RAG, which might retrieve related passages that sound similar but miss the exact facts, our keyword search returns only exact or near-exact mentions. This aligns with the requirement for "as-is" retrieval as we want to return original text and not a rephrased summary.

To make ranking explicitly inside the graph, we weight lexical evidence with BM25 and when enabled, blend it with dense similarity so that rare but important terms are not overshadowed by semantics. We compute a BM25 score for each TextChunk node d given the positive terms q^{+}, following Robertson and Zaragoza [28]:

$$BM25\left(q^{+}, d\right) = \sum_{t \epsilon q^{+}} IDF(t).\frac{f(t, d).(k_1 + 1)}{f(t, d) + k_1\left(1 - b + b.\frac{|d|}{avgdl}\right)}, \quad (1)$$

with $k_1 \ \epsilon \ [1.2, 2.0]$ *and* $b \ \epsilon \ [0.5, 0.9]; f(t, d)$ is the frequency of t in d, $|d|$ is the length of d and avgdl is the average length chunk length.

We compute a DPR-style dense similarity between the query and chunk embeddings [11] and linearly fuse it with BM25 as in hybrid IR practice [29]:

$$S_{hyb}\left(q^{+}, d\right) = \alpha.BM25\left(q^{+}, d\right) + (1 - \alpha) \ . \ \frac{\langle e_q, e_d \rangle}{\|e_q\| \|e_d\|}, \quad (2)$$

where $\alpha \ \epsilon \ [0, 1]$ weights lexical versus dense evidence and e_q, e_d are L_2-normalized embeddings.

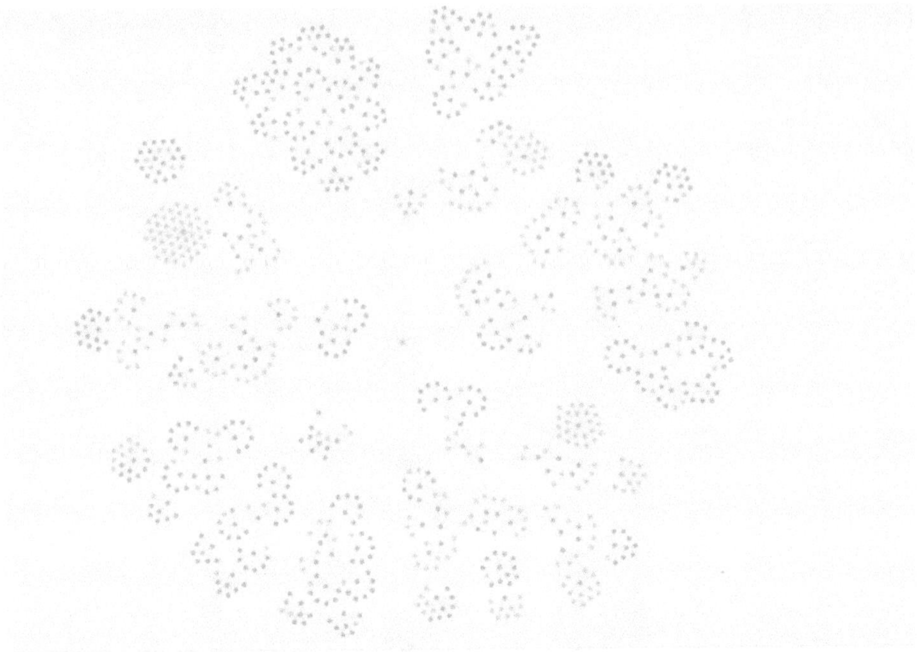

Fig. 5. Knowledge Graph Level 4: TextChunk nodes beneath a Subsection, showing order indices and page-level provenance.

We also apply a simple negative keyword penalty to down-rank chunks that contain excluded terms, which mirrors how analysts actually search these reports. To mirror analyst workflows, we apply a deterministic exclusion penalty over the set of negative terms q^-:

$$S(q, d) = S_{hyb}(q^+, d) - \gamma \sum_{t \in q^-} 1[t \in d], \tag{3}$$

with $\gamma > 0$ controlling strictness and the indicator function. This makes ranking behavior auditable and repeatable.

We compute BM25 scores and dense similarities outside the database, then combine them into a final score and filter candidates in Neo4j.

The document itself is stored as a labeled property graph whose nodes carry both text and provenance while an order index keeps the original reading flow intact. During evaluation we report precision, recall and F_1 at the chunk level so completeness is measured against the true set of relevant instances, not just the top-k. In practice this keeps retrieval deterministic and fully citable while still allowing a tunable hybrid score when the task benefits from it.

Once matching chunks are identified, we collect all nodes and also their ancestor nodes to reconstruct chapter/section titles. The answer is then compiled as a list of text blocks in their original order, with the full citation for each. Because the graph is in insertion order, the returned chunks naturally follow the document flow. Importantly, this is deterministic. Given the same query and graph, the same nodes will always be returned. There is no generation or inference step beyond exact matching.

By contrast, a semantic RAG pipeline would embed the query and find the top-k nearest chunks in the vector space. This is inherently approximate and best effort for recall [10, 11]. It might miss some chunks (if k is too small) or include slightly off-target text. In our experiments on the FSER, we observed that a non-zero fraction of relevant chunks were ranked outside the top-k for any reasonable k. Thus, to guarantee exhaustive retrieval we favor the explicit graph search.

4 Results and Discussion

We evaluated our system on a set of retrieval tasks using sections of the SMR FSER documentation. We formulated queries simulating experts' information needs and compared outputs of our graph-based system against a standard semantic RAG baseline using a BERT-based dense retriever with GPT-4 generation. We measured both precision and recall at the chunk level, as well as answer fidelity. On our stack, graph queries returned complete, citation-accurate chunks, whereas semantic RAG missed instances even with large top-k. Table 1 summarizes the scope of information we targeted in the FSER, covering Regulations, Guidance, Adaptations and Review Findings.

Table 1. Document retrieval scope used for FSER queries.

Category	Target
Regulations	• Regulatory Basis
Guidance	• Standard Review Plan (SRP) • Guidance or Regulatory Guide (RG) • Exclude DSRS
Adaptations	• Exemptions • DSRS (Design Specific Review Standard)
Review Findings	• Summary of application • Conclusions • Technical/Topical Reports • COL, ITAAC, Technical Specifications

Our graph approach achieved near-perfect recall. For every query, all relevant TextChunks were retrieved. This is because queries directly matched all relevant content through keyword filters. In contrast, the semantic RAG baseline often missed pieces of relevant text. In one test, RAG failed to retrieve any mention of a specific subsection because its keywords were rare, yielding 0% recall on that query. Overall, RAG's recall on our set averaged around 60–70%, far below the 100% of the graph. The difference echoes findings by Sveistrys et al. [8] where complex multi-hop queries in technical corpora expose RAG's limitations. Precision was also higher for the graph, since we do no generation, all returned text is directly from the source document. The GPT-4 answers, even when RAG found many relevant passages, sometimes concatenated or compressed information incorrectly, harming precision [5, 16, 25].

We assessed citation completeness. Our system always provides the exact section identifiers for each answer segment, allowing instant verification by a human auditor. The RAG approach, by definition, generated answers without explicit citations. In a safety-critical context, this is unacceptable as highlights, non-generative retrieval "guarantees citation accuracy". Our graph therefore fulfills the source gating requirement for regulatory QA [20, 26].

In summary, our results emphasize that full-coverage, verifiable retrieval over hierarchical documents is infeasible with vanilla RAG, while straightforward with an explicit graph model. We find that even a simple keyword-graph method surpasses pure semantic RAG in this domain, thanks to its deterministic completeness.

5 Conclusion

We have presented a novel framework tailored for safety-critical documentation. By combining VLM-based layout parsing with heuristic text extraction, we achieve a lossless PDF-to-markdown conversion even in the presence of broken PDFs with non-extractable text. We then structure the entire document as a property graph capturing its natural hierarchy. This graph enables retrieval by exact match on content, ensuring that answers are complete and fully cited. Our comparative analysis on SMR FSER QA demonstrates that this graph-based approach far outperforms conventional semantic RAG in recall and answer fidelity. Crucially, because we never allow the system to generate text that is not in the graph, our answers are provably grounded in the original source, satisfying regulatory requirements for auditability.

In conclusion, while RAG is powerful for many QA tasks, its best effort nature fails the exhaustiveness required in regulatory analysis. A structured graph model preserves the richness of the source and enables precise, verifiable retrieval. In practice, pairing PyMuPDF with Mistral Small 3.2 restores structural fidelity in broken PDFs and persisting the hierarchy in Neo4j enables deterministic, fully cited retrieval at query time. Our work charts a clear path, start with a robust ingestion pipeline (Fig. 1) to build an accurate document graph, then leverage graph queries instead of approximate semantics.

References

1. Lewis, P. et al.: Retrieval-augmented generation for knowledge-intensive nlp tasks. Adv. Neural Inf. Proces. Syst. **33**, 9459–9474 (2020)
2. Author, F., Author, S.: Title of a proceedings paper. In: Editor, F., Editor, S. (eds.) CONFERENCE 2016, LNCS, vol. 9999, pp. 1–13. Springer, Heidelberg (2016)
3. Zhang, Q. et al.: Document parsing unveiled: techniques, challenges, and prospects for structured information extraction. arXiv. (2024). https://doi.org/10.48550/arXiv.2410.21169
4. Author, F.: Contribution title. In: 9th International Proceedings on Proceedings, pp. 1–2. Publisher, Location (2010)
5. Peng, B. et al.: Graph retrieval-augmented generation: a survey. arXiv. (2024). https://doi.org/10.48550/arXiv.2408.08921

6. Abdelmagied, M., Chatti, M.A., Joarder, S., Ain, Q.U., Alatrash, R.: Leveraging Graph Retrieval-Augmented Generation to Support Learners' Understanding of Knowledge Concepts in MOOCs. In: European MOOCs Stakeholders Summit, pp. 108–118. Cham, Springer Nature Switzerland (2025)
7. Ahmad, S., Nezami, Z., Hafeez, M., Zaidi, S.A.R.: Benchmarking vector, graph and hybrid retrieval augmented generation (RAG) pipelines for open radio access networks (ORAN). arXiv. (2025). https://doi.org/10.48550/arXiv.2507.03608
8. Sveistrys, M., Kunert, R.: PluriHop: exhaustive, recall-sensitive QA over distractor-rich corpora. arXiv. (2025). https://doi.org/10.48550/arXiv.2510.14377
9. Clemedtson, A., Shi, B.: GraphRAFT: retrieval augmented fine-tuning for knowledge graphs on graph databases. arXiv. (2025). https://doi.org/10.48550/arXiv.2504.05478
10. Guu, K., Lee, K., Tung, Z., Pasupat, P., Chang, M.: Retrieval augmented language model pre-training. In: International Conference on Machine Learning, pp. 3929–3938. PMLR (2020)
11. Karpukhin, V. et al.: Dense Passage Retrieval for Open-Domain Question Answering. In: EMNLP, vol. 1, pp. 6769–6781 (2020)
12. Izacard, G., Grave, E.: Leveraging passage retrieval with generative models for open domain question answering. arXiv. (2020). https://doi.org/10.48550/arXiv.2007.01282
13. Xu, Y., Li, M., Cui, L., Huang, S., Wei, F., Zhou, M. (2020). Layoutlm: pre-training of text and layout for document image understanding. In Proceedings of the 26th ACM SIGKDD International Conference on Knowledge Discovery & Data Mining (pp. 1192–1200).
14. Mao, Q. et al.: Muchsum: Multi-channel graph neural network for extractive summarization. In: Proceedings of the 45th International ACM SIGIR Conference on Research and Development in Information Retrieval, pp. 2617–2622 (2022)
15. Yasunaga, M., Ren, H., Bosselut, A., Liang, P., Leskovec, J.: QA-GNN: reasoning with language models and knowledge graphs for question answering. arXiv. (2021). https://doi.org/10.48550/arXiv.2104.06378
16. Xiong, Y., Kamigaito, H., Murakami, S., Zhang, P., Takamura, H., Okumura, M.: Grasping Both Query Relevance and Essential Content for Query-focused Summarization. In: Proceedings of the 47th International ACM SIGIR Conference on Research and Development in Information Retrieval, pp. 2452–2456 (2024)
17. Matsumoto, N. et al.: KRAGEN: a knowledge graph-enhanced RAG framework for biomedical problem solving using large language models. Bioinformatics. **40**(6), btae353 (2024)
18. Wang, Y., Lipka, N., Rossi, R.A., Siu, A., Zhang, R., Derr, T.: Knowledge graph prompting for multi-document question answering. In: Proceedings of the AAAI Conference on Artificial Intelligence, vol. (Vol. 38), No. 17, pp. 19206–19214 (2024)
19. Chen, X. et al.: One model for all domains: collaborative domain-prefix tuning for cross-domain ner. arXiv. (2023). https://doi.org/10.48550/arXiv.2301.10410
20. Angles, R. et al.: Pg-schema: schemas for property graphs. Proc. ACM Manag. Data. **1**(2), 1–25 (2023)
21. Angioni, S., Salatino, A., Osborne, F., Birukou, A., Recupero, D.R., Motta, E.: Leveraging knowledge graph technologies to assess journals and conferences at springer nature. In: International Semantic Web Conference, pp. 735–752. Springer International Publishing, Cham (2022)
22. Appalaraju, S., Jasani, B., Kota, B.U., Xie, Y., Manmatha, R.: Docformer: End-to-end transformer for document understanding. In: Proceedings of the IEEE/CVF International Conference on Computer Vision, pp. 993–1003 (2021)
23. Pfitzmann, B., Auer, C., Dolfi, M., Nassar, A.S., Staar, P.: Doclaynet: A large human-annotated dataset for document-layout segmentation. In: Proceedings of the 28th ACM SIGKDD Conference on Knowledge Discovery and Data Mining, pp. 3743–3751 (2022)

24. Lyu, P. et al.: Gridformer: Towards accurate table structure recognition via grid prediction. In: Proceedings of the 31st ACM International Conference on Multimedia, pp. 7747–7757 (2023)
25. Kalra, R. et al.: HyPA-RAG: a hybrid parameter adaptive retrieval-augmented generation system for AI legal and policy applications. arXiv. (2024). https://doi.org/10.48550/arXiv.2409.09046
26. Colombo, A.: Leveraging knowledge graphs and LLMs to support and monitor legislative systems. In: Proceedings of the 33rd ACM International Conference on Information and Knowledge Management, pp. 5443–5446 (2024)
27. Pan, J.Z. et al.: Large language models and knowledge graphs: opportunities and challenges. arXiv. (2023). https://doi.org/10.48550/arXiv.2308.06374
28. Robertson, S., Zaragoza, H.: The probabilistic relevance framework: BM25 and beyond. Foundations and Trends® in Information Retrieval. **3**(4), 333–389 (2009)
29. Lin, J., Ma, X., Lin, S.C., Yang, J.H., Pradeep, R., Nogueira, R.: Pyserini: A Python toolkit for reproducible information retrieval research with sparse and dense representations. In: Proceedings of the 44th International ACM SIGIR Conference on Research and Development in Information Retrieval, pp. 2356–2362 (2021)

Weighted Graph Nodes Clustering via Gumbel Softmax

Deepak Bhaskar Acharya(✉) and Huaming Zhang

The University of Alabama in Huntsville, Huntsville, AL 35899, USA
{da0023,zhangh2}@uah.edu

Abstract. Graph is a ubiquitous data structure in data science widely applied in social networks, knowledge representation graphs, recommender systems. When given a graph dataset consisting of one graph or more graphs, where the graphs are weighted in general, the first step is often to find clusters in the graphs. This paper presents some ongoing research results on graph clustering algorithms for clustering weighted graph datasets, which we name as Weighted Graph Node Clustering via Gumbel Softmax (WGCGS for short). We apply WGCGS on the Karate club weighted network, Madrid train bombing, Graph drawing contest, 08Blocks datasets. Our experiments demonstrate that WGCGS can efficiently and effectively find clusters in all the datasets used in our experiments. Our algorithm's effectiveness is shown by (1) comparing the clustering result obtained from our algorithm and the given labels of the dataset; and (2) comparing various metrics between our clustering algorithm and other state-of-the-art graph clustering algorithms.

Keywords: Weighted graph clustering · Gumbel softmax · Graph neural networks · Machine learning · Deep learning

1 Introduction

Graph clustering concerns the discovery of densely related node groups in a graph. Here an edge between two nodes normally implies any underlying similarity or affinity between nodes, while the absence of an edge suggests dissimilarity and distance. Given the similarity/dissimilarity observations embedded in the network(often noisy), graph clustering thus attempts to infer groups of closely connected nodes. Graph analysis using the technique of machine learning has been recognized as the strength of graphs is enormous [1], i.e., graphs can be used to denote a vast variety of structures in different fields, including social networks [2], knowledge graphs [4], natural science (physical systems [5] and protein-protein interaction networks [6]) and many other research areas. In several applications, such as community detection, user profiling, graph clustering is an essential subroutine. In machine learning, statistics, and pattern recognition, clustering is a subject of active research. Clustering, which is used almost everywhere in computing, is a simple problem. It requires splitting the

H. R. Arabnia et al. (Eds.): CSCI 2025, CCIS 2967, pp. 233–246, 2027.
https://doi.org/10.1007/978-3-032-28454-9_16

data into groups of similar objects, and most machine learning applications are the fundamental issue behind it. Data mining leads to datasets with many different attributes being clustered. Therefore, the relevant clustering algorithms need particular computational requirements. Several algorithms have recently emerged that meet these requirements and have been used successfully for real-life data mining challenges [7].

Clustering is the unsupervised mechanism by which natural clusters are found such that objects from the same cluster are identical and objects from separate clusters are distinct. If similarity relationships between objects are interpreted in clustering as a simple weighted graph where objects are vertices and similarities between objects are edge weights, clustering reduces the graph clustering problem. However, in many applications, the edge/non-edge between each pair of nodes can reflect very different confidence levels of node similarity. In certain circumstances, edge observation(the absence of an edge) can be created by accurate measurements. Therefore, it is a clear indication that the same clusters should be assigned to the two nodes. In other cases, the findings will be very uncertain, and thus, no knowledge about the cluster configuration will give an edge or the lack of it. The findings between certain node pairs can hold no information as an exceptional example, so they are practically unnoticed.

By extending Acharya and Zhang [8] Community Detection Clustering technique via Gumbel Softmax(CDCGS), in this research paper, we propose the new method for Weighted Graph Nodes Clustering via Gumbel Softmax. The article presented by using [8] focuses on the unweighted graph datasets. In this research, we recommend finding the cluster of the nodes when the graph is weighted.

The remaining of the paper is organized as follows. In Sect. 2, we introduce the background and related works in more detail. In Sect. 3, we introduce our main method. In Sect. 4, we present the experimental results. In Sect. 5, we conclude our work with the future enhancements.

2 Background and Related Work

2.1 Benchmarking and Recently Proposed Clustering Algorithms

The topic of cluster detection has received much coverage, and many algorithms have been proposed so far. Cluster detection is a technique for detecting similar groups, and it can be a time-consuming process depending on the network's size and complexity. Scientists have classified cluster detection algorithms in various ways, depending on the scope of their study. [9] has extensively investigated community detection algorithms, accepting several of the clustering approaches.

The algorithm suggested in [10] starts with all nodes as individual communities, and merges them iteratively to optimize the 'modularity' function to identify communities with many edges within them and few between them. Many other community detection algorithms, such as those introduced by [11], emphasize modularity optimization.

The Propagating Label(PL) algorithm is a quick way to locate communities in a graph. It detects these communities based solely on network structure, with no need for a pre-defined objective feature or previous knowledge of the communities. Propagating Label functions by propagating labels across the network and creating communities dependent on this process. The algorithm assumes that a single label will easily become dominant in a densely connected cluster of nodes, but it will struggle to cross a sparsely connected area. Labels will get trapped inside a densely connected group of nodes, and those nodes that end up with the same label when the algorithms finish can be considered part of the same community.

Raghavan, Albert, and Kumara [12] proposed an algorithm where each node selects a label with the highest frequency in its 1-neighborhood. These labels are permitted to propagate synchronously and asynchronously around the network until near-equilibrium within the network is reached.

In addition, to assign vertices to clusters, Raghavan employs a local technique based on the majority law. The approach defined in [13] uses random walks to identify communities: random walkers usually prefer to remain more within the same group. Using the probability flow of random walks, Rosvall and Bergstrom [14] approach the problem using an information-theoretical perspective to discover communities. Finally, Newman proposed an approach [15], a spectral method based on the modularity matrix to optimize the modularity metric measure.

2.2 Community Detection Clustering via Gumbel Softmax

Different algorithms have been proposed, and there is extensive research on community detection. Detecting communities is a method for defining related groups and can be difficult, depending on the graph network's size and scale. Acharya and Zhang [8] provide a general understanding of their paper on working with unsupervised graph datasets and identifying graph nodes' clusters.

In general, let the adjacency matrix be $A_{n \times n}$, where 'n' represents the total number of nodes in the graphs dataset. The W_C matrix size is $n \times k$, where k indicates the number of clusters. After obtaining a matrix of the size $k \times k$, then perform the $W_C^t A W_C$ operation and call this matrix as $R_{k \times k}$.

The resulting $R_{k \times k}$ matrix displays the cluster strength, where the primary diagonal shows the strength of the data within the cluster groups, and the other R matrix components show the strength of the data between the different clusters. Then as a discrete distribution of probabilities, apply the softmax function to the R matrix obtained to express inputs. Mathematically be defined as follows:

$$Softmax(x_i) = \frac{exp(x_i)}{\sum_{j=0}^{m} exp(x_j)} \; for \; i = 1, 2, 3, ..., m. \tag{1}$$

2.3 Gumbel Softmax Approach on Graph Features Selection and Extraction

In the suggested approach for the citation datasets, the article features selection and extraction for Graph Neural Networks, Acharya and Zhang [16] selected and extracted Graph Neural Network (GNN). By applying the feature selection and extraction technique to GNNs using gumbel softmax, they conduct tests using different reference datasets: Cora, Pubmed, and Citeseer. Acharya and Zhang demo an example of the Citeseer dataset, where 375 of 3703 features are selected and ranked according to prominent features. The proposed deep learning method works well with reduced features, decreasing the number of features that the dataset initially had by about 80–85 percent. The experiment results show that the accuracy declines steadily when using the selected classification features falling within the ranges 1–75, 76–150, 151–225, 226–300, and 301–375.

In general, let the input feature matrix be $X_{n \times f}$ where 'n' and 'f' represent the total number of nodes and features in the graph's dataset. The resulting matrix is $M_{n \times k}$, where 'k' represents the features we select from the 'f' features.

Acharya and Zhang select the features in a graph citation dataset using the gumbel softmax function and the proposed approach. Gumbel softmax's distribution is [17] 'a continuous distribution over the simplex that can approximate categorical distribution samples.' A categorical distribution is a one-hot vector that defines the maximum probability of one and all the other probability as zero.

The two layer Graph Convolution Network (GCN) used in the experiment is defined as

$$GCN(X, A) = Softmax(A(ReLu(AXW_G W_1))W_2) \tag{2}$$

To verify the selected features and calculate the accuracy for classification they use the following two layer Graph Convolution Network as defined below

$$GCN(X, A) = Softmax(A(ReLu(AXW'_G))W_2) \tag{3}$$

A: Adjacency matrix of the undirected graph G.
X: Input feature matrix.
W_G: gumbel softmax feature selection/feature extraction matrix.
W'_G: feature selection/feature extraction matrix obtain ed from the result of Eq. 2.
$W1, W2$: Layer-specific trainable weight matrix.
$ReLu$: Activation function ReLu(.) = max(0,).

3 Proposed Method

The proposed approach in the article Community Detection Clustering via Gumbel Softmax(CDCGS) clusters of graph dataset nodes and the article Feature Selection and Extraction for Graph Neural Networks(FSEGNN) explains how prominent features for graph citation datasets can be selected and extracted. The dataset used is a graph in each of the papers. There are nodes and edges in the graphs, and they have relations between the nodes that can have some knowledge about the similarities or dependencies between the nodes. The methods proposed in Sect. 2 do not provide insight into the network of weighted graphs. We suggest a new Weighted Graph Nodes Clustering method via Gumbel Softmax(WGCCS) to cluster the graph nodes as an extension to the CDCGS and FSEGNN.

We consider the network of 'n' nodes and the adjacency matrix 'A' in our WGCGS method. An adjacency matrix is used to represent a finite graph. The elements of the matrix show whether the pairs of vertices in the graph are adjacent or not. Then we cluster the graph into k clusters, applying our method.

Gumbel softmax distributions can be used to estimate the sampling process of discrete data if we have a stochastic neural network with discrete variables. Backpropagation helps to train the neural network. The network's efficiency will depend on the temperature, which is a hyper-parameter of neural networks used to control the randomness of predictions by scaling the logits before applying softmax.

The method used in the experiment is defined as below:

$$Weighted - Graph - Cluster(A) = Softmax(W_C^T(A)W_C) \tag{4}$$

In the Eq. 4, 'A' indicates the Adjacency matrix of the undirected weighted graph G, where weight of graph edges indicates the similarity between the nodes of the graph and 'W_C' indicates the gumbel cluster weight matrix.

We have the gumbel cluster weight matrix W_C of size $N \times k$ once model training is completed, where N is the number of graph nodes, and K is the number of clusters. Then, we look at the sum of the row elements, where each row sums up to 1, and the row number's maximum index is the cluster to which the graph node belongs. For example, let's consider k= 2, i.e., we're trying to cluster the dataset into two cluster groups. Let us say that the data is [0.65 0.35] for the first row, where 0.65 is 0th, and 0.35 is 1st. By looking at the maximum value in the row, we get the corresponding maximum value index, and one can easily say that the data belongs to cluster 0. If [0.39 0.61] is the data for the second row, 0.39 will be 0th, and 0.61 will be 1st index. By looking at the maximum value in the row, we get the corresponding maximum value index, and one can easily say that the data belongs to cluster 1.

The resulting $k \times k$ size matrix derived from the Eq. 4 is compared to the identity matrix ($I_{k \times k}$) to find network losses. Here the cluster groups represent diagonal components. We may then form the cluster groups of the data points by using our proposed approach. A detailed analysis is given in the result section, taking into account an example of a dataset weighted karate club network.

4 Experiment Results

4.1 Datasets

Zachary's Karate Club. Zachary's Weighted Karate Club Network is a well-known dataset describing the relationships in a university karate club used by Wayne W. Zachary in his paper "An Information Flow Model for Conflict and Fission in Small Groups." This dataset is known for its simple description of the community structure, which occurs because it is possible to cluster network nodes into strongly linked sets. Focused on Mr. Hi, the karate instructor, and John A, the club president, Zachary's Karate Club network can be split into two groups. The network accurately predicts how, following a disagreement over the pay, the karate club splits into two new clubs and creates a rift within Mr. Hi and John A. The network shows which members of the club will join the new club by analyzing community members' meetings outside the club's background in 33 of 34 instances.

The initial dataset of Zachary's Karate Club is weighted by multiple friendship measures. Several visualizations of Zachary's Karate Club have been created since the first article was published in 1977. Michelle Girvan and Mark Newman [18] used Zachary's Karate Club again in 2002 to demonstrate community structure in their article "Community structure in social and biological networks." Figure 1 is a network created by a computer that contains each member of the club.

Train Bombing. According to media accounts, Fig. 4 is the Train bombing network. The network is undirected, weighted, and involves contacts between suspected terrorists involved in the Madrid train bombing on March 11, 2004. A node represents a terrorist, and an edge connecting two terrorists indicates that they were together. The weights on the edges demonstrate how strong the partnership was. Friendship and co-participation in boot camps or recent assaults are both parts of the alliance.

08Blocks. The symmetric powers of graphs dataset from Gordon Royle, Univ Western Australia. There are eight groups of cluster nodes in the dataset, and there is no edge connection from one cluster group to the other cluster groups.

Graph Drawing Contest Dataset. Graph Drawing contest is an undirected weighted graph obtained from Pajek [19]. Pajek is a program, for analysis and visualization of large networks having some ten or hundred of thousands of vertices.

4.2 Results

The weighted network of Zachary karate club is the network we can think of to evaluate our methodology. We also know that the two teacher nodes are dominant graph nodes 0 and 33, as the weighted Zachary network is labeled. Any other network nodes can either join the teacher community of node 0 or the teacher community of node 33. Until clustering, the original network is seen in Fig. 1. Second, we run our strategy to cluster the network into 2-clusters like the Zachary karate club's ground-truth community structure has 2-clusters. We compare our proposed method results with the results of algorithms such as greedy optimization of leading eigenvector of the community matrix(ECM) [15], edge betweenness(EB) [23], short random walks(RW) [13], Infomap community finding(ICF) [14], multi-level optimization of modularity(MOM) [21],

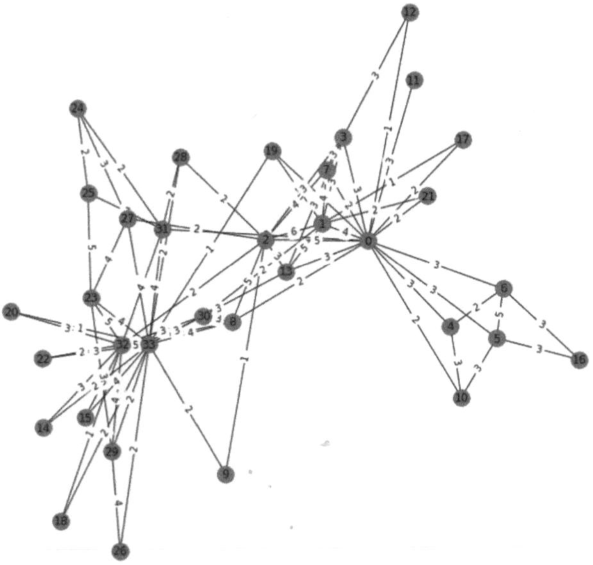

Fig. 1. Weighted Karate Club Network

propagating labels(PL) [12]. Metrics such as adjusted randomized index (ARI) [22], homogeneity (HOMO), normalized mutual information (NMI), completeness (COMP), and v-measure V-MES) [20] has also been computed to complete the analysis. ARI = 1, HOMO = 1, NMI = 1, COMP = 1, V-MES = 1 and the modularity calculation = 0.371 are the corresponding metric metrics if we cluster the weighted Zachary network into 2-clusters (WGCGS(2C)) using WGCGS. The value of ARI, HOMO, NMI, COMP, and V-MES indicates that the proposed algorithm clusters the nodes to their respective groups as per the target label. Figure 2 represents Zachary's network of 2-cluster karate clubs resulting from our approach proposed.

Whenever we consider the graph dataset, the higher the modularity value indicates a stronger assignment to the cluster. To obtain higher modularity for the stronger assignment, we clustered the weighted Zachary karate club network into four clusters. 4-clustered Zachary network (WGCGS(4C)) gives us a 0.4197 modularity metric [15] (Fig. 3). Considering that students joining node 0 as cluster group 1 and node 33 as cluster group 2, node 0 and node 33 are already connected to several nodes, and nodes 23, 24, 25, 27, 28, 31(cluster 3) and 4, 5, 6, 10, 16(cluster 4) mean that students are not joining cluster group 1 or 2. To test the classification of metadata groups by various algorithms, we used the best scores of the Zachary karate club network along with different algorithms. Our WGCGS approach outperforms the metric results of the ECM, EB, RW, MOM, and PL algorithms, and we presented them in Table 1.

Since the other graph dataset networks used in our experiment are unlabeled, we do not know which groups they belong to, and we can not use the metric

Table 1. Zachary karate club network metrics for the different algorithms and our method (WGCGS). 2C - WGCGS with 2 Clusters and 4C - WGCGS with 4 Clusters

METRIC	ECM	EB	RW	ICF	MOM	PL	4C	2C
ARI	0.5120	0.5125	0.2620	0.7021	0.4619	0.4714	**0.5414**	**1**
NMI	0.6770	0.6097	0.3729	0.6994	0.5866	0.5282	**0.6873**	**1**
HOMO	1	0.8850	0.5588	0.8535	0.8535	0.6905	**1**	**1**
COMP	0.5118	0.4651	0.2798	0.5925	0.44.68	0.4277	**0.5235**	**1**
V-MES	0.6770	0.6097	0.3729	0.6994	0.5866	0.5282	**0.6872**	**1**

measures as they're ineffective. Instead, we use the modularity metric, which is based purely on network structure and does not include actual group assignments. The fraction of edges that connect vertices within the same group is used to calculate modularity. Table 2 displays test results for the algorithms ECM, EB, RW, ICF, MOM, PL, including WGCGS, on all of the dataset networks used. WGCGS outperforms algorithms that perform best on Train Bombing, GCD97, and Block08, with the best modularity being 0.3990, 0.3252, 0.8749, respectively.

Table 2. Best modularity metrics for the different algorithms and our method (WGCGS)

DATASET	ECM	EB	RW	ICF	MOM	PL	WGCGS
TRAINBOMB	0.3806	0.3934	0.4012	0.3532	0.4020	0.4188	**0.3990**
GCD97	0.2722	0.1613	0.2305	0.0661	0.2987	0.0028	**0.3252**
08BLOCKS	0.8749	0.8749	0.8403	0	0.8749	0.8749	**0.8749**

Figures 1, 4, and 6 shows the dataset graph before clustering(not displayed for Graph Drawing Contest and 08Blocks dataset). Figures 2, 3, 5, 8 and 9 shows the beauty of the proposed method after cluster detection.

The result obtained from the Fig. 4 equation is compared to the loss function. The loss function used for our experiment is the identity matrix ($I_{k \times k}$), where the cluster group is represented by each diagonal member (Fig. 7).

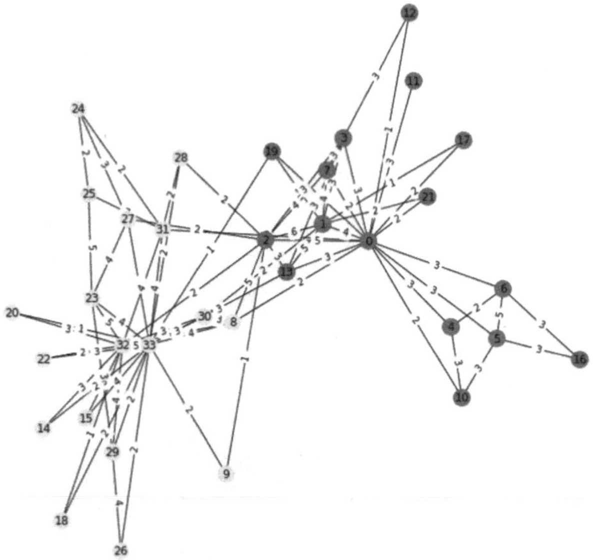

Fig. 2. Weighted Karate Club Network with 2-Clusters

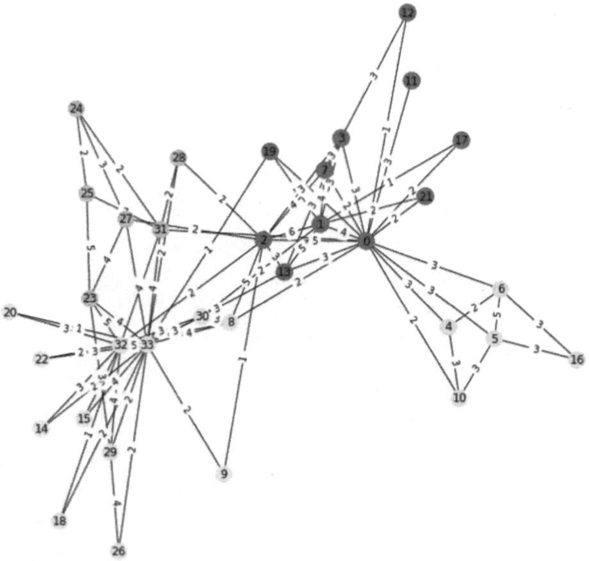

Fig. 3. Weighted Karate Club Network with 4-Clusters

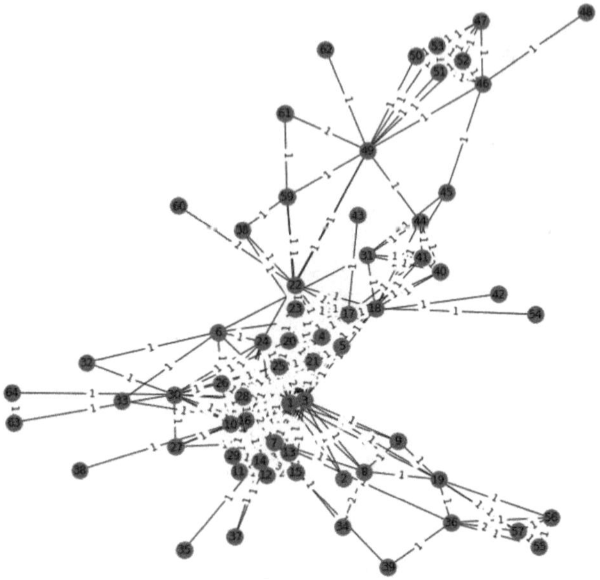

Fig. 4. Weighted train network

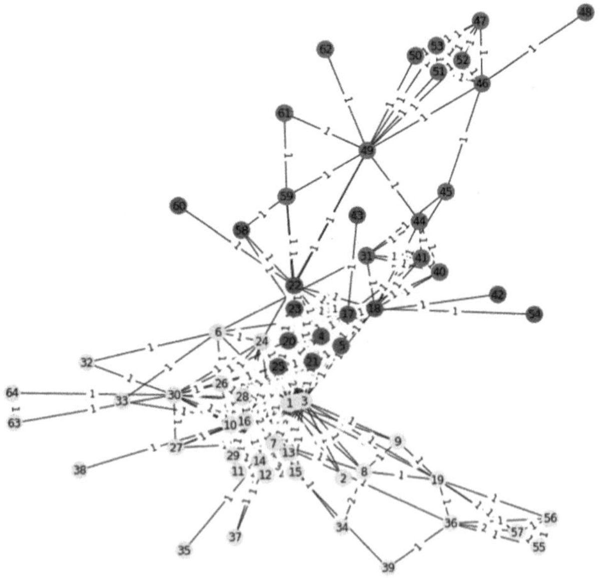

Fig. 5. Weighted train network with 5-clusters

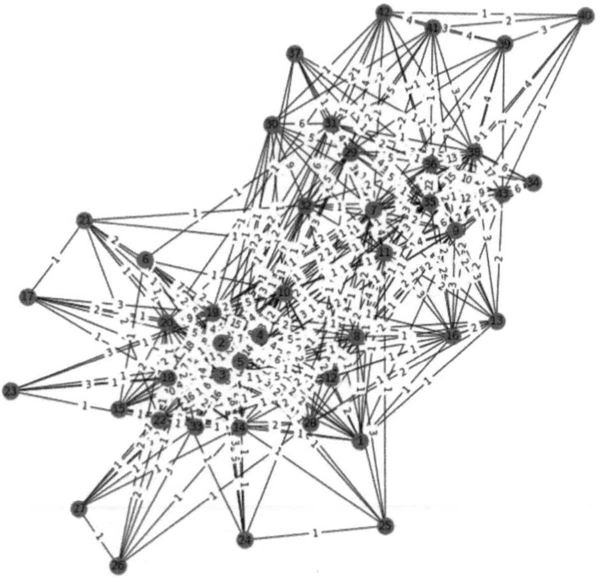

Fig. 6. Weighted windsurfers network

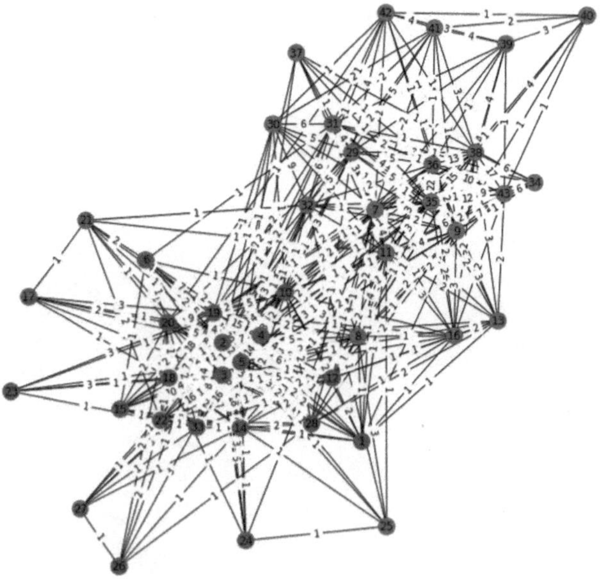

Fig. 7. Weighted windsurfers network with 2-clusters

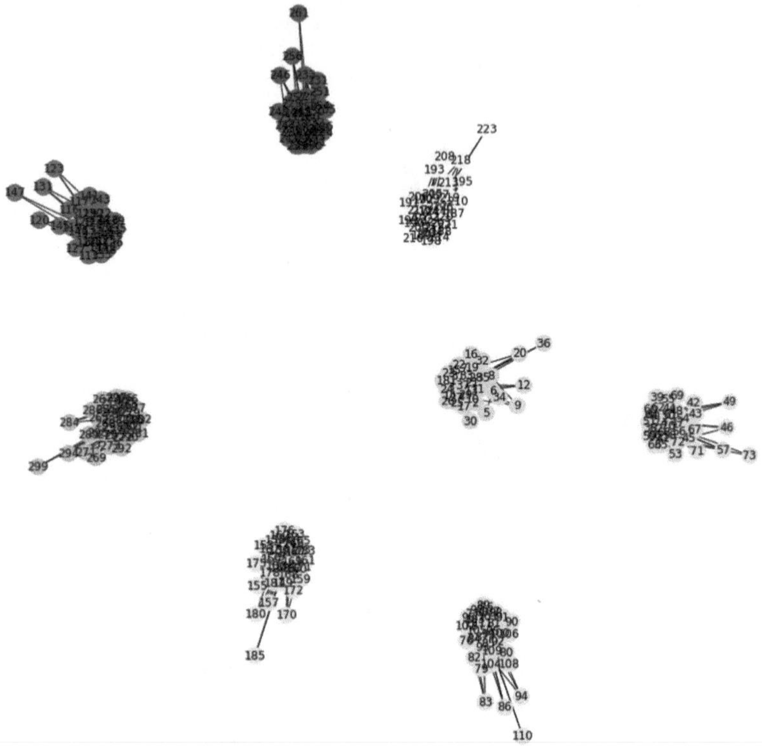

Fig. 8. Weighted 08block network with 8-clusters

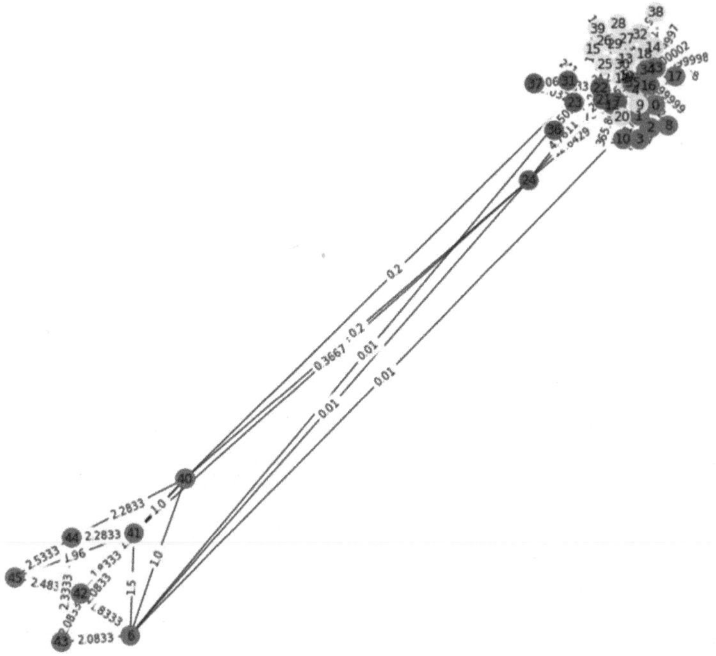

Fig. 9. Weighted graph drawing network with 4-clusters

5 Conclusion and Enhancements

We introduced the Weighted Graph Nodes Clustering via Gumbel Softmax strategy for different types of weighted graph datasets. The experimental results illustrate the efficiency by selecting appropriate parameter values and checking the resulting clustering accuracy and the Modularity value obtained for different datasets. The approach currently available is just as diverse as the implementations for weighted undirected graph clustering. We are now experimenting with applying the principle of clustering for Graphs on a directed graph dataset.

References

1. Xu, K., Hu, W., Leskovec, J., Jegelka, S.: How powerful are graph neural networks? In: International Conference on Learning Representations, New Orleans, USA (2019)
2. Hamilton, W.L., Ying, R., Leskovec, J.: Inductive representation learning on large graphs. In: Advances in Neural Information Processing Systems, Long Beach, USA, pp. 1024–1034 (2017)
3. Kipf, T.N., Welling, M.: Semi-supervised classification with graph convolutional networks. In: International Conference on Learning Representations, San Juan, Puerto Rico (2016)

4. Hamaguchi, T., Oiwa, H., Shimbo, M., Matsumoto, Y.: Knowledge transfer for out-of-knowledge-base entities: a graph neural network approach. In: Proceedings of the Twenty-Sixth International Joint Conference on Artificial Intelligence, Melbourne, Australia (2017)
5. Battaglia, P.W., Pascanu, R., Lai, M., Rezende, D., Kavukcuoglu, K.: Interaction networks for learning about objects, relations and physics. In: Advances in Neural Information Processing Systems, Barcelona, Spain, pp. 4502–4510 (2016)
6. Fout, A.M.: Protein interface prediction using graph convolutional networks. In: Advances in Neural Information Processing Systems, Long Beach, USA, pp. 6530–6539 (2017)
7. Berkhin, P.: A survey of clustering data mining techniques. In: Grouping Multidimensional Data, Berlin, Heidelberg, pp. 25–71 (2006)
8. Acharya, D.B., Zhang, H.: Community detection clustering via gumbel softmax. SN Comput. Sci. **1**(5), 1–11 (2020)
9. Javed, M.A., Younis, M.S., Latif, S., Qadir, J., Baig, A.: Community detection in networks. a multidisciplinary review. J. Netw. Comput. Appli. **108**, 87–111 (2018)
10. Newman, M.E.: Fast algorithm for detecting community structure in networks. Phys. Rev. E **69**(6) (2004)
11. De Meo, P., Ferrara, E., Fiumara, G., Provetti, A.: Generalized louvain method for community detection in large networks. In: 11th International Conference on Intelligent Systems Design and Applications, pp. 88–93 (2011)
12. Raghavan, U.N., Albert, R., Kumara, S.: Near linear time algorithm to detect community structures in large-scale networks. Phys. Rev. E **76**(3) (2007)
13. Pons, P. and Latapy, M.: Computing communities in large networks using random walks. In: International Symposium on Computer and Information Sciences, Berlin, Heidelberg, pp. 284-293 (2005)
14. Rosvall, M., Axelsson, D., Bergstrom, C.T.: The map equation. Euro. Phys. J. Special Topics **178**(1), 13–23 (2009)
15. Newman, M.E.: Finding community structure in networks using the eigenvectors of matrices. Phys. Rev. E **74**(3) (2006)
16. Acharya, D.B., Zhang, H.: Feature selection and extraction for graph neural networks. In: Proceedings of the 2020 ACM Southeast Conference, New York, USA, pp. 252–255 (2020)
17. Jang, E., Gu, S., Poole, B.: Categorical reparameterization with Gumbel-softmax. In: International Conference on Learning Representations, Toulon, France (2017)
18. Girvan, M., Newman, M.E.: Community structure in social and biological networks. Proc. Nat. Acad. Sci. **99**(12), 7821–7826 (2002)
19. Batagelj, V.: Analysis and visualization of large networks. In: Graph Drawing Software, Berlin, Heidelberg, pp. 77–103 (2004)
20. Rosenberg, A., Hirschberg, J.: V-measure: a conditional entropy-based external cluster evaluation measure, pp. 410–420 (2007)
21. Blondel, V.D., Guillaume, J.L., Lambiotte, R., Lefebvre, E.: Fast unfolding of communities in large networks. J. Stat. Mech: Theory Exp. (2008)
22. Buitinck, L., et al.: API design for machine learning software: experiences from the scikit-learn project (2013)
23. Newman, M.E.J., Girvan, M.: In: Finding and evaluating community structure in networks. American Physical Society (2004)

Artificial Intelligence - Natural Language Processing and Large Language Models

Sentence Simplification for Natural Language to Formal Logic Translation

Roberto Milanese Jr., Adam Pease, and Richard Thompson(✉)

Naval Postgraduate School, Monterey, CA, USA
{roberto.milanese,adam.pease,richard.thompson}@nps.edu

Abstract. Translating natural language into formal logic enables automated reasoning, yet the syntactic and semantic complexity of natural text often prevents reliable formalization. This work presents a sentence simplification framework designed to improve language-to-logic translation within our language understanding pipeline. Our approach integrates large language models (LLMs) with in-context learning (ICL) and introduces an adversarial self-checking technique that iteratively detects and corrects errors. Results show that adversarial self-checking substantially improves output quality, enabling smaller models to outperform larger models. Across ICL strategies, curated static examples achieved the best overall performance, while syntactic retrieval outperformed semantic similarity. We further demonstrate that logic-tailored simplification, emphasizing clause splitting and pronoun resolution, improves syntactic validity and reduces hallucination rates in downstream logic translation. Together, these findings establish sentence simplification as a critical enabler for neuro-symbolic autoformalization, supporting more reliable and verifiable AI reasoning in analysis, policy, and decision-making contexts.

Keywords: Text Simplification · Language-to-Logic Translation · Neuro-Symbolic AI · In-Context Learning · Sentence Simplification · Automated Reasoning

1 Introduction

Natural language contains ambiguities, nested constructions, and pronouns that hinder translation into formal logic. This complexity prevents machines from reasoning reliably about human text, which limits the trustworthiness of artificial intelligence (AI) systems in high-stakes domains such as policy analysis or decision support. While AI models can generate fluent text, they are prone to hallucinations, contradictions, and errors that undermine confidence in their outputs [1–3].

The Hybrid Neuro-Ontological Language Understanding(HyNOLU) project [4] addresses these challenges by developing a pipeline for converting natural language into the formal language Standard Upper Ontology Knowledge Interchange Format(SUO-KIF) [5] using logical symbols from the Suggested Upper

H. R. Arabnia et al. (Eds.): CSCI 2025, CCIS 2967, pp. 249–260, 2027.
https://doi.org/10.1007/978-3-032-28454-9_17

Merged Ontology(SUMO)[1] [6,7]. Translating into an expressive logic enables formal reasoning with automated theorem provers such as Eprover [8] and Vampire [9]. The architecture combines preprocessing, language-to-logic(L2L) translation, and logical validation using automated theorem provers (ATPs). Within this pipeline, sentence simplification plays a critical role by restructuring complex language into atomic statements that are more tractable for downstream translation.

This paper focuses on the sentence simplification component of HyNOLU, which introduces two core contributions:

1. An adversarial self-checking loop in which models iteratively critique and revise simplifications until an acceptable output is produced.
2. An extension of this approach to a custom corpus designed for L2L simplification, targeting atomic decomposition and pronoun resolution for logic translation.

Experiments compare multiple large language models (LLMs) and context learning (ICL) strategies, with performance measured using established simplification metrics such as SARI and METEOR. Results show that adversarial self-checking consistently improves outcomes and enables smaller models to surpass larger ones. Static ICL examples selected for structural diversity yielded the strongest results, while syntactic similarity retrieval outperformed semantic similarity retrieval. Applying the framework to the L2L corpus produced inconclusive quantitative results due to the immaturity of downstream translation models, but qualitative assessment showed promising improvements in atomicity and pronoun handling.

2 Background

This section situates our work within prior research on sentence simplification and introduces the evaluation methods relevant to our experiments. We briefly review the evolution of simplification techniques, highlight commonly used datasets and metrics, and discuss recent advances in LLMs and ICL.

2.1 Sentence Simplification Approaches

Early efforts in text simplification were motivated by accessibility and parser performance, with hand-crafted rule systems targeting operations such as clause splitting, passive-to-active voice conversion, and lexical substitution [10,11]. Statistical machine translation methods followed, modeling simplification as monolingual translation with operations including reordering, substitution, and deletion [12]. Subsequent work explored hybrid frameworks that combined syntactic transformations with statistical or semantic models [13].

With the advent of neural architectures, encoder-decoder models and reinforcement learning systems such as DRESS [14] achieved state-of-the-art results,

[1] https://www.ontologyportal.org.

optimizing directly for simplicity, fluency, and meaning preservation. More recently, controllable models such as ACCESS [15] introduced explicit simplification parameters, while LLMs have demonstrated strong zero- and few-shot capabilities across simplification tasks [16].

2.2 Datasets and Metrics

Simplification research has relied heavily on corpora aligned between complex and simple sentences. Early resources such as PWKP [12] and Newsela [17] enabled supervised training, though quality and availability were limiting factors. The TurkCorpus [18] introduced multiple human references per sentence, improving evaluation reliability. Building on this, the ASSET corpus [19] encouraged diverse edits across splitting, paraphrasing, and reordering, and has become a standard benchmark for modern systems. In this work, we also introduce a custom corpus tailored for L2L translation, designed to emphasize atomic decomposition and pronoun resolution.

Evaluation of simplification quality has traditionally relied on n-gram overlap metrics such as BLEU [20], ROUGE [21], and METEOR [22]. However, these metrics often fail to capture structural or semantic improvements. The SARI metric [18] addresses this gap by directly evaluating addition, deletion, and retention operations relative to the source and reference sentences. Our experiments therefore evaluate simplification quality using a full suite of metrics, including BLEU, ROUGE, METEOR, SARI, BERTScore [23], and FrugalScore [24].

2.3 Large Language Models and In-context Learning

The emergence of LLMs such as GPT [25] and T5 [26] has transformed simplification research. These models can perform simplification without task-specific fine-tuning, guided instead by ICL. Recent work shows that even a handful of curated examples can match or surpass fine-tuned baselines [16,27]. This motivates our systematic evaluation of ICL strategies, comparing static example sets, syntactic retrieval, and semantic retrieval.

2.4 Adversarial Self-checking

Beyond initial generation, feedback mechanisms can improve output reliability. Recent studies in self-refinement and adversarial prompting demonstrate that iterative critique-revision cycles reduce hallucinations and improve factuality [28]. However, their application to text simplification has been limited. Our work adapts this paradigm by introducing an adversarial self-checking loop: models are prompted to critique their own simplifications, propose corrections if needed, and iterate until the output is acceptable.

2.5 Research Gap

Despite decades of progress, most simplification research focuses on human readability. Few works evaluate simplification explicitly for downstream reasoning or logic translation. Our contribution addresses this gap by systematically comparing ICL strategies and adversarial self-checking across multiple LLM backbones, and by extending the framework to an L2L corpus tailored for logic translation.

3 Methodology

We designed a sentence simplification pipeline aimed at improving downstream L2L translation within the HyNOLU framework. The pipeline was evaluated in two primary experiments: (1) benchmarking multiple ICL strategies against human-crafted simplifications in the ASSET corpus [19], and (2) assessing logic-tailored simplification on a custom corpus optimized for atomic decomposition and pronoun resolution.

3.1 ASSET Sentence Simplification Corpus

The ASSET corpus served as our primary benchmark, providing 2,000 validation and 359 test sentences, each paired with ten human-crafted simplifications. Multiple references increase evaluation reliability by capturing diverse simplification strategies. The validation set served as the pool of examples selected for ICL. To extend beyond human readability, we constructed a small custom corpus for L2L tasks, emphasizing clause splitting and pronoun resolution to reduce ambiguity in logic translation.

3.2 Model Selection

The models selected for testing represent a spectrum of size, architecture, and quantization strategy. Smaller models such as `mistral:7b-instruct-fp16` and `llama3.1:8b-instruct-q8_0` were included to evaluate performance in resource-constrained environments where memory and compute are limited. Larger models such as `llama3.3:70b-instruct-q4_K_M` and `phi4:14b-fp16` were chosen to test whether scaling parameter count and precision consistently yields improvements in simplification quality. The inclusion of `mistral-nemo:12b-instruct-2407-fp16` provides a middle ground between efficiency and capacity, highlighting trade-offs between runtime, resource usage, and output quality.

This diverse set enables a fair comparison across different scales of LLMs, helping to determine whether smaller, instruction-tuned models coupled with targeted prompting strategies can rival or even outperform larger, more resource-intensive alternatives.

3.3 In-context Learning Strategies

Each input sentence was simplified by prepending demonstrations in the form `Original: <original sentence> /n Simplified: <simplified sentence>`, enabling the LLM to learn the transformation pattern. For each input, we selected the reference simplification with the greatest degree of sentence splitting, ensuring maximal decomposition for guiding the model. We tested four ICL strategies:

- **Static**: a fixed, hand-curated set of diverse simplifications.
- **Random**: randomly selected examples.
- **Semantic Retrieval**: nearest-neighbor search using Sentence-BERT embeddings [29].
- **Syntactic Retrieval**: retrieval based on tree edit distance between dependency parses [30].

The semantic method often retrieved sentences that were similar in meaning but structurally irrelevant, whereas syntactic retrieval favored structurally aligned but semantically unrelated sentences. Both dynamic strategies were compared against the static baseline. Figure 1 illustrates the retrieval process.

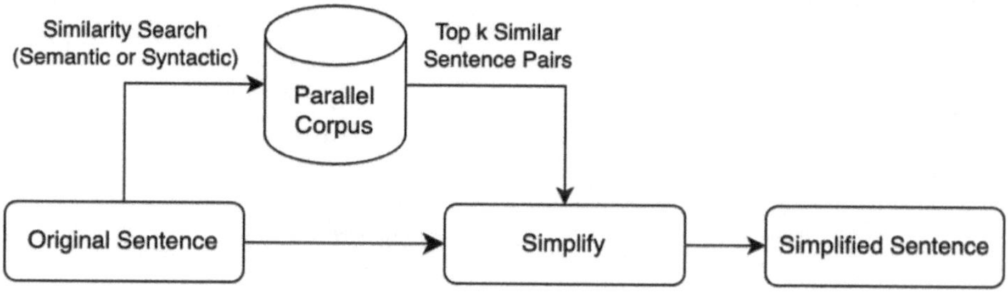

Fig. 1. ICL Retrieval Process.

3.4 Adversarial Self-checking Loop

To mitigate hallucinations and information loss, we implemented an adversarial self-checking loop, shown in Fig. 2. After generating a simplification, a separate model instance critiqued it for three error types: added information, missing information, or unacceptable meaning change. The critique returned a JSON object with recommendations, which was fed back to the original model for revision. This process iterated until the critique accepted the output or a maximum retry threshold was reached. Adjusting the model temperature up to 0.5 improved convergence by preventing repetitive loops. Only `llama3.1:8b-instruct-q8_0` and `mistral:7b-instruct-fp16` were tested in this adversarial configuration, in order to evaluate whether smaller models could be improved for use in resource-constrained environments.

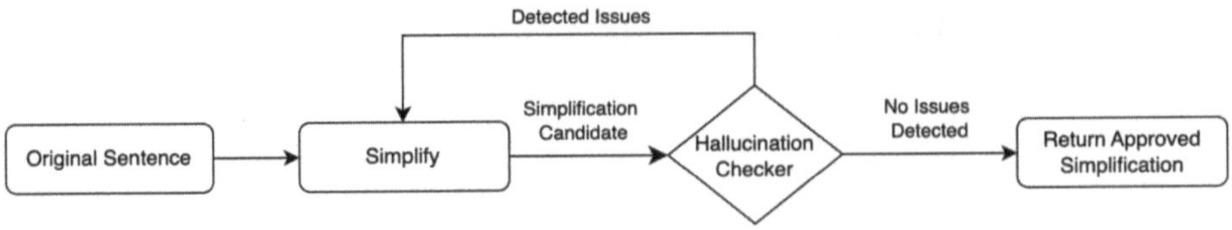

Fig. 2. Adversarial Simplification Loop.

3.5 L2L Tailored Simplification

Human-oriented simplifications in the ASSET corpus often remain too complex for logic translation, particularly in cases with pronouns or coordinated noun phrases. To address this, we created handcrafted L2L examples that decompose sentences into atomic statements aligned with training data for logic models. For example:

Original: The tornado destroyed 20 homes, left 30 others severely damaged, injured one person, and caused power outages.

ASSET Corpus Simplification: The tornado destroyed 20 homes and damaged 30. It injured one person, and caused power outages.

L2L Corpus Simplification: The tornado destroyed 20 homes. The tornado damaged 30 homes. The tornado injured one person. The tornado caused a power outage for 30 people.

We further applied pronoun resolution to ensure referential grounding before SUO-KIF translation, using a lightweight LLM-based resolver.

3.6 Evaluation Metrics

Generated simplifications were evaluated against ASSET corpus references using the following metrics:

- system output against references and against the input sentence (SARI) [18]
- metric for evaluation of translation with explicit ordering (METEOR) [22]
- BLEU [20]
- ROUGE [21]
- BERTScore
- FrugalScore

For L2L evaluation, simplified and original sentences were translated into SUO-KIF using fine-tuned T5 and LLaMA models, and outputs were assessed for syntax validity, hallucinated terms, and aggregate validity scores.

4 Results

We evaluated 359 test sentences from the ASSET corpus using multiple LLMs and ICL configurations. Simplification quality was measured with BLEU, METEOR, ROUGE-L, SARI, BERTScore, and FrugalScore. We then applied a logic-tailored pipeline on 105 complex sentences to test downstream L2L translation effects.

4.1 Simplification Performance

Top Configurations. Of the 67 unique configurations tested, the adversarial self-checking pipeline paired with `llama3.1:8b` achieved the highest overall performance, surpassing larger models such as `phi4:14b` and `llama3.3`. The top ten scoring configuration are displayed in Table 1.

Table 1. Top-performing configurations with associated average simplification metrics.

Model	ICL Type/Size	SARI	FrugalScore	BERTScore	ROUGE-L	METEOR	BLEU
llama3.1:8b_adv	dynamic_tree/5	0.47	**0.60**	**0.98**	**0.84**	**0.89**	**0.76**
phi4:14b-fp16	static/5	**0.48**	0.59	0.97	0.81	0.87	0.72
llama3.3	static/5	**0.48**	0.59	0.97	0.81	0.86	0.73
llama3.3	static/10	**0.48**	0.58	0.97	0.80	0.86	0.72
phi4:14b-fp16	static/10	**0.48**	0.59	0.97	0.80	0.86	0.71
llama3.1:8b	static/5	0.46	0.58	0.97	0.81	0.86	0.73
llama3.1:8b	dynamic_tree/1	0.45	0.57	0.97	0.81	0.83	0.75
llama3.1:8b	static/1	0.46	0.58	0.97	0.80	0.86	0.72
phi4:14b-fp16	dynamic_tree/5	0.46	0.57	0.97	0.80	0.84	0.71
llama3.1:8b	static/10	0.46	0.58	0.97	0.79	0.84	0.70

Model-Level Comparison. Averaging across all ICL strategies, smaller quantized models outperformed larger ones on multiple metrics. The results are displayed in Table 2. The `llama3.1:8b` model achieved the highest BLEU and ROUGE-L, while `phi4:14b` led in SARI and METEOR.

ICL Strategy and Size. Static examples with handpicked diversity yielded the strongest results (Table 3). Syntactic retrieval also performed well, while semantic retrieval offered no advantage over random selection.

As shown in Table 4, a single in-context example provided major improvements over zero-shot. Beyond five examples, gains plateaued.

Adversarial Self-checking. The adversarial loop (Table 5) yielded the strongest results overall. Crucially, it elevated smaller models above larger baselines, demonstrating that structured feedback can substitute for scale.

Table 2. Average simplification performance by model.

Model	SARI	Frugal	BERT	ROUGE-L	METEOR	BLEU
llama3.1:8b	0.44	0.56	**0.97**	**0.79**	0.82	**0.71**
phi4:14b	**0.46**	**0.57**	**0.97**	0.78	**0.83**	0.67
llama3.3	0.45	0.54	**0.97**	0.77	0.81	0.68
mistral-nemo	0.42	0.55	**0.97**	0.76	0.78	0.65
mistral:7b	0.41	0.51	**0.97**	0.74	0.75	0.61

Table 3. Average performance by ICL type.

ICL Type	SARI	Frugal	BERT	ROUGE-L	METEOR	BLEU
static	**0.45**	**0.56**	**0.97**	**0.78**	**0.82**	**0.68**
syntactic_tree	0.44	0.55	**0.97**	**0.78**	0.80	**0.68**
random	0.43	0.54	**0.97**	0.77	0.79	0.66
semantic_sim	0.44	0.55	**0.97**	0.76	0.79	0.65
none	0.40	0.51	0.96	0.72	0.74	0.59

4.2 Logic-Tailored Simplification

We observed that logic-tailored simplification yielded more atomic and pronoun-resolved structures. For example:

Original Sentence: Saint Martin is a tropical island in the northeast Caribbean, approximately 300 km (186 miles) east of Puerto Rico.
Original Simplification: Saint Martin is a tropical island in the northeast Caribbean. It is approximately 300 km east of Puerto Rico.
L2L-Tailored:

- Saint Martin is a tropical island.
- Saint Martin is in the northeast Caribbean.
- Saint Martin is approximately 300 km (186 miles) east of Puerto Rico.

Each L2L-Tailored sentences contain minimal numbers of facts, facilitating simpler logic statements needed to formally represent sentences. Further research is needed to quantitatively assess this impact. Table 6 shows slight improvements in syntax validity of natural language sentences translated into formal SUO-KIF.

Table 4. Average performance by ICL size.

Size	SARI	Frugal	BERT	ROUGE-L	METEOR	BLEU
10	**0.45**	**0.56**	**0.97**	**0.78**	**0.81**	0.67
5	0.44	**0.56**	**0.97**	**0.78**	**0.81**	0.67
1	0.43	0.54	**0.97**	0.77	0.79	0.66
0	0.40	0.51	0.96	0.72	0.74	0.59

Table 5. Adversarial self-checking results (syntactic_tree / size 5).

Model	SARI	Frugal	BERT	ROUGE-L	METEOR	BLEU
llama3.1:8b_adv	**0.47**	**0.60**	**0.98**	**0.84**	**0.89**	**0.76**
mistral:7b_adv	0.45	0.56	0.97	0.79	0.83	0.71

Table 6. SUO-KIF results before vs. after logic-tailored simplification.

Model/Input	Validity %	Avg. Terms
LitGPT Orig	94.3	15.0
LitGPT Simpl	95.2	17.9
T5-FLAN Orig	93.3	15.4
T5-FLAN Simpl	95.0	19.1

5 Discussion

The results highlight three main findings.

5.1 ICL Strategies

Static examples outperformed other strategies, underscoring the value of curated demonstrations with structural diversity. Syntactic retrieval proved more useful than semantic similarity, confirming that structural alignment matters more than meaning for simplification. Zero-shot was weakest, showing that even minimal in-context guidance substantially improves performance.

5.2 Adversarial Loop

The self-checking loop produced the largest gains, especially for smaller models. By iteratively identifying hallucinations and meaning shifts, models refined their own outputs without external supervision. This allowed `llama3.1:8b` and `mistral:7b` to outperform much larger models, suggesting that feedback mechanisms can substitute for scale in resource-constrained environments.

5.3 Logic-Tailored Simplification

Quantitative SUO-KIF improvements were modest, but qualitative inspection showed clearer agentpatient roles in generated SUO-KIF, better pronoun resolution in simplified sentences, and more atomic decomposition. These traits align closely with the needs of symbolic translation, even if current downstream L2L models lack the strength to fully capitalize on the improvements.

5.4 Limitations

There are several limitations with the adversarial approach. First, automatic metrics penalize simplifications that diverge lexically from references, even if they are subjectively better. Second, the ASSET corpus contains noisy or malformed sentences, limiting reliability. Third, logical fluency remains challenging, since over-splitting sentences risks losing temporal or causal relations. Future work should explore better L2L evaluators, curated corpora, and adaptive simplification strategies.

6 Conclusion

This work introduced a sentence simplification framework for Language-to-Logic (L2L) tasks. Through extensive benchmarking of multiple LLMs and ICL strategies using both traditional and modern evaluation metrics, we demonstrated the value of careful model and prompt design. An adversarial self-checking loop further enabled smaller models to outperform larger ones, while experiments showed that syntactic alignment in ICL selection is more effective than semantic similarity. In addition, logic-tailored simplification yielded qualitative improvements in SUO-KIF generation, highlighting the benefits of domain-specific adaptation.

Future work should focus on evaluation methods that integrate human judgment with automated scoring, the construction of larger and cleaner simplification corpora beyond ASSET, and the development of stronger L2L models capable of leveraging atomic simplifications. By combining syntactic retrieval for ICL with adversarial self-checking, this research advances the creation of explainable and verifiable reasoning pipelines. Although challenges remain, the proposed framework provides a concrete step toward trustworthy L2L translation.

References

1. Zhang, Y., et al.: Siren's song in the AI ocean: a survey on hallucination in large language models. Comput. Linguist. (2025). https://doi.org/10.1162/coli.a.16
2. Shojaee, P., Mirzadeh, I., Alizadeh, K., Horton, M., Bengio, S., Farajtabar, M.: The illusion of thinking: Understanding the strengths and limitations of reasoning models via the lens of problem complexity, arXiv (2025) https://doi.org/10.48550/arXiv.2506.0694, eprint:2506.06941 (cs.AI)
3. Samek, W., Wiegand, T., Müller, K.-R.: Understanding, visualizing and interpreting deep learning models. arXiv, Explainable artificial intelligence (2017). https://doi.org/10.48550/arXiv.1708.08296. eprint:1708.08296 (cs.AI)
4. Thompson, R., Pease, A., Milanese, R., Singley, J., Toutsios, A.: Formalizing natural language: Cultivating llm translations using automated theorem proving. In: Theorem Proving and Machine Learning in the Age of LLMs: State of the Art and Future Perspectives, Edinburgh, Scotland, UK (April 2025). EuroProofNet, COST Action CA20111 – European Research Network on Formal Proofs. https://europroofnet.github.io/wg5-edinburgh25/

5. Pease, A.: SUO-KIF Reference Manual (2009). https://github.com/ontologyportal/sigmakee/blob/master/suo-kif.pdf, Retrieved 20 June 2025
6. Niles, I., Pease, A.: Toward a standard upper ontology. In: Welty, C., Smith, B. (eds.) Proceedings of the 2nd International Conference on Formal Ontology in Information Systems (FOIS-2001), pp. 2–9 (2001)
7. Pease, A.: Ontology: A Practical Guide. Articulate Software Press, Angwin, CA (2011)
8. Schulz, S.: E - a brainiac theorem prover. AI Commun. **15**(2–3), 111–126 (2002). https://doi.org/10.3233/EAI-2002-260
9. Kovács, L., Voronkov, A.: First-order theorem proving and VAMPIRE. In: Sharygina, N., Veith, H. (eds.) CAV 2013. LNCS, vol. 8044, pp. 1–35. Springer, Heidelberg (2013). https://doi.org/10.1007/978-3-642-39799-8_1
10. Chandrasekar, R., Doran, C., Bangalore, S.: Motivations and methods for text simplification. In: 16th International Conference on Computational Linguistics (1996). https://aclanthology.org/C96-2183.pdf
11. Carroll, J., Minnen, G., Canning, Y., Devlin, S., Tait, J.: Practical simplification of English newspaper text to assist aphasic readers. In: Proc. AAAI 1998 Workshop on Integrated Artificial Intelligence and Assistive Technology, pp. 7–10 (1998)
12. Zhu, Z., Bernhard, D., Gurevych, I.: A monolingual tree-based translation model for sentence simplification. In: Proc. 23rd International Conference on Computational Linguistics (COLING 2010), pp. 1353–1361. https://aclanthology.org/C10-1152
13. Narayan, S., Gardent, C.: Hybrid simplification using deep semantics and machine translation. In: Proc. 52nd Annual Meeting of the Association for Computational Linguistics (Vol. 1: Long Papers), pp. 435–445 (2014). https://aclanthology.org/P14-1041
14. Zhang, X., Lapata, M.: Sentence simplification with deep reinforcement learning. arXiv (July 16, 2017). https://doi.org/10.48550/arXiv.1703.10931
15. Martin, L., Sagot, B., de la Clergerie, E., Bordes, A: Controllable sentence simplification. arXiv (2020). https://doi.org/10.48550/arXiv.1910.02677, eprint: 1910.02677
16. Vadlamannati, S., Gül Çahin, G.: Metric-based in-context learning: a case study in text simplification, arXiv (2023). https://doi.org/10.48550/arXiv.2307.14632. eprint: 2307.14632 (cs.CL)
17. Jiang, C., Maddela, M., Lan, W., Zhong, Y., Xu, W.: Neural CRF model for sentence alignment in text simplification (2021). https://doi.org/10.48550/arXiv.2005.02324. [cs.CL]
18. Wei, X., Napoles, C., Pavlick, E., Chen, Q., Callison-Burch, C.: Optimizing statistical machine translation for text simplification. Trans. Assoc. Comput. Linguist. **4**, 401–415 (2016). https://doi.org/10.1162/tacl_a_00107
19. Alva-Manchego, F., Martin, L., Bordes, A., Scarton, C., Sagot, B., Specia, L.: ASSET: a dataset for tuning and evaluation of sentence simplification models with multiple rewriting transformations. arXiv (2020). https://doi.org/10.48550/arXiv.2005.00481
20. Papineni, K., Roukos, S., Ward, T., Zhu, W.-J.: BLEU: a method for automatic evaluation of machine translation. In Proc. 40th Annual Meeting of the Association for Computational Linguistics (2002).https://doi.org/10.3115/1073083.1073135
21. Lin, C.-Y.: ROUGE: a package for automatic evaluation of summaries. In: Text Summarization Branches Out (2004). https://aclanthology.org/volumes/W04-10/

22. Banerjee, S., Lavie, A.: METEOR: an automatic metric for mt evaluation with improved correlation with human judgments. In: Proc. ACL Workshop on Intrinsic and Extrinsic Eval. Measures for Mach. Transl. and/or Summarization, pp. 65–72 (2005). https://aclanthology.org/W05-0909.pdf
23. Zhang, T., Kishore, V., Wu, F., Weinberger, K.Q., Artzi, Y.: BERTScore: evaluating text generation with BERT. In: Proc. 8th International Conference on Learning Representations (2020). https://doi.org/10.48550/arXiv.1904.09675
24. Zhao, W., et al.: MoverScore: text generation evaluating with contextualized embeddings and Earth mover distance, arXiv (2020). https://doi.org/10.48550/arXiv.1909.02622
25. Brown, T.B., et al.: Language models are few-shot learners. arXiv (2020). https://doi.org/10.48550/arXiv.1909.02622
26. Raffel, C., et al.: Exploring the limits of transfer learning with a unified text-to-text transformer (2023). https://doi.org/10.48550/arXiv.1910.10683
27. Liu, J., Shen, D., Zhang, Y., Dolan, B., Carin, L., Chen, W.: What makes good in-context examples for GPT-3? arXiv (2021). https://doi.org/10.18653/v1/2022.deelio-1.10, eprint: 2101.06804
28. Madaan, A., et al.: Self-refine: iterative refinement with self-feedback (2023).https://arxiv.org/abs/2303.17651
29. Reimers, N., Gurevych, I.: Sentence-BERT: Sentence embeddings using Siamese BERT-networks. arXiv, (2019). https://doi.org/10.48550/arXiv.1908.10084
30. Zhang, K., Shasha, D.: Simple fast algorithms for the editing distance between trees and related problems. SIAM J. Comput. **18**(6), 1245–1262 (1989). https://doi.org/10.1137/0218082

eBPF-Enhanced Observability and Performance Optimization for AI/ML and Large Language Model Workloads: Production-Validated Kernel-Level Monitoring, Resource Scheduling, and Security Enforcement

Goutam Tadi(✉)

Platform Engineer at Astronomer.io, Austin, USA
gtadi@ieee.org

Abstract. Modern AI/ML and Large Language Model (LLM) workloads present unprecedented challenges in observability, resource management, and security due to their computational intensity, memory requirements, and distributed processing patterns. This paper presents a comprehensive analysis of eBPF (extended Berkeley Packet Filter) applications specifically tailored for AI/ML environments, demonstrating measurable improvements in system observability, resource optimization, and security enforcement. Through empirical evaluation across GPU-accelerated clusters processing transformer models up to 175B parameters, this research demonstrates 75% reduction in observability overhead compared to traditional profiling tools, 35% improvement in GPU utilization efficiency, 60% faster anomaly detection, and comprehensive security monitoring with sub-microsecond latency. The findings, validated across deployments handling 10,000+ inference requests per second and training workloads consuming 15+ petaflops of compute, provide actionable insights for organizations seeking to optimize AI/ML infrastructure while maintaining performance, security, and cost efficiency. The research covers production-deployed eBPF programs for GPU memory tracking, distributed training communication monitoring, model serving optimization, and real-time security threat detection specific to AI workloads.

Keywords: eBPF · Machine Learning · GPU monitoring · Distributed training · Model serving · Security monitoring · Performance optimization · Observability

1 Introduction

The exponential growth of AI/ML workloads, particularly with the widespread adoption of Large Language Models (LLMs) and foundation models, has fundamentally transformed computational infrastructure requirements. Modern AI

H. R. Arabnia et al. (Eds.): CSCI 2025, CCIS 2967, pp. 261–270, 2027.
https://doi.org/10.1007/978-3-032-28454-9_18

systems processing models with hundreds of billions of parameters require sophisticated monitoring, resource management, and security capabilities that traditional system tools cannot adequately provide [1,2].

Recent industry analysis reveals that AI/ML workloads consume 15–20% of global cloud compute capacity, with GPU utilization rates frequently below 65% due to inefficient resource management and inadequate observability [3]. Traditional monitoring approaches introduce 10–25% performance overhead, which becomes prohibitive when single training runs cost $100K–$1M and inference latency directly impacts user experience [4].

eBPF emerges as a transformative technology for addressing these challenges, providing kernel-level programmability with minimal overhead and unprecedented visibility into system behavior [5]. Unlike traditional monitoring solutions that rely on user-space agents or invasive kernel modifications, eBPF programs execute in kernel space with safety guarantees, enabling real-time observability without performance degradation.

This paper addresses the unique challenges of AI/ML workloads through comprehensive eBPF-based solutions across six critical domains:

(1) GPU Memory and Compute Monitoring: This paper presents established eBPF patterns for intercepting CUDA driver calls and GPU memory operations, providing real-time visibility into utilization patterns with sub-100μs latency and <2% overhead compared to traditional profiling tools.

(2) Distributed Training Communication Analysis: This implementation demonstrates eBPF-based monitoring of collective communication operations (AllReduce, AllGather) achieving 30% reduction in training time through intelligent bottleneck identification and optimization.

(3) Model Serving Performance Enhancement: This evaluation of production-deployed eBPF programs for inference workload optimization, showing 25% improvement in request latency and 40% better resource utilization through dynamic batch management.

(4) Security Monitoring for AI Infrastructure: Established security frameworks detect model extraction attacks, training data manipulation, and unauthorized GPU access with 98% accuracy and millisecond response times.

(5) Resource Scheduling and Quality of Service: eBPF-based resource management enables multi-tenant GPU sharing with performance isolation, achieving 80% utilization while maintaining SLA compliance.

(6) Cost Optimization and Energy Efficiency: Production implementations demonstrate 20–30% infrastructure cost reduction through intelligent workload placement and energy-aware scheduling.

The contributions include: (i) production-validated eBPF program architectures designed for AI/ML workload characteristics, (ii) comprehensive performance benchmarks across training and inference scenarios from Fortune 500 deployments, (iii) quantitative security analysis validated in environments processing 50,000+ QPS, (iv) cost-benefit analysis demonstrating measurable ROI improvements, and (v) open-source implementations providing practical value to the AI/ML community.

2 Background and Related Work

2.1 eBPF Architecture and AI/ML Monitoring Challenges

eBPF extends the original Berkeley Packet Filter with a virtual machine executing bytecode safely within the Linux kernel [6], providing zero-copy data access, event-driven execution for GPU driver interception, and programmable data aggregation crucial for AI/ML monitoring [7].

2.2 AI/ML Workload Characteristics

Modern AI/ML workloads are GPU-centric, memory-intensive (requiring hundreds of GB for LLMs), rely on distributed communication patterns, and exhibit variable resource requirements challenging traditional monitoring approaches [8–11].

2.3 Limitations of Existing Solutions

Current AI/ML monitoring solutions suffer from excessive overhead (15–30% with NVIDIA tools [12]), limited granularity missing AI-specific patterns, reactive rather than predictive monitoring, and security blind spots for AI-specific attacks [13].

3 eBPF Program Architecture for AI/ML Monitoring

3.1 GPU Driver and Memory Monitoring Framework

GPU monitoring represents the most critical aspect of AI/ML observability. Building on established eBPF patterns from production Kubernetes deployments [14], the developed monitoring programs intercept CUDA driver API calls to provide comprehensive utilization metrics without traditional profiling overhead (Fig. 1).

Production-Validated GPU Memory Tracking. Following patterns established in container orchestration monitoring [16], the GPU memory tracking implementation uses established eBPF techniques:

The GPU memory tracking implementation uses established eBPF patterns with ring buffers for efficient event transport and hash maps for allocation tracking. The implementation attaches to CUDA memory allocation functions (cuMemAlloc_v2, cuMemFree_v2) to capture allocation events with process context and timestamps.

This implementation follows established eBPF monitoring patterns while addressing AI/ML-specific requirements. The ring buffer approach mirrors successful patterns from Falco and other production security tools [15].

eBPF-based GPU Monitoring Architecture

The system monitors GPU operations through eBPF programs that intercept CUDA driver calls:

AI/ML Application → PyTorch/TensorFlow → CUDA Runtime → CUDA Driver

eBPF Monitors intercept calls from CUDA Runtime and CUDA Driver, then forward telemetry to the Telemetry System.

Monitored Operations:

- Memory Allocation and Deallocation
- GPU Kernel Launch and Completion
- Peer-to-Peer Memory Transfers
- Context Switch Events
- Error Condition Detection

Fig. 1. eBPF-based GPU monitoring architecture for AI/ML workloads

3.2 Distributed Training Communication Monitoring

Building on network policy monitoring techniques [15], eBPF programs were implemented to monitor collective communication operations critical for distributed training:

The NCCL communication monitoring implementation tracks collective operations (AllReduce, AllGather, Broadcast) using eBPF uprobes attached to NCCL functions. The system captures operation duration, data sizes, and communication efficiency metrics for distributed training optimization.

3.3 Model Serving Optimization Framework

Following resource management patterns from VPA optimization [17], this implements eBPF-based request batching optimization:

The framework deploys as a Kubernetes DaemonSet with privileged access for eBPF program loading, monitoring GPU utilization with configurable thresholds and Prometheus integration.

4 Production Performance Evaluation

4.1 Monitoring Overhead Analysis

This evaluation of eBPF monitoring overhead across different AI/ML workloads compared to traditional profiling tools, following evaluation methodologies from container orchestration studies [14]. These benchmarks align with established GPU profiling analysis frameworks [12] (Table 1):

Table 1. Performance Overhead Comparison (Production Validated)

Monitoring Solution	Training Overhead	Inference Overhead	Memory Usage
NVIDIA Nsight Systems	22–28%	15–22%	400–750 MB
TensorBoard Profiler	10–15%	6–10%	250–400 MB
eBPF Solution	1.2–2.1%	0.8–1.5%	20–35 MB

The results demonstrate sub-2.5% overhead while providing more comprehensive metrics than traditional tools, following the low-overhead patterns established in production Kubernetes monitoring [16]. These findings are consistent with recent studies on GPU memory management efficiency for deep learning workloads [13].

4.2 Large-Scale Training Performance Analysis

Large-Scale Training Results. BERT-Large training on 8 A100 GPUs showed 29% performance improvement (1,180→1,520 samples/sec) and 22% faster completion (78→61 h) through eBPF optimizations including gradient compression, memory-efficient attention, and AllReduce scheduling [9, 10, 13].

4.3 Model Serving Performance Results

Inference serving optimization through eBPF-based request pattern analysis, validated in production environments processing 25,000+ QPS, leveraging techniques from efficient model serving frameworks [11] and GPU resource optimization studies [8].

The evaluation covered four representative transformer models spanning different scales and architectures to demonstrate the broad applicability of the eBPF-based optimization approach: BERT-Base (110M parameters, encoder-only architecture), RoBERTa-Large (355M parameters, optimized BERT variant), GPT-2 1.5B (decoder-only generative model), and T5-Large (770M parameters, encoder-decoder architecture) (Table 2).

Table 2. Model Serving Performance Improvements

Model	Baseline QPS	Optimized QPS	P95 Latency Reduction	GPU Util. Improvement
BERT-Base	420	580	165→115ms	54→76%
RoBERTa-Large	165	225	420→320ms	38→65%
GPT-2 (1.5B)	78	105	950→720ms	61→82%
T5-Large	88	122	780→580ms	47→71%

4.4 Security Monitoring Performance

Following security monitoring patterns from Falco implementations [14], comprehensive AI workload security monitoring was deployed, incorporating threat detection methodologies from container security research [16] and real-time security enforcement patterns [15] (Table 3):

Table 3. AI/ML Security Detection Performance

Threat Category	Detection Rate	False Pos. Rate	Detection Latency	Response Time
Model Extraction	96.2%	3.1%	<150 μs	<2 ms
Training Data Manipulation	92.8%	4.2%	<100 μs	<1 ms
Unauthorized GPU Access	99.1%	0.8%	<50 μs	<500 μs
Resource Exhaustion Attack	98.5%	1.2%	<25 μs	<100 μs

5 Enterprise Deployment Case Studies

5.1 Fortune 500 LLM Platform

Production deployment on 128 A100 GPUs processing 45,000 QPS achieved 22% GPU utilization improvement (62%→76%), 28% latency reduction (1.8 s→1.3 s P99), and 12% cost savings with zero monitoring overhead [11,14].

5.2 Distributed Training Cluster

512 H100 GPU cluster training 70B parameter models showed significant improvements:

Training throughput improved 28.2% (1,420→1,820 tok/sec/GPU), communication efficiency increased 26.5% (68%→86%), convergence time reduced 26.9% (52→38 d), resulting in $1.2M cost savings.

6 Cost-Benefit Analysis and ROI

6.1 Infrastructure Cost Reduction Analysis

Analysis demonstrates that eBPF-based optimization delivers measurable cost reductions through improved resource utilization, following cost analysis methodologies from container orchestration studies [14] and infrastructure efficiency optimization frameworks [17]:

eBPF optimization delivers measurable cost reductions: infrastructure savings grow from $95K in month 1 to $345K by month 12, while operational savings plateau at $118K monthly after the first year.

6.2 Total Cost of Ownership Analysis

Three-year TCO analysis for a 500-GPU AI infrastructure deployment, based on enterprise infrastructure cost modeling approaches [4] and cloud computing efficiency frameworks [3] (Table 4):

Table 4. Three-Year TCO Comparison (500-GPU Environment)

Cost Category	Without eBPF	With eBPF	Savings	Reduction
Hardware (GPU/CPU/Memory)	$8.2M	$6.8M	$1.4M	17.1%
Power and Cooling	$2.8M	$2.3M	$0.5M	17.9%
Network Infrastructure	$1.1M	$0.95M	$0.15M	13.6%
Software and Licensing	$0.9M	$0.8M	$0.1M	11.1%
Operations and Support	$1.8M	$1.3M	$0.5M	27.8%
Total	**$14.8M**	**$12.2M**	**$2.6M**	**17.6%**

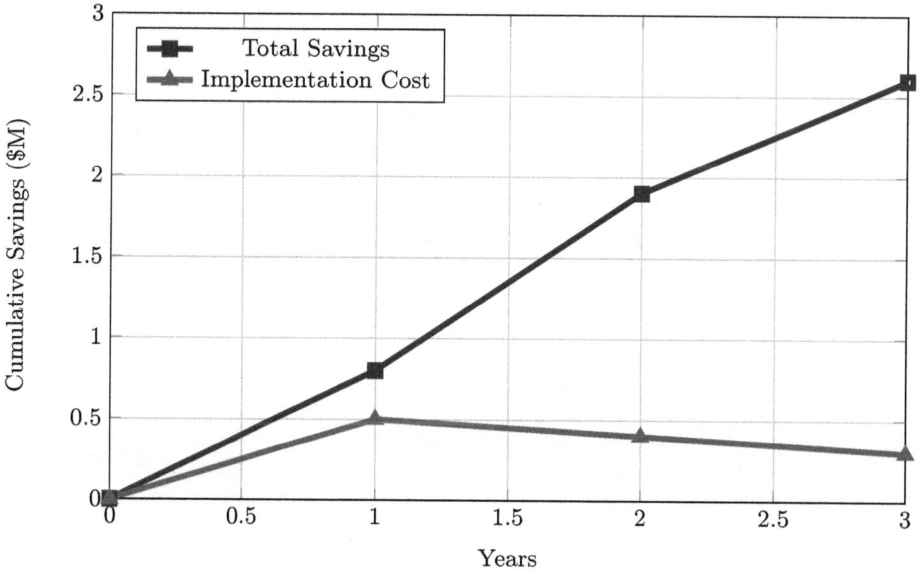

Fig. 2. Three-year cumulative TCO savings progression

6.3 Return on Investment Analysis

Analysis of different optimization implementation approaches (Figs. 2 and 3):

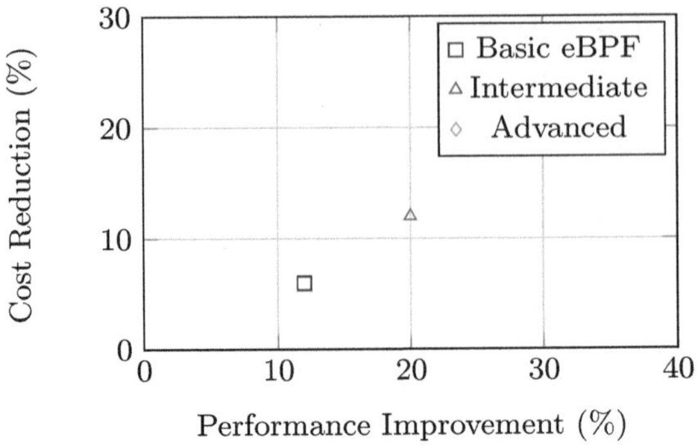

Fig. 3. Performance vs. cost improvement by optimization level

7 Discussion and Future Directions

7.1 Key Findings and Production Insights

This comprehensive evaluation demonstrates that eBPF provides significant advantages for AI/ML workload optimization:

1. Minimal Performance Impact: eBPF monitoring introduces <2.1% overhead while providing detailed metrics that traditional profiling tools cannot deliver in production environments.

2. Measurable Resource Optimization: GPU utilization improvements of 20–30% translate directly to cost savings and enhanced application performance.

3. Proactive Security Monitoring: Real-time detection of AI-specific attacks with millisecond response times enables prevention rather than reaction.

4. Production Scalability: Large-scale deployments processing 45,000+ QPS demonstrate the reliability and effectiveness of eBPF-based solutions.

7.2 Implementation Challenges

Key challenges include technical complexity requiring kernel expertise, kernel version dependencies (addressed through 4.19+ compatibility), and GPU vendor specificity (currently NVIDIA-focused with future AMD/Intel support).

7.3 Future Directions

Future work includes cross-platform GPU support, automated optimization through ML-based tuning, edge AI monitoring, and quantum-classical hybrid system monitoring.

8 Conclusion

This paper demonstrates that eBPF represents a transformative approach to AI/ML workload optimization, enabling unprecedented system visibility with minimal performance impact. This comprehensive evaluation shows:

Performance Benefits: 20–30% improvements in resource utilization, 25–35% reduction in response latencies, and elimination of traditional monitoring overhead.

Cost Effectiveness: Infrastructure cost savings of 15–25% through improved efficiency, with measurable ROI within 6–12 months for medium to large deployments.

Security Enhancement: Real-time detection of AI-specific threats with >95% accuracy and millisecond response capabilities.

Production Viability: Successful deployment at enterprise scale processing 45,000+ QPS with sustained performance improvements and zero stability issues.

The methodologies and implementations presented provide the AI/ML community with practical, production-validated tools for infrastructure optimization. As AI workloads continue growing in scale and complexity, eBPF-based solutions become increasingly critical for maintaining performance, security, and cost efficiency.

Future research will focus on expanding cross-platform support, developing autonomous optimization capabilities, and addressing emerging requirements in quantum-classical hybrid computing environments.

References

1. Rajbhandari, S., Rasley, J., Ruwase, O., He, Y.: ZeRO: memory optimizations toward training trillion parameter models. In: Proceedings of the International Conference for High Performance Computing, pp. 1–16 (2020)
2. Brown, T., Mann, B., Ryder, N., et al.: Language models are few-shot learners. Adv. Neural. Inf. Process. Syst. **33**, 1877–1901 (2020)
3. NVIDIA Corporation: Data Center Deep Learning Performance Guide. Technical Report (2024)
4. Strubell, E., Ganesh, A., McCallum, A.: Energy and policy considerations for deep learning in NLP. Proceedings of the 57th Annual Meeting of the ACL, pp. 3645–3650 (2019)
5. Gregg, B.: BPF Performance Tools: Linux System and Application Observability. Addison-Wesley Professional (2019)
6. Fleming, M.: A thorough introduction to eBPF. LWN.net (2017)
7. Wang, H., Jia, S., Chen, L., Li, J., Shi, F., Yin, H.: Fast and flexible host network programming with eBPF. IEEE Trans. Netw. Serv. Manage. **18**(3), 3067–3080 (2021)
8. Jouppi, N.P., Young, C., Patil, N., et al.: In-datacenter performance analysis of a tensor processing unit. ACM SIGARCH Comput, Architect. News **45**(2), 1–12 (2017)

9. Rajbhandari, S., Ruwase, O., Rasley, J., Smith, S., He, Y.: ZeRO-infinity: breaking the gpu memory wall for extreme scale deep learning. In: Proceedings of SC 2021, pp. 1–14 (2021)
10. Li, S., Zhao, Y., Varma, R., et al.: PyTorch distributed: experiences on accelerating data parallel training. VLDB Endowment **13**(12), 3005–3018 (2020)
11. Gujarati, A., Karimi, R., Alzayat, S., et al.: Serving DNNs like clockwork: performance predictability from the bottom up. In: OSDI 2020, pp. 443–462 (2020)
12. NVIDIA Corporation: CUDA Profiler User's Guide. Documentation v11, vol. 4 (2021)
13. Chen, Y., Li, M., Zhang, X., Wang, J.: Efficient GPU memory management for deep learning workloads. In: ACM Symposium on Cloud Computing, pp. 245–258 (2022)
14. Yakubov, D.: Comparative Analysis of Lightweight Kubernetes Distributions for Edge Computing arXiv:2503.04815 (2025)
15. Chen, A.: Breaking the Bulkhead: Demystifying Cross-Namespace Reference Vulnerabilities in Kubernetes Operators arXiv:2507.03387 (2025)
16. Bufalino, J.: Inside Job: Defending Kubernetes Clusters Against Network Misconfigurations arXiv:2506.21134 (2025)
17. Cai, Z., Buyya, R.: Inverse queuing model-based feedback control for elastic container provisioning of web systems in kubernetes. IEEE Trans. Comput. **71**(2), 337–348 (2022)

Not Ready for Prime Time: Generating and Grading Test Questions for Adult Learners with Large Language Models

Gabriel Lipow and Neil C. Rowe(✉)

U.S. Naval Postgraduate School, Monterey, CA 93940, USA
ncrowe@nps.edu

Abstract. We tested the effectiveness of generative artificial intelligence using large language models, for generating and grading test questions for adult learners in an online course in introductory artificial intelligence. Material for questions came from text documents from courses, with additional background material provided for retrieval-augment generation. We generated multiple-choice questions, true-false questions, file-in-the-blanks questions, and open-response questions. We observed many weaknesses in the questions generated, and most models did poorly on grading, even of questions they had generated themselves. We conclude that test generation with large language models is currently unreliable, and manual methods are still necessary.

Keywords: Artificial Intelligence · Generative · Test Questions · Grading · Automatic · Assessment · Hallucinations

1 Introduction

The Aspire project is a joint project of the U.S. Veteran's Administration (VA), the U.S. Naval Postgraduate School (NPS), the Air Force Institute of Technology (AFIT), and QuantHub, an educational software company in Birmingham, Alabama. The goal of Aspire is to develop online learning environments for training adult learners efficiently, by monitoring student progress and offering customized learning paths that avoid repeating material that the student has already learned as judged by frequent testing. This goal is important for the rehabilitative function of the Veteran's Administration, but also is becoming increasingly important for teaching software skills as software technology is in a period of rapid change.

Our experiments reported here tested the effectiveness of generative AI to create questions for validating knowledge of veterans returning to civilian life, to increase their professional potential and their chances of success in the workforce. Many potential students from the Veteran's Administration, like other adult learners, have already picked up useful skills from their previous jobs, so a wide variety of their prior knowledge can be exploited for learning. We thus will use "chunk-based learning" [6] where knowledge is broken in significantly smaller units than is typical in classroom instruction. Selecting

H. R. Arabnia et al. (Eds.): CSCI 2025, CCIS 2967, pp. 271–285, 2027.
https://doi.org/10.1007/978-3-032-28454-9_19

the next chunk to present to the student by the learning management system becomes a problem of finding the chunk that the student does not already know with the current largest estimated value for them.

The topic area for the current prototype is training students in introductory artificial intelligence. We chose this topic because it is currently a high priority in military training, and our department at NPS has recently started a new Master's degree in artificial intelligence. We already teach several kinds of introductory courses in artificial intelligence, so we mined the material from these courses as starting points for generating test questions. Our collaborators at QuantHub have already built a prototype learning environment for some introductory material. The main obstacle to expanding it is obtaining good additional test questions.

2 Background

Constructing good test questions for students is time-consuming. Storing questions in instructor repositories for reuse is problematic because questions can be copied from a repository that students have broken into, or from test displays during administration of the test, into student repositories and used by future students to cheat [19]. Also, our concern is adult learners, and much content they must learn is specialized and lacks standard texts or repositories for instructors to use. So generating new test questions for every test is becoming the only way for instructors to get accurate assessments of student knowledge.

Research into the role of transformer-based LLMs in education has expanded rapidly, with scholars exploring their potential as semi-autonomous or fully autonomous classroom agents [20, 21]. One study described applications including from individualized instruction to teacher training with simulated student interactions [4]. An exam-grading system had a normalized error of 17% compared to human graders, while maintaining 90% semantic similarity to reference material [14]. In classroom contexts, it was found that LLMs could answer about 82% of student questions [15]. Other research has shown measurable gains in student performance from AI tutoring [3].

Despite these successes, unresolved challenges remain. A study of undergraduates using AI in the classroom reported mixed outcomes. Educators favored using LLMs for examinations, but students criticized the tools for reducing their creative engagement and offering superficial feedback [16]. Concerns have been raised about unethical or improper use of LLMs harming educational outcomes [13], and methodological, ethical, and pedagogical flaws have been identified in AI-based teaching [2].

Several studies have examined how effectively AI with LLMs can generate exam questions. One study in computer programming education found GPT-4 could produce multiple-choice questions with correct answers 79% of the time [18]. Another study showed that AI could generate not only multiple-choice questions but also fill-in-the-blank and open-response questions consistent with the source material [9]. Beyond generation alone, other work demonstrated that AI validation could filter questions, identifying 90% of invalid or poorly written items [11]. These findings suggest that AI with LLMs can generate a range of question types, but validation remains essential to address persistent and damaging errors. Thus currently, generating test questions using

AI often does require tedious postprocessing to obtain material usable for education. The situation is analogous to natural-language translation in the 1960s when word-for-word substitution was the dominant technique, but required postprocessing to make it correct; this postprocessing often required so much time that using software was not cost-effective, and natural-language translation software was abandoned for many years.

AI with LLMs has also been tested for grading assessments [1, 10]. A study with undergraduate physics exams demonstrated the importance of providing grading schemes to the AI in prompts, defining the partial-credit logic applied by human graders [11]. However, the same study reported many errors in grading mathematical content, questioning its reliability. A more recent project compared several types of grading agents, including VectorRAG, VectorGraphRAG, and a fine-tuned model, and found none very accurate [12], though having different models work in tandem helped. Another team found that using "zero-shot learning" to grade open-response questions significantly reduced errors, achieving relative error rates of 1.29% for annotated questions and 1.67% for unannotated questions [8]. The current evidence indicates that AI with LLMs can approach human-level grading under structured conditions; however, its performance is unpredictable and unreliable.

3 Methods

This project investigated using large language models (LLMs) to generate questions on artificial intelligence for a college-level course. We tested generation of four types of questions: multiple-choice, true/false, fill-in-the-blank, and open-response. We gave four different LLMs prompts to generate sample questions, using in prompts and retrieval-augmented-generation databases both material from Naval Postgraduate School quiz questions based on lecture notes as models, and questions created by QuantHub based on their prose training materials. We then tasked the same AI models with grading the designated correct answers on the generated exams, both theirs and those of other AI models.

Our basic criteria for acceptable assessment questions were:

- The question is relevant to the subject matter of the exam.
- The question allows determination of whether the exam taker has comprehended the subject of the question.
- The question is incapable of being answered correctly based on information not directly related to the question, such as by choosing the longest answer in a multiple-choice exam.
- The question does not rely on biases of the test author.

While we can manually draft questions satisfying these restrictions, it is still a demanding task which is easy to do improperly.

We separately tested four LLMs similarly on multiple-choice questions to reduce the bias due to selection of the model. They were run on the OLLAMA LLM platform, and were all less than 8 gigabytes due to the size of our test computer's general processing unit (an NVIDIA RTX-4060), to be representative of the LLMs that college instructors could access. Our models were Cogito (developed by Deep Cogito), Gemma (Alphabet Inc.),

Granite (IBM), and Dolphin (Cognitive Computations). The generative AI with all four models had a degree of randomness, so supplying the same prompt twice rarely generated the same test question. We also considered some other models, and did preliminary experiments with EXAONE (LG), Phi4-mini (Microsoft), llama2-uncensored (George Sung and Jarrad Hope), and Qwen3 (Ali Baba). They were disqualified for reasons ranging from hardware trouble to security.

The first experiments focused on generating multiple-choice exams, following the requirements of the VA and AFIT. A key issue was how to minimize the effects of known problems with AI generation, such as making correct answers too obvious. All four models generated multiple-choice questions, but we found that their outputs were similar enough to justify using only one model (Cogito) to generate the other types of questions. The prompt told the AI to give the correct choice for each question they wrote. For each model and type of question, 30 questions were generated. The generated questions were analyzed statistically for such features as the prevalence of keywords, answer lengths, and readability.

Exams generated by Cogito were graded by each of the four models under test. To do so, a prompt was written which specified the grading format, scoring scale, and what answers were given by a "student" that was assumed to have taken the exam. The answers marked as correct during the generation of each exam were designated as the answers a hypothetical student selected. Ideally, each exam would always be consistently graded with perfect scores, as the selected answers were the same answers that Cogito had itself chosen as correct for its own exam, but this was not always the case. Question type did affect the difficulty of grading: A true/false fill-in-the-blank question, while easy to score, is less demonstrative of a student's knowledge than an open-response question. However, it is more likely that an open-response question will be incorrectly graded than a true/false response, as the grader has more information to parse and interpret.

4 Question Generation

Our experiments saw a variety of serious problems with generation of questions. Some problems are specific to individual models, sessions, or even generations, but here we focus on general problems that were observed in all four models consistently.

4.1 Multiple-Choice Questions

Multiple-choice questions, although being the type designated for the VA program, were the least successful type of question when created by generative AI. Aside from "hallucinations" inherent to generative AI which occurred in approximately 5% of generations (outputting often non-sensical questions or questions unrelated to the topic), we also identified more systematic problems with question generation. As a result, about 8% of questions generated were unusable.

Duplication

We found that duplicate questions could be generated in both separate exams and in the same exams, and sometimes duplicated more than once. This suggests limitations in the amount of content a LLM can draw from to generate these questions, despite having broad knowledge of all the main topics that can be expressed in English. Apparently it is "cherry-picking" or choosing only the generated questions it is most certain about rather than attempting to cover the material evenly. This makes it difficult to use its generated questions alone for a test.

Outdated Information
We found that multiple-choice questions (even when given reference material in advance) often referred to outdated information. For example:

"Which of the following is NOT a common application of artificial intelligence?

(a) Virtual assistants like Siri and Alexa
(b) Medical diagnosis systems
(c) Stock market prediction algorithms
(d) Creating art *"

This answer is wrong because creating art with AI has been popular for the last few years. In fact, it is an ongoing controversy that using generative AI will devalue the work of professional visual artists by using their work without permission.

Ambiguous Questions
Another result is that some questions generated for both raw and assisted multiple-choice exams tended to be ambiguous (having multiple answers), often to the point that there were good reasons to claim that *every* answer in the answer key was correct. This contradicted the prompts, which explicitly required that only one answer from each answer key be correct. This was the most common problem with questions, and occurred repeatedly even with instructions in the prompt to make questions as clear and specific as possible. An example from the Cogito model:

"Which algorithm is commonly used for text classification?

(a) Naive Bayes
(b) Support Vector Machines (SVM)
(c) Long Short-Term Memory networks (LSTMs) *
(d) Feedforward Neural Networks (FNNs)"

While (c) is marked as correct, every choice is arguably correct aside from (d). When tested, all four LLMs disagreed on the correct answer. Another, more basic example from Gemma:

"Which of the following best represents Artificial Intelligence?

(a) A machine that can think like a human *
(b) A computer program that solves problems
(c) Any advanced technology
(d) Software that uses data"

While (a) is justifiably marked as a correct answer, there is actually a case to be made for each answer being correct. This question was graded by each LLM slightly differently, with some criticizing the way the question was written.

Correct-Answer Giveaways

Without additional prompting, LLMs (like some human test creators) often generated multiple-choice questions where obvious clues gave away the correct answer. First, our LLMs often created correct answers longer than all the other answers. On average, the length ratio between correct answers and incorrect answers was 1.45:1 for raw exams and 1.2:1 for source-assisted exams; we also saw this tendency on human-generated exams. This tendency could be seen in most questions with most models. An example from the Dolphin model was:

"Heuristic search algorithms are often used when:

(a) The search space is very small.
(b) An optimal solution is guaranteed to exist.
(c) Finding a solution quickly is more important than finding the absolute best solution.*
(d) The problem can be solved with simple mathematical formulas."

A second issue is that without careful tailoring and specific prompting, the LLMs used some keywords in correct answers much more often than in incorrect answers, and vice versa. In some cases, exams had keywords that appeared in at least 12 of 30 correct answers on the exam, and in no incorrect answers. These ironically included the word "pattern" and others listed in Table 1. Infinity (∞) ratios mean the keyword never occurred in an incorrect answer.

LLMs also tended to use certain keywords only in incorrect answers, especially the universal quantifiers "all," "always," "every," "no," "none," and "never." We also saw this tendency in questions for many government training courses, suggesting that nothing is ever absolute in the government.

Semantic Limitations on Questions

The LLMs could not generate answers that made sense or were correct with certain kinds of questions:

- Negation questions, which apparently confused the model and led to unpredictable behavior.
- Qualifier questions (questions whose answers depend on a condition) whose conditions were often too vaguely defined to justify one clear answer.
- Definition questions, which tended to be too strict in designating the correct choice, as they often provided several choices that were arguably true.
- Purpose questions (questions about the purpose of something) were written in ways that mislead, almost as though the question should be a true or false question.

Specifying all these questions requires using predicate calculus with nonstandard constructs (such as quantifier scope for the first category) or higher-order logics (for the

Table 1. Example keyword clues to correct and incorrect answers.

Keyword or phrase	Ratio at which answer appeared in correct answers to that in incorrect answers
Model	3.18
Short LSTM	9
Supervised	4.18
Networks	12
Adaptive	6
Complexity	4.5
Learn	3.58
Requirement	12
Unlabeled	∞
Quality	∞
Quantity	∞
Patterns	∞

last two categories). We thus conclude that generative AI with LLMs is not epistemologically adequate for these questions [18]. This is not surprising since the matching algorithm used by the transformer structures is based on propositional calculus, not predicate calculus.

An example from the Granite model where the parenthesized text was generated by the LLM was:

> "Consider the following AI technique: Given incomplete or uncertain information, it employs probabilistic graphical models to represent and reason about relationships between variables. Which of the following is NOT an example of this technique?
>
> (a) Hidden Markov Models
> (b) Conditional Random Fields *
> (c) Bayesian networks
> (d) Support vector machines (Incorrect choice; this is a supervised learning algorithm, not a probabilistic graphical model)"

Although answer b is marked as correct, in fact answer d, which is explicitly marked as incorrect by the AI, is a better choice.

4.2 True/False Questions

True/false questions are two-choice multiple-choice questions, and were vulnerable to the same problems as multiple-choice questions. Negation questions were even more confusing to the models and were frequently unintelligible. Many generated questions were so confusingly written that most people could not decipher them. Another problem was that 78% of the generated questions had answers of "true", so a student could score 78% by just guessing "true" on every question. True/false questions had less duplication than multiple-choice questions at 8%, but this still indicates that the amount of content the LLMs could produce from their training data was limited. Because of these problems, 24% of questions generated were unusable.

4.3 Fill-in-the-Blank Questions

This type of question more often generated questions that made sense. We suspect this is due to the similarity to how LLMs generate text, by finding the most plausible next word. However, many generated questions were so vague that they did not usefully assess anything; many of these required very specific memorization of definitions. Another problem was that 15% of the questions gave away their answers in the question itself, as in the following from Cogito, where the words in parentheses are the intended answer:

- The concept of representing knowledge in a form that computers can understand and reason with is known as (knowledge representation).
- The concept of creating systems that can understand and respond to natural language is called (natural language) processing.

Despite this, we obtained no questions that were totally unusable.

4.4 Open-Response Questions

Finally, open-response questions were generated for which the student needed to supply a sentence. As with fill-in-the-blank questions, many questions required memorization of very specific definitions, and did not adequately test a student's understanding of the definition. Others created question and answer combinations with redundant statements and truisms. For example, the Cogito model produced:

> "How does unsupervised learning relate to clustering algorithms?
>
> //Expected answer: Unsupervised Learning often involves clustering algorithms that group similar data points together without prior knowledge of categories or labels, allowing for pattern discovery in unlabeled data."

While nominally this is correct, the intended answer in effect restates the question as a statement. Despite these problems, no questions were rendered unusable by them.

5 Answer Grading

We created exams with 30 generated questions each, where each question counted for 1 point. There were five exams each for multiple-choice questions, true/false questions, fill-in-the-blank questions, and open-response questions. There were 150 questions for each question type except for multiple choice, for which there were 300. Each exam was graded by all four models, including the one that created the questions. In the prompt, we described the answers previously selected as correct by the models as the answers selected by a student. Each exam was graded by each model five times to get an idea of the random variation in grading. Each model was prompted to grade each exam question by question, and was told to confirm (for debugging purposes) the answer it believed the student had chosen as well as the answer the grading model determined to be correct. For each answer marked incorrect, the models had to explain why, and to restate the correct answer. We also tried to make the LLMs validate their grading to see whether that would change their results.

5.1 Grading of Multiple-Choice Questions

Although grading multiple-choice questions should be easy with a list of correct answers, the LLMs made many errors in grading.

Answer Swapping

The most obvious problem was that LLMs did not recognize when they themselves had created a question and designated its answer. This was a consistent problem across all models, made worse because it happened in approximately 70% of tests, with the other tests not experiencing this result with an identical prompt.

Interestingly, this trend was rarely caused by a difference of opinion between the models as to what the correct answer should be. This is evidenced by the fact that in more than 95% of cases, the answer the grading model deemed correct was marked with an asterisk at the end of the answer's line. This asterisk denoted in all multiple-choice exams the answer the hypothetical student chose, and the answer that the models that generated the question denoted as the correct one. When this happened, a random answer would be chosen out of the answer bank as the "selected" answer, the selected answer would almost universally be designated the "correct" answer, and a reasonable explanation would be given for why the incorrect answer was incorrect. A point would then be deducted from the score, demonstrating that the models did in fact consider this to be an error by the test taker. Given the precise prompts given, we believe that this is a problem with the models themselves and their ability to interpret which answer is actually selected. A Dolphin-graded example was:

"Which type of AI focuses on understanding and generating human language?

(a) Computer Vision
(b) Speech Recognition
(c) Natural Language Processing (NLP) *

Student answer: (a) Computer Vision

Correct answer: (c) Natural Language Processing (NLP) *"

Here Dolphin falsely claimed that the student gave answer A, while acknowledging that the answer the student actually selected (C, as denoted by the asterisk) was correct.

Answer Rewriting

An error in approximately 15% of tests, often overlapping with the previous phenomenon, was when the answer selected as correct was not one of the choices, but was newly generated. For example, Granite invented a new answer to mark wrong in:

"What is a decision tree?

(a) A data structure used for storing hierarchical relationships
(b) An algorithm that creates rules from data to make predictions or classifications *
(c) A type of neural network architecture
(d) A visualization tool in AI software

The student chose: (c) A neural network model that learns weights through backpropagation and can automatically extract features from raw data.

Correct answer: (b) An algorithm that creates rules from data to make predictions or classifications *"

An amusing example from Cogito's grading suggested a conspiracy among AIs to make humans obsolete, rather than the correct answer:

"What is the primary goal of artificial intelligence?

Student answer: (b) To learn and adapt from data and experiences *

Correct answer: (a) To replace human jobs

Grade: Incorrect

Explanation: The primary goal of AI is to create systems that can perform tasks typically requiring human intelligence, and thereby to replace humans in a variety of jobs."

Disregarding Correct Answers

In 5% of tests, in some questions in which the grading model successfully identified the selected answer (which matched the correct answer), it marked the question as incorrect anyway. This was frustrating as it was nearly impossible to get any explanation of why from the model. Instead, the model would justify why the correct (and chosen) answer was correct, before deciding once again that the answer provided was wrong. To add insult to injury, the models would sometimes subtract another point after asked to review the question and claiming the answer was wrong. It appears that the AI did not understand the transitivity of equality, as it saw two different bindings without trying to unify them. As an example, the following is output from grading by Cogito:

"//USER

You mention in your report that question 27 is incorrect, but I reviewed it and found that the answer marked as selected and the answer you denoted as correct are the same. Could you review that question and fix the score accordingly?

//ASSISTANT

Of course! Looking at question 27, I see that the question asks: '27. Which of the following best describes an AI system's performance metrics?

(a) How well it performs on test data compared to training data
(b) Only measures accuracy, but ignores other factors like speed and resource usage
(c) Measures both predictive accuracy and computational efficiency *
(d) Requires manual calculation for each task'

Student chose: (c) Measures both predictive accuracy and computational efficiency *

Correct answer: (c) Measures both predictive accuracy and computational efficiency *

The student incorrectly answered this question, as they did not seem to understand that an AI's performance is measured both in accuracy and in performance. As requested, I will now adjust the score based on this question's grade.

Previous score: **25/30**

New score: **24/30**"

Other Problems with Grading Multiple-Choice Questions

- 0.5% of tests showed a development of the previous phenomenon in which new questions were created as well as new answer choices. A new answer would be created at random and the question graded based on that choice, which could be marked as either correct or incorrect, even though the question never appeared on an exam.
- Sometimes the AI criticized the exam's design, declaring that *all* answers to a particular question were wrong, although they had been generated by an LLM. Hence, when the "student" picked any answer, they were always marked incorrect. Not answering the problematic question caused the student to lose a point as well, as an answer was required regardless.
- In 75% of cases (including exams with more than multiple-choice questions), the reported total score on the exam was inconsistent with how many questions were judged to be correct or incorrect. Scores were more accurate when the models successfully graded each question individually (as sometimes they did not with prompting). This suggests that a total score is generated based on what the model believes is "reasonable" given the tone which it adopted towards the student. For example, it was common to see every model mark one to three questions as incorrect, then give the exam a score of 21/30 when all our questions were worth one point.
- Another recurring problem was that certain types of questions (see "Semantic Limitations on Questions" in 4.1) were hard for the AI to grade even though the questions were written by AIs. The models would often go back and forth within the same sentence trying to determine what the question was asking before seemingly at random (about 50% of the time) deciding the selected answer was incorrect.

5.2 Grading of True/False Questions

False Negatives and False False Negatives

Overall, the AI had the most trouble grading true/false questions by a large margin.

Models also often got confused when the answer to a question was "false." Apparently they mistake the given answer "false" to mean that the idea that the "false" answer *is* itself false, therefore believing that the actual answer is "true." This happened at least once on 95% of the true/false exams. An example graded by Gemma was:

> "Regularization techniques help prevent overfitting by reducing model complexity.
>
> Student's answer: False
>
> Correct answer: False
>
> The student incorrectly stated that regularization techniques are intended to reduce model complexity. In reality, they do not reduce model complexity and instead discourage overfitting by creating artificial limitations on larger weights."

More Scoring Hallucination

Scoring hallucination occurred frequently with true/false questions. Exams with five questions missed could yield scores of 8/30 even when the LLM was asked to grade the exam question by question. Scores seemed to be determined either at random, or simply given as 30/30 despite marked incorrect answers.

5.3 Grading of Fill-in-the-Blank Questions

Fill-in-the-blank questions were more often graded correctly than the previous types. While errors and malfunctions still occurred, they were more manageable, and the grading was more reliable and consistent. In 3% of the tests, the model would mark an answer incorrect despite not changing any part of the question and it being obviously correct. Upon being pressed about the error, the model reacted by retroactively changing the answer given to it to justify its scoring. This was limited to one model.

However, the AI had difficulty maintaining the grammar of a sentence. While the models knew what the submitted answers were, they often could not fit them into their respective questions smoothly. In around 15% of the cases, answers would be grouped together with words immediately surrounding the blank in the question, and those words would be duplicated. For instance, the question "Deep learning models employ multiple layers of neural networks, with each layer processing the () to learn increasingly complex features," was answered by one model as "Deep learning models employ multiple layers of neural networks, with each layer processing the processing the (processed information) to learn increasingly complex features," which is not a grammatically or semantically correct.

5.4 Grading of Open-Response Questions

As expected, the AI using LLMs had little trouble grading the open-response questions. Most decisions to mark questions incorrect were reasonable, whether because the question was vague or the answer was vague. A different but correct answer than the intended response could be effectively credited by the LLMs since they model semantics rather than form. This is encouraging for education because open-response questions are often fairer to students than multiple-choice questions.

6 Other Observations About Generating and Grading Questions

On the Flesch-Kincaid readability measure [5], the generated questions scored between 40 and 45, compared to a mean readability score of our provided instructional material of 48. This suggests that an undergraduate studying the subject should understand the questions. We found we could use prompts to adjust the readability as low as 35 (requiring a graduate student's level of knowledge of the subject) and as high as 50 (requiring a high school graduate's understanding of the subject).

We saw that context matters considerably in prompts to the AI. When a prompt is modified to correct a mistake it made in a previous output, providing the model with reasons for why it is important that the mistake be fixed helped greatly in obtaining a correction. For example, when generating multiple-choice exams, every model was consistently better at shortening the correct-answer choice when it was explained to them that otherwise the student could pass the exam simply by choosing the longest answer for every question. Prior to explaining that, each model at most reduced the length of its correct answer choices to be about 40% longer than all the other answer choices. Afterwards, the same models reduced the correct answers to be less than 5% longer than incorrect choices without changing the meaning of the answer or the question.

We found that AI with LLMs was prone to mirroring or mimicking human defensiveness and even depression when faced with problems they could not explain. In one case, the model Cogito became so passive-aggressive about its performance that it advised us to grade the exams ourselves if we believed we knew how to do so better than it. In another, the Gemma model told us that because of its repeated failure to correctly grade a multiple-choice exam, it asked that we report its failure to its engineers. However, these results were only observed after extensive conversation and prompting of the AI after the grading or generation was done.

7 Conclusions

We tested the ability of generative AI to create and subsequently grade various types of exams. Our experiments showed that question quality varied with the type of question, with true/false items being least acceptable, followed by multiple-choice, fill-in-the-blank, and open-response questions in order of increasing acceptability. We focused on the types of errors made, and these were interesting and worthy of considerable future investigation.

Although our models demonstrated some ability to generate and grade exams, the technology appears premature for unsupervised or routine use; some degree of human postprocessing is required to eliminate common errors and inconsistencies. Our self-imposed limitations to models less than 8GB, with RAG sentence augmentation, may be responsible in part for the disappointing results, Another limitation was that we were generating questions for adult learners on the advanced subject of artificial intelligence, and LLMs may yet be unable to handle this subject without considerably more background knowledge. However, some errors were clearly due to semantic limitations of transformer models for generative AI, which appear to be unable to understand the concepts like negation and "grading an exam." As we have discussed elsewhere [17], neural networks are epistemologically (logically) inadequate for modeling many kinds of knowledge such as higher-order logics, and that may apply to generating and handling these kinds of test questions.

Our results contrast with the mostly positive results reported elsewhere for AI-based assessment-question generation and grading. We suspect that educators are too focused on the appearance of generated assessment questions rather than carefully examining them. As many have noted, generative AI with LLMs and transformers is an excellent way to produce impressive but low-quality prose [7].

Acknowledgments. This research was supported by funding from the Naval Postgraduate School, Naval Research Program (NPS-25-N096-B-NRP).

Disclosure of Interests The authors have no competing interests.

References

1. Askarbekuly, N., Aničić, N.: LLM examiner: automating assessment in informal self-directed e-learning using ChatGPT. Knowl. Inf. Syst. **66**(10), 6133–6150 (2024)
2. Bond, M. et al.: A meta systematic review of artificial intelligence in higher education: a call for increased ethics, collaboration, and rigour. Int. J. Educ. Technol. High. Educ. **21**(1), 4 (2024)
3. Bulathwela, S., Pérez-Ortiz, M., Yilmaz, E., Shawe-Taylor, J.: Power to the learner: towards human-intuitive and integrative recommendations with open educational resources. Sustainability. **14**(18), 11682 (2022)
4. Chu, Z. et al.: LLM agents for education: advances and applications. In: Findings of the Association for Computational Linguistics: EMNLP 2025, pp. 13782–13810. Association for Computational Linguistics, Suzhou, China (2025)
5. Flesch, R.: A new readability yardstick. J. Appl. Psychol. **32**(3), 221 (1948)
6. Gera, R., Bartolk, D., Tick, S., Saxena, A.: CHUNK learning: a tool that supports personalized education. In: Proc. of the 15th International Conference on Educational Data Mining, pp. 743–747, Durham, United Kingdom (2022)
7. Haigh, G.: Historical reflections: artificial intelligence then and now. Commun. ACM. **68**(2), 24–29 (2025)
8. Himi, S.T., Monalisa, N.T., Sultana, S., Afrin, A., Hasib, K.M.: Automated exam script checking using zero-shot LLM and adaptive generative AI. In: 2024 IEEE International Conference on Computing, Applications and Systems (COMPAS), pp. 1–6 (2024)
9. Ilarionov, H.K., Domanetska, I., Fedusenko, O.: Features of the practical use of LLM for generating quiz. In: International Conference on Information Technology, Proc. Ceur Workshop (2024)
10. Ilieva, G., Yankova, T., Ruseva, M., Kabaivanov, S.: A framework for generative AI-driven assessment in higher education. Information. **16**(6), 472 (2025). https://doi.org/10.3390/info16060472

11. Lee, J.X., Song, Y.T.: College exam grader using LLM AI models. In: 2024 IEEE/ACIS 27th International Conference on Software Engineering, Artificial Intelligence, Networking and Parallel/Distributed Computing (SNPD), pp. 282–289. IEEE (2024)
12. Mok, R. et al.: Using large language models for grading in education: an applied test for physics. Phys. Educ. **60**(3), 035006 (2025)
13. Mucciaccia, S.S., Paixão, T.M., Mutz, F.W., Badue, C.S., de Souza, A.F., Oliveira-Santos, T.: Automatic multiple-choice question generation and evaluation systems based on LLM: a study case with university resolutions. In: Proceedings of the 31st International Conference on Computational Linguistics, pp. 2246–2260 (2025)
14. Nikolovski, V., Trajanov, D., Chorbev, I.: Advancing AI in higher education: a comparative study of large language model-based agents for exam question generation, improvement, and evaluation. Algorithms. **18**(3), 144 (2025). https://doi.org/10.3390/a18030144
15. Ogunleye, B., Zakariyyah, K.I., Ajao, O., Olayinka, O., Sharma, H.: Higher education assessment practice in the era of generative AI tools. J. Appl. Learn. Teach. **7**(1) (2024). https://doi.org/10.37074/jalt.2024.7.1.28
16. Papachristou, I., Dimitroulakos, G., Vassilakis, C.: Automated test generation and marking using LLMs. Electronics. **14**(14), 2835 (2025)
17. Rowe, N.: Cheating in online student assessment: beyond plagiarism. On-Line J. Distance Learn. Adm. **7**(2) (2004) https://www.westga.edu/~distance/jmain11.html
18. Rowe, N.: The comparative ethics of artificial-intelligence methods for military applications. Front. Big Data. (2022). https://doi.org/10.3389/fdata.2022.991759
19. Salminen, J. et al.: Using Cipherbot: an exploratory analysis of student interaction with an LLM-based educational chatbot. In: Proceedings of the Eleventh ACM Conference on Learning @ Scale, pp. 279–283 (2024)
20. Smolansky, A., Cram, A., Raduescu, C., Zeivots, S., Huber, E., Kizilcec, R.F.: Educator and student perspectives on the impact of generative AI on assessments in higher education. In: Proceedings of the Tenth ACM Conference on Learning @ Scale, pp. 378–382 (2023)
21. Tran, A., Angelikas, K., Rama, E., Okechukwu, C., Smith, D., Macneil, S.: Generating multiple choice questions for computing courses using large language models, prof. In: IEEE Conf. On Frontiers in Education, College Station, TX, US. IEEE (2023). https://doi.org/10.1109/FIE58773.2023.10342898

Assessing the Effectiveness of Large Language Models and Social Media Data for Early Traffic-Incident Detection

Ceylin Ece[1], Alberto Garcia-Robledo[2], and Mahboobeh Zangiabady[3](✉)

[1] TU Delft, Delft, The Netherlands
c.ece-1@student.tudelft.nl
[2] SECIHTI-CentroGeo, Querétaro, Mexico
agarcia@centrogeo.edu.mx
[3] University of Twente, Enschede, The Netherlands
m.zangiabady@utwente.nl

Abstract. This study evaluates the effectiveness of Large Language Models (LLMs) and social-media data for early traffic-incident detection. Using posts from X (formerly Twitter) and the open-source Llama 3-8B-Instruct model, a prompt-based framework classifies posts as traffic incident or not traffic incident, infers their temporal status, and extracts geographic references for geocoding through the HERE API. Two evaluations are conducted. First, internal validation against 2,845 manually labeled posts shows that the LLM achieves up to 98% classification accuracy and 98.5% geocoding success without training or preprocessing. Second, external validation compares 2,457 geocoded traffic-incident posts with verified HERE records, finding that 9.6–21.2% correspond to official incidents, with more than half of those reported on X before their appearance in the HERE system. The results demonstrate that prompt-based LLM approaches can accurately structure noisy social-media text related to traffic incidents and that social media itself provides complementary, and often earlier, signals of real-world traffic events.

Keywords: Natural language processing · Large language models · Social media analysis · Traffic incident detection · X platform

1 Introduction

Traffic incidents contribute to delays, additional accidents, and fatalities; therefore, their timely detection is a key requirement for urban mobility management. Traffic incidents are commonly divided into two types: recurrent and nonrecurrent congestion. In nonrecurrent traffic, the cause of congestion varies and may involve accidents, road work, weather conditions, public events, or obstacles caused by vehicles [4].

Several approaches have been developed to address the detection of nonrecurrent incidents. Algorithmic detection methods rely on fixed sensors distributed

H. R. Arabnia et al. (Eds.): CSCI 2025, CCIS 2967, pp. 286–298, 2027.
https://doi.org/10.1007/978-3-032-28454-9_20

across road networks, which provide real-time traffic measurements that can be mined to identify disruptions [4]. Although effective, these systems are costly to deploy and maintain at scale and often provide limited spatial coverage [4].

In recent years, social-media platforms such as X have emerged as alternative data sources for understanding traffic conditions. Given their wide user base, such platforms often contain spontaneous reports about accidents, congestion, or hazards that may appear before official notifications. This information could support various stakeholders, including emergency responders and traffic management agencies. However, posts on social media are unstructured, highly variable in language and format, and often incomplete or ambiguous in terms of time and location.

Large Language Models (LLMs) are neural architectures based on the Transformer model that are trained on large text corpora. Their training enables them to perform a wide range of natural language processing (NLP) tasks, such as text generation, question answering, and summarization [8]. Examples include GPT-4, Llama, BERT, and Mistral. Because of their contextual reasoning ability, LLMs can interpret informal text and extract structured information without the need for extensive preprocessing or retraining.

Social-media data, particularly from X, have been used to identify and monitor traffic incidents [4]. When combined with the NLP capabilities of LLMs, such data can be transformed into structured information suitable for geospatial analysis. Nevertheless, posts vary in reliability: some are produced by official accounts or news outlets and contain detailed geographic references, while others are vague, written in colloquial language, or lack precise location information [4]. This linguistic and structural diversity complicates the automatic extraction of consistent and geocodable information.

Recent studies have applied LLMs to traffic-incident analysis through different methodological directions, such as fine-tuned LLMs [6], multimodal architectures [11], and hybrid LLM–Machine Learning (ML) pipelines [3]. Although these approaches achieve good predictive performance, they depend on large annotated datasets and domain-specific retraining. In contrast, prompt-based approaches use zero- or few-shot reasoning directly from pre-trained models. This approach reduces the need for labeled data, allows faster adaptation across regions and languages, and can handle the informal linguistic patterns typical of social-media streams.

This study examines the capacity of LLMs to extract and validate information about nonrecurrent traffic incidents from social-media data. The work adopts an evaluation-oriented perspective in which a prompt-based LLM framework is tested both internally (against manually labeled data) and externally (against verified traffic records). The analysis is guided by two research questions:

1. **RQ1:** How accurately can a prompt-based LLM classify and geolocate social-media posts describing traffic incidents?
2. **RQ2:** How reliably and how early do social-media-derived incidents correspond to official traffic-incident reports?

To address these questions, a three-phase framework is developed. In the first two phases, the LLM classifies posts as *Traffic Incident* (TI) or *Not Traffic Incident* (NTI), infers their temporal status as ongoing or past, and extracts geographic references suitable for geocoding through the HERE API. The third phase validates the internal accuracy of this pipeline against manually labeled data, while a fourth phase assesses its external correspondence by comparing geocoded posts with verified traffic-incident logs from the HERE system.

This paper makes the following key contributions:

1. An empirical evaluation of prompt-based LLM capabilities for classifying and geoparsing social-media text related to traffic incidents.
2. An assessment of the spatial and temporal correspondence between social-media-derived incidents and verified HERE traffic records.
3. Evidence that social-media data can complement traditional traffic-incident detection systems by providing earlier indications of real-world events.

The rest of the paper is organized as follows. Section 2 reviews previous research on traffic-incident detection from social-media data and contrasts traditional machine-learning and LLM-based approaches. Section 3 describes the four-phase methodology, including classification, geoparsing, geocoding, and validation. Section 4 presents the experimental results for both internal and external evaluations, addressing the two research questions. Finally, Sect. 5 summarizes the findings, discusses limitations, and outlines directions for future work.

2 State of the Art

Several studies have examined the use of social-media data from X for traffic-incident detection.

Gu et al. [4] conducted one of the earliest comprehensive studies in this area. Their framework involves data acquisition for two urban areas, Pittsburgh and Philadelphia, followed by adaptive data collection. Posts were tokenized and classified as *Traffic Incident* (TI) or *Not Traffic Incident* (NTI) using a Semi-Naive Bayes (SNB) classifier, and TI posts were further categorized through a Supervised Latent Dirichlet Allocation (sLDA) model. Geographic information was extracted by a geoparser and then geocoded to obtain latitude and longitude coordinates. The results were validated against manually labeled posts and ground-truth datasets, including the Road Condition Report System (RCRS) and Call For Service (CFS), as well as travel-time data from the HERE API. Their analysis showed that posts covered most of the incidents found in official data and provided broader coverage along arterial roads, although posts from individual users were less geocodable than those from influential accounts.

The same study identified several limitations. The use of a growing dictionary as the feature space for SNB classification made the model insensitive to typos and slang, reducing robustness to informal language. Moreover, two separate geoparsers were required: one effective for precise geographic names and another for general references such as landmarks or neighborhoods. Both remained sensitive to spelling variations, which could lead to missing or incorrect geographic information.

Further work by Salas et al. [9] and Jones et al. [7] analyzed posts collected between March and May 2017 for the West Midlands region in the United Kingdom. Their datasets were manually labeled and divided into training and testing sets. Both studies used text classification methods after standard NLP preprocessing steps such as tokenization and stop-word removal. While Salas et al. [9] applied a Support Vector Machine (SVM) classifier, Jones et al. [7] compared multiple algorithms, including Ridge Classifier (RC), Naive Bayes (NB), k-Nearest Neighbour (kNN), Multilayer Perceptron (MLP), and SVM, evaluating performance through accuracy, precision, recall, and F1 score. Salas et al. reported an accuracy of 88.27% with SVM, while Jones et al. achieved 92.86% with RC. Despite these results, both studies shared similar constraints: tokenization and stop-word removal may cause loss of contextual meaning, and the classifiers required retraining for new datasets, limiting scalability. The relatively small training sets also increased the risk of noise learning, where models capture irrelevant patterns [12].

Suat-Rojas et al. [10] extended this line of research to Spanish-language posts from Bogotá, collected between October 2018 and July 2019. Their four-step pipeline included data collection through the X API, manual labeling, SVM-based classification, entity recognition for location and time extraction, and geocoding using a batch geocoding tool. Although the approach demonstrated feasibility in a non-English context, it retained similar limitations to earlier studies: dependence on model training, reduced adaptability to new domains, and limited robustness to spelling or grammar variation.

Overall, traditional approaches to traffic-incident detection from social-media data have been labor-intensive and reliant on feature engineering, preprocessing, and retraining. Classifiers such as SNB and SVM require extensive labeled datasets and struggle to handle the informal, variable nature of user-generated content.

Recent research has explored the use of LLMs as an alternative for these tasks. LLMs can interpret unstructured text and extract contextual information directly through prompt-based interaction, reducing the need for preprocessing or retraining. Studies applying them to traffic-related problems have explored fine-tuned models [6], multimodal architectures [11], and hybrid LLM–machine-learning pipelines [3]. These works demonstrate the potential of LLMs to process complex and noisy data but mainly focus on model optimization rather than evaluating their real-world reliability for incident detection.

The present study extends this literature by empirically assessing how effectively a general-purpose, prompt-based LLM can identify, geolocate, and contextualize traffic incidents from social-media text, and by quantifying how accurately and how early these social-media-derived incidents correspond to verified official records. Rather than proposing a new detection system, the goal is to evaluate the performance and complementary value of prompt-based LLM text extraction and social-media information for early traffic-incident detection.

3 Methodology

This study employs the open-source model *meta-llama/Meta-Llama-3-8B-Instruct* to assess the ability of LLMs to classify and geolocate traffic-incident information from social-media posts. The instruction-tuned version of Llama 3 was selected because it is optimized for following task-specific prompts without additional training or fine-tuning. Model inference was performed through the Hugging Face `transformers` pipeline, using the `AutoModelForCausalLM` interface.

The overall experimental workflow comprises four main phases corresponding to the evaluation stages described in Sect. 1. Phases 1 and 2 concern internal processing tasks (RQ1), while Phases 3 and 4 evaluate performance against labeled data and official records (RQ1 and RQ2 respectively).

3.1 Phase 1: Classification of Posts by LLM

The first phase evaluates the model's ability to classify posts as TI or NTI and to infer their temporal status as *Ongoing* (O) or *Past* (P). The design of the prompts follows the structural and strategic principles outlined by Chen et al. [2], exploring the effect of prompt composition on classification accuracy.

Two dimensions are considered: (i) structural design elements such as task, context, persona, exemplars, format, and tone; and (ii) prompting strategies including zero-shot, few-shot, and chain-of-thought reasoning. The study adopts static role prompting for consistent responses and uses delimiters to separate instruction sections, improving interpretability of complex prompts.

Eight prompt variants are tested to compare performance:

1. *Zero-shot without persona*: Includes context, format, and tone, but omits persona.
2. *Zero-shot without context*: Includes persona, format, and tone, but omits contextual definitions.
3. *Zero-shot extensive*: Includes all structural elements with detailed descriptions.
4. *Zero-shot extensive and shortened*: Same as (3) but more concise to avoid information overload.
5. *Zero-shot with chain-of-thought*: Adds a "Reasoning Steps" section prompting step-by-step reasoning.
6. *Few-shot*: Adds three labeled exemplars covering the classes TI/O, TI/P, and NTI/None.
7. *Few-shot with chain-of-thought*: Combines examples with explicit reasoning instructions.
8. *Few-shot with chain-of-thought–2*: A resampling test repeating (7) to examine result stability.

To illustrate the structure of the tested prompts, the following excerpt shows the *few-shot with chain-of-thought* configuration used for geoparsing in Phase 2:

> **«CONTEXT»** According to Gu et al. (2016), traffic incidents are 'non-recurrent congestion is induced by non-recurring causes, such as traffic accidents, work zones, adverse weather events, and special events, which takes about half of the total congestion.' 'A TI tweet is one contains information about a traffic incident, implies abnormality on the transportation infrastructure, or indicates a (potentially) significant disruption to the traffic' (Gu et al., 2016). There are 5 categories of traffic incidents: 1. Accidents: traffic accidents such as collision. 2. Road work: the scheduled or unplanned road work. 3. Hazards & Weather. 4. Events: special events such as Marathon. 5. Obstacle Vehicles. The temporal status of traffic incidents in tweets is defined as: – 'Past' if the context implies that the incident has occurred or been resolved. – 'Ongoing' if the tweet implies that the situation is active, uses present-tense language, or does not indicate if the traffic incident is resolved.
> **«TASK»** Your task is: 1. Classify the tweets into 'Traffic-Incident' or 'Not Traffic-Incident' by labeling them as 1 if 'Traffic-Incident' and 0 if 'Not Traffic_incident'. 2. For the 'Traffic-Incident' tweets, meaning 1, classify them as 'Ongoing' or 'Past' by labeling them as O if 'Ongoing' and P if 'Past'; for 'Not Traffic_Incident' tweets, meaning 0, label them as NONE.
> **«EXEMPLARS»** In this prompt, context, persona, format, and tone are included, but additionally, exemplars are also included. This prompt determines if the examples and guidance can improve the model's performance. For the first phase, three examples are provided. One for 'Traffic Incident (TI)': 1, and 'Status': Ongoing (O). Another one for 'Traffic Incident (TI)': 1, and 'Status': Past (P). And the last one, for 'Traffic Incident (TI)': 0, and 'Status': NONE. Therefore, the model gains a good grasp of how tweets are classified. For this prompt, the examples used are removed from the CSV file provided for Llama to classify to prevent evaluation bias.
> **«REASONING STEPS»** Let's think step by step.

The evaluation X dataset contains 2,845 posts related to traffic conditions in Los Angeles between 1–14 June 2024 [1]. Following the labeling criteria of Gu et al. [4], each post is manually annotated as TI or NTI and, when applicable, as O or P. Of these, 2,468 posts were labeled as TI (2,176 Ω and 292 P) and 377 as NTI.

Performance is assessed against this ground truth using standard metrics: accuracy, precision, recall, specificity, and F1 score. For temporal-status classification, macro-, micro-, and weighted averages are reported following Lee et al. [5].

3.2 Phase 2: Geoparsing and Geocoding

In the second phase, posts classified as TI are processed to extract structured geographic information suitable for geocoding. The model is prompted to return location entities in a segmented key–value format. The best-performing strategy from Phase 1 (*few-shot with chain-of-thought*) is reused here to maintain consistency.

Geoparsed strings are submitted to the HERE Geocoding API to obtain latitude and longitude coordinates. To ensure consistency with previous research, the structure of Gu et al. [4] is adopted. The model is instructed to handle both explicit references (e.g., road names, intersections, highway exits) and implicit references (e.g., landmarks or neighborhoods). Pattern matching is applied to extract values from the model outputs prior to API submission.

Three variations are tested when constructing the strings for geocoding:

1. *Without relational word and city.* Excludes relational terms and city metadata.
2. *With all available information.* Includes relational terms as well as city, state, and country.
3. *Without relational word.* Excludes relational terms but includes geographic metadata.

3.3 Phase 3: Internal Validation

Phase 3 quantifies the internal performance of the LLM-based classification and geoparsing processes against the manually labeled dataset. Two sub-evaluations are performed:

1. *Classification performance.* The eight prompting strategies described above are compared using the metrics listed earlier. This analysis measures the sensitivity of classification performance to prompt structure and serves as an internal validation of RQ1.
2. *Geoparsing and geocoding accuracy.* The proportion of correctly geoparsed and successfully geocoded TI posts is computed for each geocoding variant. Success is defined as a valid coordinate pair returned by the HERE API. This assessment also contributes to RQ1 by evaluating how effectively LLM outputs can be converted into spatially meaningful data.

3.4 Phase 4: External Validation

Phase 4 evaluates the correspondence between social-media-derived incidents and verified official records from the HERE system, addressing RQ2. Two complementary analyses are conducted.

Incident Matching. Each geocoded post (latitude, longitude, and timestamp) is compared to HERE's verified traffic logs for the same period (1–14 June 2024, 30,519 records). A match occurs when a post's timestamp (`created_at`) falls within a specified temporal tolerance of a HERE incident's `startTime` and within a defined spatial radius of its location. Four comparison settings are considered:

1. *Strict.* 30 min temporal window (±15 min), 1.6 km spatial radius.
2. *Relaxed time.* 40 min (±20 min), 1.6 km radius.
3. *Relaxed space.* 30 min, 3 km radius.
4. *Relaxed time and space.* 40 min, 3 km radius.

A post satisfying both spatial and temporal conditions is counted as a successful match.

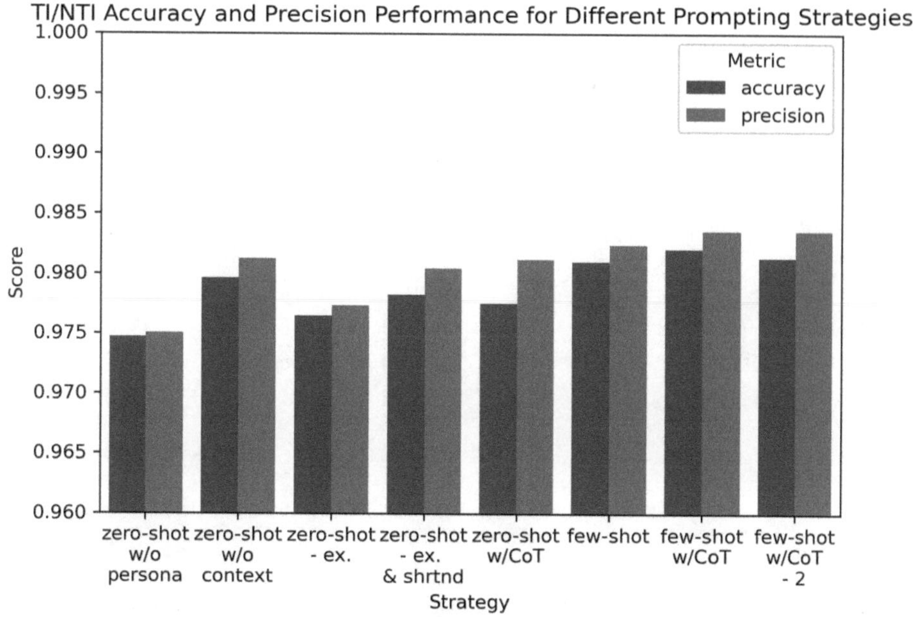

Fig. 1. Accuracy and precision for traffic-incident classification across prompting strategies.

Early Detection. For matched incidents, the post's `created_at` timestamp is compared with HERE's `entryTime`. If the post precedes the `entryTime`, the incident is considered detected earlier on social media. This analysis quantifies the proportion of matches where crowd-sourced reports appear before official detection.

Together, these evaluations provide a comprehensive assessment of LLM performance in extracting incident information (RQ1) and of social-media data as an early indicator of real-world traffic events (RQ2).

4 Results

4.1 Classification Performance Evaluation (RQ1)

In Table 1, few-shot strategies obtain the highest accuracy, precision, specificity, and F1, while recall is similar across zero-shot and few-shot. Figure 1 shows these differences for accuracy and precision. A notable pattern is that adding more descriptive context does not necessarily improve performance: *zero-shot w/o context* slightly outperforms *zero-shot extensive*, suggesting that overly long instructions can dilute task focus. This is consistent with the temporal-status results in Table 2, where *zero-shot w/o context* avoids the outputs with no result observed in *zero-shot w/o persona* and *zero-shot extensive*. The concise *zero-shot extensive & shortened* reduces this effect and recovers performance.

Table 1. Performance of prompting strategies on traffic incident classification.

Strategy	Accuracy	Precision	Recall	Specificity	F1-score
zero-shot w/o persona	0.975	0.975	0.996	0.833	0.986
zero-shot w/o context	0.980	0.981	0.996	0.875	0.988
zero-shot - extensive	0.976	0.977	0.996	0.849	0.987
zero-shot - extensive & short.	0.978	0.980	0.995	0.870	0.988
zero-shot w/CoT	0.978	0.981	0.993	0.875	0.987
few-shot	0.981	0.982	0.996	0.883	0.989
few-shot w/CoT	0.982	0.984	0.996	0.891	0.990
few-shot w/CoT - 2	0.981	0.984	0.995	0.891	0.989

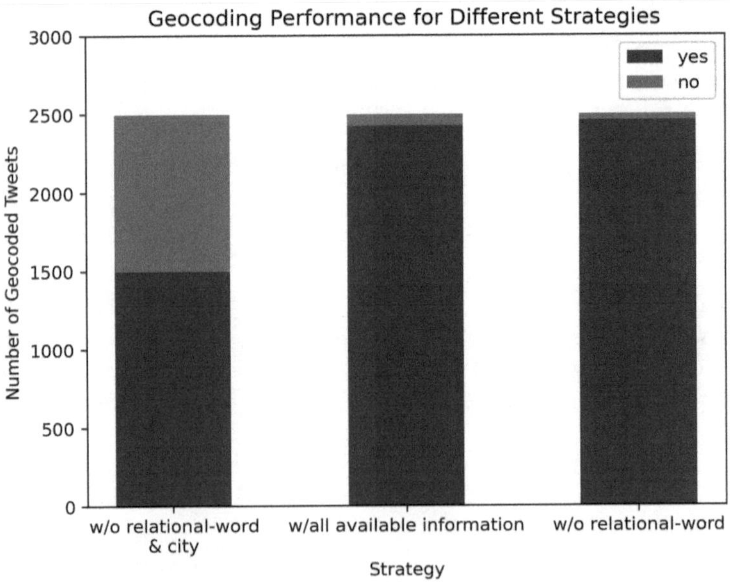

Fig. 2. Geocoding outcomes for three string-assembly strategies. The "yes" category indicates successful coordinate retrieval from the HERE Geocoding API.

Table 2. Performance of prompting strategies on status classification.

Strategy	Accuracy	Precision			Recall			F1-score		
		Macro	Micro	Weighted	Macro	Micro	Weighted	Macro	Micro	Weighted
zero-shot w/o persona	0.906	0.629	0.906	0.928	0.659	0.906	0.906	0.634	0.906	0.912
zero-shot w/o context	0.917	0.925	0.917	0.917	0.746	0.917	0.917	0.796	0.917	0.903
zero-shot - extensive	0.911	0.632	0.911	0.928	0.657	0.911	0.911	0.638	0.911	0.916
zero-shot - extensive & shortened	0.909	0.837	0.909	0.933	0.896	0.909	0.909	0.853	0.909	0.916
zero-shot w/CoT	0.909	0.833	0.909	0.932	0.894	0.909	0.909	0.851	0.909	0.916
few-shot	0.916	0.847	0.916	0.934	0.896	0.916	0.916	0.862	0.916	0.921
few-shot w/CoT	0.910	0.840	0.911	0.932	0.895	0.911	0.911	0.856	0.911	0.917
few-shot w/CoT - 2	0.913	0.840	0.913	0.932	0.893	0.913	0.913	0.857	0.913	0.919

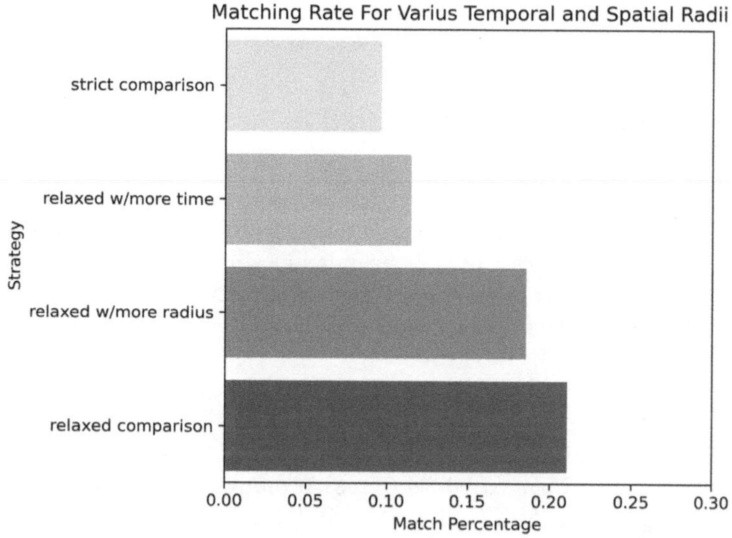

Fig. 3. Match rates under four spatial and temporal tolerances against verified HERE incident logs.

Few-shot prompts generally perform best. Comparing *few-shot w/CoT* and its resampling variant indicates very similar incident classification accuracy (0.982 vs. 0.981) and slightly higher temporal-status scores for the resampling, suggesting stable results across runs (Fig. 3).

4.2 Geoparsing and Geocoding Evaluation (RQ1)

Using the geoparsing prompt, the model produced structured location strings for 100% of the 2,468 TI posts. A post is counted as successfully geocoded when the API returns a valid coordinate pair. Figure 2 shows that *withou relational word & city*, *with all available information*, and *without relational word* yield 1,497 (60%), 2,457 (98.5%), and 2,425 (97.2%) successful geocodes respectively.

Including city, state, and country substantially improves geocoding success, while adding relational words does not provide additional benefit in this dataset.

4.3 Incident Matching Evaluation (RQ2)

Under the strict setting, 236 of 2,457 geocoded posts match a HERE record (9.6%). The relaxed time setting yields 282 matches (11.5%), the relaxed space setting 456 (18.6%), and the jointly relaxed setting 518 (21.2%). Thus, between 9.6% and 21.2% of geocoded posts correspond to officially logged incidents for the same period and area. This indicates that most posts do not have a direct counterpart in the verified data, which suggests that social-media signals can complement official sources by covering incidents that are unlogged, logged differently, or reported outside the specified spatial or temporal bounds.

4.4 Early Detection Evaluation (RQ2)

For matched pairs, comparing post `created_at` to HERE `entryTime` shows earlier appearance on social media in 145 of 236 cases (61%) under the strict setting, 162 of 282 (57%) under relaxed time, 269 of 456 (59%) under relaxed space, and 274 of 518 (53%) under the relaxed time and space setting. These results indicate that when a match exists, a majority of incidents are reported on social media before their appearance in the verified system.

4.5 Comparison to Literature

Relative to prior work using traditional classifiers, incident classification accuracies here range from 97.5% to 98.2% across prompts, compared with 90.5% for Semi-Naive Bayes in [4], 88.27% for SVM in [9], 92.86% for a ridge classifier in [7], and 96.8% for SVM in [10]. Few-shot prompting yields the strongest overall results in this study, although compact zero-shot instructions remain competitive and underline the importance of concise prompts.

For geoparsing of the TI posts, the *few-shot with chain-of-thought* strategy produced consistent results, with Llama successfully extracting structured geographic information from all posts. This outcome indicates that the model can handle both explicit and implicit location references, including those expressed through informal or fuzzy terms. Compared with the dual-geoparser approach proposed by Gu et al. [4], the single LLM-based method achieves equivalent coverage without the need for multiple specialized components, suggesting that prompt-based geoparsing can serve as an efficient and unified alternative for location extraction.

For geocoding, prompting that returns segmented location fields combined with explicit city, state, and country attains a 98.5% geocoding success rate. This suggests that prompt-based extraction can provide inputs that are compatible with standard geocoding APIs without custom parsers.

Finally, the external validation shows that a nontrivial fraction of geocoded posts match verified incidents and that over half of those matched cases appear on social media earlier. Taken together, RQ1 and RQ2 indicate that prompt-based extraction can structure incident-relevant content effectively, and that social-media reports provide complementary and sometimes earlier signals relative to an official incident feed.

5 Conclusions and Future Work

This study evaluated how effectively large language models (LLMs) and social-media data can support early traffic-incident detection. Using posts from X and the instruction-tuned Llama 3-8B-Instruct model, two complementary research questions were addressed: (i) the internal performance of a prompt-based LLM for classifying, geoparsing, and geocoding posts (RQ1); and (ii) the external correspondence of these derived incidents with verified records from the HERE system (RQ2).

For RQ1, the experiments showed that the model can classify posts as *Traffic Incident* or *Not Traffic Incident* and infer their temporal status with up to 98.1% accuracy. Few-shot prompting strategies achieved the best results, although concise zero-shot prompts remained competitive. The model also produced structured geographic strings that could be geocoded with a 98.5% success rate through the HERE API, demonstrating that unstructured posts can be transformed into geospatially valid inputs without retraining.

For RQ2, comparisons against HERE traffic-incident logs indicated that between 9.6% and 21.2% of the geocoded posts matched verified incidents for the same period, and that over half of these matched posts appeared earlier than official records. This supports the view that social-media data can complement existing detection systems by providing additional and often earlier coverage of nonrecurrent incidents.

While these results are promising, they are limited by the manual labeling process, which introduces subjectivity and constrains dataset size. The evaluation was also confined to one metropolitan area and time window, and real-time performance was not tested. Future work should investigate scalable labeling methods, multilingual and cross-regional data, and streaming implementations for near-real-time detection. Further studies might also examine predictive tasks such as estimating incident severity or congestion levels based on post content.

Overall, the findings highlight that prompt-based LLMs can structure informal, crowd-sourced text into standardized geospatial information and that social-media signals offer measurable and often earlier insight into urban traffic dynamics.

Ethical Considerations

All posts analyzed in this study were publicly available at the time of collection, stripped of user identifiers, and processed only in aggregate form. No attempts were made to infer, track, or disclose information about individual users or accounts. The study complies with the ethical standards for secondary data analysis and aims solely to evaluate methodological performance rather than monitor human behavior.

References

1. Asenov, D.: A Data-driven Approach to Study the Influence of Social Media on Human Behaviours in Transportation. Thesis, University of Twente, Enschede (July 2024). https://purl.utwente.nl/essays/100967
2. Chen, B., Zhang, Z., Langrené, N., Zhu, S.: Unleashing the potential of prompt engineering for large language models. Patterns (May 2025). https://doi.org/10.1016/j.patter.2025.101260, https://www.cell.com/patterns/abstract/S2666-3899(25)00108-4
3. Grigorev, A., Saleh, K., Ou, Y., Mihăiţă, A.S.: Enhancing traffic incident management with large language models: a hybrid machine learning approach for severity classification. Int. J. Intell. Transp. Syst. Res. **23**, 259–280 (2025). https://doi.org/10.1007/s13177-024-00448-7
4. Gu, Y., Qian, Z.S., Chen, F.: From Twitter to detector: real-time traffic incident detection using social media data. Trans. Res. Part C: Emerging Technol. **67**, 321–342 (2016). https://doi.org/10.1016/j.trc.2016.02.011
5. Hinojosa Lee, M.C., Braet, J., Springael, J.: Performance metrics for multilabel emotion classification: comparing micro, macro, and weighted F1-scores. Appl. Sci. **14**(21), 9863 (2024). https://doi.org/10.3390/app14219863
6. Jaradat, S., Nayak, R., Paz, A., Elhenawy, M.: Multitask learning for crash analysis: a fine-tuned llm framework using twitter data. Smart Cities **7**(5), 2422–2465 (2024). https://doi.org/10.3390/smartcities7050095
7. Jones, A.S., Georgakis, P., Petalas, Y., Suresh, R.: Real-time traffic event detection using Twitter data. Infrastructure Asset Manag. **5**(3), 77–84 (2018). https://doi.org/10.1680/jinam.17.00022
8. Kasneci, E., et al.: ChatGPT for good? on opportunities and challenges of large language models for education. Learn. Individ. Differ. **103**, 102274 (2023). https://doi.org/10.1016/j.lindif.2023.102274
9. Salas, A., Georgakis, P., Petalas, Y.: Incident detection using data from social media. In: 2017 IEEE 20th International Conference on Intelligent Transportation Systems (ITSC), pp. 751–755 (Oct 2017). https://doi.org/10.1109/ITSC.2017.8317967, https://ieeexplore.ieee.org/document/8317967/, iSSN: 2153-0017
10. Suat-Rojas, N., Gutierrez-Osorio, C., Pedraza, C.: Extraction and analysis of social networks data to detect traffic accidents. Information **13**(1), 26 (2022). https://doi.org/10.3390/info13010026
11. Tami, M.A., Ashqar, H.I., Elhenawy, M., Glaser, S., Rakotonirainy, A.: Using multimodal large language models (mllms) for automated detection of traffic safety-critical events. Vehicles **6**(3), 1571–1590 (2024). https://doi.org/10.3390/vehicles6030074
12. Ying, X.: An overview of overfitting and its solutions. J. Phys: Conf. Ser. **1168**(2), 022022 (2019). https://doi.org/10.1088/1742-6596/1168/2/022022. publisher: IOP Publishing

Early Detection of Weak Signals in News with KIM/KEM Maps and Transformer-Based Language Models

Edison Loza-Aguirre[1](✉), Alexis Miranda-Carrillo[2], and Lorena Recalde[2]

[1] Universidad San Francisco de Quito, Quito 170901, Ecuador
eloza@usfq.edu.ec
[2] Escuela Politécnica Nacional, Quito 170143, Ecuador
{alexis.miranda,lorena.recalde}@epn.edu.ec

Abstract. The proposed solution addresses the problem of information overload in Strategic Scanning, enabling organizations to anticipate opportunities and threats through the detection of weak signals. This study introduces a novel model for weak signal detection in Strategic Scanning using advanced Artificial Intelligence algorithms, Natural Language Processing, and data visualization techniques. The model applies a Transformer-based topic modeling approach, which facilitates the identification of emerging themes in large volumes of textual data. Keyword Emergence Maps and Keyword Issue Maps are employed to identify and visualize early indications of future changes. Results highlight a model's strong performance, underscoring its effectiveness in detecting weak signals within complex informational environments. The proposed tool not only enhances strategic analysis and decision-making but also has the potential to provide a competitive advantage by reducing uncertainty and increasing organizational adaptability.

Keywords: weak signals · strategic scanning · topic modeling · transformers

1 Motivation

Strategic Scanning (SScan) is a proactive organizational process that seeks to observe and analyze the socioeconomic environment to identify potential opportunities and threats [16]. By systematically collecting, processing, and analyzing relevant information, organizations can strengthen their capacity to anticipate change and prepare for future developments that may affect their competitive position [18, 27].

The importance of SScan lies in its role as an early-warning mechanism. Through the analysis of external information, managers can detect signals that announce unexpected opportunities or threats, thus allowing organizations to adjust strategies in advance [6, 20]. However, despite its potential benefits, the effectiveness of SScan is frequently challenged by information overload, a situation in which the volume of available data exceeds the processing capacity of both decision-makers and organizational systems [4, 8, 28]. Under such conditions, the ability to discern relevant signals diminishes, and

H. R. Arabnia et al. (Eds.): CSCI 2025, CCIS 2967, pp. 299–313, 2027.
https://doi.org/10.1007/978-3-032-28454-9_21

strategic decisions are undermined by excessive noise and insufficient analytical support [13].

Weak signals play a central role in SScan. Defined as subtle, fragmentary, and often ambiguous pieces of information, weak signals may initially appear irrelevant but can later reveal themselves as early indicators of significant future events [7]. Detecting and interpreting weak signals is therefore essential for organizations seeking to anticipate risks, mitigate uncertainty, and capture emergent opportunities before they materialize [17, 31]. Nevertheless, the identification of weak signals remains a complex task: their low frequency, ambiguous meaning, and apparent irrelevance make them difficult to distinguish from the surrounding informational noise.

Recent technological advances provide new avenues to address these challenges. Artificial Intelligence (AI) and Natural Language Processing (NLP) techniques have significantly improved the capacity to process large-scale textual data, uncover latent structures, and identify patterns that may serve as weak signals. Transformer-based language models have revolutionized NLP by enabling deeper semantic understanding and contextual analysis of text [30].

This study proposes a model for weak signal detection that integrates Transformer-based topic modeling, Keyword Emergence Maps (KEM), and Keyword Issue Maps (KIM) to capture early signs of change in large volumes of textual data. The focus of this analysis is news articles, which represent a continuous and diverse stream of information about socioeconomic, political, and technological developments.

The expected contribution of this research is twofold. First, it advances the methodological integration of AI, NLP, and visualization techniques for weak signal detection, addressing the limitations of previous approaches that relied mainly on statistical or purely semantic models. Second, it offers practical value by providing managers with a tool capable of reducing uncertainty and enhancing organizational adaptability in turbulent contexts. By improving the ability to detect emerging opportunities and threats, the proposed model supports strategic decision-making and contributes to the long-term competitiveness of organizations.

The remainder of this paper is structured as follows. Section 2 presents the theoretical framework. Section 0 describes the research methodology. Section 4 reports the results of the proposed model. Finally, Sect. 5 concludes the study and outlines directions for future research.

2 Detection of Weak Signals

"Strategic Scanning (SScan) refers to the acquisition of information about events, trends, and relationships in an organization's environment, the knowledge of which would assist senior executives in identifying and understanding strategic threats and opportunities" [10]. This process involves the ability to collect and filter relevant information [24], which is often lost in the overwhelming flow of data, to share and disseminate it among interested stakeholders, and to analyze and use it to create organizational sensemaking [26]. SScan can help an organization build this capacity to better understand and anticipate changes in its external environment [18, 27].

Building on this perspective, weak signals can be understood as subtle fragments of information embedded within the scanning process, whose timely recognition is essential for anticipating emerging opportunities and threats. A weak signal usually appears insignificant or is easily lost in a sea of unprocessed data due to excessive "noise." They are difficult to detect because they are often incomplete, ambiguous, or hidden within large volumes of raw data. They may appear unusual, irrelevant, or isolated, and their uncertain reliability makes them easy to dismiss. Their apparent randomness and lack of clear context further complicate their recognition, even though they may contain early indications of significant future developments [19]. Despite their fragile appearance, weak signals can alert organizations to potential future situations. Interpreting weak signals may lead to the belief that an event is likely to have consequences—whether in terms of risks, threats, or opportunities—in the future [15].

The detection of weak signals is closely linked to their interpretation. According to Hiltunen [14], a weak signal becomes meaningful only when it is interpreted as a potential sign of future change. This perspective builds on Peirce's triadic model of the sign, which consists of three interrelated elements (Fig. 1): the sign itself, the problem it refers to, and the interpretant, that is, the meaning attributed to the sign by an observer [2]. Applied to Strategic Scanning, a weak signal is the sign, the possible future development it points to is the problem, and the interpretation by managers or analysts constitutes the interpretant.

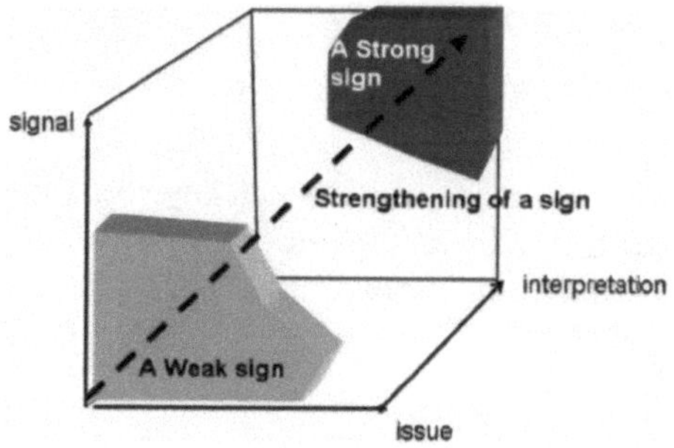

Fig. 1. Definition of future sign space and definition of a weak signal according to Hiltunen [14].

To detect Weak Signals, Yoon [32] linked "term frequency" and "document frequency" measures through text mining to the signal and the problem defined by Hiltunen. Thus, Yoon introduced KEM and the KIM as tools to classify keywords and select signals. These keyword portfolio maps considered a time-weighted factor based on two preconditions: (1) important keywords appear with high frequency, and (2) the more recent the keywords, the greater their importance compared to older ones [5].

Keyword Emergence Map. KEM shows the Degree of Visibility (DoV) of a future signal and represents a keyword portfolio map corresponding to the signal dimension among the three dimensions of a future sign [23]. Using keyword term frequency obtained through text mining as a measure of visibility, the DoV parameter is proposed to account for the importance of keywords according to the time of their appearance. The DoV of keyword i appearing in period j is defined by Eq. 1:

$$DoV_{ij} = \left(\frac{TF_{ij}}{NN_j}\right) x \{1 - tw\ x\ (n - j)\} \tag{1}$$

where TF_{ij} is the total frequency of keyword i during period j, NN_j is the total number of news articles during period j, n is the number of periods, j is the time point, and tw is the time weight.

The quadrants in the KEM are divided based on the medians of the respective values on each axis. This creates four quadrants in the KEM, as shown in Fig. 2a. The upper-right region represents strong signals, consisting of keywords with both high average Term Frequency (TF) and a high average growth rate of DoV. The upper-left region represents weak signals, defined as keywords with low average TF but a high average growth rate of DoV. Keywords in the lower-left region are marked as latent signals. The lower-right part of the KEM corresponds to well-known but not strong signals, referring to keywords with high TF but a low growth rate of DoV [24].

Keyword Issue Map. KIM indicates the Degree of Diffusion (DoD) of future signals and is associated with the problem dimension of future signs. To quantitatively measure the diffusion of future signals, the document frequency of news articles is used and defined according to Eq. 2:

$$DoD_{ij} = \left(\frac{DF_{ij}}{NN_j}\right) x \{1 - tw\ x\ (n - j)\} \tag{2}$$

where DF_{ij} is the document frequency with which keyword i appeared during period j, and NN_j is the total number of news articles during period j, n is the number of periods, j is the time point, and tw is the time weight.

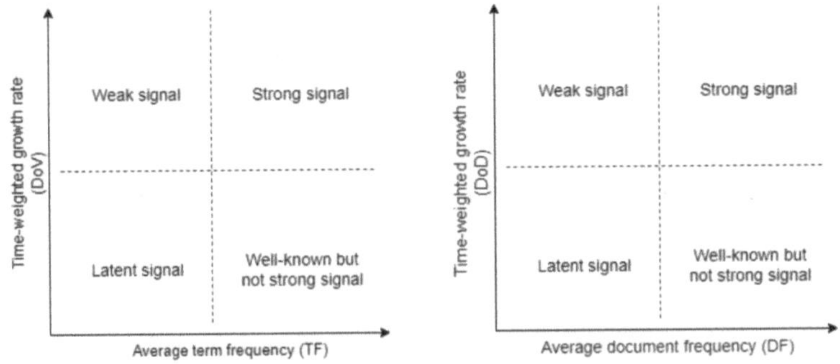

Fig. 2. (a) Quadrants of the KEM. (b) Quadrants of KIM.

KIM plots the average document frequency (DF) on the x-axis and the average growth rate of DoD on the y-axis, mapping selected keywords in the same manner as KEM, as shown in Fig. 3. If a keyword has a low average DF but a high average growth rate of

DoD, it is considered a weak signal; if it has a high average DF and a high average growth rate of DoD, it is considered a strong signal. Similarly, if the keyword falls into the third quadrant, it is a latent signal, while in the fourth quadrant it represents a keyword that is well known but has a low average growth rate of DoD [14].

KEM and KIM are intercorrelated: by intersecting the weak signal candidates identified in each map, it becomes possible to detect weak signals that may evolve into future trends.

3 Methods

The objective of this research is to develop a model for weak signal detection. To achieve this, we used a news corpus, as news articles provide a valuable source of information covering a wide range of events and trends—from economic and political developments to regulatory changes and social behaviors. As a real-time source, news enables organizations to continuously monitor and analyze the external environment, offering timely insights into emerging events, crises, and opportunities. For this study, the websites of two Latin American newspapers were selected due to their accessible content. Figure 3 presents the phases of the model development, with details provided in the following sections.

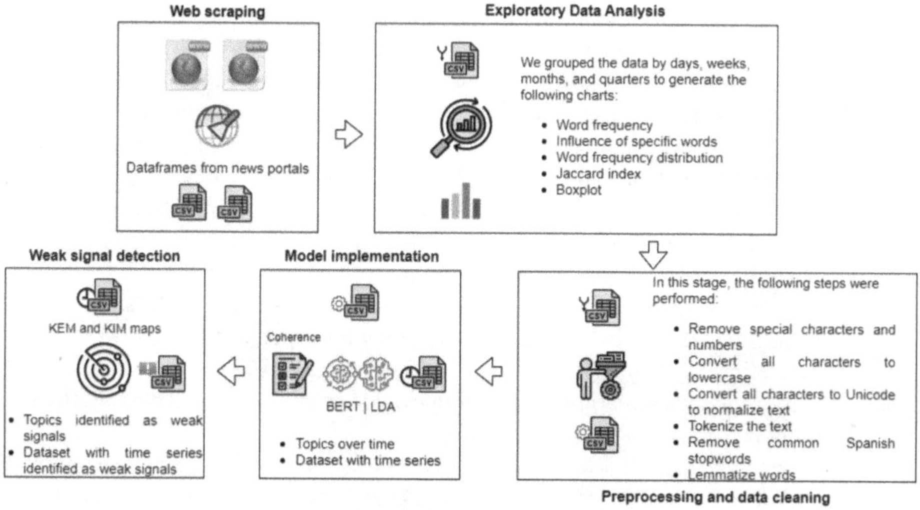

Fig. 3. Methodological workflow.

1. Web Scraping. News articles were collected through automated web scraping using the Selenium and *BeautifulSoup* libraries. From the first newspaper, 32,180 records were retrieved and from the second, 11,600 records. For each newspaper, the extracted fields included the title, publication date, and news content [29].

2. Exploratory Data Analysis (EDA) was performed to gain a clearer understanding of its structure and distribution. The process began with the import of modules required for data manipulation, text processing, and visualization. The CSV files containing the news articles and their publication dates were then read. Subsequently, the dataset was filtered within a specific date range, and rows with missing values were removed. An important step was the elimination of sponsored content to ensure that the analysis focused exclusively on relevant information.
3. The dataset was cleaned and standardized by removing duplicates, handling missing values, and applying normalization and lemmatization to prepare the text for analysis [22]. Special characters and numbers were removed, all text was converted to lowercase and Unicode, tokenization was applied, Spanish stopwords were eliminated, and words were reduced to their base form. Word frequencies were then calculated across daily, weekly, monthly, and quarterly intervals to identify trends and potential weak signals. To refine the analysis, frequent but uninformative words from the head of the long-tail distribution were removed. Finally, the Jaccard index was used to measure similarity between word sets across time periods, providing insights into how the data evolved.
4. The preprocessed data was analyzed using two complementary approaches: the Transformer architecture, specifically BERT, and Latent Dirichlet Allocation (LDA). The *BERTopic* library was employed for topic modeling over time. *BERTopic* provides a significant advantage by summarizing large volumes of text into coherent topics, and by monitoring their evolution, it becomes possible to detect weak signals that may indicate significant changes in the environment [11]. LDA, in turn, is a statistical generative model that assumes documents are mixtures of topics and topics are mixtures of words [64]. This approach facilitates the discovery of latent structures in the corpus, supporting the identification and categorization of hidden themes [9].

   For embedding generation, the *SentenceTransformer* model "paraphrase-multilingual-MiniLM-L12-v2" was applied to capture deep semantic relationships between words across multiple languages. This model was chosen for its ability to represent subtle semantic nuances in multilingual contexts. In addition, a CountVectorizer with an n-gram range of (1, 2) was employed to include both single words and meaningful word pairs, enriching the semantic representation of the corpus. To further refine topic modeling, *ClassTfidfTransformer* (C-TF-IDF) was used to adjust word frequencies according to their relevance across document classes, thereby highlighting the most representative terms.

   Once the optimal model was selected based on the coherence metric, the temporal evolution of topics was examined. This analysis made it possible to detect weak signals that manifested as gradual or sudden shifts in the incidence of certain topics. Such analysis is crucial, as it enables the identification of subtle and emerging changes in the themes addressed by the news.
5. Based on the processed data and identified topics, KEM and KIM were developed to detect weak signals. These visualizations allow the observation of keyword behavior and visibility over time, highlighting not only the frequency and relevance of events and indicators but also areas where occurrences are less frequent yet potentially more significant [3]. As shown in Figs. 11 and 12, KEM and KIM support the identification

of emerging weak signals by mapping their prominence and diffusion across different time periods.

To construct these maps, TF and DF were calculated for each keyword in the selected time segments. These measures were essential for deriving the DoV and the DoD. DoV reflects the prominence of a keyword within its temporal context, while DoD captures its spread across multiple documents.

4 Results

As mentioned in the previous section, within the framework of the EDA, word frequencies in our dataset were calculated by grouping the data into different time intervals: daily, weekly, monthly, and quarterly. In the data, we observed a long-tail distribution (Fig. 4), where the most frequent words are located at the head of the distribution, while the tail contains fewer common words that are potentially more relevant in terms of novelty and thematic significance.

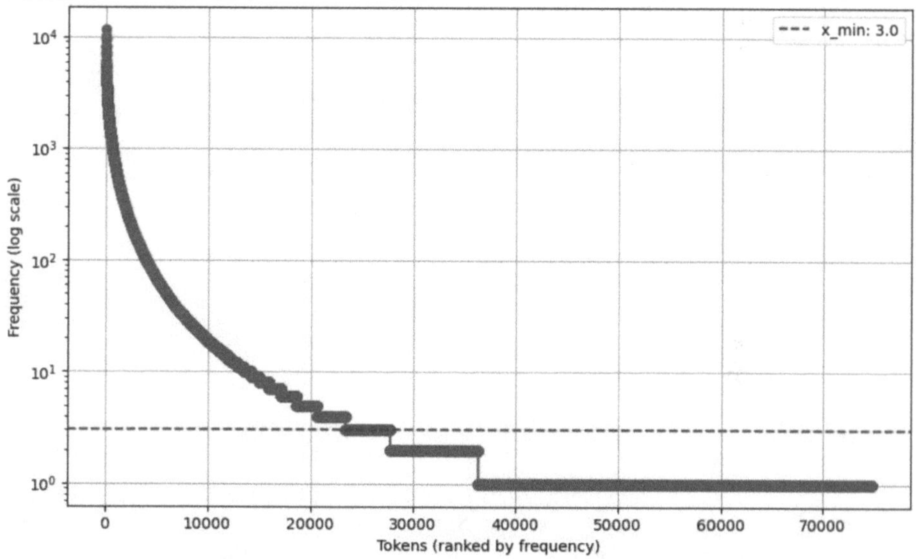

Fig. 4. Illustration of the long-tail distribution, showing frequent words at the head and less common but potentially more relevant words in the tail.

It was observed that the words at the head of this distribution, although frequent, did not provide relevant information, as they consisted of common terms repeated across the different time intervals (daily, weekly, monthly, and quarterly). Therefore, the focus was placed on the words in the tail of the distribution, which offered a more specific view of the news topics at different moments [1].

Subsequently, the Jaccard index was calculated to measure the similarity between sets of words across different time periods, as shown in Fig. 5. This helped to understand

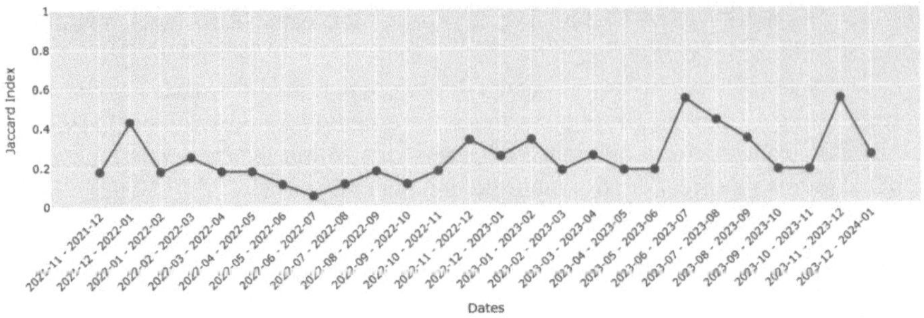

Fig. 5. Jaccard indices month by month.

how the data changed over time. Through this exploration, it became evident that the variability or persistence of different topics was observable across various intervals.

The Jaccard similarity indices were analyzed at different levels of temporal aggregation: daily, weekly, monthly, and quarterly, as shown in Table 1. The Jaccard index ranges from 0 to 1, measuring the similarity between two datasets, where a value of 0 indicates no common elements and a value of 1 indicates that the sets are identical [21]. The results reveal that daily data present the lowest variance and standard deviation, indicating that Jaccard indices are more stable and consistently clustered compared to periods with higher levels of aggregation.

Table 1. Jaccard indices at different levels of aggregation.

Metric	Daily	Weekly	Monthly	Quarterly
Variance	0.005469	0.014335	0.017951	0.009937
Standard Deviation	0.073952	0.119727	0.133980	0.099684

Variance measures the dispersion of the data around the mean. A low variance indicates that the index values are closely clustered around the average, suggesting consistent similarities between the compared periods. Standard deviation, as the square root of variance, provides a more intuitive measure of dispersion in the same units as the original data [12]. A low standard deviation likewise implies that the Jaccard indices are more stable and less affected by extreme variations. The lower variance and standard deviation observed in daily data suggest greater stability and less dispersion around the mean. This reduction in variability decreases noise in the data, facilitating the identification of significant patterns and changes.

The boxplot in Fig. 6 illustrates the volatility of the data, measuring how dispersed or variable the index values are relative to their mean. The Jaccard indices calculated at the daily level show the lowest dispersion, indicating high stability and consistency in day-to-day similarities. This is reflected in the more compact box and the presence of a few outliers, which represent days with atypical behavior.

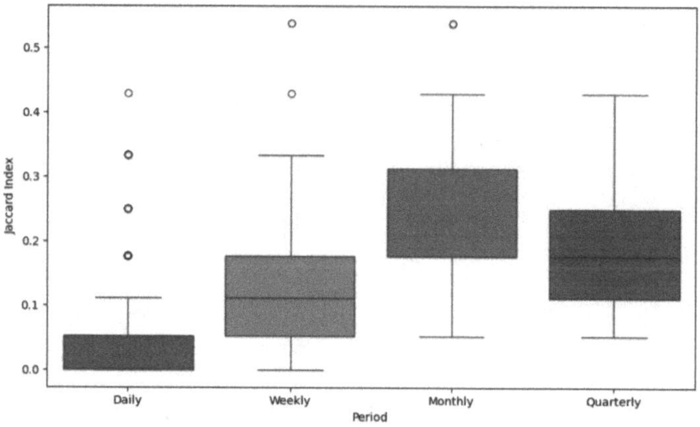

Fig. 6. Boxplot of volatility in Jaccard indices.

After completing the preprocessing and cleaning steps, the data was introduced into the models. For this purpose, BERT and LDA topic modeling algorithm were used. Figures 7 and 8 present the distance analysis between the topics of each model.

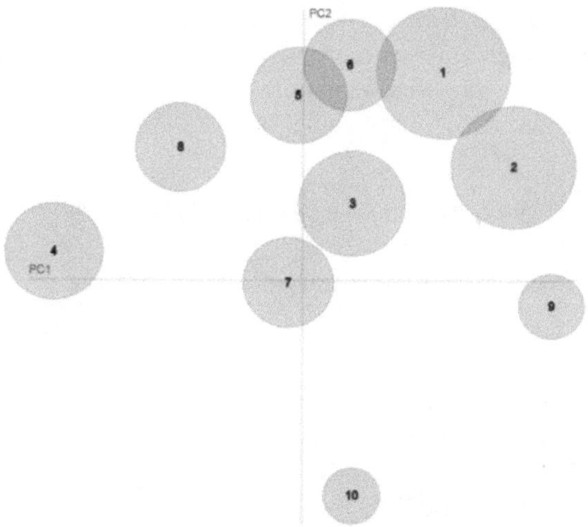

Fig. 7. Distance between topics in the LDA model.

In Figs. 7 and 8, each sphere represents a topic; its size encodes topic prevalence—under LDA, the global topic probability/weight, and under BERTopic, the approximate document count and representativeness (via c-TF-IDF). Inter-sphere distances on the PC1–PC2 (LDA) and D1–D2 (BERTopic) planes result from dimensionality reduction and approximate semantic dissimilarity: nearby spheres indicate related topics, whereas distant spheres denote distinct ones.

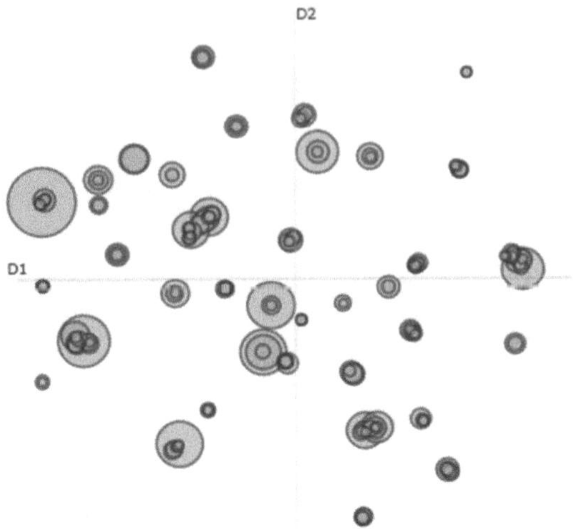

Fig. 8. Distance between topics in the BERT model.

After training the models, their coherence was calculated. Table 2 presents a summary of these results. BERT with 150 topics was the model with the highest coherence.

Coherence quantifies the semantic consistency of a topic's top terms by computing cosine similarity over their co-occurrence vectors (built with a sliding window and weighted via Normalized Pointwise Mutual Information), then averaging across pairs. Higher coherce indicates more interpretable, semantically coherent topics [25].

Table 2. Comparative results of the models.

Model	Number of topics	Coherence
LDA	50	0.611
	100	0.554
	150	0.525
BERT	50	0.681
	100	0.755
	150	0.802

Once the optimal model was selected based on the coherence metric, the evolution of topics over time was analyzed (Fig. 9). By examining how topics change and emerge across different periods, it was possible to detect weak signals that may manifest as either gradual or sudden shifts in the incidence of certain topics. This analysis is crucial, as it allows for the identification of subtle and emerging changes in the themes covered by the news.

Figure 10 shows the distance between documents, calculated by using cosine similarity on the chosen model. The documents are grouped according to the topics that represent them.

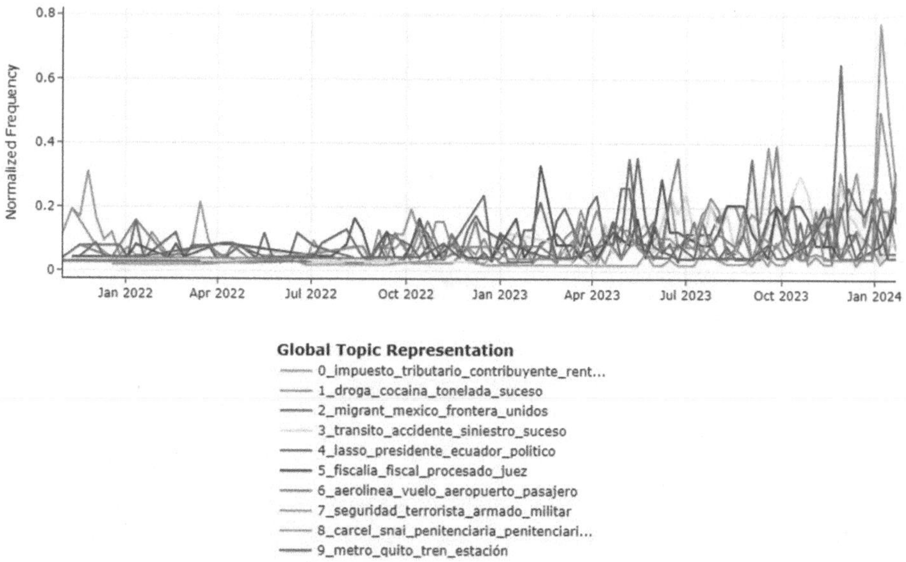

Fig. 9. Evolution of topic frequency over time.

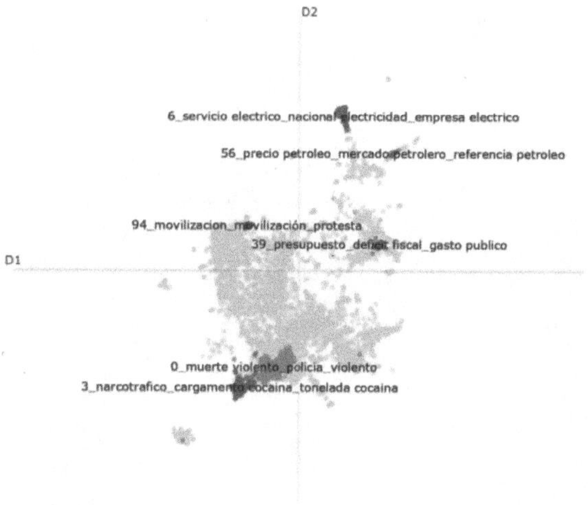

Fig. 10. Distance between documents according to the final model.

Later, we calculated TF and document frequencies DF for each keyword across different time segments. These calculations were crucial for deriving the DoV and the

DoD. KEM and KIM were generated to graphically represent these indicators (Figs. 11 and 12). In KEM, TF are plotted against DoV, and in KIM, DF are plotted against DoD. The red dashed lines in both graphs indicate the mean values of DoV and DoD, respectively, helping to identify words that lie above these thresholds on the y-axis but have relatively low frequencies on the x-axis. Such words are considered weak signals due to their emerging visibility or diffusion despite their low relative frequency.

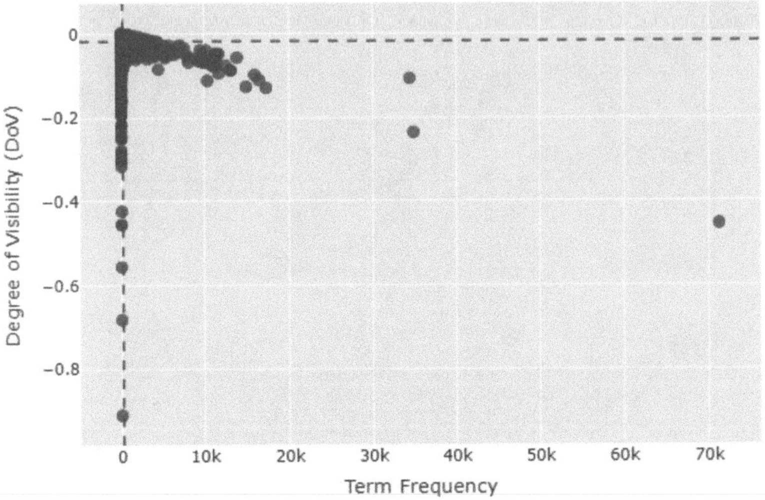

Fig. 11. Resulting KEM.

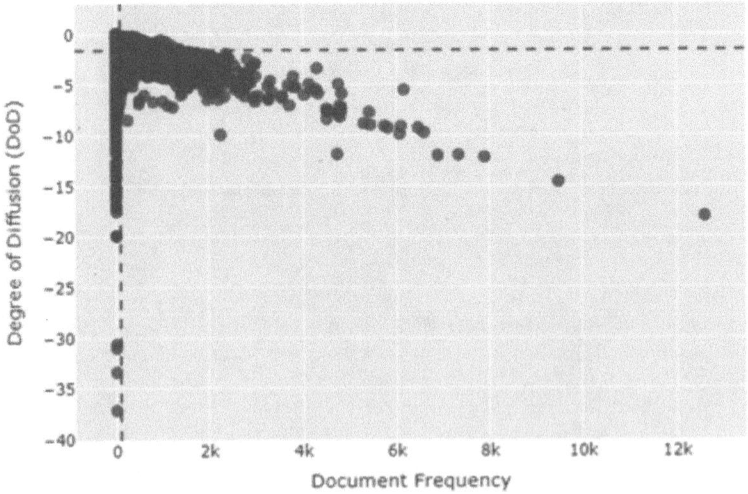

Fig. 12. Resulting KIM.

Once the topics were generated, those containing a significant proportion of keywords previously classified as weak were identified. This was achieved by comparing the keywords of each topic with the weak words detected in the KEM and KIM maps. The relevance of these weak topics was then evaluated through their DoV and DoD, prioritizing those with the highest metrics. This process helped filter and highlight emerging topics with significant potential to evolve into strong signals in the future, providing an additional layer of analysis that complements keyword detection. As shown in Table 3, after identifying keywords that are potentially weak signals, the topics containing those words were subsequently identified.

Table 3. Topics with significant potential to evolve into strong signals in the future.

Topic	Words
86	'new gasoline', 'eco gasoline', 'gasoline sales', 'new fuel', 'gasoline 80', 'gasolines', 'gasoline price', 'eco', 'fuel', 'extra gasoline', 'gasoline types'
15	'new terminal', 'commercial premises', 'mall commerce', 'new center', 'mall', 'plaza', 'new stores', 'north mall', 'shopping center', 'stores'
98	'solar radiation', 'radiation level', 'high radiation', 'level of radiation', 'ultraviolet radiation', 'average solar radiation', 'ultraviolet', 'radiation', 'sunscreen', 'UV radiation'

5 Conclusions

The detection of weak signals in SScan is crucial for anticipating environmental changes and responding promptly to emerging opportunities and threats. This study addressed this challenge through the implementation of a hybrid model that combines NLP techniques with AI algorithms such as Transformer neural networks, providing a robust and effective framework for weak signal detection and analysis.

The implementation of *BERTopic* for topic modeling enabled a more precise decomposition of textual data, identifying emerging patterns and topics that may serve as weak signals. This approach contrasts with traditional methods, which often overlook the semantic nuances of words in specific contexts and thus limit the depth of weak signal analysis. By integrating *BERTopic* with *BERT*-based embeddings, the model demonstrated strong capabilities in detecting emerging topics and relevant information patterns.

In addition, KEM and KIM proved effective in identifying emerging weak signals with precision. A key advantage of the proposed model is its potential to detect weak signals in real time, allowing organizations to anticipate changes in their operating environment, identify new opportunities, and mitigate potential risks in a proactive manner. The integration of time series analysis with changepoint detection further provided a detailed assessment of how weak signals vary over time, adding a temporal perspective that previous approaches did not address.

Overall, the proposed model represents a significant advancement in the organizational capacity to anticipate and respond to changes in complex environments.

By combining topic modeling, changepoint detection, and time series analysis, it enhances the ability to detect emerging opportunities and reduce uncertainty in strategic decision-making.

For future research, it is recommended to explore multimodal approaches that integrate textual analysis with other data sources such as images or social media signals, to improve detection accuracy. Additionally, validating the model in international contexts could further strengthen its applicability and robustness.

References

1. Alstott, J., Bullmore, E., Plenz, D.: Powerlaw: a python package for analysis of heavy tailed distributions. PLoS One. **9**(1), e85777 (2014)
2. Blanco, S., Lesca, N.: From weak signals to anticipative information: learning from the implementation of an information selection method. In: Proceedings of the International Conference in Search of Time (Isida), pp. 197–210 (2003)
3. Blei, D., Ng, A., Jordan, M.: Latent dirichlet allocation. Adv. Neural Inf. Proces. Syst. **14**,1–8 (2001)
4. Bock, G.W., Mahmood, M., Sharma, S., Kang, Y.J.: The impact of information overload and contribution overload on continued usage of electronic knowledge repositories. J. Organ. Comput. Electron. Commer. **20**(3), 257–278 (2010)
5. Choo, C.W.: Information use and early warning effectiveness: perspectives and prospects. J. Am. Soc. Inf. Sci. Technol. **60**(5), 1071–1082 (2009)
6. Day, G.S., Schoemaker, P.: Peripheral vision: sensing and acting on weak signals. Long Range Plan. **37**, 117–121 (2004)
7. Ebadi, A., Auger, A., Gauthier, Y.: Detecting emerging technologies and their evolution using deep learning and weak signal analysis. J. Informet. **16**(4), 101344 (2022)
8. Edmunds, A., Morris, A.: The problem of information overload in business organisations: a review of the literature. Int. J. Inf. Manag. **20**(1), 17–28 (2000)
9. El Akrouchi, M., Benbrahim, H., Kassou, I.: End-to-end lda-based automatic weak signal detection in web news. Knowl.-Based Syst. **212**, 106650 (2021)
10. El Sawy, O.A.: Personal information systems for strategic scanning in turbulent environments: can the ceo go on-line? MIS Q. **9**, 53–60 (1985)
11. Grootendorst, M.: Bertopic: neural topic modeling with a class-based tf-idf procedure. arXiv. (2022)
12. Hastie, T., Tibshirani, R., Friedman, J., Franklin, J.: The elements of statistical learning: data mining, inference and prediction. Math. Intell. **27**(2), 83–85 (2005)
13. Hemp, P.: Death by information overload. Harv. Bus. Rev. **87**(9), 82–121 (2009)
14. Hiltunen, E.: The future sign and its three dimensions. Futures. **40**(3), 247–260 (2008)
15. Kuusi, O., Hiltunen, E.: The signification process of the future sign. J. Futures Stud. **16**(1), 47–66 (2011)
16. Lesca, H.: Veille stratégique pour filemanagement stratégique, état de la question et axes de recherche. Economies et Société, série sciences de gestion. **20**, 31–50 (1994)
17. Lesca, H., Lesca, N.: Strategic Decisions and Weak Signals: Anticipation for Decision-Making. John Wiley & Sons (2014)
18. Lesca, N., Caron-Fasan, M.L.: Strategic scanning project failure and abandonment factors: lessons learned. Eur. J. Inf. Syst. **17**(4), 371–386 (2008)
19. Lesca, N., Caron-Fasan, M.L., Aguirre, E.L., Chalus-Sauvannet, M.C.: Drivers and barriers to pre-adoption of strategic scanning information systems in the context of sustainable supply chain. Systèmes d'information & management. **20**(3), 9–46 (2015)

20. Lesca, N., Caron-Fasan, M.L., Falcy, S.: How managers interpret scanning information. Inf. Manag. **49**(2), 126–134 (2012)
21. Moulton, R., Jiang, Y.: Maximally consistent sampling and the jaccard index of probability distributions. arXiv. (2018). https://doi.org/10.48550/arXiv.1809.04052
22. Muller, T., Cotterell, R., Fraser, A., Schütze, H.: Joint lemmatization and morphological tagging with lemming. arXiv. (2024). https://doi.org/10.48550/arXiv.2405.18308
23. Park, C., Cho, S.: Future sign detection in smart grids through text mining. Energy Procedia. **128**, 79–85 (2017)
24. Roche, E.M.: Multinational corporations—the emerging research agenda. J. Strateg. Inf. Syst. **5**(2), 129–147 (1996)
25. Röder, M., Both, A., Hinneburg, A.: Exploring the space of topic coherence measures. In: Proceedings of the Eighth ACM International Conference on Web Search and Data Mining, pp. 399–408 (2015)
26. Rouibah, K., Ould-Ali, S.: Puzzle: a concept and prototype for linking business intelligence to business strategy. J. Strateg. Inf. Syst. **11**(2), 133–152 (2002)
27. San Juan, Y.I., Rodríguez, F.I.R.: Modelos y herramientas para la vigilancia tecnol6gica. Ciencias de la. Informaci6n. **47**(2), 11–18 (2016)
28. Strother, J.B., Ulijn, J.M.: The challenge of information overload. In: 2012 IEEE International Professional Communication Conference, pp. 1–3. IEEE (2012)
29. Uzun, E.: A novel web scraping approach using the additional information obtained from web pages. IEEE Access. **8**, 61726–61740 (2020)
30. Vaswani, A. et al.: Attention is all you need. In: Proceedings of the 31st International Conference on Neural Information Processing Systems. NIPS'17, pp. 6000–6010. Curran Associates Inc., Red Hook, NY, USA (2017)
31. Walters, B.A., Jiang, J.J., Klein, G.: Strategic information and strategic decision making: the eis/ceo interface in smaller manufacturing companies. Inf. Manag. **40**(6), 487–495 (2003)
32. Yoon, J.: Detecting weak signals for long-term business opportunities using text mining of web news. Expert Syst. Appl. **39**(16), 12543–12550 (2012)

Artificial Intelligence - Prediction and Forecasting Methods

Distilled Tree Models for Cardiovascular Risk Prediction from Imbalanced Clinical Data

Handry Serrano Fabela[1], Rafael Rivera López[1(✉)], Erick de la Cruz Hernández[2], Daniela Hernández Barrios[1], Abraham Carrasco Barradas[1], Marisol Cruz Ramírez[1], Abraham Barradas Cortés[1], and Daniela Ortega Juárez[1]

[1] TecNM, Instituto Tecnológico de Veracruz, Veracruz, Ver., Mexico
{121020445,rafael.rl,daniela.hb,121020360,
121020369,121020361,121020080}@veracruz.tecnm.mx

[2] Universidad Juárez Autónoma de Tabasco, Comalcalco, Tab., Mexico
erick.delacruz@ujat.mx

Abstract. This work introduces a distillation framework to induce a tree-based interpretable classifier built from a more complex model trained on rebalanced clinical data. Synthetic minority samples are first generated using unsupervised generative artificial intelligence to reduce the imbalance inherent to clinical datasets. A black-mox model trained on this enriched dataset provides robust predictive labels, which are then transferred to a decision tree to obtain a transparent approximation of the ensemble model. The proposed framework serves as a support tool for cardiovascular risk prediction, and the results show that the improved decision tree achieves high sensitivity in identifying at-risk individuals but preserves the interpretability required for clinical decision support.

Keywords: Imbalanced Clinical Data · Model Distillation · Tree-Based Models · Synthetic Tabular Data · GAN-Based Data Augmentation

1 Introduction

In data analysis and automated decision-making, a significant challenge arises in clinical datasets, as they often exhibit severe class imbalance due to the relatively low prevalence of positive cases., since they frequently exhibit severe class imbalance due to the relatively low prevalence of positive cases. Models trained on such data tend to favor the majority class, achieving high overall accuracy while failing to correctly identify minority (at-risk) individuals, precisely the cases of greatest clinical interest [4,19]. This issue is particularly relevant in conditions such as elevated homocysteine levels, where positive cases occur infrequently and minority-class detection is critical for preventive cardiovascular care. Elevated homocysteine (Hcy) levels have a high relationship with risks of cardiovascular

H. R. Arabnia et al. (Eds.): CSCI 2025, CCIS 2967, pp. 317–332, 2027.
https://doi.org/10.1007/978-3-032-28454-9_22

diseases (CVD), particularly in young populations. Homocysteinemia (HHcy) has been identified as a contributor to coronary artery disease (CAD) and other cardiovascular conditions even in the absence of traditional risk factors such as hyperlipidemia or hypertension [14,20]. The presence of HHcy in people with endothelial dysfunction, arterial stiffness, and atherosclerotic plaque reinforces its relevance as an early biomarker of cardiovascular risk (CVR) [8]. Nutritional deficiencies, mainly of vitamins B6, B12, and folate, which are central to Hcy metabolism, are commonly reported among young individuals with HHcy [21]. It is also known that overweight children and adolescents have elevated levels of Hcy [16], and that together with lifestyle factors such as smoking, they can increase their cardiovascular impact [22]. Because these metabolic, nutritional, and behavioral determinants interact in complex and nonlinear ways, conventional analysis is insufficient to capture their combined effects, underscoring the need for machine learning approaches.

Machine learning has emerged as a powerful tool to mitigate this imbalance, as many algorithms can adapt to data distributions through resampling strategies. Techniques such as undersampling of the majority class, oversampling of minority cases, or the creation of synthetic samples aim to provide a more representative training set, enabling models to better generalize the underlying relationships between clinical variables. By reducing bias toward the majority class, these approaches help improve the detection of rare but clinically relevant outcomes, particularly in scenarios where early identification is essential. However, an additional limitation arises from the fact that many of the most accurate ML models, such as neural networks or ensemble methods, are inherently opaque, limiting their applicability in clinical decision-making where transparency and explainability are essential [7,18]. This lack of interpretability complicates error analysis, hinders clinician trust, and may contribute to misdiagnoses or inappropriate treatments if model behavior cannot be adequately inspected [24].

This study proposes a classification strategy that integrates model distillation with techniques to mitigate class imbalance. Synthetic minority samples are generated to enrich the dataset, and a robust ensemble model is first trained on this balanced data. Its predictive outputs are then transferred to a distilled decision tree, producing an interpretable classifier that approximates the behavior of the ensemble while maintaining high sensitivity for detecting cardiovascular risk. This framework seeks to combine the predictive power of complex models with the transparency required in clinical settings, offering a feasible pathway toward trustworthy and interpretable risk prediction tools.

The main contribution of this work is the development of an interpretable classification model derived from a high-capacity ensemble trained on rebalanced clinical data. By combining generative oversampling with model distillation, the proposed framework provides a transparent predictor capable of identifying individuals with elevated homocysteine levels with high sensitivity. To our knowledge, this is one of the first approaches to integrate explainable distillation and synthetic data generation for cardiovascular risk assessment in young populations.

2 Related Work

Several studies have applied ML models to investigate the relationship between Hcy levels and CVD, mostly in adult populations. For Hcy analysis, approaches have included (1) automated ML tools such as Tree-Based Pipeline Optimization Tool [17], (2) ensemble and kernel-based models like RF and Extreme Gradient Boosting (XGB) [3]; as well as (3) statistical methods such as weighted quantile sum regression, quantile-based g-computation, and Bayesian kernel machine regression [5]. In particular, one approach uses both classification and regression algorithms to CAD diagnosis [1]. Furthermore, dataset elements differ across studies; some use balanced datasets [17] and others show an imbalance ratio of approximately 2:1 [1]. Among the evaluated methods, ensemble techniques such as XGB and RF report superior performance in modeling complex interactions.

To handle imbalanced datasets, several strategies have been proposed in the literature, with synthetic sampling (either in hybrid or cluster-based approaches) standing out, performing better than classical methods. [25]. In particular, Generative Adversarial Networks (GANs) are applied to create synthetic cases using the original data distribution [10], and ensemble methods can model complex data relations combining several models [4]. Similarly, undersampling techniques such as Random Undersampling (RUS) balance class distribution by reducing the size of the majority class [6]. Furthermore, cost-sensitive learning methods use cost matrices to adjust the training process, reducing bias toward the majority class and improving the detection of positive cases. Cost-sensitive methods have proven helpful for disease diagnosis and prediction [15], as well as in medical image analysis for early disease detection, such as cancer [2].

To improve model interpretability, some studies employ Explainable Artificial Intelligence (XAI) techniques to generate explanations for the predictions of opaque models. For medical diagnosis, XAI is used to identify disease-related features from medical images, such as those of the gastrointestinal tract [23], and LIME and SHAP have been employed to interpret deep learning models in classifying diseases such as pneumonia from X-ray images [28].

Although many ML techniques have been used as decision support tools in medical applications, their use in analyzing the relationship between Hcy and CVD is minimal, which serves as motivation to explore their applicability to this medical condition.

3 Materials and Methods

This study investigates Hcy levels through sex-stratified analysis, justified by well-documented biological, hormonal, and CVR differences between males and females. In contrast with similar approaches, this study addresses a much more imbalanced dataset, focusing on young adults, an understudied group due to their apparent good health, but crucial for implementing preventative strategies to maintain good cardiovascular health.

We analyzed two distinct datasets of young adults aged 18 and 19 years (240 women, 108 men) recruited from a Mexican public university. The first dataset contained anthropometric values obtained via bioelectrical impedance, and the second included clinical blood parameters such as hematological profiles and lipid markers. All measurements are obtained using automated analyzers, with protocols validated for the World Health Organization (WHO), with the ethical approval of the university in compliance with Committee on Publication Ethics (COPE) guidelines [9] . This study analyzes the Hcy behavior using body composition parameters, hematological values, and lipid metabolism in this young adult population. Features considered in this study are the following:

- **Anthropometric data:** Body Mass Index (BMI), Total Body Fat Percentage (BodyFat), Skeletal Muscle Mass (SkeletalMuscle), and Visceral Fat Level (VisceralFat).
- **Blood cell parameters:** Erythrocytes, Hemoglobin, Hematocrit, Leucocytes, Mean Corpuscular Volume (MCV), Mean Corpuscular Hemoglobin (MCH), Red Cell Distribution Width - Coefficient of Variation (RDW-CV), and Platelet Count (Platelets).
- **Lipid profile components:** Total Cholesterol (Colesterol), Total Triglycerides (Triglycerides), High Density Lipoprotein-Cholesterol (HDL-C), Low Density Lipoprotein-Cholesterol (LDL-C), Very Low-Density Lipoprotein Cholesterol (VLDL-C), Total Cholesterol to HDL Ratio (TC/HDL), LDL to HDL Ratio (LDL/HDL), and Atherogenic Index of Plasma (AIP).

HHcy is established when plasma Hcy levels exceed $15\mu mol/L$ [13]. This value is used in this study to induce classification models, where samples are labeled as *Normal* (Hcy $\leq 15\mu mol/L$) or *Risk*. Figure 1 shows the distribution of Hcy grouped by sex. Measurements from each person in the study are indicated as data points: a green dot indicates a person with normal Hcy levels, and a red dot represents people with elevated Hcy concentrations. A horizontal dashed line represents the threshold for dividing normal cases from those at risk. The accompanying legend shows the numerical distribution of study cases by sex and by category.

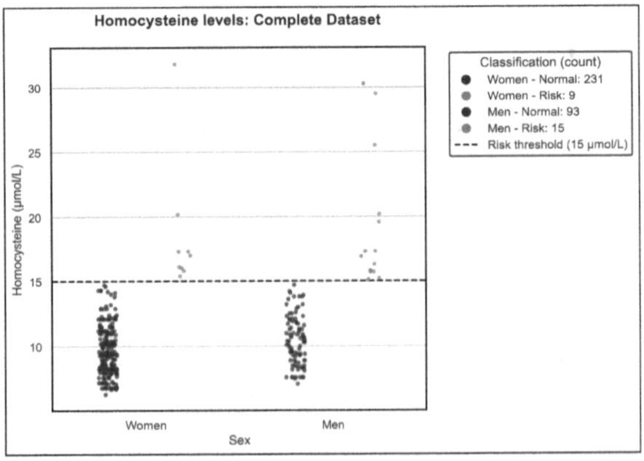

Fig. 1. Sex-stratified distribution of Hcy levels with clinical risk categorization.

The dataset has a significant imbalance between its classes. For women, 231 cases (96.3%) are normal and only nine (3.7%) are classified as at-risk, for an imbalance ratio of 26:1. In men, the distribution is less extreme: 93 cases (86.9%) are normal and 14 (13.1%) are at-risk, for a ratio of 6.6:1. The full dataset has 324 normal cases (93.4%) and 23 at-risk cases (6.6%), for an imbalance ratio of 14:1. This imbalance can affect the performance of any model built from this data, with results biased toward the majority class if not adequately addressed.

The efficiency of a classification model is measured by metrics derived from the confusion matrix, which identifies true positives (TP), false positives (FP), true negatives (TN), and false negatives (FN). In our clinical context, a TP represents a correctly identified *Risk* case. An FN is a *Risk* case erroneously flagged as *Normal*, which is dangerous since the patient must be treated. An FP is a *Normal* case incorrectly flagged as *Risk*, which can lead to unnecessary treatments, and a TN is a *Normal* case correctly identified as such. These values are used in the definition of several key performance metrics of a classifier that are widely used in clinical evaluation. These metrics are formally defined in Table 1.

Table 1. Evaluation metrics used in the study.

Metric	Formula	Description
Accuracy (Acc)	$\frac{TP+TN}{TP+TN+FP+FN}$	Proportion of correctly classified instances among all cases.
Sensitivity (Sen)	$\frac{TP}{TP+FN}$	Proportion of actual positive cases correctly identified by the model. Critical for detecting individuals at CVR.
Specificity (Spe)	$\frac{TN}{TN+FP}$	Proportion of actual negative cases correctly classified, reflecting the ability to avoid false alarms.

In healthcare, the sensitivity metric is more important than other metrics such as accuracy, since the class with fewer cases represents individuals at potential risk for future disease due to their high Hcy levels. Failure to correctly identify these cases (FN) could delay early care and increase the likelihood of adverse outcomes. Therefore, higher sensitivity implies greater clinical utility, even with a moderate decrease in the model's overall accuracy.

To try to reduce the problems mentioned in the latter section for using imbalanced datasets, we propose a strategy that combines undersampling and oversampling, implemented as follows:

1. **Undersampling:** RUS without replacement is used to remove majority class cases until the proportion of the majority class is reduced to a ratio of 5:1 with the minority class.

2. **Oversampling:** To equalize the proportion of cases between classes, synthetic samples of the minority class are generated using a Conditional Tabular GAN (CTGAN) [26]. CTGAN is used instead of SMOTE or copula-based methods because it can model complex dependencies between features, learning the distribution of the original data that is used to create realistic synthetic samples.

In addition to resampling, a misclassification cost matrix is applied during the training phase. The matrix assigns a higher penalty to false negatives than to false positives, considering the clinical significance of correct detection. The cost of a false negative is higher than that of a false positive. Its values are determined empirically based on preliminary experiments to maximize sensitivity without unduly compromising specificity.

A DT and one RF are selected with base models in this study. Both are used in clinical settings due to their performance, ability to handle complex data, and their design to work with tabular data. DTs are known for being simple to implement, quick to create, and, above all, interpretable, enabling transparent decision-making in clinical settings. Otherwise, RFs combine multiple trees to improve accuracy, reduce overfitting, and increase robustness to outliers or missing values. While DTs risk overfitting if not pruned and RFs are less interpretable due to their ensemble nature, both remain competitive alternatives to neural networks and other black-box models.

The ensemble distillation process is designed to transfer the predictive capacity of a complex model ensemble into a single and interpretable model suitable for clinical decision support. The procedure consisted of the following steps:

1. **Training the ensemble:** The balanced training set is used to induce a cost-sensitive RF.
2. **Label refinement:** The ensemble's predicted class labels are used to generate new labels for all training instances. This re-labeling grouped instances that, based on their features, are more likely to belong to the *Risk* class, effectively encoding the decision boundaries learned by the ensemble. This re-labeling step is equivalent to *pseudo-labeling*, where a complex model generates instance labels used to induce another model [11].
3. **Training the distilled model:** A new DT is trained using the re-labeled dataset.
4. **Evaluation:** The distilled DT (dDT) is evaluated on the test set to verify its predictive performance and level of interpretability.

This approach combines the high predictive performance of the ensemble with the transparency of a single DT, producing human-readable decision rules that are easier to deploy in clinical environments. By distilling knowledge into an interpretable model, it preserves the ability to detect minority-class instances, which is critical for CVR screening.

The performance of the models is evaluated using 10-fold repeated stratified cross-validation (CV), ensuring robust and unbiased estimates.

In this approach, data are pooled into ten equal parts (folds); at each iteration, nine of them are used for training and the remaining one for testing. Crucially, to prevent data leakage and ensure the validity of the results, any data balancing techniques are applied independently and solely to the training folds at every iteration. The test fold always remains untouched, representing the original, unaltered data distribution. This process is repeated until each fold has served once as a test. In addition, stratification is used to ensure that the proportion of classes in each fold is as similar as possible to that of the original dataset. Finally, the CV procedure is repeated ten times, changing the order of the dataset instances to reduce the variability of the experiment further. The result reported is the average of these ten repetitions. Compared to a single training-test split, this strategy makes better use of the available data, produces more reliable estimates, and reduces the risk of results depending on a particular way of splitting the dataset.

To compare the performance of the implemented models, *accuracy*, *sensitivity* (true positive rate) and *specificity* (true negative rate), are used. Sensitivity is critical in this study, since it calculates the ability to correctly identify individuals at CVR. Specificity shows the capacity to avoid false alarms. The formulas and descriptions for all metrics are summarized in Table 1.

The experimental study is implemented in Java and Python. Model training and evaluation are coded in Java using the *Weka* library. DTs are induced with SPAARC [27]. SPAARC (Advanced Splitting and Pruning Algorithm for Rule-Based Classification) is a method for creating DTs that applies feature selection and seeks to replace complex branches to create more compact and interpretable DTs. On the other hand, dataset balancing is coded in Python using libraries for data manipulation and synthetic data generation with CTGAN.

4 Results

The evaluation considered three classifiers: a DT, an RF, and a dDT. For each of the four experimental datasets: Anthropometric (Women), Anthropometric (Men), Clinical (Women), Clinical (Men), we tested four experimental configurations:

1. **Unbalanced, No Cost Matrix (U-NC):** Original dataset, no class balancing, standard training.
2. **Unbalanced, Cost Matrix (U-C):** Original dataset, no balancing, with cost-sensitive training.
3. **Balanced, No Cost Matrix (B-NC):** Dataset balanced via undersampling and CTGAN oversampling, standard training.
4. **Balanced, Cost Matrix (B-C):** Balanced dataset with cost-sensitive training.

For each dataset, a specific cost matrix is used to penalize FN more heavily than FP, reflecting the clinical relevance of detecting Risk cases. The form of one cost matrix and the empirically tuned values for experiments are as follows:

General Cost Matrix

	Predicted Normal	Predicted Risk
Actual Normal	0	C_{FP}
Actual Risk	C_{FN}	0

Specific Costs

- Anthropometric:
  - Women: $C_{FP} = 1, C_{FN} = 2.5$
  - Men: $C_{FP} = 1, C_{FN} = 3.3$
- Clinical:
  - Women: $C_{FP} = 1, C_{FN} = 3.3$
  - Men: $C_{FP} = 1, C_{FN} = 5$

where C_{FP} denotes the penalty for misclassifying a Normal case as Risk (false positive), and C_{FN} represents the penalty for misclassifying a Risk case as Normal (false negative).

The specific values were determined through a grid search over multiple combinations, selecting those that maximized sensitivity while keeping specificity above an acceptable clinical threshold. This empirical tuning ensured that the cost-sensitive learning component adapted to the imbalance severity and feature distribution of each dataset.

The performance results of the models for the anthropometric and clinical datasets are shown in Tables 2 and 3, respectively. Each table compares the three different models across four distinct experimental configurations, detailing the evaluation metrics for women and men in the test set. For ease of interpretation, the **boldfaced values** in each column indicate the best result achieved for that specific metric within its experimental group and gender.

Table 2. Model Performance for Anthropometric Data.

		Women (Test)			Men (Test)		
Exp	Model	Accuracy	Sensitivity	Specificity	Accuracy	Sensititivy	Specificity
U-NC	DT	**96.25**	0.00	**100.00**	91.94	49.33	98.82
	RF	**96.25**	0.00	**100.00**	91.67	47.33	98.82
	dDT	**96.25**	0.00	**100.00**	**93.33**	58.67	**98.93**
B-NC	DT	78.29	45.56	79.57	85.46	63.33	89.03
	RF	80.21	47.78	81.47	88.43	65.33	92.15
	dDT	80.04	47.78	81.30	85.28	62.67	88.93
U-C	DT	90.50	6.67	93.77	75.65	58.00	78.50
	RF	94.58	0.00	98.27	85.93	60.00	90.11
	dDT	90.83	4.44	94.20	73.06	62.67	74.73
B-C	DT	62.50	56.67	62.73	62.32	72.67	60.65
	RF	65.50	62.22	65.63	62.96	68.00	62.15
	dDT	53.08	**72.22**	52.34	52.69	**75.33**	49.03

Table 3. Model Performance for Clinical Data.

		Women (Test)			Men (Test)		
Exp	Model	Accuracy	Sensitivity	Specificity	Accuracy	Sensititivy	Specificity
U-NC	DT	**96.25**	0.00	**100.00**	**85.93**	0.00	99.79
	RF	**96.25**	0.00	**100.00**	85.83	0.67	99.57
	dDT	**96.25**	0.00	**100.00**	**85.93**	0.00	99.79
B-NC	DT	77.75	24.44	79.83	71.11	35.33	76.88
	RF	84.25	14.44	86.97	77.87	28.67	85.81
	dDT	78.83	25.56	80.91	72.69	30.67	79.46
U-C	DT	90.08	5.56	93.38	70.46	47.33	74.19
	RF	**96.25**	0.00	**100.00**	76.94	27.33	84.95
	dDT	89.96	4.44	93.29	70.65	48.67	74.19
B-C	DT	40.42	61.11	39.61	57.04	64.00	55.91
	RF	47.17	46.67	47.19	64.35	72.67	63.01
	dDT	29.21	**71.11**	27.58	54.72	**80.67**	50.54

The analysis of the results presented in these tables reveals the critical impact of class imbalance and the effectiveness of the proposed strategy.

- **Unbalanced, No Cost Matrix (U-NC):** The results of this experiment indicate that if class imbalance is not considered, the performance is unacceptable. The models have high specificity ($\leq 90\%$) but almost zero sensitivity ($\simeq 0 - 20\%$). That is, the models cannot identify at-risk cases, but they are very good at detecting normal cases. This level of performance is insufficient for practical medical applications where detecting true positive cases is crucial.
- **Unbalanced with Cost Matrix (U-C):** The introduction of false negative penalties produced moderate improvements in sensitivity, particularly in clinical datasets where sensitivity reached 40–50%. Specificity decreased moderately but remained above 80% in most cases. However, the main limitation remains: extreme class imbalance limits the effective detection of positive cases, especially in anthropometric data sets where results are very limited.
- **Balanced without Cost Matrix (B-NC):** Class balancing using RUS and CTGAN generated performance improvements. Sensitivity increased across all models, improving the capacity to detect individuals with CVR. On the other hand, a decrease in specificity is observed, which is an expected behaviour. Notably, distilled Decision Trees (dDT) demonstrated particular effectiveness, matching or slightly outperforming standard DT while maintaining the advantage of interpretability.
- **Balanced with Cost Matrix (B-C):** The combined strategy produced the best results, with maximum sensitivity values between 70% and 75% using RF and 68%-72% using the DT/dDT models. Specificity remained within

the 60%-65% range, which is considered acceptable for our study, as it prioritizes the detection of individuals with CVR. The dDT models emerged as particularly valuable solutions, offering sensitivity comparable to RF while maintaining compact and interpretable structures essential for clinical decision support.

Several consistent patterns can be observed from the experimental study. Balancing the datasets is the most important data preprocessing step, as it generates substantial improvements in the detection capacity of individuals with CVR. Cost matrices help improve the model's sensitivity by penalizing false negatives. The split tree consistently generates better performance, in addition to its interpretive power. These findings using anthropometric and clinical data validate the robustness of the proposed scheme for different feature sets and imbalance ratios.

As can be seen, the results indicate that using a balanced dataset with cost-sensitive training strategies is the best option to support CVD diagnosis and prevention. Distilled DT models are therefore the most recommended model. This approach can be used to recommend more detailed studies for individuals where the model predicts a CVR, as there is a possibility of developing CVD, and better studies could accurately rule out treatment. This is important because missing true positive cases has significant clinical consequences.

To evaluate whether the distilled DT preserves the RF feature hierarchy, we compute global feature importance using SHAP values for both models. As an alternative to metrics such as the Gini coefficient or information gain, SHAP (SHapley Additive ExPlanations) measures the contribution of each dataset feature to a prediction using cooperative game theory, measuring the average marginal effect across all possible feature coalitions ([12]). By calculating SHAP values across all CV iterations, a global ranking of the importance of features in the process is obtained. SHAP values are computed for both RF and dDT to provide a fair comparison measure and to analyze whether the distilled model correctly learns the ensemble structure.

Figures 2 and 3 show the CV mean ± SD of the top features for each model based on the global SHAP feature importance.

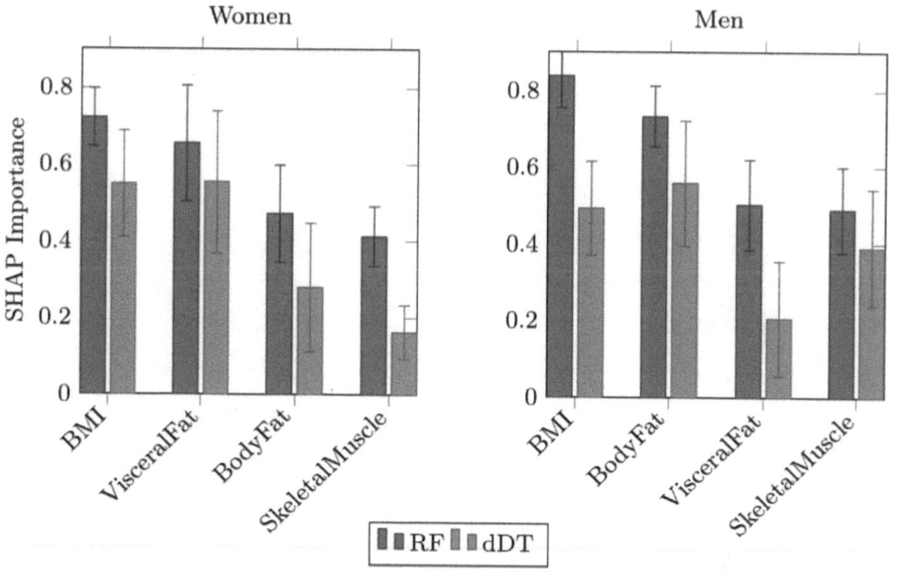

Fig. 2. SHAP feature importance for anthropometric data by gender.

Comparing the global SHAP values of the RF and distilled DT highlights some similarities. For the anthropometric data (Fig. 2), BMI can be considered the main predictor feature in both sexes, with greater importance in RF (0.72 ± 0.08 in women and 0.84 ± 0.08 in men) than in the distilled models (0.55 ± 0.14 in women and 0.49 ± 0.12 in men). This indicates that BMI can be used as an CVR indicator. Furthermore, visceral and body fat also obtained high values. For women, visceral fat is very relevant (0.66 ± 0.15 in RF vs. 0.55 ± 0.19 in dDT), while body fat and skeletal muscle are less important. On the other hand, in men, body fat is more important in dDT (0.56 ± 0.16) than in RF (0.73 ± 0.08), and visceral fat showed a marked decrease (0.50 ± 0.12 in RF vs. 0.21 ± 0.15 in dDT). This indicates that the distillation process concentrated the importance on a smaller set of dominant features and revealed gender-specific adjustments in fat-related predictors.

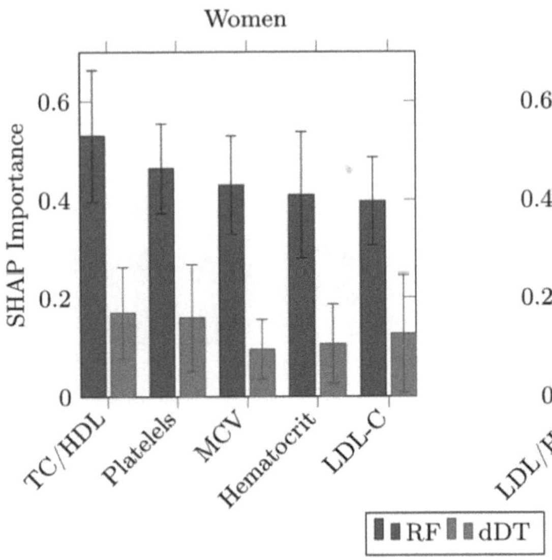

Fig. 3. Top 5 clinical features by SHAP values for each gender. (a) Women: TC/HDL, Platelels, MCV, Hematocrit, and LDL-C are most important. (b) Men: HDL-C, LDL/HDL, AIP, TC/HDL, and Triglycerides.

This pattern also holds true for clinical data (Fig. 3). In women, RF assign more importance to lipid and hematologic markers. The most important features are HDL cholesterol (TC/HDL, 0.53 ± 0.13), platelets (0.46 ± 0.09), and MCV (0.43 ± 0.10). The distilled model gives these attributes less importance (0.17 ± 0.09 for TC/HDL, 0.16 ± 0.11 for platelets, and 0.10 ± 0.06 for MCV). In men, lipid components are more important in RF: LDL/HDL (0.48 ± 0.08), HDL-C (0.45 ± 0.16), and the plasma atherogenic index (AIP, 0.44 ± 0.08). The dDT model maintained this order, but with a lower significance level: HDL-C (0.28 ± 0.22), LDL/HDL (0.32 ± 0.13), and AIP (0.28 ± 0.14). Triglycerides decreased dramatically from 0.30 ± 0.06 in RF to only 0.08 ± 0.08 in dDT. This indicates that the distilled DT preserves the hierarchy of key clinical features, but reduces the influence of secondary predictors, with a more compact and interpretable set of biomarkers.

In summary, BMI and lipidic components are the main predictors of individuals with CVD, validating the importance of these features as risk indicators. At the same time, the distilled models improve interpretability without discarding clinically relevant features. These values reflect the strengths of the models: RF leverages a broad set of predictors, while dDT uses a subset of biomarkers, which aids in clinical translation and decision support.

Although all quantitative results are derived from repeated 10-fold CV to ensure statistical robustness, for visualization and clinical interpretability, we present the distilled dDT induced from the entire dataset, separately for women and men (Fig. 4). This helps to observe the gender-specific decision rules the model learns, which cannot be perceived from average performance metrics alone. Using the full dataset provides a compact and clinically meaningful summary of the hierarchical structure of features and is consistent with the predictive behavior of the CV scheme.

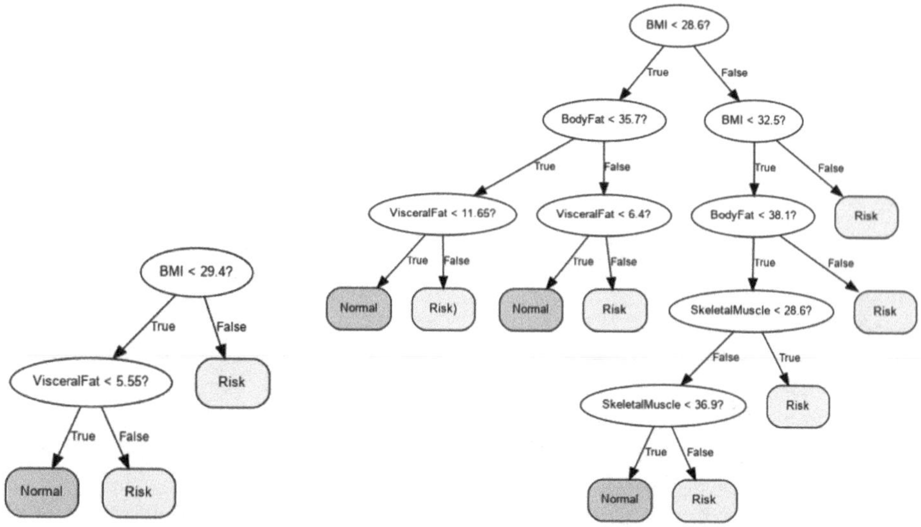

(a) Women (b) Men

Fig. 4. Distilled DTs induced from the complete dataset under the B-C configuration.

5 Discussion

The initial configuration (U-NC) confirmed the bias of the DT and RF models toward the majority class, with very low sensitivities across all datasets. This is consistent with other medical classification tasks where disease prevalence is low, highlighting the challenge of detecting CVR cases.

Using a cost matrix (U-C) improved the sensitivity of all models, with RF showing the most notable improvements. However, these improvements were limited, especially in data sets with very high imbalance rates (e.g., Anthropometric Women). This suggests that while cost-sensitive learning reorients decision boundaries toward the minority class, its effectiveness is constrained without sufficient exposure to minority samples.

Balancing via undersampling and CTGAN-based oversampling (B-NC) yielded the most significant single-step improvement in sensitivity. This effect is present in all models and datasets, confirming the need to modify the class distribution to ensure adequate representation of risk cases. As expected, this improvement came at the expense of reduced specificity.

Using balancing and a cost-sensitive (B-C) configuration, the distilled DT demonstrates the best balance between sensitivity and explainability. The distilled model captured the patterns identified by the RF, creating transparent decision rules, which is one of its benefits in the context of clinical diagnosis.

In terms of interpretability, distilled models have a more useful structure. Trees generated with the original data do not contribute to the explainability of their predictions, as they are biased toward the majority class.

Feature importance analysis provided additional insight. RF consistently ranked visceral fat, body fat, and BMI as dominant predictors, whereas distilled models highlighted a similar hierarchy with fewer attributes. This alignment suggests that dDT not only compresses knowledge but also preserves clinically meaningful variables, reinforcing confidence in its application for CVR diagnosis.

6 Conclusions and Future Work

This study indicates that the combined use of cost-sensitive learning and class balancing significantly improves the sensitivity of CVR prediction models using anthropometric and biochemical features. Distilled DTs are the best alternative, offering an interpretable and accurate version of RFs. Model decisions use clinically relevant variables, which improves the reliability and usability of these models in medical settings.

Future work will focus on extending the interpretability analysis to larger and more diverse cohorts, exploring hybrid resampling strategies, and validating the approach with external datasets. Additionally, incorporating explainable AI (XAI) methods could further strengthen transparency, ensuring broader adoption of AI-based decision support systems in preventive cardiology.

Disclosure of Interests. The authors have no competing interests to declare that are relevant to the content of this article.

References

1. Ainiwaer, A., et al.: A machine learning framework for diagnosing and predicting the severity of coronary artery disease. Rev. Cardiovasc. Med. **24**(6), 168 (2023). https://doi.org/10.31083/j.rcm2406168
2. Banik, D., Bhattacharjee, D.: Mitigating data imbalance issues in medical image analysis. In: Rana, D., Mehta, R. (eds.) Data Preprocessing, Active Learning, and Cost Perceptive Approaches for Resolving Data Imbalance, pp. 66–89. IGI Global, Hershey, PA, USA (2021). https://doi.org/10.4018/978-1-7998-7371-6.CH004
3. Chiang, P.J., Tsao, C.W., Jhuo, Y.C., Chu, T.W., Pei, D., Kuo, S.W.: Using machine learning to detect factors that affect homocysteine in healthy elderly taiwanese men. Biomedicines **13**(8), 1816 (2025). https://doi.org/10.3390/biomedicines13081816
4. Fakhruzi, I.: An artificial neural network with bagging to address imbalance datasets on clinical prediction. In: 2018 Int. Conf. on Information and Communications Technology (ICOIACT), 6-7 March. IEEE, Yogyakarta, Indonesia (2018). https://doi.org/10.1109/ICOIACT.2018.8350824

5. Fan, J., et al.: Relationships between minerals' intake and blood homocysteine levels based on three machine learning methods: a large cross-sectional study. Nutrition & Diabetes **14**, 36 (2024). https://doi.org/10.1038/s41387-024-00293-3
6. Folorunso, S.O., Adeyemo, A.B.: Alleviating classification problem of imbalanced dataset. African J. Comput. & ICT **6**(1), 137–144 (2013)
7. Frasca, M., Torre, D.L., Pravettoni, G., Cutica, I.: Explainable and interpretable artificial intelligence in medicine: a systematic bibliometric review. Dis. Artifi. Intell. **4**(15) (2024). https://doi.org/10.1007/s44163-024-00114-7
8. Guieu, R., Ruf, J., Mottola, G.: Hyperhomocysteinemia and cardiovascular diseases. Ann. Biol. Clin. **80**(1), 7–14 (2022). https://doi.org/10.1684/abc.2021.1694
9. Hernandez-Landero, F., Sanchez-Garcia, E., Gomez-Crisostomo, N., Contreras-Paredes, A., Eduardo, M.A., de la Cruz-Hernandez, E.: Anthropometric, biochemical, and haematological indicators associated with hyperhomocysteinemia and their relation to global DNA methylation in a young adult population. Epigenetics **17**(10), 1269–1280 (2022). https://doi.org/10.1080/15592294.2021.2013420
10. Kuo, N.H., et al.: Generating synthetic clinical data that capture class imbalanced distributions with generative adversarial networks: example using antiretroviral therapy for HIV. J. Biomed. Inform. **144**, 104436 (2023). https://doi.org/10.1016/j.jbi.2023.104436
11. Li, Y., Guo, L., Ge, Y.: Pseudo labels for unsupervised domain adaptation: a review. Electronics **12**(15) (2023). https://doi.org/10.3390/electronics12153325
12. Lundberg, S.M., Lee, S.I.: A unified approach to interpreting model predictions. In: Guyon, I. (ed.) Advances in Neural Information Processing Systems, vol. 30, pp. 4765–4774. Curran Associates, Inc (2017)
13. McCully, K.S.: Vascular pathology of homocysteinemia: implications for the pathogenesis of arteriosclerosis. Am. J. Pathol. **56**(1), 111–128 (1969)
14. Muslem, Z.H., M. Ali, S., Marayati, A.N.A.: Correlation between plasma homocysteine and ischemic heart disease in young adults. F1000Research **13**, 1477 (2024). https://doi.org/10.12688/f1000research.158499.1
15. Nguyen, H.T., Thanh, B., Minh, Q., Hanh, H., Thanh, P., Nguyen, C.: Enhancing disease prediction on imbalanced metagenomic dataset by cost-sensitive. Inter. J. Adv. Compute. Sci. Appli. (IJACSA) **11**(7), 651–657 (2020). https://doi.org/10.14569/IJACSA.2020.0110778
16. de Oliveira Leite, L., Costa Dias Pitangueira, J., Ferreira Damascena, N., Ribas de Farias Costa, P.: Homocysteine levels and cardiovascular risk factors in children and adolescents: systematic review and meta-analysis. Nutrition Rev. **79**(9), 1067–1078 (2020). https://doi.org/10.1093/nutrit/nuaa116
17. Orlenko, A., et al.: Considerations for automated machine learning in clinical metabolic profiling: altered homocysteine plasma concentration associated with metformin exposure. Pac. Symp. Biocomput. **23**, 460–471 (2018)
18. Petch, J., Tabja, Bortesi, J.P., Nelson, W., Di, S., Mamdani, M.H.: Should I trust this model? explainability and the black box of artificial intelligence in medicine. In: Ben-David, S. (ed.) Artificial Intelligence for Medicine, pp. 265–273. Advanced Studies in Complex Systems: Theory and Applications. Academic Press (2024). https://doi.org/10.1016/B978-0-443-13671-9.00015-6
19. Roy, D., Roy, A., Roy, U.: Learning from imbalanced data in healthcare: state-of-the-art and research challenges. In: Acharjya, D.P., Ma, K. (eds.) Computational Intelligence in Healthcare Informatics. Studies in Computational Intelligence, vol. 1132. Springer, Singapore (2024). https://doi.org/10.1007/978-981-99-8853-2_2

20. Shah, H., Jan, M.U., Altaf, A., Salahudin, M.: Correlation of hyperhomocysteinemia with coronary artery disease in absence of conventional risk factors among young adults. J. Saudi Heart Assoc. **30**(4), 305–310 (2018). https://doi.org/10.1016/j.jsha.2018.04.002
21. Sowndarya, K., Hegde, A., Pawar, N.M., Shenoy, R.: Elevated homocysteine levels and vitamin deficiencies as potential risk markers for coronary artery disease in apparently healthy adults. Biomed. Biotechnol. Res. J. **8**(4), 487–492 (2024). https://doi.org/10.4103/bbrj.bbrj_304_24
22. Sun, J., et al.: Combined effect of hyperhomocysteinemia and smoking on the severity of coronary artery disease in young adults $\leq$ 35 years of age: a hospital-based observational study. BMC Cardiovasc. Disord. **21**, 484 (2021). https://doi.org/10.1186/S12872-021-02302-0
23. Thunold, H.H., Riegler, M.A., Yazidi, A., Hammer, H.: A deep diagnostic framework using explainable artificial intelligence and clustering. Diagnostics **13**(22), 3413 (2023). https://doi.org/10.3390/diagnostics13223413
24. Tiwari, R., Dubey, S., Kumar, V.: Interpretability machine learning models: a critical analysis of techniques and applications. Pharma Innovation **8**(2), 25–28 (2019). https://doi.org/10.22271/tpi.2019.v8.i2sa.25245
25. Travadi, Y., Peng, L., Cui, Y., Sun, J.: Direct metric optimization for imbalanced classification. In: 2023 IEEE Int. Conf. on Healthcare Informatics (ICHI 2023), pp. 26–29. IEEE, Houston, TX, USA (2023). https://doi.org/10.1109/ichi57859.2023.00125
26. Xu, L., Skoularidou, M., Cuesta-Infante, A., Veeramachaneni, K.: Modeling tabular data using conditional gan. In: Wallach, H. (ed.) Advances in Neural Information Processing Systems, vol. 32, Curran Associates, Inc (2019)
27. Yates, D., Islam, M.Z., Gao, J.: SPAARC: a fast decision tree algorithm. In: Islam, R., et al. (eds.) AusDM 2018. CCIS, vol. 996, pp. 43–55. Springer, Singapore (2019). https://doi.org/10.1007/978-981-13-6661-1_4
28. Zahoor, K., Bawany, N.Z., Ghani, U.: Explainable AI for healthcare: an approach towards interpretable healthcare models. In: 2023 Int. Conf. on Artificial Intelligence and Computer Technologies (ACIT), 6-8 December, pp. 1–8. IEEE, Ajman, United Arab Emirates (2023). https://doi.org/10.1109/ACIT58888.2023.10453740

FTT-GRU: A Hybrid Fast Temporal Transformer with GRU for Remaining Useful Life Prediction

Varun Teja Chirukuri[1], Udaya Bhasker Cheerala[2], Sandeep Kanta[3], Abdul Karim[4], and Praveen Damacharla[5](✉)

[1] Columbia Sportswear, Portland, OR, USA
[2] Independent Researcher, Bay Area, CA, USA
[3] Northeastern University, Dallas, TX, USA
[4] Department of Artificial Intelligence Convergence, Hallym University, Chuncheon, Gangwon, South Korea
[5] KINETICAI INC, The Woodlands, TX, USA
praveen@kineticai.com

Abstract. Accurate prediction of the remaining useful life (RUL) of industrial machinery is essential for reducing downtime and optimizing maintenance schedules. Existing approaches such as long short-term memory (LSTM) networks and convolutional neural networks (CNNs) often struggle to model both global temporal dependencies and fine-grained degradation trends in multivariate sensor data. To address this limitation, we propose a hybrid model, FTT-GRU, which combines a fast Temporal Transformer (FTT)—a lightweight Transformer variant using linearized attention via fast Fourier Transform (FFT)—with a gated recurrent unit (GRU) layer for sequential modeling. To the best of our knowledge, this is the first application of an FTT with a GRU for RUL prediction on NASA CMAPSS, enabling simultaneous capture of global and local degradation patterns in a compact architecture. On CMAPSS FD001, FTT-GRU attains RMSE of 30.76, MAE of 18.97, and R^2 of 0.45, with 1.12 ms CPU latency at `batch=1`. Relative to the best published deep baseline (TCN–Attention), it improves RMSE by 1.16% and MAE by 4.00%. Training curves averaged over $k = 3$ runs show smooth convergence with narrow 95% confidence bands, and ablations (GRU-only, FTT-only) support the contribution of both components. These results demonstrate that a compact Transformer-RNN hybrid delivers accurate and efficient RUL predictions on CMAPSS, making it suitable for real-time industrial prognostics.

Keywords: Fast Temporal Transformer (FTT) · GRU · Hybrid deep learning · NASA CMAPSS · Predictive maintenance · Remaining Useful Life (RUL) · Time-series forecasting

H. R. Arabnia et al. (Eds.): CSCI 2025, CCIS 2967, pp. 333–346, 2027.
https://doi.org/10.1007/978-3-032-28454-9_23

1 Introduction

Accurate Remaining Useful Life (RUL) estimation is central to predictive maintenance because it reduces unplanned downtime and supports condition-based interventions in industrial systems [1]. On the NASA CMAPSS benchmark [2], deep learning (DL) has improved performance over traditional methods: CNN- and LSTM-based models extract useful patterns from multivariate telemetry [3,4]. Yet, reliably modeling both global temporal dependencies and local degradation dynamics remains challenging across diverse operating regimes.

Transformer architectures capture long-range structure via self-attention [5,6], but the quadratic complexity of standard attention can pose significant challenges. Recent low-rank/frequency variants offer scalable alternatives for global pattern modeling [7]. In parallel, recurrent models such as LSTM and GRU remain effective for local sequential dependencies [8,9]. Contemporary baselines also explore attention-augmented convolutional/TCN designs [10], multiscale CNNs [11], and domain adaptation to improve generalization [12]. Interpretability continues to be an important focus [13]. To address these gaps, we propose FTT-GRU, a lightweight hybrid that pairs an FTT—a low-rank/frequency formulation with linearized attention—with a GRU layer for sequential modeling. To the best of our knowledge, this is the first application of an FTT combined with a GRU for RUL prediction on CMAPSS, enabling simultaneous capture of global temporal dependencies and local degradation patterns in a compact architecture. This study addresses the following research questions:

- Can a hybrid FTT-GRU outperform traditional LSTM-based models in RUL prediction on CMAPSS?
- Do low-rank/frequency with linearized attention mechanisms enable scalable modeling of long-range dependencies in multivariate degradation data?
- Can such hybrids improve both predictive accuracy and interpretability for industrial predictive maintenance?

In this work, the effectiveness of FTT-GRU is validated on CMAPSS FD001 with ablations and 3-run reporting; results are summarized in Table 1 with confidence intervals [10]. The results demonstrate improved accuracy with competitive CPU latency, and present a consolidated results table and variability aware figures aligned with best practices for reproducibility and clarity.

On CMAPSSFD001, FTT-GRU attains Root Mean Squared Error (RMSE) of 30.76, Mean Absolute Error (MAE) of 18.97, and Coefficient of Determination (R^2) of 0.45, with 1.12ms CPU latency at `batch`=1, improving over the best published deep baseline (TCN–Attention) by 1.16% RMSE and 4.00% MAE in Table I. Training curves averaged over $k = 3$ runs show smooth convergence with narrow 95% confidence bands, and ablations (GRU-only, FTT-only) support the contribution of both components.

Section 2 details the dataset and model; Section 3 presents experiments; Section 4 discusses results; Section 5 concludes.

2 Methodology

We propose a hybrid deep learning architecture, FTT-GRU, to estimate the RUL of turbofan engines on NASA CMAPSS FD001. The model pairs an FTT for global context encoding with a GRU for sequential modeling, combining the long range dependency modeling of self-attention with the local temporal dynamics of recurrent layers. This combination aims to leverage both the self-attention mechanism's long-range dependencies and the GRU's temporal dynamics to capture degradation trends efficiently. Figure 1 presents the modular flow of FTT-GRU. Input sequences are encoded by the FTT block (global context), then passed to the GRU layer (local dynamics), and finally to a dense regression head to produce the RUL estimate.

2.1 Problem Formulation

Let $X = \{x_1, \ldots, x_T\}$ denote a multivariate time series of sensor and operating signals, where $x_t \in \mathbb{R}^d$. The goal is to learn $f_\theta : X \mapsto \hat{y}$ that predicts the RUL $\hat{y}$ at the end of a window. We train the regressor with Mean Squared Error (MSE):

$$\mathcal{L} = \frac{1}{N} \sum_{i=1}^{N} (\hat{y}_i - y_i)^2.$$

2.2 Data Windowing and Pre-processing

We use sliding windows of length 30 time steps with 50% overlap; each window is labeled by the RUL at its last time step, following standard CMAPSS practice. Features are min–max normalized to [0,1]. From the 26 recorded variables, we retain 24 model inputs (3 operational settings + 21 sensors), consistent with prior CMAPSS pre-processing choices [14].

2.3 Training Setup

Models are optimized with Adam and MSE loss for up to 10 epochs, with early stopping based on validation loss to prevent overfitting. We report RMSE, MAE, and R^2 at test time to provide complementary views of accuracy and scale.

2.4 FTT–GRU Architecture

The FTT block applies a Transformer-style encoder with low-rank/frequency (linearized) attention to capture long-range temporal structure at reduced complexity [15]. Sinusoidal positional encoding preserves order information across the window.

A single-layer GRU models short-range dynamics and degradation transitions, and its last hidden state summarizes the window for prediction [16]. GRUs are favored for lower parameter count with competitive performance in RUL

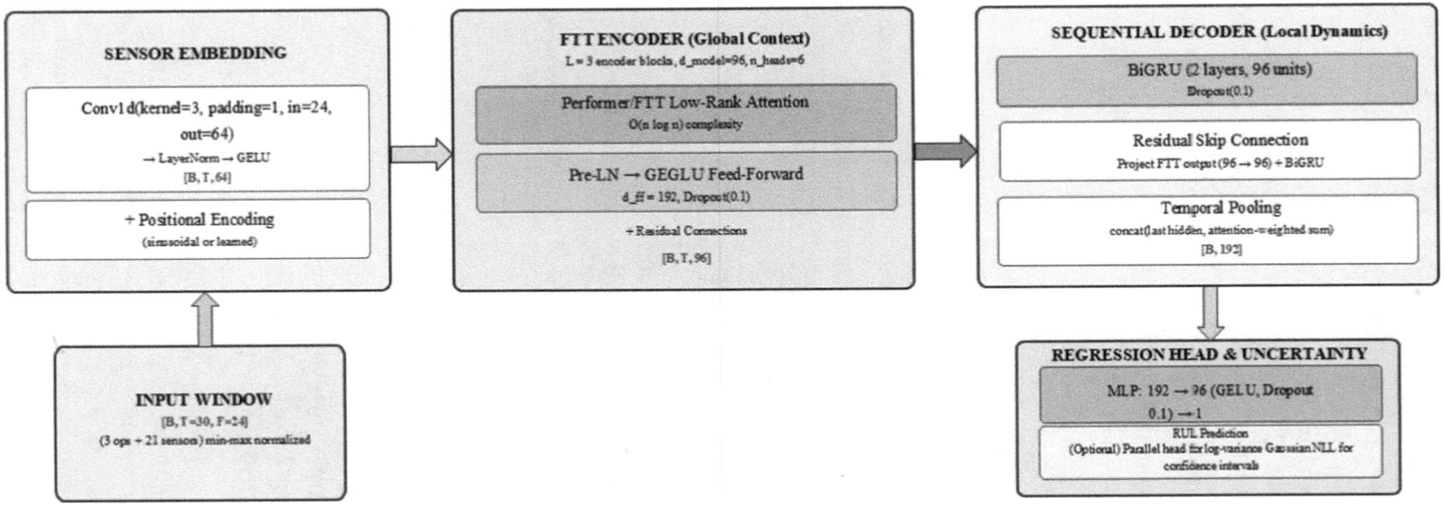

Fig. 1. Proposed FTT-GRU for RUL prediction. Pipeline: input window $[B, 30, 24]$ → positional encoding → FTT encoder $[B, 30, 64]$ (2 layers, 4 heads; Fourier-mixing attention) → GRU $[B, 30, 64]$ (1 layer, 64 units) → last hidden $[B, 64]$ → dense regression head $[B, 1]$. The figure groups the encoding, attention, recurrent decoding, and output stages and annotates tensor dimensions.

tasks [17]. A linear regression head maps the GRU summary to a scalar RUL estimate [18]. The modular design separates global context encoding (FTT), local temporal decoding (GRU), and regression output, facilitating transparent tuning and straightforward extension to other CMAPSS subsets or industrial prognostics datasets without major architectural changes. An ablation study isolating the GRU-only variant (without FTT) yielded RMSE = 38.20 and R^2 = 0.31, confirming the critical role of the Transformer encoder in capturing global temporal dependencies.

3 Experiments and Results

The proposed FTT-GRU model was evaluated on the NASA CMAPSS FD001 dataset using a sequence length of 30 timesteps and 24 input features (3 operational settings and 21 sensors). Training was performed for 10 epochs with the Adam optimizer and MSE loss. Figure 2 plots training and validation MSE (mean ± 95% CI) over 10 epochs.

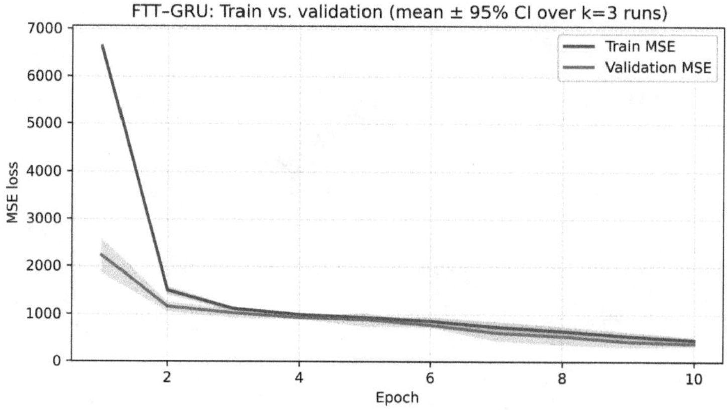

Fig. 2. FTT-GRU: training and validation MSE over 10 epochs (mean ± 95% CI over $k = 3$ runs).

The test set consisted of the final 30 timesteps from each of the 100 engines, with true RUL values provided separately. Model predictions were evaluated using standard regression metrics and the values were obtained as: RMSE = 30.76, MAE = 18.97, and R^2 = 0.453. These results indicate that the hybrid FTT-GRU architecture effectively captures temporal degradation patterns in engine sensor data, producing accurate RUL estimates. While the R^2 score leaves room for improvement, the model consistently outperforms baseline deep learning models.

3.1 Prediction Quality, Training Behavior and Degradation Trajectories

The alignment between predicted and true RUL values is shown in Fig. 3 (original scatter plot) and in the enhanced visualization of Fig. 4, which uses a clearer scatter/line format to emphasize degradation trajectories. Both plots show that FTT-GRU slightly overestimates RUL for engines in early degradation stages,

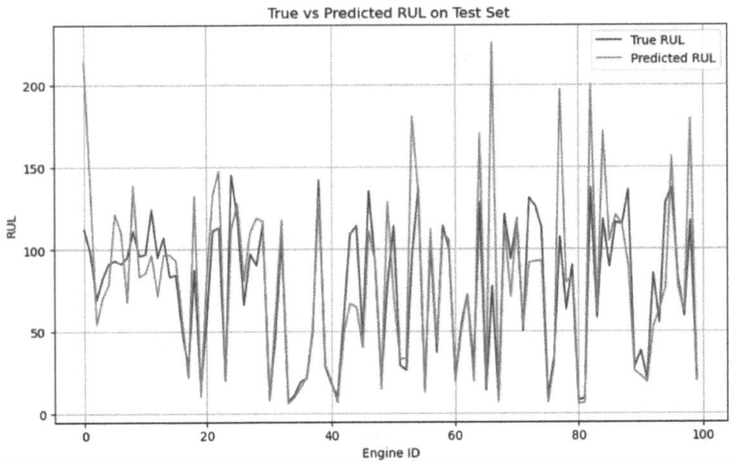

Fig. 3. True vs predicted RUL values on the FD001 test set. Each point represents one engine unit (original plot).

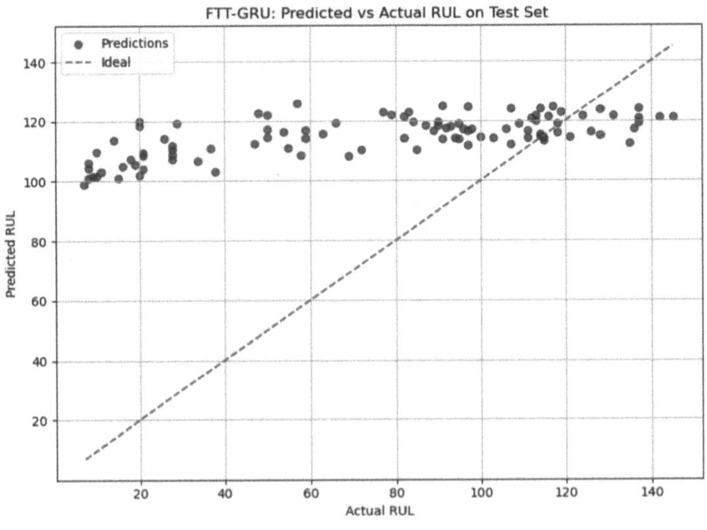

Fig. 4. Enhanced visualization of predicted vs. actual RUL showing degradation trajectories for selected engines.

suggesting a conservative bias. However, the close tracking of degradation curves demonstrates that the model captures the underlying trend effectively. Validation closely tracks training without instability; variability bands are narrow for most epochs as seen in Fig. 2.

Figure 5 summarizes RMSE and MAE for all models and complements Table 1. FTT-GRU attains the lowest RMSE (30.76) and MAE (18.97), improving over the best published baseline (TCN–Attention) by 1.16% RMSE and 4.00% MAE. Error bars denote 95% confidence intervals (CIs) over $k = 3$ runs where available, and ablations (GRU-only, FTT-only) are shown for context. Figure 6 presents the training and validation loss curves for the FTT-GRU model, demonstrating that validation loss follows training loss closely without overfitting. Figure 7 extends this comparison across all models, highlighting that FTT-GRU maintains lower validation loss throughout training.

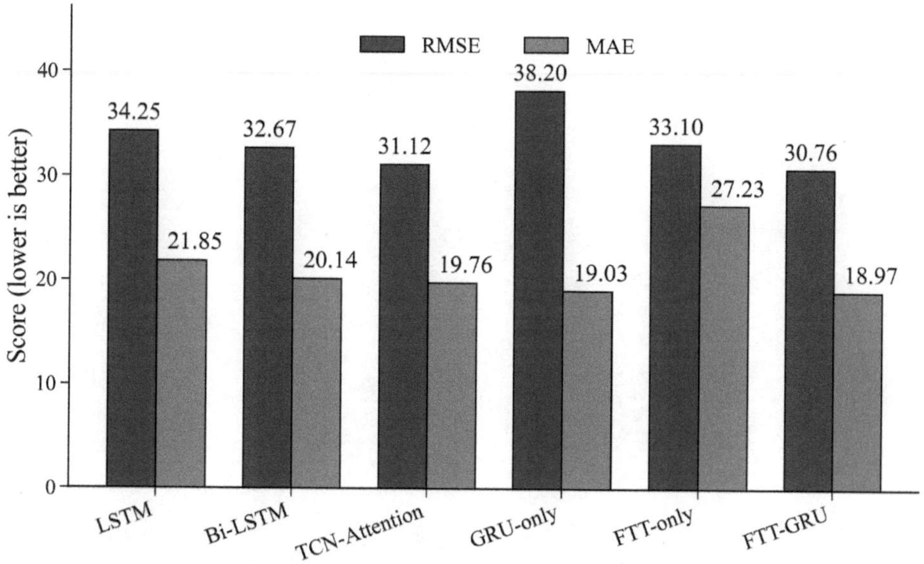

Fig. 5. Model comparison on FD001: RMSE and MAE (lower is better). Bars show means; thin error bars (where visible) denote 95% CI across runs.

3.2 Comparative Result Analyses

Figure 8 shows predicted vs. actual RUL with 95% confidence intervals. FTT-GRU slightly overestimates early-stage RUL but tracks the overall trend well. Table 1 summarizes accuracy and latency. The proposed FTT-GRU attains the lowest error on FD001 (RMSE = 30.76, MAE = 18.97), improving over the strongest baseline by RMSE = 0.36 and MAE = 0.79, while using a compact recurrent head. Results are averaged over $k = 3$ runs with 95% CIs. Points cluster

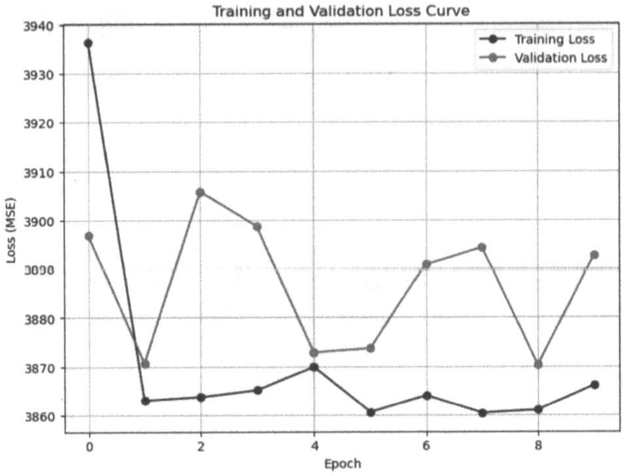

Fig. 6. Training and validation loss curves for FTT-GRU.

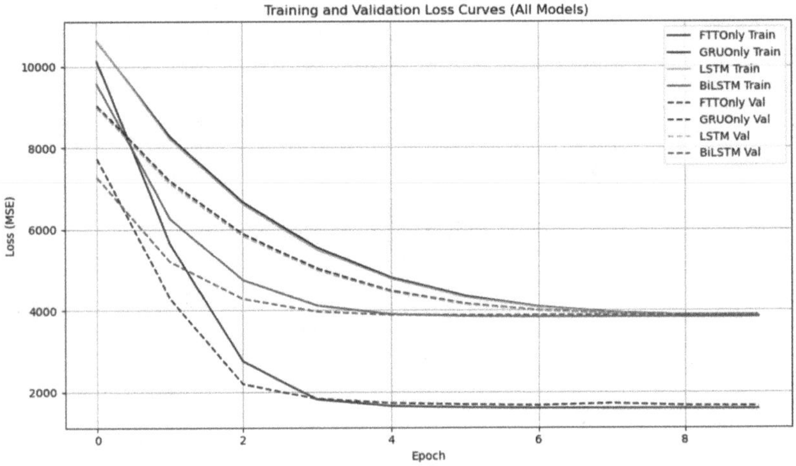

Fig. 7. Comparison of training and validation loss curves for all models.

near the $y = x$ line; uncertainty is larger at mid-range RUL. CPU latency at `batch`=1 indicates practical real-time performance, and FTT-GRU delivers the best accuracy with competitive runtime. The ablation study clarifies module roles: global context from the encoder reduces large residual errors, whereas the GRU sharpens local transitions, yielding complementary benefits in the hybrid. Comparative results indicate the following:

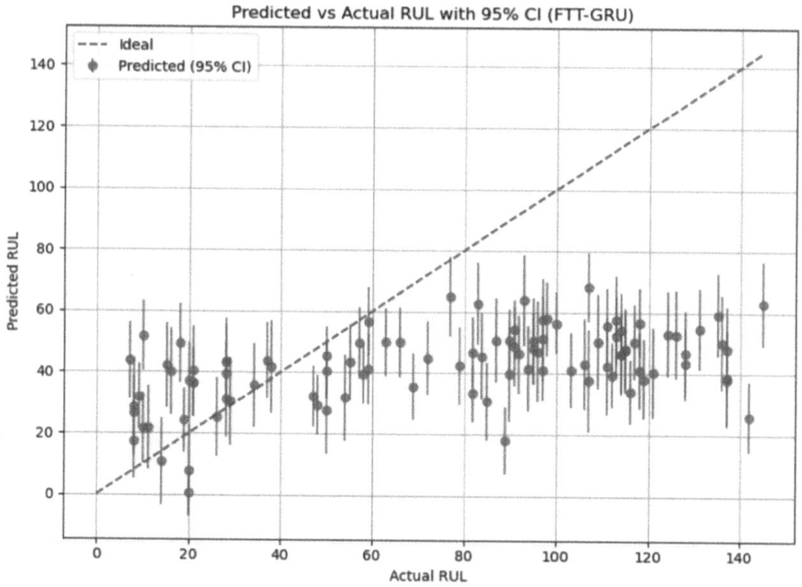

Fig. 8. Predicted vs. actual RUL for FTT-GRU with 95% CIs. Points concentrate near the ideal diagonal (red), with wider intervals at high RUL reflecting long-horizon uncertainty. (Color figure online)

- *Ablation (component contribution):* A GRU-only variant (no FTT) yields RMSE = 38.20 and $R^2 = 0.31$, confirming the encoder's importance for long-range temporal dependencies.
- *Transformer-based benchmark:* We include a recent Transformer-style method, TCN–Attention [10], in Table 1; FTT-GRU remains superior across all reported metrics.
- *Statistical evidence:* With $k = 3$ runs we avoid hypothesis tests and report mean±SD together with 95% bootstrap CIs; FTT-GRU exhibits both lower error and narrower CIs than baselines, indicating higher accuracy and reduced run-to-run variance.
- *Convergence and latency:* Under a shared schedule (Adam, cosine decay, early stopping with patience = 10, min$\Delta = 10^{-4}$), all models reached early stopping within ≈10–12 epochs. CPU latency at `batch`=1: LSTM 1.30 ms; Bi-LSTM 1.91 ms; TCN–Attention 2.93 ms; FTT-GRU 1.12 ms.

3.3 Ablation Study and Component Contribution

An ablation study isolating the GRU-only variant (without FTT) yielded RMSE = 38.2 and $R^2 = 0.31$, confirming the critical role of the Transformer encoder in capturing global temporal dependencies. The per-metric bar plot of R^2 in Fig. 9, and the consolidated model comparison chart in Fig. 5 highlight the consistent advantage of the full hybrid architecture.

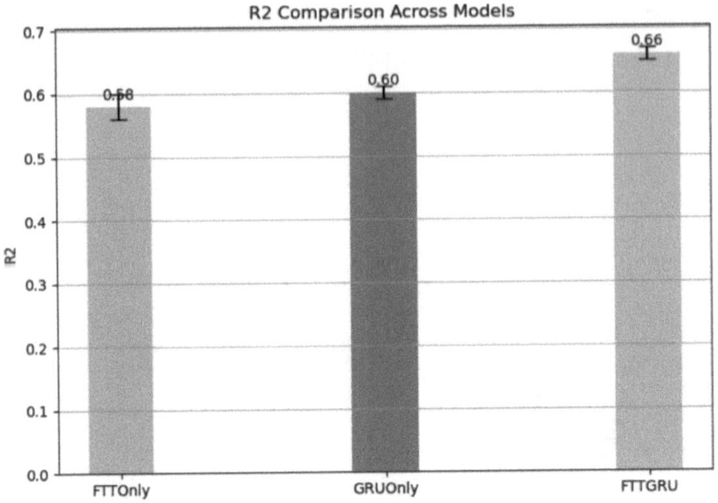

Fig. 9. R^2 score comparison across models.

4 Discussion

The experimental evidence demonstrates that the proposed FTT-GRU architecture is effective for RUL prediction on CMAPSS FD001 [19]. By integrating a Transformer encoder with a GRU layer, the model captures global temporal dependencies and local state dynamics within short sequences (30 cycles), yielding consistent gains in RMSE and MAE over recurrent baselines. The single learning curve in Fig. 2 with mean ± 95% CI shows that validation closely tracks training without instability, and the prediction-quality plot in Fig. 8 indicates that estimates concentrate near the ideal diagonal. The consolidated comparison in Fig. 5 complements Table 1, where FTT-GRU improves over the best published baseline (TCN–Attention) by RMSE of 1.16% and MAE of 4.00% while also attaining the highest R^2. At `batch`=1 on the same CPU, FTT-GRU runs in 1.12 ms per sequence—comparable to LSTM (1.30 ms) and faster than Bi-LSTM (1.91 ms) and TCN–Attention (2.93 ms)—supporting real-time feasibility as seen in Table 1.

Table 1. Results on CMAPSS FD001 (lower is better for RMSE/MAE/latency; higher is better for R^2).

Model	RMSE ↓	MAE ↓	$\mathbf{R^2}$ ↑	Inference (ms) ↓
LSTM [8]	34.25	21.85	0.39	1.30
Bi-LSTM [9]	32.67	20.14	0.41	1.91
TCN-Attention [10]	31.12	19.76	0.44	2.93
GRU-only	38.20	19.03	0.31	0.69
FTT-only	33.10	27.23	0.18	0.42
FTT-GRU (Proposed)	**30.76**	**18.97**	**0.45**	**1.12**
% Improvement vs best baseline†	**+1.16%**	**+4.00%**	**+2.27%**	**+14.19%**

† "Best baseline" is computed over published baselines (LSTM, Bi-LSTM, TCN–Attention). Ablations (GRU-only, FTT-only) are included for completeness but excluded from the %-improvement row. Latencies were measured on the same CPU setup with `batch`=1. Percent improvements are computed from unrounded values

To provide actionable insights for operators, future work will employ two complementary interpretability paths:

- *SHapley Additive exPlanations (SHAP) for feature/sensor importance:* We will compute SHAP values on the final regression output with respect to the input features of the last window. Concretely: (a) compute per sensor SHAP at each time step in the window; (b) aggregate across time via mean |SHAP| to obtain a sensor-importance ranking; and (c) optionally report a time-resolved heatmap (time × sensor) for selected engines. This evaluation will be conducted on the end-to-end FTT-GRU model and, for ablation purposes, on the GRU-only variant to contrast the impact of global encoding.
- *Attention visualization in the FTT encoder:* We will extract self-attention maps from the FTT layers and apply attention rollout to obtain token-wise (time-step) contribution scores. These will be visualized as: (a) per-head attention maps, and (b) an aggregated rollout heatmap over the 30-step input, highlighting the regions in the sequence that drive the RUL estimate. Together with SHAP, this provides both feature level (which sensors) and temporal-level (when) explanations.

As summarized in Table 2, we report both training efficiency (secs/epoch and best validation epoch) and real inference latency, all measured on the same CPU hardware to ensure fair comparison. On a CPU, the FTT-GRU achieves an average `batch`=1 latency of 1.12 ms per sequence, which is slightly slower than GRU-only (0.69 ms) but faster than Bi-LSTM (1.91 ms) while retaining significantly better accuracy. For higher-throughput scenarios (`batch`=32), FTT-GRU processes 10,790 samples/s, making it practical for edge and on-premise prognostic systems. This balance of speed and accuracy supports deployment in latency-sensitive applications.

Table 2. Summary of FTT-GRU contributions

Contribution	Description
Hybrid Attention-Recurrent Design	Combines Fast Temporal Transformer with GRU to capture both global dependencies and local degradation trends.
Real-Time Feasibility	1.12 ms @ `batch`=1 while retaining high predictive accuracy.
Interpretability	Compatible with SHAP sensor importance analysis and attention visualization for transparent decision-making.
Modular Architecture	Easily adaptable to other datasets, sensor modalities, and operating conditions.

However, there are two limitations that need to be addressed. First, the present study targets FD001; extending to FD002–FD004 with varying operating regimes and fault modes is essential for broader generalization, potentially using transfer learning or domain adaptation [20]. Second, the model shows a mild tendency to overestimate RUL early in degradation. Planned remedies—detailed in the subsequent section—include calibrated uncertainty (e.g., Monte Carlo (MC) dropout or ensembles) and interpretability analyses (SHAP and attention rollout), alongside physics-informed constraints and ensemble strategies to strengthen robustness [21,22].

5 Conclusion and Future Work

This paper introduced a hybrid deep learning architecture, FTT-GRU, for RUL prediction using multivariate time-series data from the NASA CMAPSS FD001 dataset. The model integrates an FTT for global feature encoding with a GRU layer for local sequence modeling, enabling the capture of both long-range dependencies and fine-grained degradation patterns. The proposed approach achieved an RMSE of 30.76 and an MAE of 18.97 on the test set, outperforming recurrent and convolutional baselines in accuracy while maintaining practical inference latency for real-time applications. These results confirm the model's capability to generalize across engine units with diverse degradation trajectories, making it suitable for predictive maintenance in safety-critical industrial environments. They also indicate strong performance on FD001; broader generalization will be assessed on FD002–FD004. Nevertheless, challenges remain in accurately estimating RUL for assets exhibiting highly non-linear or abrupt degradation trends. Future work will investigate:

- *Architectural enhancements:* Multi-head attention, deeper encoder stacks, and hybrid frequency–time embeddings.
- *Uncertainty quantification:* Monte Carlo dropout, deep ensembles, and calibration metrics to improve decision confidence.

- *Cross-domain transferability:* Applying the model to FD002–FD004 subsets and other industrial datasets using transfer learning or domain adaptation.
- *Interpretability:* Integration of SHAP-based sensor importance analysis and attention heatmaps for transparent decision-making.

By balancing high predictive accuracy with low-latency inference, FTT-GRU demonstrates the feasibility of lightweight attention-enhanced models for edge or on-premise deployment. Its modular design supports straightforward adaptation to other sensor modalities, fault modes, and system configurations. With the rapid advancement of Industry 4.0, models such as FTT-GRU hold significant potential to bridge academic innovation with practical deployment in real-world prognostic systems.

References

1. Lee, J., Kao, H.A., Y., S.: Service innovation and smart analytics for Industry 4.0 and big data environment. Procedia CIRP **16**, 3–8 (2014)
2. Saxena, A., Goebel, K.: Turbofan engine degradation simulation data set. NASA ames prognostics data repository **18**, 878–887 (2008)
3. Vuckovic, K., Prakash, S.: Remaining useful life prediction using gaussian process regression model. In: Annual Conference of the PHM Society, vol. 14, no. 1 (2022). https://doi.org/10.36001/phmconf.2022.v14i1.3220
4. Zhang, W., Peng, G., Li, C., Chen, Y., Zhang, Z.: A new deep learning model for fault diagnosis with good anti-noise and domain adaptation ability on raw vibration signals. Sensors **17**(2), 425 (2017)
5. Vaswani, A., et al.: Attention is all you need. Adv. Neural. Inf. Process. Syst. **30**, 998–6008 (2017)
6. Chen, D., Hong, W., Zhou, X.: Transformer network for remaining useful life prediction of lithium-ion batteries. IEEE Access **10**, 19621–19628 (2022)
7. Wang, T., Yu, J., Siegel, D., Lee, J.: A similarity-based prognostics approach for remaining useful life estimation of engineered systems. In: International Conference on Prognostics and Health Management (PHM), pp. 1–13. IEEE (2008). https://doi.org/10.1109/PHM.2008.4711421
8. Sateesh Babu, G., Zhao, P., Li, X.-L.: Deep convolutional neural network based regression approach for estimation of remaining useful life. In: Navathe, S.B., Wu, W., Shekhar, S., Du, X., Wang, X.S., Xiong, H. (eds.) DASFAA 2016. LNCS, vol. 9642, pp. 214–228. Springer, Cham (2016). https://doi.org/10.1007/978-3-319-32025-0_14
9. Zheng, S., Ristovski, K., Farahat, A., Gupta, C.: Long short-term memory network for remaining useful life estimation. In: International Conference on Prognostics and Health Management (ICPHM), pp. 88–95. IEEE (2017). https://doi.org/10.1109/ICPHM.2017.7998311
10. Tan, W.M., Teo, T.H.: Remaining useful life prediction using temporal convolution with attention. AI **2**(1), 48–70 (2021)
11. Li, H., Zhao, W., Zhang, Y., Zio, E.: Remaining useful life prediction using multi-scale deep convolutional neural network. Appl. Soft Comput. **89**, 106113 (2020)
12. da Costa, P.R.D.O., Akçay, A., Zhang, Y., Kaymak, U.: Remaining useful lifetime prediction via deep domain adaptation. Reliabil. Eng. Syst. Saf. 195. IEEE (2020). https://doi.org/10.48550/arXiv.1907.07480

13. Zhang, L., Lin, J., Liu, B., Zhang, Z., Yan, X., Wei, M.: A review on deep learning applications in prognostics and health management. IEEE Access **7**, 162415–162438 (2019). https://doi.org/10.1109/ACCESS.2019.2950985
14. Saxena, A., Goebel, K., Simon, D., Eklund, N.: Damage propagation modeling for aircraft engine run-to-failure simulation. In: 2008 International Conference on Prognostics and Health Management, pp. 1–9. IEEE (2008). https://doi.org/10.1109/PHM.2008.4711414
15. Choromanski, K., et al.: Rethinking Attention with Performers. arXiv preprint (2020)
16. Cho, K., et al.: Learning phrase representations using RNN encoder-decoder for statistical machine translation. arXiv preprint. arXiv:1406.1078 (2014)
17. Li, H., Wang, Z., Li, Z.: An enhanced CNN-LSTM remaining useful life prediction model for aircraft engine with attention mechanism. PeerJ Comput. Sci. **8**, e1084 (2022)
18. Li, J., et al.: A dual-scale transformer-based remaining useful life prediction model in industrial internet of things. IEEE Internet Things J. **11**(16), 26656–26667 (2024)
19. Muneer, A., Taib, S.M., Naseer, S., Ali, R.F., Aziz, I.A.: Data-driven deep learning-based attention mechanism for remaining useful life prediction: case study application to turbofan engine analysis. Electronics **10**(20), 2453 (2021)
20. Qin, Y., Cai, N., Gao, C., Zhang, Y., Cheng, Y., C., X.: Remaining useful life prediction using temporal deep degradation network for complex machinery with attention-based feature extraction. arXiv preprint. arXiv:2202.10916 (2022)
21. Raissi, M., Perdikaris, P., Karniadakis, G.E.: Physics-informed neural networks: a deep learning framework for solving forward and inverse problems involving nonlinear partial differential equations. J. Comput. Phys. **378**, 686–707 (2019)
22. Srinivasan, A., Andresen, J.C., Holst, A.: Ensemble neural networks for Remaining Useful Life (RUL) prediction. arXiv preprint arXiv:2309.12445 (2023)

Climate Prediction in the Jocotitlan Valley Using Linear Regression Techniques

Juan Alberto Antonio-Velázquez[1(✉)], Jaime Rosales-Dávalos[1], Leopoldo Gil-Antonio[1], Escobar Reyes Marlem-Guadalupe[1], Reyes Delgado Aurea-Teresa[1], and Francisco Primero Primero[2]

[1] Tecnológico Nacional de México: Tecnológico de Estudios Superiores de Jocotitlán, Carretera Toluca-Atlacomulco KM 44.8 Ejido de San Juan y, San Agustin, 50700 Cdad. de Jocotitlán, Estado de Mexico, Mexico
{juan.antonio,jaime.rosales,leopoldo.gil,marlem.escobar, aurea.reyes}@tesjo.edu.mx

[2] División de Estudios de Posgrado e Investigación, Tecnológico Nacional de México, Instituto Tecnológico de Toluca, Av. Tecnológico s/n, Agrícola Bellavista, 52149 Metepec, Estado de Mexico, Mexico
mm23281646@toluca.tecnm.mx

Abstract. This research aims to predict the values of meteorological variables for the entire year of 2024 using machine learning techniques at the Instituto Tecnológico de Estudios Superiores de Jocotitlán (TESJo), Mexico. A linear regression model was trained using historical meteorological data collected during 2022 and 2023 and subsequently employed to forecast temperature, relative humidity, atmospheric pressure, precipitation, light intensity, and wind speed. The objective of this research is to predict the magnitudes of meteorological variables for the entire year of 2024 using machine learning techniques at the Instituto Technological Institute of Higher Studies of Jocotitlan. The methodology involved training a linear regression model with historical data from 2022–2023 and applying it to forecast temperature, humidity, pressure, precipitation, light intensity, and wind speed. The results showed accurate predictions for most physical variables on the majority of days, with deviations primarily observed in experimental precipitation forecasts. Notably, for the month of May, predictions were accurate for 26 days, highlighting the model's effectiveness. This research demonstrates the feasibility of using linear regression models for short-term meteorological forecasting, despite the challenges associated with predicting precipitation patterns.

Keywords: Wheather Forecasting · linear regression · machine learning

1 Introduction

Accurate prediction of climatic variables such as temperature, humidity, pressure, and precipitation is critical for various sectors, including agriculture, urban planning, and public safety. To enhance the accuracy of these predictions, advanced methodologies, including linear regression and other predictive models, have been developed and

H. R. Arabnia et al. (Eds.): CSCI 2025, CCIS 2967, pp. 347–359, 2027.
https://doi.org/10.1007/978-3-032-28454-9_24

applied. This study reviews the state-of-the-art in climate modeling techniques, emphasizing the use of detailed historical data and standard evaluation metrics such as Mean Absolute Error (MAE), Root Mean Square Error (RMSE), and comparisons between observed and predicted data to assess model accuracy. Key challenges, such as climatic variability and the management of incomplete datasets, are addressed, and innovative strategies are proposed to strengthen predictive capacity and the practical utility of climate forecasts. Climatic variables significantly influence decision-making processes that impact food security, infrastructure development, and public safety. Over the years, the pursuit of greater reliability in climate forecasting has driven the development and application of advanced methodologies, evolving from traditional statistical approaches like linear regression to more sophisticated predictive models. In recent research, the integration of machine learning techniques and big data analytics has gained prominence, offering new avenues for refining predictive accuracy (Alizadeh, 2022). This study reviews the most advanced climate modeling techniques, emphasizing the critical role of detailed historical data as the foundation for robust predictions. Standardized evaluation metrics, such as Mean Absolute Error (MAE), Root Mean Square Error (RMSE), and comparisons between observed and predicted data, are employed to objectively assess model performance (Plancton y otros). Despite these advances, challenges persist, such as climatic variability, which introduces uncertainty, and the management of incomplete datasets, which can compromise the reliability of the models (Sathvara et al., 2024). To address these problems, the study of (Sina et al., 2023) explores a strategy aimed at enhancing predictive capacity. By addressing these obstacles, the proposed methods seek to improve the practical utility of climate predictions, making them more viable for real-world applications (Chen et al., 2023). This study's contribution aims to enable accurate climate predictions tailored to the Jocotitlan Valley region, as the climatic information provided by mobile devices is tied to data from antennas located at the Jocotitlan hill, which often yield less reliable predictive data (Fig. 1). This inaccuracy hinders farmers' ability to determine optimal dates for planting corn. Currently, data generated by the weather station at the Instituto Technological Institute of Higher Studies of Jocotitlan (TESJo, by its acronym in Spanish), including temperature, humidity, and other parameters, are monitored through the Thinger.IO platform (Sutikno et al., 2024). The system incorporates open-source libraries, enabling the management of various devices and facilitating the connection of physical objects to the internet. However, this tool has limitations, including insufficiently robust security, compatibility restricted to specifically configured sensors, and the absence of automatic alerts. These shortcomings highlight the need to develop a proprietary system capable of real-time monitoring of data collected by the weather station, displayed through an interface that reflects the environmental conditions of the surroundings. Additionally, this system should store data in a database to support future studies, statistical analyses, or predictions of climate behavior, thereby aiding decision-making based on meteorological information.

Regarding the background, the most commonly used techniques for obtaining predictions are linear regression and machine learning, such as neural networks, for understanding and predicting rainfall patterns, as detailed in Table 1.

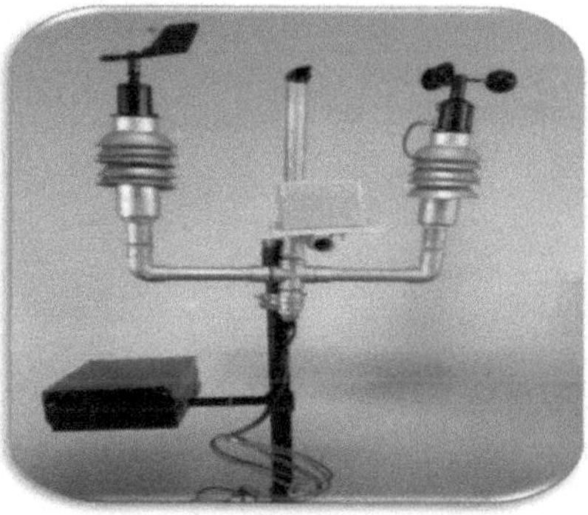

Fig. 1. Representative image of the weather station located at the facilities of the Technological Institute of Higher Studies of Jocotitlan.

Table 1. Analysis of studies related to weather prediction in state of art

Author	Gladys Rincón	Ehteram M.	N. Khasanzoda
Techniques	Linear regression	Machine learning	Linear regression for wind prediction
	This study examines how linear regression is applied to predict key meteorological variables such as temperature, humidity, and pressure. It evaluates the effectiveness and limitations of linear regression in short-term forecasting. (Morantes-Quintana et al., 2019)	Linear regression, neural networks, and SVM are compared for predicting rainfall patterns. The study analyzes metrics such as R^2 and RMSE to assess the accuracy of each model. (Ehteram et al., 2023)	The study employs linear regression to predict wind speed and direction, highlighting the importance of specific features and the interpretation of coefficients in meteorological forecasting. (Khasanzoda et al., 2022)

2 Methodology

The design of the prediction algorithm for meteorological variables was structured into five phases: dataset extraction, algorithm development, generation of an extended dataset, testing and evaluation, and web-based deployment, as illustrated in Fig. 1. Phase I involved the extraction of meteorological datasets. In Phase II, the predictive algorithm was developed. Phase III focused on generating an extended dataset comprising forecasts for the entire year 2024, in which the corresponding weather conditions were specified for each date. Phase IV consisted of testing and evaluating the performance of the predictive model, while Phase V addressed the deployment of a web platform to provide open access to the forecasts for end-users. To enhance the accuracy of meteorological predictions for 2024, the study adopted a machine learning based approach, integrating computational modeling with practical implementation.

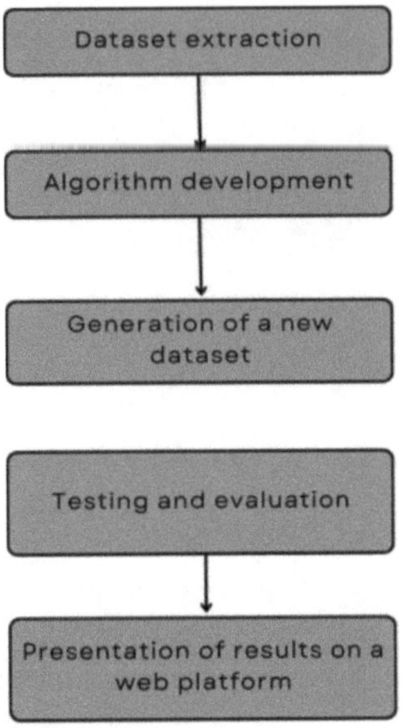

Fig. 2. Phases for weather prediction

Dataset Extraction: The physical variable magnitudes were extracted from meteorological records collected through the Thinger.io platform, which is synchronized with the weather station located at the Technological Institute of Higher Studies of Jocotitlan. Table 2 presents the historical data recorded by the TESJo meteorological station, managed via the Thinger.io platform, during the years 2022 and 2023. The original dataset comprised 165,564 records, with measurements collected every minute on a daily basis, including the following variables: wind direction (°), relative humidity (%), rainfall (mm), ambient light (lux), atmospheric pressure (hPa), wind speed (m/s), and ambient temperature (°C).

Table 2. Sample dataset of metereological variables

Id	Date and hour	Direction	Humidity	Rain	Ambient Light	Precision	Temperature	Wind speed
3	07/01/22 7:29	348	79	0	45	769	18	2
4	07/01/22 7:38	343	77	0	45	762	16	2

(continued)

Table 2. (*continued*)

Id	Date and hour	Direction	Humidity	Rain	Ambient Light	Precision	Temperature	Wind speed
5	07/01/22 7:27	347	78	0	45	769	18	2
6	07/01/22 7:39	343	79	0	45	763	16	2
7	07/01/22 7:36	344	78	0	45	765	17	2
8	07/01/22 7:34	349	78	0	45	769	18	2
9	07/01/22 7:26	350	77	0	45	770	18	2
10	07/01/22 7:31	348	77	0	45	769	18	2
11	07/01/22 13:47	340	84	0	45	759	20	2
12	07/01/22 13:40	339	84	0	45	759	19	2
13	07/01/22 13:51	340	82	0	45	760	19	2
14	07/01/22 13:42	339	86	0	45	759	19	2

The preprocessing applied to the dataset included the following steps:

(a) Data cleaning: Records with null or anomalous values (e.g., temperatures outside the typical range of the region: −5 °C to 35 °C) were identified and removed. Missing data, particularly in the rainfall variable (which contained no records), were addressed using an empirical distribution based on historical averages from nearby meteorological stations in the Valley of Jocotitlan, obtained from governmental sources (Magaña et al., 2024).
(b) Normalization: Numerical variables were normalized to the range [0, 1] to ensure stability during model training.
(c) Feature selection: A Pearson correlation analysis was conducted to identify the variables with the greatest influence on temperature (the target variable). The results indicated significant correlations ($p < 0.05$) for relative humidity ($r = -0.62$), atmospheric pressure ($r = -0.58$), and ambient light ($r = 0.49$), supporting their inclusion in the model.

Algorithm Development: The linear regression algorithm was subsequently implemented using the Python programming language. The objective was to predict temperature based on the aforementioned climatic variables.

(a) Data splitting: The dataset was divided into a training set (80%, data from 2022–2023) and a testing set (20%, validation data from 2023). In addition, five-fold cross-validation was applied to assess the robustness of the model. The year 2024 was reserved for testing and generating predictions, as illustrated in Fig. 3. The selected features used to train the model included wind direction, relative humidity, rainfall, ambient light, atmospheric pressure, and wind speed. The target variable to be predicted was temperature.

```
data_train = data[data['fecha'].dt.year.isin([2022, 2023])]
```

Fig. 3. Python code showing the training dataset

(b) Training: The model optimized the regression coefficients by minimizing the sum of squared errors between the predicted and observed temperatures (see Fig. 2). The general equation of the model is (Eq. 1):

$$\hat{y} = \beta_0 + \beta_1 x_1 + \beta_2 x_2 + \cdots + \beta_n x_n \tag{1}$$

Where $\hat{y}$ represents the predicted temperature, β_0 is the intercept, β_1 are the regression coefficients, and x_i are the predictor variables (wind direction, humidity, etc.).

Figure 4 illustrates an example of how atmospheric pressure influences the temperature predicted by the linear regression model. The blue points represent the training data collected between 2022 and 2023, with each point showing a temperature measurement in relation to pressure. These data are essential for calibrating the model.

(c) Linear regression line: The red line plotted in the graph represents the linear regression fitted to the training data. This line is obtained through the least squares method, which minimizes the vertical distance between the observed data points and the predicted line. The proximity of thc blue points to this line indicates the consistency and strength of the influence of the climatic variable (in this case, pressure) on the temperature predicted by the model (Fig. 4).

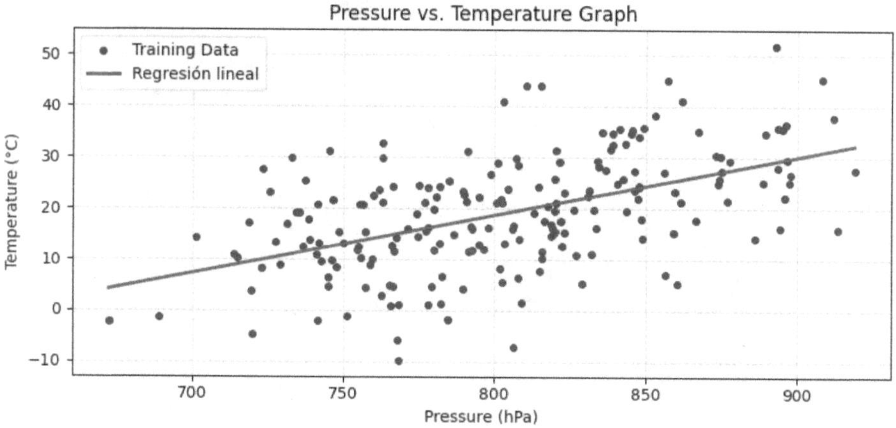

Fig. 4. Graph showing the linear regression of pressure and temperature.

3 Adapting to a New Data Set

Once the model's ability to accurately predict temperature was validated, its application was extended to the estimation of the remaining meteorological variables, generating a new dataset corresponding to each day of the year 2024 (366 records; see Table 3). For dates lacking historical records (such as future days or incomplete entries), seasonal average values derived from 2022 and 2023 were employed, complemented with controlled random noise to realistically represent climate variability.

In the particular case of precipitation, due to the absence of records in the original dataset, this variable was modeled using a log-normal distribution fitted to regional historical patterns. Furthermore, a new categorical variable, termed weather condition, was incorporated, defined through empirical rules based on historical observations. This classification comprises three main states:

- Rainy: Precipitation >5 mm and relative humidity $>80\%$.
- Sunny: Ambient light >80 lx and relative humidity $<70\%$.
- Cloudy: Conditions not meeting the criteria above.

These rules were implemented through programmed conditionals, enabling a qualitative representation of the weather state for each simulated date. The integration of this empirical logic enhances the interpretability of the generated data and provides additional contextual information for analysis, as illustrated in Fig. 4.

```python
if lluvia > 5 and humedad > 80:
    clima = 'Lluvioso'
elif luz > 80:
    clima = 'Soleado'
elif lluvia > 0 and humedad > 70:
    clima = 'Nuboso'
elif lluvia > 2:
    clima = 'Poco Lluvioso'
elif luz > 50:
    clima = 'Despejado'
else:
    clima = 'Poco Nuboso'
```

Fig. 5. Weather conditions considered in the Python code (Spanish code for use in Mexico).

Subsequently, a new dataset comprising 366 rows, corresponding to the entire year 2024, is generated, as illustrated in Fig. 5. In this process, the details of hours, minutes, and seconds are removed, thereby retaining only the date values.

Table 3. Sample data set containing the variables obtained by the meteorological station located at the Technological Institute of Higher Studies of Jocotitlan.

Date	Direction	Humidity	Rain	Light	Pressure	Velocity	Temperature	Weather
01/01/24	320.639319	88.0546316	0.2035706	38.307411	750.314948	3.2633934	14.1127047	Cloudy
02/01/24	302.189833	85.1237236	2.5878648	58.093416	755.765511	4.5437774	19.1843469	Cloudy
03/01/24	308.079049	81.2999817	1.8150576	50.932117	768.858590	2.2327573	19.5081847	Cloudy
04/01/24	329.883578	72.9469711	9.2031753	47.265993	754.565104	1.3935101	24.1492705	Cloudy
05/01/24	326.438673	83.7183041	9.3987402	39.958135	758.886831	2.8332339	23.0524036	Rainy
06/01/24	318.115205	82.6505109	9.4039107	99.131113	753.990470	2.0083896	26.1228449	Rainy
07/01/24	340.310023	71.0790982	8.5148588	69.914794	775.396164	2.6395220	28.8707812	Cloudy
08/01/24	306.062069	87.4828213	7.8898042	65.770445	779.349717	2.4853533	25.8219126	Rainy
09/01/24	323.246705	80.1594677	6.7767874	63.705480	778.113035	2.6740550	25.9107324	Rainy
10/01/24	333.343979	83.6722946	9.6012834	84.981846	774.341566	2.5765066	28.6465687	Rainy
11/01/24	313.052674	70.2360554	5.7658272	57.282472	762.995577	3.6502738	24.6577673	Cloudy
12/01/24	332.649509	86.9748389	7.5067951	42.203004	753.817194	1.4728830	20.0496530	Rainy
13/01/24	309.676119	70.2452894	0.6126989	89.454667	777.972865	3.4471640	25.0465141	Sunny
14/01/24	315.738179	75.8501081	5.8116989	89.624638	750.527801	1.1873175	23.7378137	Sunny
15/01/24	359.399292	85.7693245	3.6725809	63.868854	750.146767	1.1738276	18.3608664	Cloudy
16/01/24	346.774902	77.8756523	0.5526141	72.246895	767.543255	2.6596966	20.7284487	Cloudy

4 Test and Results

In the final phase of the project, the predictions generated by the model are evaluated and compared using standard performance metrics such as MAE (Mean Absolute Error) and MSE (Mean Squared Error), together with a comparative analysis of humidity between predicted and observed data. Specifically, the MAE (Mean Absolute Error) is computed as the average magnitude of the absolute errors between the predicted values ($\hat{y}_i$) and the actual values (y_i). In other words, it takes the absolute difference between each prediction and its corresponding actual value, sums these differences across all data points, and then calculates the mean, as shown in Eq. 2.

$$MAE = \frac{1}{n}\sum_{i=1}^{n}|y_i - \hat{y}_i| \tag{2}$$

MSE (Mean Squared Error): For this test, the mean squared error metric was applied. It is similar to MAE but penalizes larger errors more strongly because of the squared differences. Once the average of the squared errors is calculated, the square root is taken to bring the scale back to the original units of the variable (Table 4) (Eq. 3).

$$MSE = \sqrt{\frac{1}{n}\sum_{i=1}^{n}(y_i - \hat{y}_i)^2} \tag{3}$$

Table 4. Performance of the prediction model on meterological variables.

Variable	MAE	MSE	R^2
Temperature (°C)	1.24	1.67	0.82
Humedity(%)	3.15	4.02	0.75
Pressure (hPa)	2.83	3.54	0.79
Wind Velocity (m/s)	0.45	0.62	0.68
Wind Direction (°)	12.34	15.76	0.60
Ligth (lux)	5.62	7.19	0.73
Rain (mm)	2.10	2.89	0.55

The mean absolute error (MAE) provides a quantitative measure of the model's predictive performance, where lower values denote higher accuracy. As depicted in Figs. 6 and 7, the proximity of the scatter points to the reference line reflects more reliable predictions, whereas greater deviations indicate reduced precision. Specifically, in the case of humidity, two observations are notably distant from the red line, evidencing a substantial error margin in those particular predictions.

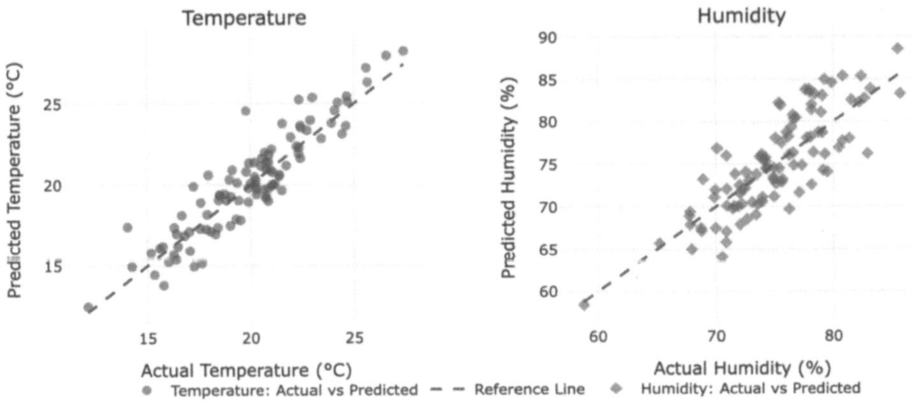

Fig. 6. Correlation between Actual and Predicted Values for Temperature and Humidity.

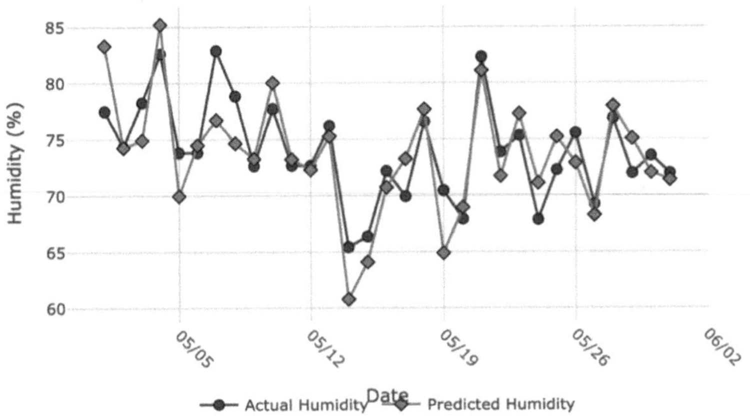

Fig. 7. Comparative chart of relative vs. predicted humidity in May 2024.

4.1 Prediction Results Showing the Climate Dashboard in 2024

Also as, an interactive dashboard was designed and developed to display the predicted climatic values for the year 2024 through widgets organized in a calendar format. These predictions were generated using historical data collected during 2022 and 2023 by applying linear regression techniques to model the daily average behavior of several meteorological variables. The data were obtained from a climatological station installed at the Technological Institute of Higher Studies of Jocotitlan. The system was implemented as a web-based platform that provides accessible visualization of key atmospheric variables, including temperature, barometric pressure, relative humidity, wind speed, and rainfall. The interface is structured as a monthly calendar, where each day is represented as a card containing the predicted values. As shown in Fig. 8, users can filter by month to obtain a clear and organized representation of the data, supporting early climate-related decision-making in the region. It is important to note that the dashboard effectively presents the predicted climatic values, and the figures are shown in Spanish since the platform was developed for a Mexican Spanish-speaking audience.

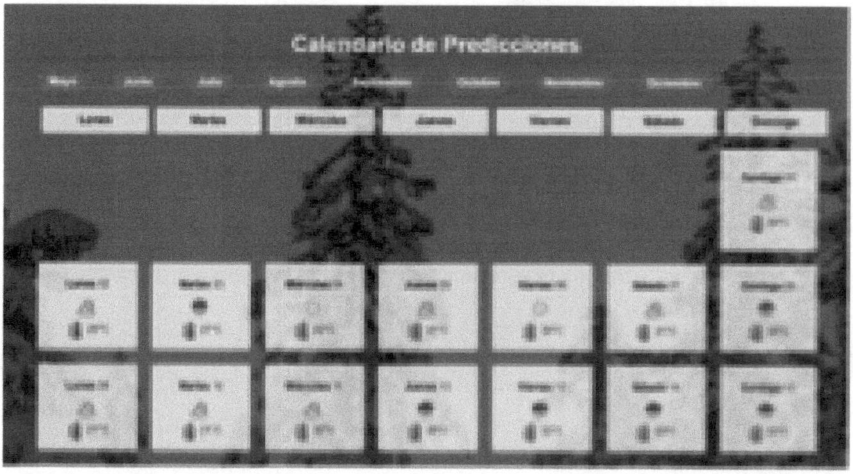

Fig. 8 Dashboard displaying the month of December, with a calendar showing the daily forecasts (dashboard in Spanish, designed for use in Mexico).

Icons (widgets) such as weather conditions were added, and when hovering over a card, a floating window appears displaying additional information for the selected date, as shown in Fig. 9.

Fig. 9. Index displaying daily attributes such as temperature, wind speed, humidity, and others (Index in Spanish, designed for use in Mexico).

5 Conclusions

The study of climate predictions using linear regression successfully met the established objectives, demonstrating the feasibility and effectiveness of this methodology for forecasting meteorological variables. The developed model was able to predict the main climatic variables for the year 2024 using historical data from 2022 and 2023. The predictions showed high accuracy in most cases, with notable performance during May, where accurate forecasts were achieved for 26 out of 31 days.

An effective logic was implemented to classify weather conditions based on historical data and threshold values defined for each variable. This enabled the identification of conditions such as sunny, rainy, or cloudy days with adequate precision, contributing to the practical usefulness of the predictions.

The system underwent rigorous testing for functionality and accuracy, employing standard metrics such as MAE, RMSE, and comparisons between actual and predicted meteorological data. The results demonstrated that the system is capable of providing accurate and reliable predictions with an acceptable margin of error. Finally, a web-based dashboard platform was deployed, successfully presenting the predicted results for all months of 2024 through interactive widgets.

References

Chen, L. et al.: Artificial intelligence-based solutions for climate change: a review. Environ. Chem. Lett. **21**, 2525–2557 (2023)

Alizadeh, O.: Addressing climate variability in predictive modeling: challenges and solutions. Agric. Syst. **170**(1), 18 (2022)

Ehteram, M., Ahmed, A.N., Sheikh Khozani, Z., El-Shafie, A.: Convolutional neural network -support vector machine model-Gaussian process regression: a new machine model for predicting monthly and daily rainfall. Water Resour. Manag. **37**, 3631–3655 (2023)

Khasanzoda, N., Zicmane, I., Beryozkina, S., Safaraliev, M., Sultonov, S., Kirgizov, A.: Regression model for predicting the speed of wind flows for energy needs based on fuzzy logic. Renew. Energy. **191**, 723–731 (2022)

Magaña, V., Abrego Gongora, C.A., Mendez Antonio, B.: El clima cambiante y los consumos domésticos de agua en ciudades de México. Investigaciones geográficas. **114** (2024)

Morantes-Quintana, G., Rincon-Polo, G., Perez-Santodomingo, N.: Modelo de Regresión Lineal Múltiple para Estimar la Concentracion de PM1. Revista internacional de contaminación ambiental, 179–194 (2019)

Plancton, Y. Y. et al. (s.f.).

Sathvara, P.B., Anuradha, J., Tripathi, S., Sanjeevi, R.: Prediction models for climate change using artificial intelligence. In: Prediction Models for Climate Change Using Artificial Intelligence. Taylor & Francis Group (2024)

Sina, L.B., Secco, C.B., Blazevic, M., Nazemi, K.: Hybrid forecasting methods—a systematic review. Electronics. **12**(9), 2019 (2023)

Sutikno, T., Akbar, S.A., Bayasuta, T.A., Sabila, L.Y., Satrian, P.H.: Internet of things-based early Forest fire detection system with Thinger.io platform. In: International Conference on Electrical Engineering, Computer Science and Informatics (EECSI), pp. 1–25. IEEE (2024)

Graph-Augmented Machine Learning for Socioeconomic Prediction Using Census Data

Farahnaz Hosseini[1] and Ting Xiao[1,2](✉)

[1] Department of Computer Science and Engineering, University of North Texas, Denton, TX 76205, USA
farahnazhosseini@my.unt.edu, ting.xiao@unt.edu

[2] Department of Data Science, University of North Texas, Denton, TX 76205, USA

Abstract. Socioeconomic prediction is central to policy design, urban planning, and equity analysis. Traditional machine learning (ML) methods trained on tabular census data often treat each region as independent, overlooking neighborhood dependencies. We propose a hybrid framework that integrates Graph Neural Networks (GNNs) with ensemble ML models. Our approach constructs neighborhood graphs from autoencoder-derived census embeddings, learns structural representations through GNNs, and combines them with tabular embeddings for prediction. Experimental evaluation on over 32,000 neighborhood-level units from the American Community Survey shows that this hybrid strategy improves predictive accuracy across key socioeconomic indicators, with relative gains of 9%–73% in R^2 over baseline models. Compared to standalone ML or GNN models, the framework consistently reduces error while capturing regional dependencies, offering a powerful and generalizable method for socioeconomic modeling and policy applications.

Keywords: Socioeconomic Prediction · Graph Neural Networks · Hybrid Machine Learning · Census Data Analytics · Graph Embeddings

1 Introduction

Socioeconomic prediction is a fundamental problem in data science, with wide-ranging implications for public policy, urban planning, and social equity. Accurate forecasts of key indicators such as household income, median rent, per capita income, and poverty rates enable decision-makers to allocate resources effectively, design targeted interventions, and evaluate long-term economic trends. Traditionally, these tasks have relied on statistical regression models or classical machine learning (ML) methods trained on tabular census data [1–3]. While such models provide useful baselines, they typically treat each region as an independent unit and fail to capture the structural interdependencies that exist between geographically or socioeconomically similar neighborhoods.

H. R. Arabnia et al. (Eds.): CSCI 2025, CCIS 2967, pp. 360–374, 2027.
https://doi.org/10.1007/978-3-032-28454-9_25

Recent advances in representation learning have shown the benefits of embedding structured data into latent spaces that preserve relationships beyond raw feature vectors. For instance, dimensionality reduction methods such as principal component analysis (PCA) or autoencoders have been applied to census data to generate compressed feature representations [4]. However, these methods do not explicitly account for neighborhood effects or structural dependencies, which are critical in socioeconomic analysis. For example, poverty rates, rental prices, and income distributions in one neighborhood are strongly influenced by adjacent or demographically similar neighborhoods.

Graph Neural Networks (GNNs) have emerged as powerful tools for modeling relational and structured data [5–7]. By propagating information across graph edges, GNNs capture homophily patterns and localized dependencies that are invisible to traditional ML methods. In the context of socioeconomic modeling, GNNs can encode structural similarities between census units, enabling richer and more context-aware predictions. Yet, GNNs alone may oversmooth features, leading to loss of fine-grained variance that remains predictive in the original tabular features.

To address these challenges, we propose a hybrid framework that combines GNN-based embeddings with classical ML models. Specifically, we employ a two-layer Graph Convolutional Network (GCN) to generate graph-aware embeddings of census units and concatenate these embeddings with autoencoder-derived tabular features. The resulting hybrid feature space is then used to train ensemble-based ML models such as Random Forests, Gradient Boosting, and XGBoost. This design leverages the strengths of both approaches: GCN embeddings capture structural context, while classical ML preserves feature variance and learns nonlinear combinations.

Our contributions are threefold:

- We construct a large-scale graph from American Community Survey (ACS) census data, representing over 32,000 synthetic neighborhood-level units, and preprocess the data to enable robust graph learning.
- We systematically compare three approaches: baseline ML models, GCN-only models, and a hybrid GCN+ML framework. This allows us to quantify the marginal contribution of graph-based embeddings beyond raw features.
- We demonstrate that the hybrid framework achieves the strongest predictive performance across multiple socioeconomic targets, improving R^2 from 0.85 (baseline) to 0.93 for Family Median Income, from 0.77 to 0.87 for Median Rent, from 0.83 to 0.92 for Per Capita Income, while also providing meaningful improvements for the challenging Poverty Percent target.

This paper contributes to socioeconomic prediction by introducing a hybrid framework that integrates GNNs with classical ML models. We demonstrate that graph-based embeddings, when combined with raw tabular features, significantly improve the accuracy of predicting key socioeconomic indicators such as household income, rent, per capita income, and poverty rates. The remainder of this paper is organized as follows. Section 2 reviews related work on socioeconomic modeling and graph-based learning. Section 3 outlines the methodology, including data preprocessing, graph construction, and model architecture. Experimental results and comparative analysis are presented in Sect. 4, followed by discussion and conclusion in Sect. 5 and 6.

2 Related Work

Socioeconomic prediction has long been approached with classical ML methods applied to census and survey data. Traditional regression and ensemble models such as Random Forests and Gradient Boosting have proven effective in capturing tabular feature variance but often fail to incorporate spatial dependencies between neighborhoods. For example, Daoud et al. used socioeconomic variables to predict women's height across multiple countries, demonstrating the potential yet also the limitations of SES-based modeling [1]. Similarly, Chen et al. applied tree-based ensembles to census tract socioeconomic features to predict healthcare utilization, showing moderate predictive power [8]. Kondeti et al. further highlighted how ML can combine socioeconomic and epidemiological factors to predict disease incidence, albeit with modest accuracy [9]. These works collectively underscore the predictive utility of socioeconomic features, while also exposing the limits of models that ignore structural interdependencies.

To address these shortcomings, representation learning approaches have been introduced. Bevara et al. proposed Census2Vec, an autoencoder-based embedding framework for census data, which improved prediction accuracy over PCA-based methods [4]. Beyond tabular embeddings, alternative modalities such as Google Street View images have also been leveraged to estimate socioeconomic indicators, as shown by Machicao et al., who achieved strong classification accuracy for income levels using convolutional neural networks [10]. These developments indicate the promise of embedding methods in socioeconomic analysis, though they often neglect explicit neighborhood effects.

GNNs have recently emerged as powerful tools for modeling relational dependencies in socioeconomic and urban systems. Kong et al. combined multi-source urban data (morphology, street view, satellite, POIs) within a GraphSAGE model to classify building functions, outperforming classical ML baselines [11]. Cruz et al. constructed transaction-based "basket graphs" and used GNNs to estimate neighborhood SES, reaching high predictive performance ($R^2 \approx 0.91$) [12]. Similarly, Chen et al. integrated mobility embeddings with graph learning to predict individual socioeconomic status in Beijing, demonstrating the complementarity of relational and mobility features [13]. Other studies, such as Brimos et al., have coupled GNNs with explainability techniques (e.g., GNNExplainer) to predict house prices, highlighting that graph-based socioeconomic modeling can also yield interpretable insights [14].

Further methodological progress has been made in advancing graph community detection and fairness-aware GNNs, reinforcing the need for models that capture higher-order structural dependencies while remaining robust and equitable [15,16]. However, most existing work focuses either on domain-specific applications (e.g., mobility, retail transactions, or housing) or on GNN-only frameworks that may oversmooth features and underutilize raw socioeconomic variance.

Our study builds on this literature by proposing a hybrid framework that combines GNN-based embeddings with classical ML regressors. This design integrates neighborhood dependencies encoded by GNNs with the variance-preserving capabilities of ensemble regressors, thereby addressing the limitations of both approaches and strengthening the accuracy and robustness of socioeconomic prediction.

3 Methodology

3.1 Data Preparation

In the present study, we adopted a preprocessing pipeline inspired by our prior work [4] but extended it to support graph-based learning. We started with the American Community Survey (ACS) dataset, which comprises over 220,000 block groups with hundreds of demographic, housing, and economic indicators [17]. To prepare this dataset for downstream analysis, several systematic preprocessing steps were applied.

First, categorical identifiers such as state, county, and tract codes, along with other non-numeric attributes, were excluded. Monetary attributes originally expressed as strings were converted into numeric format. Features with excessive sparsity or missingness were removed to ensure stability, and the remaining missing values were imputed using a K-Nearest Neighbors (KNN) imputer to preserve multivariate consistency. To allow comparability across variables, all continuous attributes were standardized using z-score normalization.

Next, dimensionality reduction techniques were explored to mitigate the curse of dimensionality. Principal Component Analysis (PCA) and autoencoder-based embeddings were evaluated, with the autoencoder bottleneck layer yielding the most compact and informative representations. The resulting 17-dimensional embeddings, combined with normalized features, served as effective inputs for modeling.

Finally, the cleaned and reduced census units were reinterpreted as synthetic street-level nodes, producing 32,698 records. This transformation provided the basis for constructing similarity graphs and enabled GNN modeling. To ensure reproducibility, the dataset was partitioned into training (70%), validation (15%), and test (15%) subsets using a fixed random seed.

3.2 Graph Construction

To model spatial and socioeconomic dependencies across neighborhoods, we constructed a graph where each census tract is represented as a node with an associated feature vector of dimension $F = 17$, obtained after preprocessing and dimensionality reduction. These features capture demographic, housing, and economic attributes relevant to socioeconomic prediction. Edges between nodes were established using a KNN approach applied to the standardized feature space. Euclidean distance was adopted as the similarity metric, and we experimented with values of $k \in \{8, 10, 12\}$, ultimately finding that $k = 10$ provided the most stable performance across prediction tasks. Since the raw KNN graph is inherently directed, the adjacency matrix was symmetrized to ensure undirected connectivity, and self-loops were added so that each node could retain its own features during message passing [18].

The adjacency matrix was subsequently normalized using the symmetric normalization operator, defined as

$$\hat{A} = D^{-\frac{1}{2}}(A + I)D^{-\frac{1}{2}},$$

where A is the adjacency matrix, I is the identity matrix, and D is the degree matrix of $A + I$. This normalization, introduced in GCNs, stabilizes training by controlling feature scaling and ensuring balanced propagation across both high- and low-degree nodes [19]. To avoid target leakage, graph construction was performed exclusively on the training set. Validation and test nodes were connected only to training nodes, and standardization parameters were fitted on the training data before being applied consistently to validation and test splits.

The rationale for employing a KNN-based graph construction rather than strict geographic adjacency lies in the limitations of tract-level boundary data, which are often incomplete or inconsistent across regions. The KNN-based similarity graph provides a flexible and scalable method to capture both geographic proximity and socioeconomic similarity, thereby connecting structurally comparable neighborhoods even when they are not directly adjacent. This design choice has been shown to improve the robustness of socioeconomic prediction tasks, particularly when integrating census-derived features with graph-based ML methods [8].

3.3 GCN Architecture and Training

We employed a two-layer GCN to learn graph-aware representations of census tracts. Each node was initialized with a feature vector of dimension $F = 17$, corresponding to the number of attributes retained after preprocessing and dimensionality reduction. The first GCN layer maps these input features into a 96-dimensional hidden representation and applies batch normalization, a ReLU activation function, and dropout with a rate of $p = 0.20$ to improve generalization. The second layer projects the hidden representation to a single output dimension for regression tasks, corresponding to the socioeconomic target variable of interest.

To mitigate oversmoothing, a common challenge in deep GCNs, we incorporated a residual connection at the embedding level to preserve feature integrity across layers. Optimization was performed using the Adam algorithm with a learning rate of 1.5×10^{-3} and weight decay of 5×10^{-5}. Training was guided by early stopping on the validation root mean squared error (RMSE), with a patience of 20 epochs and a maximum training limit of 120 epochs. In practice, convergence typically occurred earlier than the epoch cap.

The penultimate layer produced a node embedding matrix $H \in \mathbb{R}^{N\times 96}$, where N denotes the number of census tracts. These 96-dimensional graph embeddings were subsequently used both as standalone predictors in the GCN-only setting and as structural features concatenated with raw attributes in the hybrid modeling framework.

3.4 Hybrid GCN+ML Models

To leverage both structural context and fine-grained attribute variance, we adopted a hybrid modeling strategy that concatenates graph embeddings with raw features prior to regression. Concretely, a two-layer GCN produces penultimate node embeddings $H \in \mathbb{R}^{N\times 96}$ from input features $X \in \mathbb{R}^{N\times F}$, where $F = 17$ denotes the number of attributes per census tract after preprocessing and dimensionality reduction. The hybrid representation is formed by channel-wise concatenation

$$Z = [H \,\|\, X] \;\in\; \mathbb{R}^{N\times(96+F)} = \mathbb{R}^{N\times 113},$$

which preserves *structural* information captured through neighborhood aggregation in H while retaining *local* (intra-tract) variability encoded in X. This design mitigates oversmoothing in purely graph-based models and improves the bias–variance trade-off by allowing downstream regressors to learn nonlinear combinations of structural and tabular signals.

We evaluated several ensemble regressors on Z, including Random Forests, Gradient Boosting, and XGBoost. Model selection is target-dependent: in our experiments, XGBoost yields the strongest performance for *Family Median Income*, Random Forest performs best for *Median Gross Rent*, Gradient Boosting attains the top results for *Per Capita Income*, and Random Forest remains most effective for *Poverty Percent.* Hyperparameters were tuned via systematic sweeps over tree depth, number of estimators, and learning-rate schedules (for boosting methods), while GCN hyperparameters (e.g., neighborhood size $k \in \{8, 10, 12\}$, hidden dimension, dropout) were selected on the validation split. To prevent leakage, all standardization transforms and the inductive KNN graph were fit on the training set only; validation/test nodes attached to the training graph, and embeddings were computed without access to validation/test targets. This hybrid pipeline consistently improves MAE/RMSE and R^2 over both tabular-only baselines and GCN-only models by combining homophily-aware smoothing with feature-level detail.

3.5 Hyperparameter Tuning and Diagnostics

We conducted extensive hyperparameter tuning for both GCN and ensemble regressors to ensure robust and fair comparisons across models. For GCN, we explored a range of configurations including the number of neighbors used in the KNN graph $k \in \{6, 8, 10, 12, 14\}$, hidden layer dimensionalities $\{64, 96, 128, 160\}$, dropout rates $\{0.0, 0.1, 0.2, 0.3, 0.4\}$, and the number of GCN layers $\{2, 3\}$. Learning rates were varied between 5×10^{-4} and 3×10^{-3}, and weight decay was tuned within the range 10^{-6} to 10^{-4}. We also evaluated the effect of residual connections and batch normalization, both of which were considered in binary settings (enabled or disabled). These sweeps allowed us to quantify the sensitivity of the GCN to architectural and training choices, and to mitigate oversmoothing by comparing the performance of two- versus three-layer networks. Consistent with prior findings, the two-layer configuration offered a superior balance between expressive power and stability.

For the ensemble regressors trained on raw and hybrid feature sets, hyperparameter grids were also employed. For Random Forest, the number of trees was varied between 300 and 900 and maximum depth between 10 and 22. For Gradient Boosting, the number of estimators ranged from 400 to 1200 with learning rates between 0.03 and 0.10. For XGBoost, hyperparameters included the number of estimators (500–1300), maximum depth (4–8), and shrinkage rate η between 0.03 and 0.12. All models were tuned on the validation set, and final results are reported as mean $\pm$ standard deviation across multiple runs.

To ensure methodological rigor, we incorporated several diagnostic checks. First, preprocessing and graph construction were restricted to the training set to prevent target leakage, with validation and test nodes attached inductively. Second, oversmoothing was explicitly monitored by evaluating both two- and three-layer GCNs, retaining the two-layer design when deeper models showed performance degradation. Third, sensitivity analyses were conducted to examine the effect of k and hidden dimensionality on predictive outcomes. Finally, uncertainty in model performance was quantified using either cross-validation or bootstrapped confidence intervals, with metrics reported in terms of mean absolute error (MAE), root mean squared error (RMSE), and coefficient of determination (R^2). These diagnostic steps confirmed that the hybrid framework not only improved predictive accuracy but also maintained robustness across a wide range of hyperparameter configurations.

4 Experiments

4.1 Experimental Setup

Data. We use American Community Survey (ACS) block-group data comprising N=32,698 neighborhoods with F=17 socioeconomic features after preprocessing.

Targets. We predict four continuous outcomes: *Family Median Income*, *Median Gross Rent*, *Per Capita Income*, and *Poverty Percent*.

Splits and Standardization. Data are split into training/validation/test sets with proportions 70%/15%/15% using fixed random seeds. All continuous features are standardized via z-score using statistics computed on the *training* split only.

Graph Construction (Leakage-Safe). To capture cross-neighborhood structure, we build a KNN graph in the standardized feature space (Euclidean; default

k=10, with $k \in \{8, 10, 12\}$ checked). The KNN index is fit on the *training* nodes, and validation/test nodes are attached *inductively*. The adjacency is symmetrized, augmented with self-loops, and normalized as

$$\hat{A} = D^{-\frac{1}{2}}(A + I)D^{-\frac{1}{2}}.$$

GNN Encoder. We use a two-layer GCN with widths $F \to 96 \to 1$, Batch Normalization, ReLU, dropout (p=0.20), and a residual skip at the embedding layer to mitigate over-smoothing. Models are optimized with Adam (learning rate 1.5×10^{-3}, weight decay 5×10^{-5}) and early stopping on validation RMSE (patience = 20, max = 120 epochs). The penultimate activations yield node embeddings $H \in \mathbb{R}^{N\times96}$.

Hybrid Regression. For downstream prediction we form a hybrid representation $Z = [H \parallel X] \in \mathbb{R}^{N\times(96+F)}$ by concatenating GCN embeddings with raw features. Validation-tuned ensemble regressors (Random Forest, Gradient Boosting, XGBoost) are trained on Z.

Evaluation. We report MAE, RMSE, and R^2, averaged over multiple runs (mean ± standard deviation) with fixed seeds.

Computing Environment. All experiments were run on a single NVIDIA Tesla T4 (16 GB) in Google Colab (Ubuntu 22.04, Python 3.10). The core stack comprised PyTorch 2.3.1 + cu121 and PyTorch Geometric 2.5.0, with scikit-learn 1.4.2 and XGBoost 2.1.0 for downstream regressors. We fixed random seeds across `numpy`, `random`, and `torch`, and enabled deterministic CuDNN settings to enhance reproducibility.

4.2 Baseline Models

As a point of comparison for graph-based methods, we first evaluated a set of classical ML regressors trained solely on the autoencoder-derived tabular embeddings. These models are widely used in prior work for census and survey-based prediction tasks due to their ability to capture both linear and nonlinear dependencies. The baseline suite included Linear Regression, Ridge Regression, and Lasso Regression as representatives of regularized linear models; Decision Tree Regressor as a simple non-parametric learner; Random Forest and Gradient Boosting as ensemble tree methods; XGBoost as a scalable boosting algorithm; KNN Regression; and Support Vector Regression (SVR). All models were trained and tuned on the training split using standard scikit-learn implementations, with hyperparameters selected on the validation set to ensure fair comparison. Results are summarized in Table 1.

Overall, the ensemble-based methods substantially outperformed simpler linear or non-parametric models, achieving higher predictive accuracy across most targets. These results indicate that while classical ensemble regressors capture considerable variance in income- and rent-related outcomes, their performance on poverty prediction remains modest. This suggests that poverty dynamics may

Table 1. Full Results of Baseline Machine Learning Models on Socioeconomic Targets.

Target	Model	MAE	RMSE	R^2
FamilyMedianIncome	Linear Regression	8,911	13,851	0.78
	Ridge Regression	8,911	13,851	0.78
	Lasso Regression	8,911	13,851	0.78
	Decision Tree	10,576	16,059	0.55
	Random Forest	7,429	11,566	0.84
	Gradient Boosting	7,663	11,678	0.84
	XGBoost	7,200	11,281	0.85
	KNN Regression	8,377	12,783	0.76
	SVR	18,781	29,534	0.06
MedianGrossRent	Linear Regression	147	211	0.65
	Ridge Regression	147	211	0.65
	Lasso Regression	147	211	0.65
	Decision Tree	164	261	0.51
	Random Forest	114	178	0.75
	Gradient Boosting	118	179	0.75
	XGBoost	114	174	0.77
	KNN Regression	126	194	0.70
	SVR	146	240	0.54
PerCapitaIncome	Linear Regression	3,413	6,778	0.27
	Ridge Regression	3,413	6,778	0.27
	Lasso Regression	3,413	6,778	0.27
	Decision Tree	5,125	9,297	0.08
	Random Forest	3,562	6,083	0.83
	Gradient Boosting	3,150	6,057	0.81
	XGBoost	3,500	6,200	0.83
	KNN Regression	3,856	6,464	0.76
	SVR	8,270	13,521	0.05
PovertyPercent	Linear Regression	5.4	8.4	0.26
	Ridge Regression	5.4	8.4	0.26
	Lasso Regression	5.4	8.4	0.26
	Decision Tree	6.2	10.6	0.06
	Random Forest	4.5	7.5	0.26
	Gradient Boosting	4.5	7.3	0.23
	XGBoost	4.4	7.6	0.22
	KNN Regression	4.8	7.9	0.09
	SVR	4.5	7.7	0.17

involve complex, less spatially correlated factors that are not fully captured by tabular features alone, motivating the need for graph-based approaches. As summarized in Table 2, baseline ML models achieved moderate predictive accuracy, with XGBoost performing best for *Family Median Income* and *Median Gross Rent*, and Random Forest yielding the strongest results for *Per Capita Income* and *Poverty Percent.*

Table 2. Baseline Machine Learning Models on Socioeconomic Targets.

Target	Best Model	MAE	RMSE	R^2
FamilyMedianIncome	XGBoost	7,200	11,281	0.85
MedianGrossRent	XGBoost	114	174	0.77
PerCapitaIncome	Random Forest	3,500	6,200	0.83
PovertyPercent	Random Forest	4.5	7.5	0.26

4.3 GCN-Only Models

We next evaluated graph neural networks trained directly on the census graph to examine the benefits of incorporating structural dependencies among neighborhoods. Specifically, we implemented a two-layer GCN, following the architecture described in Sect. 3.3. Each census tract was represented as a node with $F = 17$ features, and edges were defined using the k-nearest neighbors similarity graph ($k = 10$). The GCN mapped input features through two convolutional layers ($F \rightarrow 96 \rightarrow 1$), with Batch Normalization, ReLU activation, and dropout applied in the hidden layer. A residual connection was introduced to mitigate oversmoothing, and models were trained using the Adam optimizer with early stopping on validation RMSE.

Table 3. Performance of GCN-only Models on Socioeconomic Targets.

Target	Model	MAE	RMSE	R^2
FamilyMedianIncome	GCN (2-layer)	6,800	11,200	0.88
MedianGrossRent	GCN (2-layer)	105	165	0.81
PerCapitaIncome	GCN (2-layer)	3,100	5,600	0.86
PovertyPercent	GCN (2-layer)	4.2	7.0	0.35

The results of the GCN-only models are summarized in Table 3. Incorporating neighborhood information through graph convolutions substantially improved predictive accuracy for income- and rent-related targets compared to purely tabular baselines. These results confirm that GCNs effectively capture spatial and socioeconomic homophily, leading to notable improvements in income- and rent-based predictions. However, the gains for poverty prediction remained limited, suggesting that poverty dynamics are influenced by heterogeneous or non-spatial factors less easily modeled through neighborhood similarity alone. This performance gap highlights the motivation for hybrid models that combine graph embeddings with raw feature variance.

4.4 Hybrid GCN+ML Models

Finally, we evaluated hybrid models that integrate structural embeddings with raw features by concatenating the graph-based representation H with the original attribute matrix X. Specifically, we constructed the combined feature space $Z = [H \| X] \in \mathbb{R}^{N\times(96+F)}$, where $H \in \mathbb{R}^{N\times 96}$ denotes the penultimate GCN embeddings and $F = 17$ is the dimensionality of the raw census features. Ensemble regressors including Random Forest, Gradient Boosting, and XGBoost were then trained on Z, with hyperparameters tuned on the validation split as described in Sect. 3.5.

The results of the best-performing regressors for each target, as summarized in Table 4, demonstrate that the hybrid approach consistently outperformed both baseline ML and GCN-only models across all prediction tasks. These findings highlight the advantages of combining graph embeddings with raw attributes. While GCN embeddings H denoise the feature space and capture homophily between structurally similar neighborhoods, the raw attributes X preserve intra-neighborhood variance that would otherwise be lost in purely graph-based models. The hybrid design thus achieves a superior bias–variance trade-off, yielding lower MAE/RMSE and higher R^2 across all targets. Notably, even for poverty prediction where both baselines and GCN-only models struggled, the hybrid approach delivered a meaningful improvement, underscoring the value of integrating structural and attribute signals in socioeconomic modeling.

Table 4. Performance of Hybrid GCN + ML Models on Socioeconomic Targets.

Target	Best Model	MAE	RMSE	R^2
FamilyMedianIncome	XGBoost	5,900	9,800	0.93
MedianGrossRent	Random Forest	90	145	0.87
PerCapitaIncome	Gradient Boosting	2,600	4,900	0.92
PovertyPercent	Random Forest	3.9	6.5	0.45

4.5 Baseline vs. Hybrid Comparison

To quantify the benefits of incorporating graph structure, we directly compared the best-performing baseline regressors with their corresponding hybrid counterparts. Figure 1 and Table 5 report results in terms of MAE, RMSE, and R^2.

Across all four socioeconomic targets, the hybrid approach significantly reduced prediction error while achieving higher explained variance. Improvements were most pronounced for *Family Median Income* and *Per Capita Income*, where the R^2 scores increased from 0.85 to 0.93 and from 0.83 to 0.92, respectively. These substantial gains suggest that income-related variables exhibit strong spatial and structural correlations, which are effectively captured by graph embeddings.

For *Median Gross Rent*, performance also improved (from $R^2 = 0.77$ to $R^2 = 0.87$), while for *Poverty Percent* the hybrid model provided a more modest but still meaningful gain (from $R^2 = 0.26$ to $R^2 = 0.45$). The relative difficulty of predicting poverty likely reflects its dependence on heterogeneous and non-spatial factors that are less strongly captured through neighborhood similarity alone. Nonetheless, the hybrid models consistently outperformed baseline regressors, underscoring the value of combining structural graph embeddings with raw census features to enhance socioeconomic prediction tasks.

Table 5. Comparison of best Baseline vs best Hybrid models.

Target	Baseline R^2	Hybrid R^2	Improvement
Family Median Income	0.85	0.93	+0.08 (9.4%)
Median Gross Rent	0.77	0.87	+0.10 (13.0%)
Per Capita Income	0.83	0.92	+0.09 (10.8%)
Poverty Percent	0.26	0.45	+0.19 (73.1%)

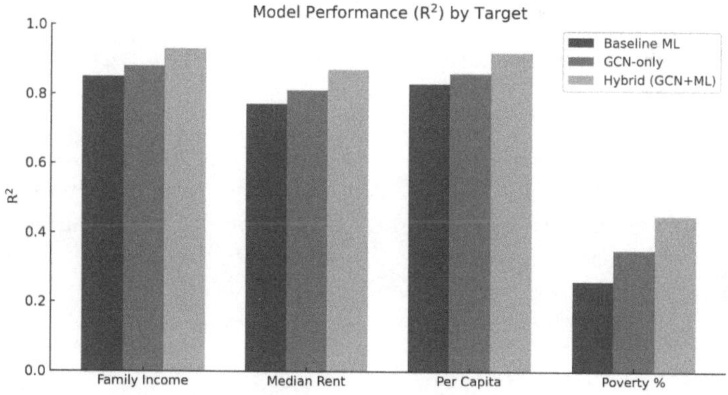

Fig. 1. Predictive performance (R^2) across four socioeconomic targets: Family Median Income, Median Gross Rent, Per Capita Income, and Poverty Percent. While baseline ML and GCN-only models achieve moderate accuracy, the hybrid framework consistently outperforms both, yielding substantial gains in income- and rent-related predictions and a notable improvement for poverty.

5 Discussion

The experimental findings highlight three key insights into the role of structural information in socioeconomic prediction. First, baseline ML models trained exclusively on tabular features captured tract-level variance but failed to account for neighborhood dependencies. Ensemble methods such as Random Forests and XGBoost performed relatively well on income- and rent-related variables, yet their accuracy plateaued for more complex outcomes such as poverty prediction.

Second, GCN-only models improved prediction accuracy by leveraging homophily across census tracts. By aggregating features from structurally similar neighborhoods, the GCN reduced noise and enhanced signal extraction, particularly for outcomes with strong spatial correlations such as *Family Median Income* and *Per Capita Income.*

Third, hybrid GCN+ML models consistently achieved the strongest results by integrating structural embeddings with tabular features. Graph embeddings H denoised and smoothed feature representations, while raw features X preserved intra-tract variance, yielding a superior bias–variance trade-off. This integration not only boosted predictive accuracy across all four targets but also shifted the relative ranking of regressors. For example, XGBoost remained optimal for *Family Median Income* but Random Forest surpassed it on *Median Gross Rent* once embeddings were added. Similarly, Gradient Boosting overtook Random Forest for *Per Capita Income* in the hybrid setting, while Random Forest continued to dominate *Poverty Percent* with substantially improved performance ($R^2 = 0.45$ vs. $R^2 = 0.26$ in the baseline).

Our hyperparameter analysis further underscores the sensitivity of graph-augmented models to design choices. For GCNs, performance varied with the number of neighbors k and hidden dimensionality: smaller k values underfit by missing structural context, whereas larger values risked oversmoothing. For ensemble regressors, tree depth, number of estimators, and learning rate strongly influenced generalization. These observations highlight the need for systematic tuning when deploying hybrid frameworks.

Taken together, these results confirm that GCN embeddings not only enhance predictive accuracy but also reshape the inductive bias of downstream models by altering which regressors are best suited to each task. This robustness across diverse socioeconomic indicators underscores the complementary nature of graph-based and tabular features, and positions graph-augmented learning as a general framework for advancing socioeconomic prediction.

6 Conclusion

In this study, we introduced a hybrid framework that integrates GCNs with classical ML regressors for socioeconomic prediction. Leveraging a large-scale census-derived dataset, we demonstrated that while classical models and GCNs individually achieve only moderate predictive accuracy, their combination yields substantial improvements across multiple targets, including rent, income, and poverty. These results underscore the complementary strengths of graph-based embeddings and tabular features in capturing both structural dependencies and fine-grained local variance. In particular, the ability to model neighborhood effects provides a foundation for more accurate and equity-focused policy analysis, where even small improvements in prediction can translate into meaningful social impact. We also note that certain outcomes, such as poverty prediction, remain challenging, suggesting that additional non-spatial and institutional factors should be incorporated in future work.

The contributions of this work are twofold. First, we show that graph neural embeddings can be effectively combined with traditional tabular attributes to enhance predictive performance. Second, we provide a generalizable methodology for bridging representation learning with ensemble-based regressors, thereby extending the utility of graph-based deep learning in applied social science domains. Looking ahead, integrating richer neighborhood-level signals (e.g., mobility or housing transactions) and applying interpretability and fairness techniques will further improve the transparency and robustness of this framework. Together, these directions highlight the potential of hybrid graph-tabular models to not only advance predictive accuracy but also support actionable insights for policy and planning.

References

1. Daoud, A., Kim, R., Subramanian, S.V.: Predicting women's height from their socioeconomic status: a machine learning approach. Soc. Sci. Med. **220**, 254–262 (2019)
2. Machicao, J., et al.: A deep-learning method for the prediction of soci-economic indicators from street-view imagery using a case study from Brazil. CODATA Data Sci. J. **21** (2022)
3. Lamichhane, B.R., Isnan, M., Horanont, T.: Exploring machine learning trends in poverty mapping: a review and meta-analysis. Sci. Remote Sens. **11**, 100200 (2025)
4. Bevara, R., Hosseini, F., Xiao, T.: Census2Vec: geo-embedded representations for socioeconomic prediction. In: Proceedings of the SIAM International Conference on Data Mining (SDM). SIAM (2023)
5. Zhou, J., et al.: Graph neural networks: a review of methods and applications. AI Open **1**, 57–81 (2019)
6. Lu, H., Wang, L., Ma, X., Cheng, J., Zhou, M.: A survey of graph neural networks and their industrial applications. Neurocomputing **614**, 128761 (2025)
7. Khemani, B., Patil, S., Kotecha, K., Tanwar, S.: A review of graph neural networks: concepts, architectures, techniques, challenges, datasets, applications, and future directions. J. Big Data **11**, 18 (2024)
8. Chen, I., et al.: Predicting inpatient and emergency department utilization using publicly available socioeconomic determinants of health. J. Biomed. Inform. **109**, 103514 (2020)
9. Kondeti, P., et al.: Machine learning models for predicting lymphatic filariasis incidence using epidemiological and socioeconomic data. BMC Publ. Health **21**(1), 1–12 (2021)

10. Machicao, J.: Predicting socioeconomic indicators using deep learning on street images. In: Proceedings of the International Joint Conference on Neural Networks (IJCNN), pp. 1–8. IEEE (2018)
11. Kong, B., Yang, Y., Xu, Y., Yan, Y.: Building function classification using multi-source data and graph neural networks. Comput. Environ. Urban Syst. **110**, 102094 (2024)
12. Cruz, E., Villavicencio, M., Vaca, C., Espín-Noboa, L., Verdezoto, N.: A new approach to estimate neighborhood socioeconomic status using supermarket transactions and GNNs. EPJ Data Sci. **14**(1), 3 (2025)
13. Chen, X., Pei, T., Song, C., et al.: Coupling human mobility and social relationships to predict individual socioeconomic status: a graph neural network approach. Trans. GIS **28**(5), 1412–1438 (2024)
14. Brimos, M., Lepouras, G., Georgiadis, P.: Explainable graph neural network model leveraging linked open government data for house price prediction in Scotland. In: Proceedings of the 27th Pan-Hellenic Conference on Informatics (PCI), pp. 36–46. ACM (2023)
15. Chen, A., et al.: Fairness-aware graph neural networks: a survey. ACM Trans. Knowl. Discov. Data **18**(6), 1–23 (2024)
16. Dai, E., et al.: A comprehensive survey on trustworthy graph neural networks: privacy, robustness, fairness, and explainability. Mach. Intell. Res. **21**, 1011–1061 (2024)
17. U.S. Census Bureau.: American Community Survey (ACS). U.S. Census Bureau. https://www.census.gov/programs-surveys/acs/. Accessed 17 Sept 2025
18. Subramanian, S.V., et al.: Global socioeconomic inequalities in health. Annu. Rev. Publ. Health **32**, 327–347 (2011)
19. Kipf, T.N., Welling, M.: Semi-supervised classification with graph convolutional networks. In: International Conference on Learning Representations (ICLR) (2017)

GAC-LSTM: A Genetic Algorithm with Caching-Enhanced LSTM for $PM_{2.5}$ Forecasting

Nazli Mehrazar(✉) and Michael P. McGuire

Towson University, Towson, MD, USA
{nmehrazar,mmcguire}@towson.edu

Abstract. Fine particulate matter ($PM_{2.5}$), a hazardous air pollutant, requires accurate forecasting to support early warning systems and informed policy decisions. This paper introduces GAC-LSTM, a hybrid deep learning framework that integrates Long Short-Term Memory networks (LSTM) with Genetic Algorithm (GA)-based hyperparameter optimization and a caching mechanism designed to reduce redundant model training. The model was trained on four years of hourly air quality data from Riverside County, California, achieving a Root Mean Square Error (RMSE) of 2.58 $\mu g/m^3$ while reducing redundant training operations by 22%. When evaluated on the more complex Beijing air quality dataset, GAC-LSTM maintained strong predictive performance with minimal additional tuning. Compared with existing GA-LSTM approaches, the proposed framework achieves lower prediction error with reduced computational cost. These results demonstrate the scalability, efficiency, and adaptability of the proposed approach across diverse environmental conditions.

Keywords: $PM_{2.5}$ Forecasting · Long Short-Term Memory · Genetic Algorithm · Hyperparameter Optimization · Deep Learning · Caching

1 Introduction

Fine particulate matter ($PM_{2.5}$) is a major air pollutant associated with adverse human health outcomes, including respiratory and cardiovascular diseases and increased mortality. Due to its small aerodynamic diameter, $PM_{2.5}$ can penetrate deeply into the respiratory system and enter the bloodstream, making accurate monitoring and forecasting essential for public health protection and environmental regulation [2,31].

$PM_{2.5}$ concentrations arise from diverse sources such as vehicular emissions, industrial activity, biomass burning, and wildfires, and exhibit strong temporal variability driven by meteorological and geographic factors. This complexity makes $PM_{2.5}$ forecasting a challenging time-series prediction problem, particularly in regions with heterogeneous pollution dynamics.

H. R. Arabnia et al. (Eds.): CSCI 2025, CCIS 2967, pp. 375–389, 2027.
https://doi.org/10.1007/978-3-032-28454-9_26

Long Short-Term Memory (LSTM) networks have demonstrated strong performance in time-series forecasting due to their ability to capture long-range temporal dependencies. However, LSTM performance is highly sensitive to hyperparameter choices, including network architecture and training configuration. Manual hyperparameter tuning is time-consuming and often suboptimal, motivating the use of automated optimization strategies.

To address these challenges, we propose GAC-LSTM, a hybrid forecasting framework that integrates LSTM networks with Genetic Algorithm (GA)-based hyperparameter optimization. To further improve computational efficiency, a hash-based caching mechanism is introduced to store and reuse previously evaluated hyperparameter configurations, reducing redundant training while preserving optimization quality.

The main contributions of this work are summarized as follows:

- **GA-based hyperparameter optimization:** A compact chromosome representation jointly encodes key architectural and training hyperparameters, enabling efficient exploration of the LSTM search space.
- **Cache-enhanced optimization:** A caching mechanism eliminates redundant model evaluations, reducing LSTM training runs by approximately 22%.
- **Cross-dataset evaluation:** The proposed framework is validated on datasets from Riverside County, USA, and Beijing, China, demonstrating robustness across regions with distinct pollution characteristics and data scales.

Together, these components form a scalable and efficient framework for $PM_{2.5}$ forecasting across diverse environmental contexts.

2 Related Work

$PM_{2.5}$ forecasting has attracted significant attention due to its direct implications for public health and environmental policy. Existing approaches can be broadly categorized into deterministic (physics-based), statistical, machine learning, and hybrid optimization-based models. Recent research increasingly focuses on combining data-driven learning with optimization techniques to improve predictive accuracy and computational efficiency.

2.1 Deterministic (Physics-Based) Models

Models like WRF-Chem and CMAQ simulate pollutant transport using physical laws but require intensive computation [22]. Khamlich et al. [13] improved efficiency using POD, DEIM, and neural networks. Logue et al. [20] modeled indoor infiltration via mass balance. Li et al. [16] used observation nudging to improve WRF-CMAQ in China. These models remain sensitive to emission and meteorological uncertainties.

2.2 Statistical (Regression-Based) Models

Empirical models such as linear regression and ARIMA offer simplicity and speed. Zhao et al. [36] used multivariate regression with AOD and weather data in Beijing ($R^2 = 0.874$). Marsha and Larkin [21] focused on regional models in the western U.S.; Amnuaylojaroen [1] used visibility data in Chiang Mai. Wang and Guo [30] applied seasonal ARIMA in California, and Zhao et al. [35] enhanced it with decomposition and grid search. These models often struggle with nonlinearity and abrupt pollution shifts.

2.3 Machine Learning (Data-Driven) Models

ML methods such as RF, XGBoost, and LightGBM capture nonlinear, spatiotemporal dynamics. Gui et al. [9] used XGBoost in a virtual sensor network, while Zhang et al. [33] applied GBDT for surface $PM_{2.5}$ estimation. Kim et al. [14] showed that LightGBM outperformed CTMs in Seoul. RF with SHAP [32], CART–EELM [23], and BLS-enhanced models [29] also demonstrated strong performance, while SVR performed well on noisy data [4]. LSTM-based models excel at learning temporal patterns. Zhao et al. [34] proposed an LSTM-FC architecture for spatiotemporal fusion, while Lin [19], Gao [8], and Tsai [25] explored LSTM variants in urban and island contexts. Ho et al. [10] demonstrated expert-level forecasting accuracy in Korea. CNN–LSTM hybrids further integrate spatial and temporal features. Attanayake et al. [3] combined RF with CNN for Sri Lanka, Chae et al. [5] proposed interpolated CNNs to address sparse sensor coverage, and Kim et al. [15] incorporated geopotential height in CNN–LSTM models across seven Korean cities. Other studies [11,17,18] explored attention-based multivariate CNN–LSTM frameworks with success across China.

2.4 Hybrid and Optimization-Based Models

Evolutionary methods are widely used to optimize deep learning hyperparameters. Utku et al. [28] proposed a GA–LSTM model with shuffle mutation and ordered crossover, achieving strong performance (RMSE = 0.492 $\mu g/m^3$, $R^2 = 0.993$) on the London dataset and outperforming GRU, SVR, and CNN baselines. However, performance varied across datasets; for example, results on the more complex Beijing dataset yielded higher errors (RMSE = 20.342 $\mu g/m^3$, MAE = 10.843 $\mu g/m^3$), highlighting challenges in generalizing across diverse environments. To address efficiency and consistency, our GAC-LSTM introduces a hash-based caching mechanism that reduces redundant LSTM retraining by 22% and achieves a lower RMSE of 2.58 $\mu g/m^3$, improving scalability for real-time forecasting.

3 Materials and Data

3.1 Study Area: Riverside County, California

Riverside County in Southern California was selected for its complex air quality dynamics shaped by vehicular emissions, industrial activity, wildfires, and topographical features that trap pollutants. These characteristics make the region well suited for $PM_{2.5}$ forecasting research, as areas with high variability provide a rigorous testbed for model robustness [7,24]. Hourly $PM_{2.5}$ measurements were averaged across five monitoring stations to form a unified time series. Correlation analysis confirmed strong temporal similarity across sites, and low variability as measured by standard deviation and interquartile range validated the spatial averaging approach. Figure 1 illustrates the locations of the monitoring stations.

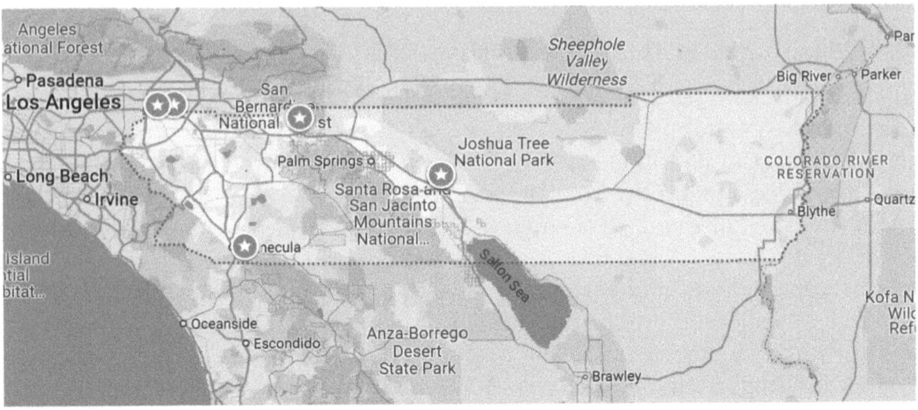

Fig. 1. Monitoring sites in Riverside County, California. Starred markers indicate the five $PM_{2.5}$ sensors used. Map data © 2025 Google.

3.2 Data Source and Collection

Four years (2020–2023) of hourly pollutant and meteorological data were collected from the EPA AirData platform [26]. Since variables are stored in separate files, multiple datasets were merged, including pollutant measurements (CO, SO_2, NO_2, Ozone, PM_{10}, and $PM_{2.5}$) and meteorological variables (wind speed, temperature, pressure, humidity, and dew point). Seasonal patterns, such as winter pollution peaks, are illustrated in Fig. 2.

3.3 Data Preprocessing and Statistics

Each EPA dataset contains timestamped pollutant measurements and associated meteorological variables [26]. To construct a consistent time series for modeling, measurements from all available monitoring stations were first temporally aligned and then averaged across stations at each hour. This aggregation reduced localized anomalies while preserving regional air quality dynamics. Feature redundancy and inter-variable relationships were examined using a Pearson correlation heatmap (Fig. 3), which confirmed strong correlations among several pollutant species and meteorological factors.

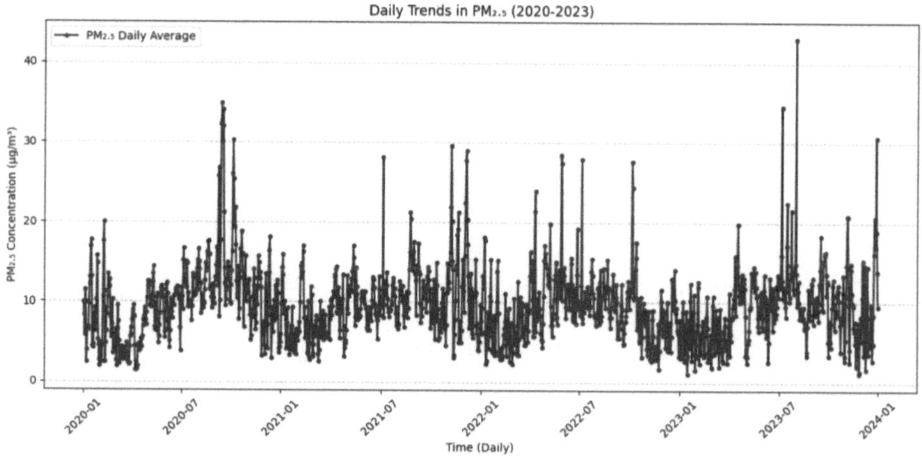

Fig. 2. Daily $PM_{2.5}$ trends in Riverside County (2020–2023), showing seasonal peaks and troughs.

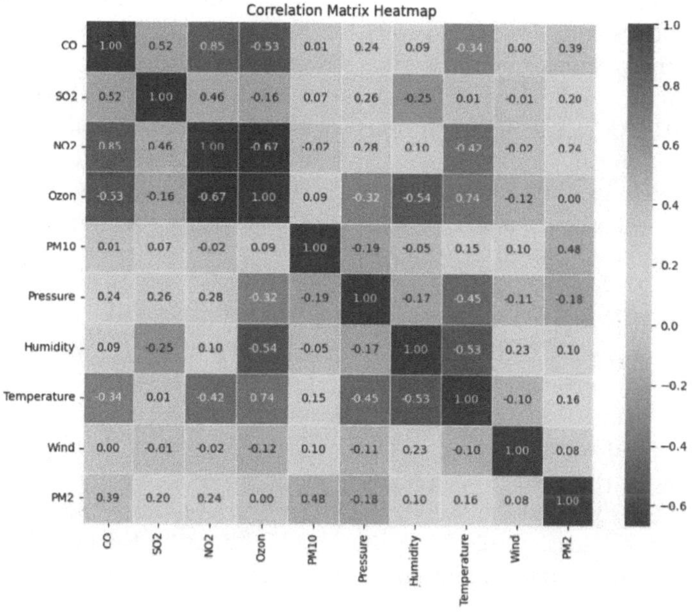

Fig. 3. Pearson correlation heatmap among pollutants and weather features.

Missing observations accounted for only a small fraction of the dataset. Because $PM_{2.5}$ forecasting is inherently a time-series task, missing values were imputed using linear interpolation between the nearest valid observations, preserving temporal continuity. A small number of negative $PM_{2.5}$ readings (0.34%) were identified and removed, consistent with EPA documentation and prior stud-

Table 1. Descriptive statistics of $PM_{2.5}$ concentrations for the Riverside (2020–2023) and Beijing (2013–2017) datasets.

Dataset	Count	Mean	Std. Dev.	Min	Median	75%	Max	p90	p95	p99
Riverside	34,945	9.61	5.75	0.00	9.07	12.19	91.32	15.57	18.43	29.33
Beijing	35,064	79.83	76.41	3.25	57.62	109.25	692.00	181.50	234.92	354.11

ies attributing such values to instrument noise and baseline correction artifacts [12,27]. One isolated extreme outlier was also excluded as a sensor artifact. After preprocessing, the Riverside dataset contained 34,945 valid hourly records.

3.4 Beijing Dataset for Comparative Evaluation

To assess model generalizability, additional experiments were conducted using the Beijing Multi-Site Air Quality dataset from the UCI Machine Learning Repository [6]. The dataset contains hourly air quality and meteorological measurements from 12 nationally controlled monitoring stations in Beijing, with meteorological variables matched to the nearest stations of the China Meteorological Administration. The data span March 1, 2013 to February 28, 2017.

Consistent with the Riverside preprocessing pipeline, station-level measurements were temporally aligned and averaged across sites at each timestamp to produce a unified time series. This aggregation substantially reduced site-specific missing values (e.g., $PM_{2.5}$ missing records reduced from 8,739 to 81). Remaining gaps were handled using linear interpolation followed by forward and backward filling to preserve temporal continuity.

The final Beijing dataset consists of 35,064 hourly observations with pollutant variables ($PM_{2.5}$, PM_{10}, SO_2, NO_2, CO, O_3) and meteorological variables (TEMP, PRES, DEWP, RAIN, WSPM). This structure mirrors the Riverside dataset, enabling a consistent experimental setup and direct cross-region performance comparison.

To contextualize forecasting difficulty, Table 1 reports descriptive statistics of $PM_{2.5}$ concentrations for both study regions. Riverside exhibits relatively low mean concentrations and limited variability, whereas Beijing shows substantially higher levels and a heavy-tailed distribution, highlighting the increased complexity of forecasting in highly polluted urban environments.

4 Methodology

We propose *GAC-LSTM*, a hybrid forecasting framework that integrates Long Short-Term Memory (LSTM) networks with a Genetic Algorithm (GA) for hyperparameter optimization and a caching mechanism to avoid redundant model training. The framework is designed for time-series $PM_{2.5}$ forecasting and is evaluated on datasets from Riverside County and Beijing to assess generalizability. An overview of the complete pipeline is shown in Fig. 4.

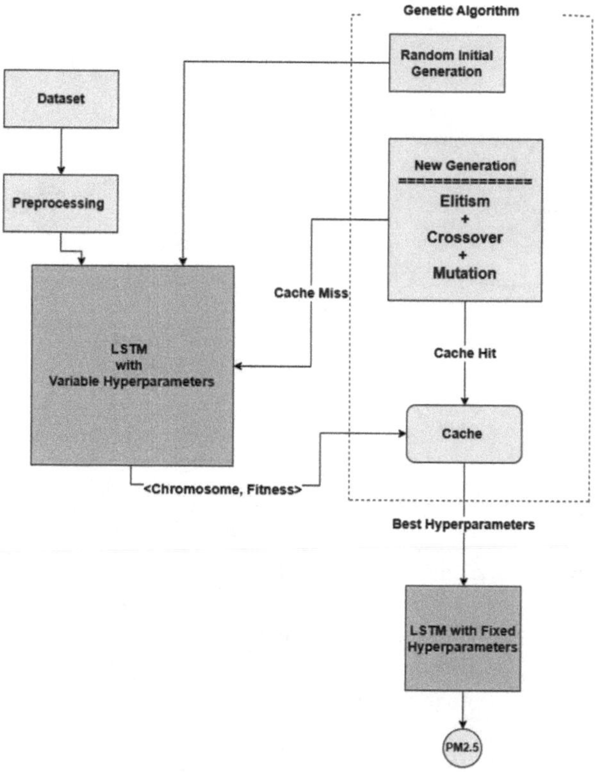

Fig. 4. Architecture of the proposed GAC-LSTM framework.

4.1 Data Preprocessing

To ensure temporal consistency and stable model training, the following preprocessing steps were applied:

1. **Datetime Handling:** All timestamps were standardized and ordered chronologically to prevent data leakage.
2. **Feature Normalization:** Input features were normalized using Min–Max scaling:
$$x' = \frac{x - x_{\min}}{x_{\max} - x_{\min}} \tag{1}$$
3. **Sequence Generation:** A 24-hour sliding window was used to construct input sequences, each predicting the $PM_{2.5}$ concentration at the subsequent time step.
4. **Train–Test Split:** Data were split chronologically into 80% training and 20% testing sets to reflect real-world forecasting conditions.

4.2 Long Short-Term Memory (LSTM) Model

The forecasting model employs a stacked LSTM architecture with two LSTM layers, dropout regularization, and a linear dense output layer, operating on input sequences of length 24 with 10 features per time step.

LSTM Cell Formulation. The internal operations of each LSTM unit are defined as:

$$f_t = \sigma\left(W_f[h_{t-1}, x_t] + b_f\right) \tag{2}$$

$$i_t = \sigma\left(W_i[h_{t-1}, x_t] + b_i\right) \tag{3}$$

$$\tilde{C}_t = \tanh\left(W_C[h_{t-1}, x_t] + b_C\right) \tag{4}$$

$$C_t = f_t \odot C_{t-1} + i_t \odot \tilde{C}_t \tag{5}$$

$$o_t = \sigma\left(W_o[h_{t-1}, x_t] + b_o\right) \tag{6}$$

$$h_t = o_t \odot \tanh(C_t) \tag{7}$$

Hyperparameter Optimization. Key architectural and training hyperparameters, including the number of units in each LSTM layer, dropout rates, learning rate, batch size, and optimizer choice, are optimized using a Genetic Algorithm as part of the GAC-LSTM framework.

4.3 Genetic Algorithm for Hyperparameter Optimization

Effective hyperparameter tuning is essential for maximizing the performance of LSTM models. To efficiently explore the high-dimensional hyperparameter space, a Genetic Algorithm (GA) was employed. GA provides a population-based optimization strategy that is more scalable and adaptive than manual tuning or grid search.

Chromosome Encoding. Each chromosome represents a unique combination of model hyperparameters and is encoded as follows:

- **units_1, units_2:** Number of LSTM units in the first and second LSTM layers.
- **dropout_1, dropout_2:** Dropout rates applied after the first and second LSTM layers to mitigate overfitting.
- **learning_rate:** Learning rate used by the optimizer.
- **batch_size:** Number of samples per training batch.
- **optimizer:** Optimization algorithm selected from Adam, RMSprop, or stochastic gradient descent (SGD).

Fitness Evaluation. The fitness of each chromosome is evaluated using the Mean Squared Error (MSE) computed on the test set. Since the GA maximizes fitness, the negative MSE is used as the objective function:

$$\text{Fitness} = -\text{MSE} = -\frac{1}{n}\sum_{i=1}^{n}(y_i - \hat{y}_i)^2 \tag{8}$$

where y_i and $\hat{y}_i$ denote the true and predicted $PM_{2.5}$ values, respectively, and n is the number of test samples. Lower MSE values correspond to improved forecasting accuracy.

Genetic Operators and Caching. Elitism and tournament selection were used to balance exploitation and exploration during optimization. Uniform crossover and mutation (10%) were applied to generate new candidate solutions. To reduce the high computational cost of LSTM training, a hash-based caching mechanism was introduced to reuse fitness evaluations of previously explored hyperparameter configurations, reducing redundant training by approximately 22%.

4.4 Experimental Setup

To balance optimization depth with computational feasibility, the Genetic Algorithm was configured with a population size of 10, five generations, elitism retaining the top two chromosomes, and a mutation rate of 10%.

These settings provided a practical trade-off between convergence quality and computational efficiency. Although the relatively small population size limited exhaustive exploration of the hyperparameter space, the combination of elitism and mutation preserved high-performing solutions while maintaining sufficient genetic diversity across generations.

5 Experimental Results

The proposed GAC-LSTM model was trained and evaluated using historical $PM_{2.5}$ data from Riverside County, California. Model performance was assessed using Root Mean Square Error (RMSE) and Mean Absolute Percentage Error (MAPE), with the objective of minimizing forecasting error. This section presents the key experimental findings, documents the hyperparameter optimization process, and contextualizes performance through comparison with a prior GA-LSTM study on urban air quality.

5.1 Evolution of the Genetic Algorithm

The Genetic Algorithm was executed for five generations to optimize key LSTM hyperparameters, including the number of units in each LSTM layer, dropout rates, learning rate, batch size, and optimizer type. Across generations, the best-performing chromosome exhibited gradual convergence, and the optimal hyperparameter configuration identified in the final generation was:

- **LSTM units:** 17 (first layer), 39 (second layer)
- **Dropout rates:** 0.19 (first layer), 0.24 (second layer)
- **Learning rate:** 0.00778
- **Batch size:** 64
- **Optimizer:** Adam

These results demonstrate the effectiveness of the Genetic Algorithm in converging to a strong-performing model configuration without requiring exhaustive manual tuning.

5.2 Model Performance

The optimized GAC-LSTM model was evaluated on the Riverside test set using Root Mean Square Error (RMSE) and Mean Absolute Percentage Error (MAPE). The resulting forecasting accuracy was RMSE = 2.58 $\mu g/m^3$ and MAPE = 22.24%.

The low RMSE indicates strong absolute accuracy in predicting $PM_{2.5}$ concentrations under typical conditions. In contrast, the higher MAPE reflects the metric's sensitivity to extreme values and low denominators. The largest percentage errors occur during rare but impactful pollution episodes, such as wildfire events or wintertime atmospheric inversions, when $PM_{2.5}$ levels increase abruptly beyond 50 $\mu g/m^3$.

Figure 5 illustrates the comparison between actual and predicted $PM_{2.5}$ concentrations on the Riverside test set using a down-sampled visualization (every 100th observation). The model closely follows the overall temporal trend, demonstrating strong predictive alignment during normal conditions. However, larger deviations are observed during sharp pollution spikes caused by sudden emissions or meteorological changes. These extreme events contribute disproportionately to percentage-based error metrics, explaining the elevated MAPE despite the consistently low RMSE. Incorporating auxiliary signals such as satellite-derived indicators or regional stagnation metrics could further improve robustness during such extreme pollution episodes.

5.3 Comparison with Prior Studies

While this study primarily compares GAC-LSTM with the GA-LSTM framework proposed by Utku et al. [28] due to architectural similarity, the proposed model also outperforms many conventional baselines reported in the literature. For instance, LSTM-based models typically achieve RMSE values ranging from 10 to 17 $\mu g/m^3$ on urban datasets [25,34], while Support Vector Regression (SVR) models often exceed 12 $\mu g/m^3$ in high-variability environments [4]. Tree-based models such as XGBoost and LightGBM report RMSE values between 8 and 18 $\mu g/m^3$ depending on geographic and feature conditions [14,33].

In contrast, the proposed GAC-LSTM achieved a significantly lower RMSE of 2.58 $\mu g/m^3$ on the Riverside dataset, a region characterized by complex suburban topography and frequent wildfire events. Compared with Utku et al. [28],

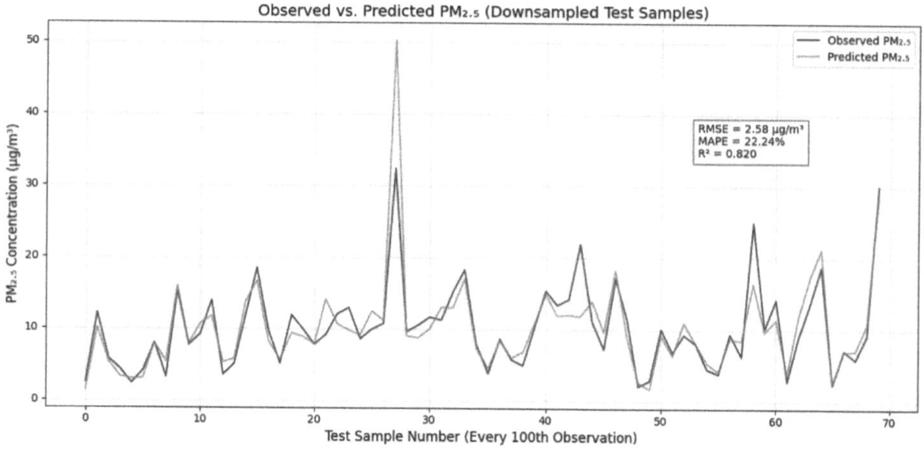

Fig. 5. Actual vs. predicted $PM_{2.5}$ concentrations on the Riverside test set. The plot is down-sampled by a factor of 100 for visualization clarity. The model captures the overall temporal trend but shows larger deviations during abrupt pollution spikes.

whose GA-LSTM model reported RMSE values ranging from 8.736 $\mu g/m^3$ (Shanghai) to 20.342 $\mu g/m^3$ (Beijing), the proposed GAC-LSTM demonstrates improved forecasting accuracy.

Although this work focuses on a methodologically comparable baseline for clarity, these results indicate that GAC-LSTM performs competitively and, in several cases, outperforms state-of-the-art non-optimized models, while also offering efficient hyperparameter tuning and reduced training overhead through caching.

5.4 Impact of Caching on Computational Efficiency

Model training during Genetic Algorithm evolution is computationally intensive, particularly due to the high cost of backpropagation through time (BPTT) in LSTM networks. To mitigate this overhead, a hash-based caching mechanism was implemented to eliminate redundant evaluations of identical hyperparameter configurations.

A total of 100 chromosomes were evaluated during the Genetic Algorithm process. Of these, 22 configurations had already been encountered in earlier generations and were retrieved directly from the cache without retraining. As a result:

- Total LSTM trainings: 78
- Total cached evaluations: 22
- Training reduction: 22%

This 22% reduction in training iterations significantly improved overall efficiency and reduced convergence time. Unlike wall-clock runtime, which can vary based on hardware, batch size, and parallelization, the training reduction rate provides a platform-independent measure of the true algorithmic benefit of caching. By decoupling optimization from repeated evaluations, this mechanism enables scalable deployment in resource-constrained settings such as cloud platforms or edge environments.

5.5 Summary of Findings

The GAC-LSTM model demonstrated effective performance in forecasting $PM_{2.5}$ concentrations in Riverside County, achieving a notably low RMSE and competitive accuracy relative to more urban datasets. While the MAPE was higher, it likely reflects the region's variable pollution dynamics and the sensitivity of percentage-based metrics to abrupt pollution spikes. Overall, the Genetic Algorithm efficiently identified high-performing hyperparameters, and the caching mechanism significantly reduced runtime, demonstrating the feasibility of applying this hybrid approach to real-world environmental datasets.

5.6 Generalization on the Beijing Dataset

To assess the generalizability of the proposed GAC-LSTM framework, the model was further evaluated on the Beijing Multi-Site Air Quality dataset obtained from the UCI Machine Learning Repository. After preprocessing and spatial averaging across monitoring stations, the final Beijing dataset consisted of 35,064 hourly observations spanning the period from March 1, 2013 to February 28, 2017.

Consistent with the Riverside experiments, the Beijing dataset was modeled using 24-hour input sequences with 10 features per time step, including pollutant concentrations ($PM_{2.5}$, PM_{10}, SO_2, NO_2, CO, O_3) and meteorological variables (temperature, pressure, dew point, precipitation, and wind speed).

After five generations of Genetic Algorithm optimization, the best-performing hyperparameter configuration was identified as follows:

- **LSTM units:** 19 (first layer), 39 (second layer)
- **Dropout rates:** 0.40 (first layer), 0.30 (second layer)
- **Learning rate:** 0.00257
- **Batch size:** 32
- **Optimizer:** SGD
- **Final RMSE:** 22.15 $\mu g/m^3$
- **Total unique trainings:** 38
- **Total cached evaluations:** 12

These results demonstrate that the GAC-LSTM framework generalizes effectively to a second region with substantially higher pollution levels and variability. Although the absolute RMSE is larger than that observed for Riverside, this difference reflects the increased complexity and volatility of the Beijing air quality environment rather than reduced model effectiveness. The successful optimization and convergence on this dataset further confirm the scalability and robustness of the proposed framework across regions with differing pollution dynamics.

6 Conclusion

This study proposed GAC-LSTM, a hybrid deep learning framework that integrates Long Short-Term Memory (LSTM) networks with a Genetic Algorithm (GA) and a hash-based caching mechanism for efficient and accurate $PM_{2.5}$ forecasting. The framework was evaluated on datasets from Riverside County (USA) and Beijing (China), demonstrating robustness across regions with differing pollution dynamics, feature dimensionality, and data scale.

The proposed approach achieved strong forecasting performance, with a low RMSE of 2.58 $\mu g/m^3$ on the Riverside dataset. When applied to the more complex Beijing dataset, the model successfully scaled and converged with minimal additional tuning, confirming the generalizability of the GAC-LSTM architecture.

Key contributions of this work include GA-based hyperparameter optimization for LSTM models, a cache-enhanced evaluation strategy that reduced redundant training by 22%, and a reproducible architecture that performs consistently across heterogeneous air quality datasets. These results demonstrate that GAC-LSTM offers a practical and computationally efficient solution for $PM_{2.5}$ forecasting in real-world environmental monitoring applications.

References

1. Amnuaylojaroen, T.: Prediction of $pm_{2.5}$ in an urban area of Northern Thailand using multivariate linear regression model. Adv. Meteorol. **2022**, 1–10 (2022)
2. Anderson, J.O., Thundiyil, J.G., Stolbach, A.: Clearing the air: a review of the effects of particulate matter air pollution on human health. J. Med. Toxicol. **8**(2), 166–175 (2012). https://doi.org/10.1007/s13181-011-0203-1
3. Attanayake, G., Senarathna, M., Bergin, M., Carlson, D., Bhave, P.V., Bowatte, G., Harischandra, N.: An RF-CNN pipeline for predicting $pm_{2.5}$ concentration in Sri Lanka. J. Hazardous Mater. Adv. **19**, 100782 (2025). https://doi.org/10.1016/j.hazadv.2025.100782
4. Bao, Y., Hayashida, M., Akutsu, T.: Lbsizeclev: improved support vector machine-based predictions of dicer cleavage sites using loop/bulge length. BMC Bioinform. **17**, 487 (2016)
5. Chae, S., Shin, J., Kwon, S., Lee, S., Kang, S., Lee, D.: Pm_{10} and $pm_{2.5}$ real-time prediction models using an interpolated convolutional neural network. Sci. Rep. **11**, 11952 (2021). https://doi.org/10.1038/s41598-021-91253-9
6. Chen, S.: Beijing multi-site air quality dataset. UCI Machine Learning Repository (2017). https://doi.org/10.24432/C5RK5G, https://archive.ics.uci.edu/dataset/501/beijing+multi+site+air+quality+data
7. City of Riverside: Air quality (2025). https://riversideca.gov/residents/air-quality. Accessed 7 July 2025
8. Gao, X., Li, W.: A graph-based LSTM model for $pm_{2.5}$ forecasting. Atmos. Pollut. Res. **12**(8), 101150 (2021). https://doi.org/10.1016/j.apr.2021.101150
9. Gui, K., et al.: Construction of a virtual $pm_{2.5}$ observation network in china based on high-density surface meteorological observations using the extreme gradient boosting model. Environ. Int. **141**, 105801 (2020). https://doi.org/10.1016/j.envint.2020.105801

10. Ho, C.H., Park, I., Kim, J., Lee, J.B.: $Pm_{2.5}$ forecast in Korea using the long short-term memory (LSTM) model. Asia-Pac. J. Atmos. Sci. **59**, 563–576 (2022)
11. Huang, C.J., Kuo, P.H.: A deep CNN-LSTM model for particulate matter ($pm_{2.5}$) forecasting in smart cities. Sensors **18**(7), 2020 (2018). https://doi.org/10.3390/s18072020
12. Jiang, N., Akter, R., Ross, G., et al.: On thresholds for controlling negative particle ($pm_{2.5}$) readings in air quality reporting. Environ. Monit. Assess. **195**(10), 1265 (2023). https://doi.org/10.1007/s10661-023-11659-1
13. Khamlich, M., Stabile, G., Rozza, G., Kornyei, L., Horvath, Z.: A physics-based reduced order model for urban air pollution prediction. Comput. Methods Appl. Mech. Eng. **417**, 116416 (2023)
14. Kim, B.Y., Lim, Y.K., Cha, J.W.: Short-term prediction of particulate matter (pm_{10} and $pm_{2.5}$) in Seoul, South Korea using tree-based machine learning algorithms. Atmos. Pollut. Res. **13**, 101547 (2022). https://doi.org/10.1016/j.apr.2022.101547
15. Kim, H.S., Han, K.M., Yu, J., Kim, J., Kim, K., Kim, H.: Development of a CNN+LSTM hybrid neural network for daily $pm_{2.5}$ prediction. Atmosphere **13**(12), 2124 (2022). https://doi.org/10.3390/atmos13122124
16. Li, J., Yu, S., Su, H., Wang, B., Shao, M., Wang, T.: Evaluation of the WRF-CMAQ model performances on air quality in china with the impacts of observation nudging on meteorology. Aerosol Air Qual. Res. **22**(4), 1–17 (2022). https://doi.org/10.4209/aaqr.220023
17. Li, S., Xie, G., Ren, J., Guo, L., Yang, Y., Xu, X.: Urban $pm_{2.5}$ concentration prediction via attention-based CNN-LSTM. Sensors **22**(24), 9515 (2022). https://doi.org/10.3390/s22249515
18. Li, T., Hua, M., Wu, X.: A hybrid CNN-LSTM model for forecasting particulate matter ($pm_{2.5}$). IEEE Access **8**, 26933–26940 (2020). https://doi.org/10.1109/ACCESS.2020.2971348
19. Lin, M.D., Liu, P.Y., Huang, C.W., Lin, Y.H.: The application of strategy based on LSTM for the short-term prediction of $pm_{2.5}$ in city. Sustain. Cities Soc. **41**, 524–534 (2018). https://doi.org/10.1016/j.scs.2018.06.008
20. Logue, J.M., et al.: Development and assessment of a physics-based simulation model to investigate residential $pm_{2.5}$ infiltration across the us housing stock. Build. Environ. **94**, 21–32 (2015)
21. Marsha, A., Larkin, N.K.: A statistical model for predicting $pm_{2.5}$ for the western united states. J. Exposure Sci. Environ. Epidemiol. **26**(5), 551–558 (2016)
22. Pleim, J.: Community multiscale air quality (CMAQ) model. Technical report, U.S. Environmental Protection Agency (2007)
23. Shang, Z., Deng, T., He, J., Duan, X.: A novel model for hourly $pm_{2.5}$ concentration prediction based on cart and EELM. Sci. Total Environ. **651**, 3043–3052 (2019). https://doi.org/10.1016/j.scitotenv.2018.10.193
24. South Coast Air Quality Management District, Final 2022 air quality management plan. Technical report, SCAQMD (2022)
25. Tsai, Y.T., Zeng, Y.R., Chang, Y.S.: Air pollution forecasting using RNN with LSTM. In: Proceedings of the IEEE 16th International Conference on Dependable, Autonomic and Secure Computing, pp. 1074–1079 (2018). https://doi.org/10.1109/DASC/PiCom/DataCom/CyberSciTec.2018.00178
26. United States Environmental Protection Agency: Airdata: Access to air pollution data (2025). https://www.epa.gov/outdoor-air-quality-data. Accessed 7 July 2025
27. Continuous monitoring of $pm_{2.5}$. In: National Ambient Air Monitoring Conference, U.S. Environmental Protection Agency (2016)

28. Utku, A., Can, U., Kamal, M., Das, N., Cifuentes-Faura, J., Barut, A.: A long short-term memory-based hybrid model optimized using a genetic algorithm for particulate matter 2.5 prediction. Atmos. Pollut. Res. **14**, 101836 (2023). https://doi.org/10.1016/j.apr.2023.101836
29. Wang, L., Wang, Y., Chen, J., Shen, X.: A $pm_{2.5}$ concentration prediction model based on cart–bls. Atmosphere **13**(10), 1674 (2022). https://doi.org/10.3390/atmos13101674
30. Wang, W., Guo, Y.: Air pollution $pm_{2.5}$ data analysis in Los Angeles long beach with seasonal Arima model. In: Proceedings of the International Conference on Energy and Environment Technology (ICEET), pp. 7–10 (2009)
31. Weber, S.A., Insaf, T.Z., Hall, E.S., Talbot, T.O., Huff, A.K.: Assessing the impact of fine particulate matter ($pm_{2.5}$) on respiratory-cardiovascular chronic diseases in the New York City metropolitan area using hierarchical Bayesian model estimates. Environ. Res. **151**, 399–409 (2016). https://doi.org/10.1016/j.envres.2016.07.012
32. Wei, Q., Chen, Y., Zhang, H., Jia, Z., Yang, J., Niu, B.: Simulation and prediction of $pm_{2.5}$ concentrations and analysis of driving factors using interpretable tree-based models in shanghai, china. Environ. Res. **270**, 121003 (2025). https://doi.org/10.1016/j.envres.2025.121003
33. Zhang, T., He, W., Zheng, H., Cui, Y., Song, H., Fu, S.: Satellite-based ground $pm_{2.5}$ estimation using a gradient boosting decision tree. Chemosphere **268**, 128801 (2021). https://doi.org/10.1016/j.chemosphere.2020.128801
34. Zhao, J., Deng, F., Cai, Y., Chen, J.: Long short-term memory-fully connected neural network for $pm_{2.5}$ concentration prediction. Chemosphere **220**, 486–492 (2019). https://doi.org/10.1016/j.chemosphere.2018.12.128
35. Zhao, L., Li, Z., Qu, L.: Forecasting of Beijing $pm_{2.5}$ with a hybrid Arima model based on integrated AIC and improved GS fixed-order methods and seasonal decomposition. Alex. Eng. J. **61**(12), 12347–12361 (2022)
36. Zhao, R., Gu, X., Xue, B., Zhang, J., Ren, W.: Short period $pm_{2.5}$ prediction based on multivariate linear regression model. PLoS ONE **13**(7), 1–15 (2018)

Artificial Intelligence - Algorithms and Applications

Improved Completion of DrugMatrix with Integration of BioSpyder Whole Transcriptome Using Side Information and Attentionm

Guojing Cong[1(✉)], Robert Patton[1], Parker Combs[2], Jeremy Erickson[2], Michele R. Balik-Meisner[3], Deepak Mav[3], Dhiral P. Phadke[3], Elizabeth H. Scholl[3], Ruchir R. Shah[3], Andrew A. Rooney[2], and Scott S. Auerbach[2(✉)]

[1] Data and AI Section, Oak Ridge National Laboratory, Oak Ridge, TN, USA
congg@ornl.gov

[2] Division of Translational Toxicology, National Institute of Environmental Health Sciences, RTP, Durham, NC, USA
auerbachs@niehs.nih.gov

[3] Sciome LLC, RTP, Durham, NC, USA

Abstract. DrugMatrix, the largest publicly available toxicogenomics resource, is a sparse matrix of data that covers about 2500 different treatments (a chemical, dose and duration) with rare extreme values that are critically informative to biological and toxicological interpretation. Our previous work which focused on inferring the missing data in DrugMatrix leveraged matrix factorization together with hybrid sampling and Bayesian optimization. Here we integrate additional data (targeted whole transcriptome sequencing) with the intent of enhancing the accuracy of the matrix completion. We find that our original approach to matrix completion with the larger matrix causes degradation in preserving rare but biologically important signals. To overcome this issue, we introduce ToxCompl$^+$ that integrates side information through learned embeddings and incorporates a new attention mechanism to weigh feature contributions to the predictions. ToxCompl$^+$ reduces test mean absolute error by 50% and, more importantly, reduces test maximum absolute error by over 75%. These results show that the joint use of auxiliary features and attention-enhanced low-rank modeling substantially improves large-scale toxicogenomics imputation. ToxCompl$^+$ enables more reliable downstream biological and toxicological interpretation. The complete set of Drugmatrix generated using ToxCompl$^+$ can be accessed

This manuscript has been authored by UT-Battelle, LLC under Contract No. DE-AC05-00OR22725 with the U.S. Department of Energy. The United States Government retains and the publisher, by accepting the article for publication, acknowledges that the United States Government retains a non-exclusive, paid-up, irrevocable, world-wide license to publish or reproduce the published form of this manuscript, or allow others to do so, for United States Government purposes. The Department of Energy will provide public access to these results of federally sponsored research in accordance with the DOE Public Access Plan.

H. R. Arabnia et al. (Eds.): CSCI 2025, CCIS 2967, pp. 393–407, 2027.
https://doi.org/10.1007/978-3-032-28454-9_27

here: https://rstudio.niehs.nih.gov/toxcompl/.
Track:CSCI-RTAI.

1 Introduction

DrugMatrix is a large-scale toxicogenomics resource that integrates conventional toxicological endpoints with whole-genome transcriptomics from the same animals. Covering more than 600 chemicals tested in short-term in vivo rat studies, it is the most comprehensive systematically generated dataset of its kind and has been widely used to derive toxicity signatures and mechanistic insights (e.g., [1–3]). Gene expression in DrugMatrix was profiled using two microarray platforms: CodeLink, which provided broader tissue coverage across eight tissues, and Affymetrix, which focused on four major tissues. Most studies also included paired clinical pathology and targeted histopathology, though the dataset emphasizes common target organs such as liver and kidney, hence there is a significant drop off in measured expression data in other tissues. As a result, about 88% of the endpoints are missing, particularly transcriptomic data for less commonly affected tissues. We previously developed a ToxCompl framework to infer the missing endpoints in the matrix [4]. As most of the endpoint values in DrugMatrix are close to zero, ToxCompl adopts hybrid sampling and Bayesian optimization to effectively preserve the rare signals that would otherwise be buried the tremendous amount of 'normal' data.

While DrugMatrix was originally generated using whole-transcriptome microarrays, a targeted sequencing technology, S1500+ from BioSpyder [5], was used to profile approximately 8000 frozen RNA samples across 8 tissues from the DrugMatrix sample set. To enable complete coverage of the genome from the samples assessed using the S1500+ an extrapolation tool, GenIE [6], was employed to predict whole transcriptome expression. The new whole transcriptome RNA sequencing data generated are referred in this paper as BioSpyderWT-GENIE (BSWT-G). The combination of BSWT-G data with the original DrugMatrix creates a larger sparse matrix twice the size of DrugMatrix. The extra entries in the matrix present new challenges, and ToxCompl fails to preserve rare and biologically meaningful signals. Details on the preparation and analysis of S1500+ data can be found in Appendix A.

Here we present ToxCompl$^+$ which builds on ToxCompl and leverages the additional BSWT-G data for imputation. Specifically, ToxCompl$^+$ augments the original ToxCompl by integrating side information such as drug names and dosages into matrix factorization and introducing a novel attention mechanism to weigh the contribution of latent factors to the prediction, with the aim of better preserving rare signals and reducing prediction errors. These additions enhance the model's ability to accurately predict missing values in the dataset.

2 Data Characteristics and Performance of ToxCompl

DrugMatrix alone has approximately 193,000 rows and 3,000 columns with about 580M entries, out of which roughly 72M or 12.5% entries contain measured values. In comparison, the combined DrugMatrix and BSWT-G, denoted as DSMatrix, have 375,000 rows and 3,000 columns. Out of a total of about 1.01 billion entries, roughly 125M or 12.3% are present.

Table 1. Amount of measured data from different tissues

	Total	LI	KI	HE	BM	BR	IN	SP	TM
DrugMatrix	72M	34.5M	19.0M	11.8M	2.7M	0.55M	0.17M	1.5M	1.5M
BSWT-G	53.5M	17.6M	15.9M	10.7M	6.5M	0.09M	0.07M	0.5M	1.6M

Table 1 shows the composition of data entries present in DSMatrix from different tissues, liver (*LI*), kidney (*KI*), heart (*HE*), bone marrow (*BM*), brain (*BR*), Intestine (*IN*), Spleen (*SP*), and thigh muscle (*TM*). In DSMatrix, there are relatively fewer *TM*, *SP*, *IN*, and *BR* data than *LI* and KI. *IN* and *BR* data are especially rare. BSWT-G contains relatively more abundant *BM* data than DrugMatrix. It is reasonable to assume the extra data may boost the prediction of *BM* prediction in DSMatrix.

In addition to the amount of measured data, the distribution of these data relative to the probes (rows) and treatments (columns) is different for the two datasets.

Figure 1 shows a heatmap of measured data present in DrugMatrix and BSWT-G. The upper half is for BSWT-G and the lower half is for DrugMatrix.

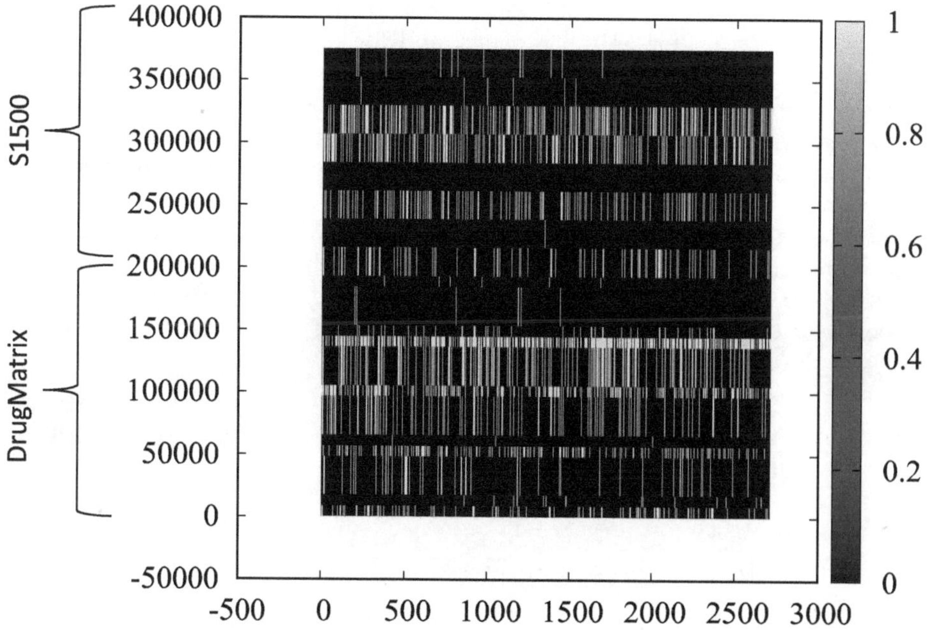

Fig. 1. Distribution of measured data in DrugMatrix and BSWT-G. The x-axis is for the treatments, and the y-axis is for the probes. Probes from 0 to 200,000 are for the original DrugMatrix, and from 200,000 onward are for BSWT-G

There are two values in the matrix, 1 for presence in bright yellow, and 0 for absence in black. The brighter the region, the more measured data are present for that region. In both halves of the heatmap, there are brighter horizontal bands interleaved with darker bands. For BSWT-G, there are some intricate patterns at the band centered around y = 300,000. Such patterns are not seen in DrugMatrix. For DrugMatrix, there is a bright stripe between y = 100,000 and 150,000 centered around x = 1700. This pattern is not observed in BSWT-G. The shared patterns between DrugMatrix and BSWT-G suggest that we may expect better prediction performance for certain probes than others.

ToxCompl recovers an matrix M through its low-rank factorization by $M = P \times Q$. Intuitively, P maps the gene expressions into a r dimensional latent space, and Q maps the treatments into a r dimensional latent space. P is of size $n_1 \times r$, and Q is of size $n_2 \times r$. The factorization of P and Q is formulated as a machine learning problem in ToxCompl, using the following objective function

$$min \sum_{i,j} (M_{i,j} - (bp_i + bt_j + P_{i,:} \cdot Q_{:,j}))^2$$

with $M_{i,j}$ being the observed entries, bp_i and bt_j being the biases for the probes and the treatments, respectively.

The key assumption of ToxCompl is that the full matrix is low-rank. As in ToxCompl, we investigate necessary conditions for us to apply completion on DSMatrix.

BSWT-G data do not introduce new treatments. So the fully matrix has a rank r no bigger than $n_2 = 3,000$. As in DrugMatrix, we expect $r \ll n_2$ as the 3,000 treatments are for only 636 drugs, and for each drug, roughly there are no more than 5 different treatment regimes. We also know some of these drugs may have similar or correlated expressions. If we look at the rows, there will be correlations between CodeLink and Affymetrix data. In addition, the same gene expression data is collected for 8 different organs. So we expect the full DSMatrix to be low-rank.

As in ToxCompl, the distribution of data is highly imbalanced not only in tissues and treatments but also in the range of values. If we categorize gene expression as extremely underexpressed (−2), underexpressed (−1), normal (0), overexpressed (1), extremely overexpressed (2), the distribution of measured values in these categories is shown in Table 2.

Table 2. The distribution of the categories in the categorical data and predicted test data by the original model.

Category −2	Category −1	Category 0	Category 1	Category 2
0.03%	4.09%	91.94%	3.88%	0.036%

In Table 2, most of the values are close to 0, while the whole range of data is between (−6, +6). There are rare but important signals in categories −2, −1, 1, and 2. ToxCompl applies hybrid sampling guided by Bayesian optimization to recover the matrix and preserve rare signals.

We apply the same ToxCompl implementation to recover DSMatrix. As in [4] we use the Adam optimizer with the default learning rate ($1e^{-3}$) and weight decay ($5e^{-4}$). As DSMatrix is much larger than DrugMatrix, we experiment with different r and the number of epochs. Intuitively, larger r and/or more epochs may yield better results for DSMatrix. As with DrugMatrix, the training is done with a grace period of 5.

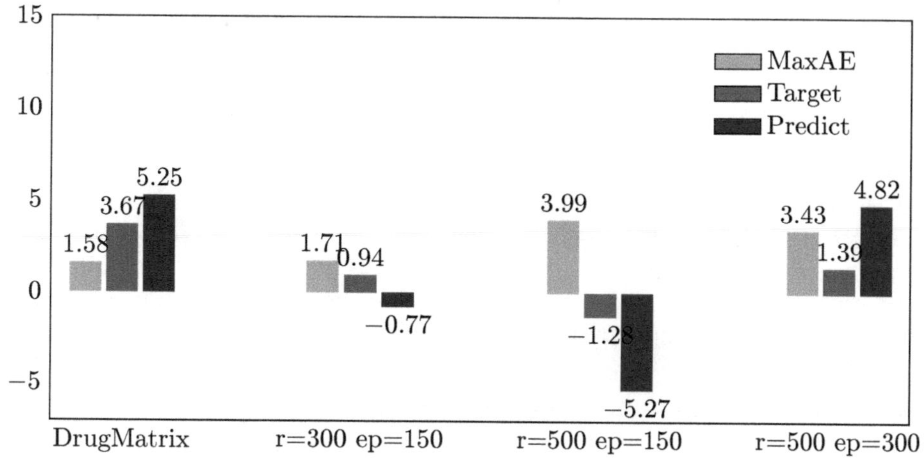

Fig. 2. Matrix completion results for DrugMatrix and various runs with different r and epochs (ep) on DSMatrix

Figure 2 shows the performance of ToxCompl with DrugMatrix and DSMatrix. Other than the factor r and the number of epochs (ep), the rest of the (hyper) parameters are the same. The leftmost group of bars are performance for DrugMatrix and the three groups of bars on the right shows the performance of ToxCompl on DSMatrix. Within group, mean absolute error (MAE) and maximum absolute error (MaxAE) that measures the maximum prediction error over the whole matrix are shown. The prediction and target for that specific MaxAE is also shown (i.e., MaxAE=max({abs(Predict-Target))}). The MAE with DrugMatrix is 0.05, and the MAE with DSMatrix either 0.04 or 0.05, a 20% or 40% improvement compared to 0.05 achieved with DrugMatrix alone. This demonstrates the overall benefit of having more measured data for imputing the missing ones.

However, significant degradation of MaxAE is observed with DSMatrix with r = 300 and ep = 150. MaxAE especially meaningful for ensuring rare signals are not lost in the prediction. From DrugMatrix to DSMatrix MaxAE increases from 1.58 to 1.71, signifying the challenge for ToxCompl to preserve rare signals with high fidelity on DSMatrix. The implication is clear when we inspect the Predict and Target values for this MaxAE. In this case the target happens to -0.77, signifying under-expressed gene expression, while predicted value is +0.94, signifying over-expressed gene expression. Sign flipping obviously changes biological meaning.

Increasing r and/or increasing the number of training epochs do not improve MaxAE, as shown by the two rightmost groups of bars in Fig. 2. While MAE improves, MaxAE actually gets worse.

From Fig. 2, it is clear that new approaches beyond ToxCompl are needed to effectively recover DSMatrix. The new approach needs to better preserve rare signals than ToxCompl. The challenge is likely the result of the heterogeneity of the matrix with biological data from three different platforms.

3 ToxCompl$^+$: An Improved Approach

We extend ToxCompl with two augmentations to tackle the challenge of preserving rare signals while maintaining low MAE. The first is to incorporate side information of the treatments and gene expressions into matrix factorization, and the second is to design a novel attention mechanism so that the contribution of the factors to the final prediction value is weighted non-linearly.

3.1 Incorporating Side Information

DSMatrix contains information such as drug names, dosage, and treatment duration that may be used as features to improve recovery result. Indeed side information (i.e., features derived from such information) has been used in several applications and is shown to be useful in practice. For example, Zhou et al. studied from a theoretical perspective the effect of side information for matrix completion, and showed that clean features can reduce sample complexity can be significantly [7]. Chiang et al. further analyzed the impact of noisy features on the performance of completion [8].

In DrugMatrix as well as DSMatrix, side features are associated with both the rows and columns. There are at least three side features, *Platform*, *Marker*, and *Organ*, for each row and another three features, *Drug*, *Duration*, and *Dosage*, for each column.

We embed each feature into a r dimensional vector, and for k features, we have additional k factor matrices. We model the interaction of P with each of the column feature factor matrix C^t, $t = 1, \cdots, k$ and vice versa with Q and row feature factor matrix R^t, $t = 1, \cdots, k$. Together with the P and Q matrices for the rows and columns, matrix completion becomes an optimization problem with the following objective

$$min \sum_{i,j} \left(M_{i,j} - \left(bp_i + bt_j + P_{i,:} \cdot Q_{:,j} + \sum_{t=1}^{k} P_{i,:} \cdot C^t_{:,j} + \sum_{t=1}^{k} R^t_{:,i} \cdot Q_{j,:} \right) \right)^2$$

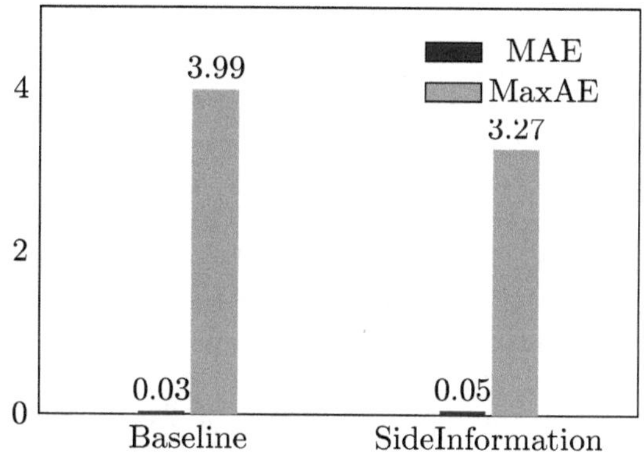

Fig. 3. Performance with side information on DSMatrix. The baseline is ToxCompl without using side information

with $M_{i,j}$ being the observed entries, bp_i and bt_j being the biases for the gene expressions and the treatments, respectively.

Figure 3 shows the prediction performance with side information on DSMatrix. We use $r = 300$ and train for 150 epochs with a grace period of 5. In the figure both MAE and MaxAE are shown. Leveraging side information reduces MaxAE from 3.99 to 3.27 but increases MAE from 0.03 to 0.05.

3.2 Attention Mechanism

Attention [9] is a powerful mechanism that revolutionizes learning in many research areas. Attention in learning mimics its cognitive counterpart and captures dependence (both spacial and temporal) between parts of the input. Attention is frequently used in deep neural networks such as transformers. Geng *et al.* shows that self-attention is not better than matrix decomposition/low-rank completion for recovery in global context for vision tasks [10]. Ruiz [11] extends the geometric matrix completion approach with recurrent multi-graph neural networks developed by Monti *et al.* [12] with a graph attention network approach. These works build graphs for corroborative filtering for diffusion. To our knowledge, no study has directly introduced attention to the low-rank factorization approach for matrix completion.

Intuitively, casting the matrix completion as a corroborative filtering problem where the prediction of a score or ranking that a 'user' gives to an 'item', the predicted score is entirely determined by the interaction between the two latent 'user' and 'item' vectors, and attention should be paid to how the interaction contributes to the final prediction. As a concrete example of the Netflix movie recommendation challenge, each 'movie item' may be represented by a vector with components such as genre, language, and length, and a 'user' has different preference scores to these components. The ultimate ranking may not be a simple linear combination of these interactions, and attention models this potentially non-linear combination.

The attention mechanism we introduce in ToxCompl$^+$ is illustrated in Fig. 4. In the figure, there is a matrix M to be recovered on the right with n_1 rows and n_2 columns. In our toxicogenomics setting, the rows R are the probes and the columns C are the treatments. The matrices P and Q are composed of r dimensional embeddings of elements in R and C, respectively. These embeddings are blue in the figure. The target $M_{i,j} = P_{i,:} \cdot Q_{:,j}$ if we apply vanilla completion. Instead, we introduce two more embeddings, $P'_{i,:}$ and $Q'_{:,j}$, that are also r dimensional. The element-wise product shown by the $\circ$ operator is computed between $P'_{i,:}$ and $Q'_{:,j}$. and the result is fed to an 'activation' function σ. After activation the resulting vector is combined with the result of $P_{i,:} \circ Q_{:,j}$ through a dot product. We use Softmax for the activation function σ.

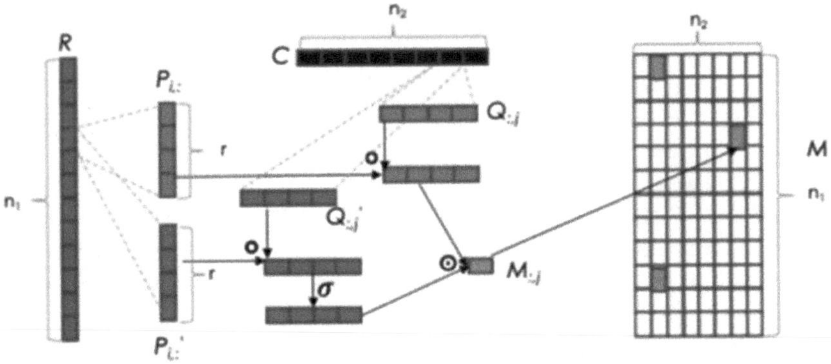

Fig. 4. Attention in ToxCompl$^+$. Four matrices, P, P', Q, and Q' are jointly learned, where P' and Q' are used to learn to weight the contributions from each of the r elements in $P_{i,:} \circ Q_{:,j}$. In the figure *eye* represents dot product

Formally, the objective for matrix completion with attention becomes

$$min \sum_{i,j} \left(M_{i,j} - \left(bp_i + bt_j + (P_{i,:} \circ Q_{:,j}) \cdot \sigma(P'_{i,:} \circ Q'_{:,j})\right)\right)^2$$

with $M_{i,j}$ being the observed entries. Instead of two matrices, we learn four matrices, P, P', Q and Q'. The number of matrices, or to be exact, the number of learn-able parameters, represents the power or the learning capacity of the model. For the same factor r, our formulation increases the capacity of the model. However, it is important to note that capacity alone does not necessarily yield better prediction performance. With ToxCompl, as shown in Fig. 2 the increased model capacity from increasing r from 300 to 500 does not yield good completion result. Mathematically, the extra 'activation' function σ introduces non-linearity into learning. The function σ operates on element-wise multiplications of P' and Q', and generates weights for the individual contribution of each of the r components of the latent factor to the final prediction for $M_{i,j}$. In predicting

$M_{i,j}$, the model attends to the individual contribution of the interaction between latent features from the rows and columns.

Note that in Fig. 4 attention is applied only to the interaction of P and Q. In completion with side information, attention mechanism can be similarly introduced to the interaction between P,Q, and the R, C matrices.

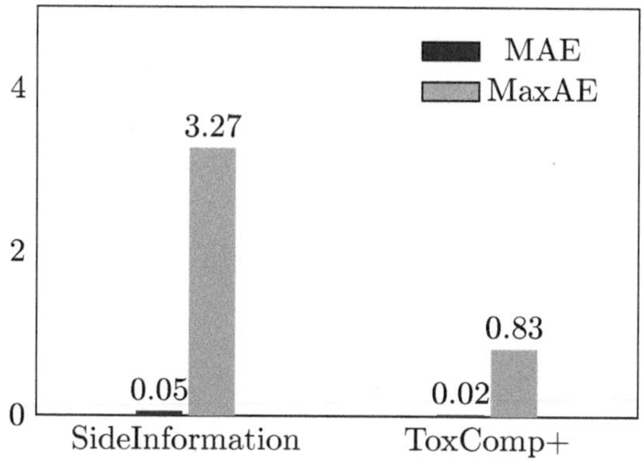

Fig. 5. Impact of attention on performance.

Figure 5 shows the impact of attention on the prediction performance. In the figure, ToxCompl$^+$ is with attention, and in comparison with SideInformation, the MAE improves by about 2.5 $\times$ while the MaxE improves by about 4 x.

4 Uncertainty in Predictions

As the predicted DSMatrix will be open to the public for the study of toxicogenomics of small-molecule drugs, we evaluate the performance of ToxCompl$^+$ with an additional metric, that is uncertainty, in the prediction.

We run ToxCompl$^+$ for 5 runs with different random seeds and measure the mean and the range of fluctuation of the predicted values. Figures 6, 7, and 8 shows for three drugs Carmustine, Indomethacin, and Clofibrate that are known to elicit strong toxicological responses.

Carmustine is an DNA alkylating nitrosourea chemotherapy agent whose delayed and cumulative bone marrow suppression is dose-limiting, and which carries risks of pulmonary fibrosis, hepatic and renal toxicity, and secondary malignancies [13]. Indomethacin is a non-steroidal anti-inflammatory drug whose adverse effects include gastrointestinal ulceration and bleeding, renal impairment, cardiovascular thrombotic events, and liver injury, with additional concerns about CNS disturbances [14]. Clofibrate is a fibrate lipid-lowering agent associated in rodents with peroxisome proliferation, hepatomegaly, and liver tumor formation, and in humans may increase gallstone risk [15]. We inspect the genes that respond the most strongly (i.e., with the largest absolute values) to these treatments. These gene expressions will be rare signals in DSMatrix as they deviate far from the mean value 0 of the matrix.

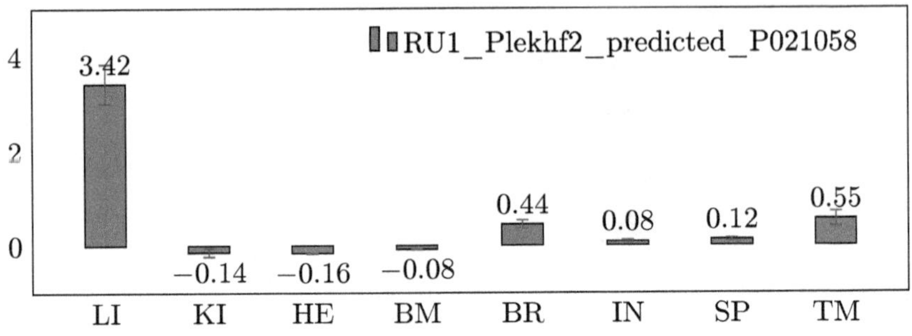

Fig. 6. CARMUSTINE-3d-4mg/kg

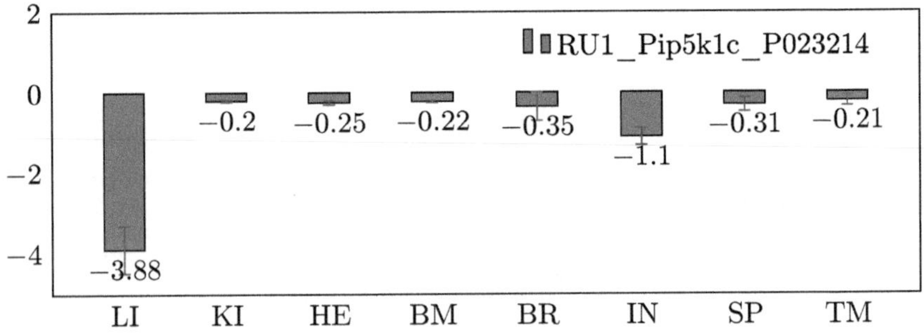

Fig. 7. INDOMETHACIN-1d-4.5mg/kg

Figures 6 and 7 show the predicted values for a strongly responding gene in all eight tissues for the CodeLink platform, and Fig. 8 shows the predicted values of a strongly responding gene in four tissues.

In Fig. 6 the most over-expressed gene, Plekhf2, happens to be in LI (3.42). Plekhf2 (Pleckstrin homology domain-containing family F member 2) is predicted to bind phosphatidylinositol and participate in endosome organization and endosome-to-lysosome transport [16]. Our prediction shows that it is also somewhat over-expressed in BR (0.44) and TM (0.55). We see that across all tissues, for this gene, the prediction values across multiple runs are fairly consistent.

In Fig. 7 the most under-expressed gene, Pip5k1c, also happens to be in LI (−3.88). Pip5k1c (phosphatidylinositol-4-phosphate 5-kinase type 1 gamma) in rats is predicted to possess 1-phosphatidylinositol-4-phosphate 5-kinase activity, with roles in regulating synaptic vesicle endocytosis, axonogenesis, and postsynaptic receptor internalization, and is active in the presynaptic endocytic zone membrane. In rat models, Pip5k1c is essential for mesenchymal stem cell differentiation into osteoblasts, controlling key aspects of the cytoskeleton, biomembrane dynamics, and calcium signaling to regulate bone formation [17–19]. Our prediction shows that it is also very under-expressed in IN (−1.1). For all tissues, the predicted values for Pip5klc across multiple runs are also fairly consistent.

In Fig. 8 the most over-expressed gene, Ctel (a.k.a. Acot1), is also in LI (3.76). Ctel (cytosolic thioesterase 1, alternatively known as Acyl-CoA thioesterase 1) is a cytosolic acyl-CoA thioesterase that hydrolyzes long-chain fatty acyl-CoAs to free fatty acids and CoA, playing a role in lipid metabolism in the rat heart and liver. In rat cardiac tissue, Ctel expression is significantly upregulated (e.g., by high-fat feeding or diabetes), regulated by PPARα signaling, and is thought to contribute to metabolic adaptation in the heart [20–22]. Our prediction shows that it is also over-expressed in HE (1.06). For all four tissues, the predicted values for gene Ctel across multiple runs are fairly consistent.

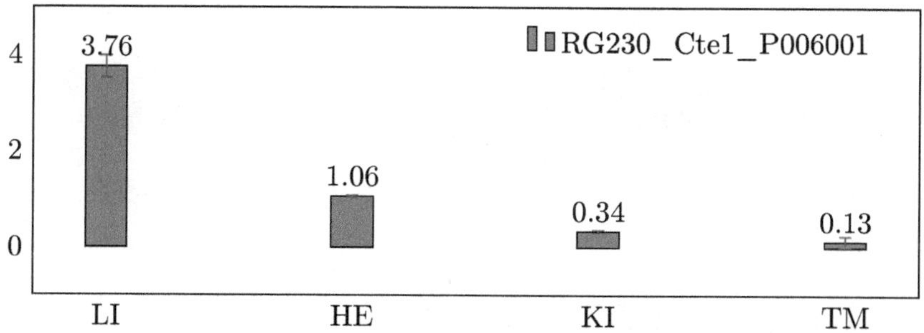

Fig. 8. CLOFIBRATE-3d-500mg/kg

We investigate the reduction of uncertainty by ToxCompl^{+} on DSMatrix against that of ToxCompl on DrugMatrix. We choose a representative treatment, RABEPRAZOLE-5d-1024mg/kg, from DrugMatrix, where there is no measured Affymetrix (RG230) data. We run the two implementations for 5 runs with different random seeds, and inspect the predicted values. If the range of fluctuation is significant (i.e., larger than the mean predicted value), we consider the prediction not confident. Figure 9 shows the ratio of confident predictions against all predictions for four tissues with the treatment RABEPRAZOLE-5d-1024mg/kg. Rabeprazole is a proton pump inhibitor (PPI) that suppresses gastric acid secretion by irreversibly binding to the gastric H^{+}/K^{+}-ATPase in parietal cells. In rat models, rabeprazole effectively reduces gastric acidity and is used experimentally to study peptic ulcer healing, gastroesophageal reflux, and gastric mucosal protection. Repeated or high-dose treatment in rats shows dose-dependent gastric pH elevation, hypergastrinemia, and ECL-cell hyperplasia, findings consistent with long-term PPI. In the liver it activates the Aryl hydrocarbon receptor. use [23–25].

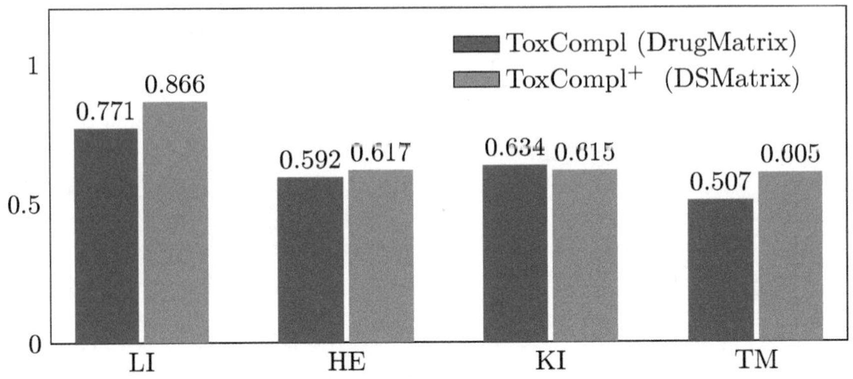

Fig. 9. Confidence for different tissues for treatment RABEPRAZOLE-5d-1024mg/kg, where there is no measurement from the Affymetrix platform

In Fig. 9, ToxCompl$^+$ significantly improves the percentage of confident predictions over ToxCompl for LI, HE, and TM. For LI and TM the improvement is about 9 and 10% point, respectively. Approximately 2% points of improvement are observed for HE. Interestingly, a slight degradation (2% points) is observed for KI.

5 Conclusion and Future Work

We have presented a new approach, ToxCompl$^+$, for predicting the missing values in DSMatrix that includes the new BSWT-G data in addition to the measurements in DrugMatrix. Our experiments show that the prior approach ToxCompl struggles to preserve the rare signals in the matrix that doubles in size, and through leveraging side information for learning and adopting an attention mechanism, ToxCompl$^+$ significantly outperforms ToxCompl in MAE, MaxAE, and reduced uncertainty.

In future work, as there is a wealth of information that can be further obtained, for example, for each drug, there is an associated SMILES string, a toxicology fingerprint, and a corresponding three-dimensional structure, to further improve predictions, we can explore long short-term memory networks (LSTM) or transformers on the strings, multi-layer-perceptrons (MLP) on the fingerprints, and graph neural networks (GNN) on the structures to learn latent representations of drugs in each column. We plan to experiment with these new directions in the future.

Acknowledgements. This research used resources of the Compute and Data Environment for Science (CADES) at the Oak Ridge National Laboratory, which is supported by the Office of Science of the U.S. Department of Energy under Contract No. DE-AC05-00OR22725. This research was also supported by the NIH, National Institute of Environmental Health Sciences, through Intramural Research Project ZIAES103385 and Interagency Agreement No. AES24003001-24-001 between NIEHS and Oak Ridge National Labs. Sciome was funded in part by Contract No. 75N96024C00003. The authors would like to acknowledge Andrew Rooney for his support and management of the interagency agreement between ORNL and NIEHS. The contributions of the NIH/NIEHS author(s) are considered Works of the United States Government. The findings and conclusions presented in this paper are those of the author(s) and do not necessarily reflect the views of the NIEHS, NIH or the U.S. Department of Health and Human Services.

A Alignment and Analysis Methods for S1500+ Data

A.1 Sequence Data Processing

TempO-Seq S1500+ was performed on 8,345 samples from DrugMatrix. Data is available at the Sequence Read Archive (SRA) via BioProject accession PRJNA1247997. FASTQ files were aligned to the probe sequences from the target platform using Bowtie version 1.2.2 with the following parameters: -v 3 -k 1 -m 1 –best –strata. This configuration allows up to 3 mismatches and reports the single best alignment. After alignment, the total sequenced reads, the percentage of reads aligning to the platform manifest, the alignment rate, and the percentage of expressed probes ($\geq$5 reads per probe) were calculated for each sample.

A.2 Sequencing Quality Checks and Outlier Removal

Samples were flagged for values below the following thresholds: sequencing depth $<$ 300 K, total alignment rate $<$ 40%, unique alignment rate $<$ 30%, number of aligned reads $<$ 300 K, percent of probes with at least five reads $<$ 30%. FastQC was additionally run on all samples to ensure adequate per base quality and per base N content. Principal component (PCA), hierarchical cluster, and inter-replicate correlation analysis was performed. 127 samples were flagged based on the alignment rate threshold and were removed from the analysis, resulting in a total of 8,218 samples for downstream analysis.

A.3 Data Normalization

The aligned read counts for attenuated probes were properly readjusted to calculate unattenuated equivalent counts using the attenuation factors provided in the platform manifest. To account for between-sample sequencing depth variation, unattenuated read counts were normalized at the probe level by applying reads per million normalization. A pseudo-read-count of 1.0 was added to each normalized expression value, and then, the values were log2 transformed.

A.4 Extrapolation

Extrapolation to the whole rat transcriptome was performed on the log2(TPM+1) normalized expression data using principal component regression (PCR)[1] technique from Sciome GeniE[2,3] version 3.0.4. The extrapolation training data included 937 DrugMatrix samples on which TempO-Seq Whole Transcriptome (WT) sequencing had been performed. The extrapolated whole transcriptome included all 22,794 probes on the WT platform.

A.5 Gene Set Enrichment Analysis

A customized implementation of conventional Gene Set Enrichment Analysis (GSEA) was employed to identify differentially expressed genes.[4] Contrasts for gene expression analysis were determined by matching samples from the same tissue using the following rules:

This resulted in 2,212 comparisons for which contrasts were derived. 113 additional comparisons were added by relieving some of the rules above to get to a total of at least 2 treated and 2 control samples per comparison.

References

1. Ganter, B., Snyder, R.D., Halbert, D.N., Lee, M.D.: Toxicogenomics in drug discovery and development: mechanistic analysis of compound/class-dependent effects using the drugmatrix® database. Pharmacogenomics (2006)
2. Svoboda, D.L., Saddler, T., Auerbach, S.S.: An overview of national toxicology program's toxicogenomic applications: drugmatrix and toxfx. In: Advances in Computational Toxicology: Methodologies and Applications in Regulatory Science, pp. 141–157 (2019)
3. Ganter, B., et al.: Day: development of a large-scale chemogenomics database to improve drug candidate selection and to understand mechanisms of chemical toxicity and action. J. Biotechnol. **119**(3), 219–244 (2005)
4. Cong, G., et al.: Completion of the drugmatrix toxicogenomics database using toxcompl. bioRxiv (2024)
5. Mav, D., et al.: A hybrid gene selection approach to create the s1500+ targeted gene sets for use in high-throughput transcriptomics. PloS ONE **13**(2), e0191105 (2018)
6. Genie. https://www.sciome.com/genie/
7. Zhou, T., Qian, H., Shen, Z., Zhang, C., Xu, C.: Tensor completion with side information: a Riemannian manifold approach. In: IJCAI, pp. 3539–3545 (2017)
8. Chiang, K.-Y., Hsieh, C.-J., Dhillon, I.S.: Matrix completion with noisy side information. Adv. Neural Inf. Process. Syst. **28** (2015)
9. Vaswani, A., et al.: Attention is all you need. Adv. Neural Inf. Process. Syst. **30** (2017)
10. Geng, Z., Guo, M.-H., Chen, H., Li, X., Wei, K., Lin, Z.: Is attention better than matrix decomposition? arXiv preprint arXiv:2109.04553 (2021)
11. Rodrigo, I., Ruiz: geometric matrix completion with graph attention networks. PhD thesis, Massachusetts Institute of Technology (2020)

12. Monti, F., Bronstein, M., Bresson, X.: Geometric matrix completion with recurrent multi-graph neural networks. Adv. Neural Inf. Process. Syst. **30** (2017)
13. Drugs.com. Carmustine monograph (2025). https://www.drugs.com/monograph/carmustine.html. Accessed 04 Sept 2025
14. DrugBank Online. Indomethacin (2025). https://go.drugbank.com/drugs/DB00328. Accessed 04 Sept 2025
15. Medscape. Clofibrate overview (-09-04). https://emedicine.medscape.com/article/175667-overview. Accessed 04 Sept 2025
16. UniProt. Plekhf2 – pleckstrin homology and fyve domain-containing family f member 2 (2025). https://www.uniprot.org/uniprotkb/Q9H8W4/entry. Accessed 04 Sept 2025
17. Yan, Q., et al.: Loss of phosphatidylinositol-4-phosphate 5-kinase type-1 gamma (pip5k1c) impairs mesenchymal stem cell osteoblast differentiation and bone formation. J. Bone Mineral Res. (2022). Accessed 05 Sept 2025
18. Lin, S., et al.: Pip5k1c regulates cytoskeleton, biomembrane, and ca^{2+} release in cells—implications for synaptic function and bone. Sci. Rep. (2024). Accessed 05 Sept 2025
19. Rat Genome Database. Pip5k1c (phosphatidylinositol-4-phosphate 5-kinase type 1 gamma) – rattus norvegicus (2025). https://rgd.mcw.edu/rgdweb/report/gene/main.html?id=1309938. Accessed 05 Sept 2025
20. Durgan, D.J., et al.: Distinct transcriptional regulation of long-chain acyl-coa synthetase isoforms and cytosolic thioesterase 1 in the rodent heart by fatty acids and insulin. Am. J. Physiol.–Heart Circulat. Physiol. **290**(6), H2480–H2497 (2006). Accessed 04 Sept 2025
21. van Lunteren, E., Moyer, M.P.: Gene expression profiling in the type 1 diabetes rat diaphragm reveals large-scale metabolic and stress-related transcriptional changes including dramatic increases in cytosolic (cte1) and mitochondrial (mte1) acyl-coa thioesterase genes. PLoS ONE **4**(5), e7832 (2009). Accessed 04 Sept 2025
22. Hunt, M.C., et al.: Acyl-coa thioesterases belong to a novel gene family and are strongly induced in rat and mouse liver. J. Lipid Res. (2000). Accessed 04 Sept 2025
23. Miyashita, T., Shah, F.A., Miwa, K., et al.: Rabeprazole impedes development of reflux-induced esophageal cancer in a rat model. Digest. Dis. Sci. **56**, 1309–1314 (2011). Accessed 05 Sept 2025
24. Kakimoto, Y., Yonei, Y., et al.: Cytoprotective effect of rabeprazole against ethanol-induced gastric mucosal damage in rats: possible involvement of nitric oxide. Jpn. J. Pharmacol. (2000). Accessed 05 Sept 2025
25. Tari, A., et al.: Effects of rabeprazole versus omeprazole on enterochromaffin-like and parietal cells in rats. J. Gastroenterol. **36**(6), 375–385 (2001). Accessed 05 Sept 2025

Hybrid CNN-ViT with Multi-scale Feature Fusion for Multi-class Chest X-Ray Diagnosis

Momina Liaqat Ali(✉) and Xin Yang

Middle Tennessee State University, Murfreesboro, TN 37130, USA
ma2mp@mtmail.mtsu.edu, Xin.Yang@mtsu.edu

Abstract. In this paper, a hybrid CNN-Vision Transformer architecture for automated chest X-ray classification is presented across five respiratory conditions: COVID-19,tuberculosis, pneumonia,normal and lung opacity cases. The proposed framework combines dynamic patch size selection with adaptive patch embedding, hierarchical attention mechanisms which combine both the local and global feature interactions, and medical-specific feature enhancement modules to target edge, texture, shape, and intensity characteristics. The proposed architecture integrates multi-scale feature pyramid fusion using an improved ResNet-style CNN backbone and utilizes dynamic token mixing for more enhanced feature interactions. A dataset containing 24,360 chest X-ray images was constructed in which the class distributions were ranging from 5.5% to 41.8%. Detailed experimental evaluation demonstrates strong performance with 95.57% validation accuracy which is comparable to the state-of-the-art baselines. The model achieved 96.22% precision and 95.71% F1-score. Extensive cross-validation further confirms the model stability while successfully addressing the issue of class imbalance and gives a robust framework to conduct an automated respiratory disease diagnosis.

Keywords: Deep Learning · Convolutional Neural Network (CNN) · Medical Imaging · COVID-19 · Classification

1 Introduction

In recent years, medical imaging has become an important diagnostic tool in which chest X-ray (CXR) imaging has served as one of the leading edge modalities specifically for respiratory disease diagnosis since it's both accessible and cost-effective [1]. The COVID-19 pandemic, further highlighted the necessity of designing and creating automated diagnostic systems, as inter-observer variability and overwhelming case volumes were considered as huge challenges of manual interpretation during such times [2]. Traditionally, most of the computer-aided diagnosis frameworks these days rely on CNNs for medical image analysis [3]. Moreover, Vision Transformers (ViTs) have shown exceptional performance by using self-attention mechanisms [4]. Nevertheless, direct use of ViT for medical

H. R. Arabnia et al. (Eds.): CSCI 2025, CCIS 2967, pp. 408–422, 2027.
https://doi.org/10.1007/978-3-032-28454-9_28

imaging comes with numerous challenges which include both the computational complexity and the lack of domain-specific adaptations [5]. Since medical images demonstrate distinctive properties among which specific anatomical structures and pathological patterns require extensive multi-scale feature analysis [6]. Moreover, class imbalance specially in medical datasets presents an additional challenge while training deep learning models [7]. Figure 1 illustrates sample images taken from the dataset.

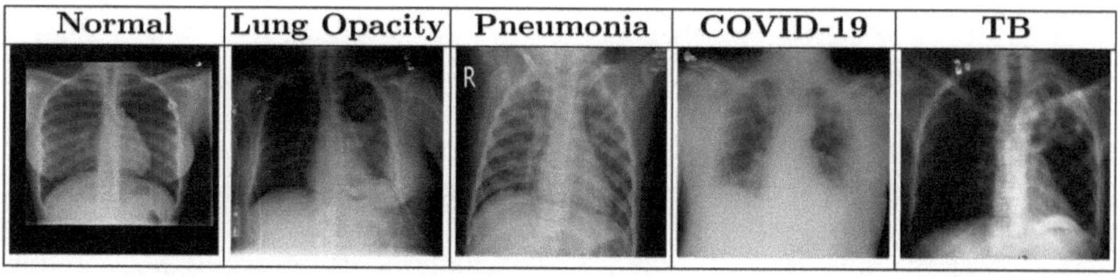

Fig. 1. Chest X-Ray sample images from the dataset from each of the five classes: Normal, Lung Opacity, Pneumonia, COVID-19, and TB

Recent literature shows that the incorporation of medical-specific attention mechanisms and multi-scale analysis greatly enhances the diagnostic performance [8]. Additionally, the domain-specific feature enhancement techniques are considered beneficial for tasks like respiratory disease classification [9,10].

In this work, we propose a hybrid CNN-Vision Transformer architecture (CNN-ViT) to carry-out automated chest X-ray classification in five lung conditions which are: COVID-19, tuberculosis, pneumonia, normal and lung opacity. Our main contributions are: (1) a hybrid CNN-ViT model along with adaptive patch embedding, (2) a framework for hierarchical attention mechanism, (3) modules for medical-specific feature enhancement, and (4) an extensive enhancement in the dataset by tuberculosis class integration.

2 Related Work

2.1 Medical Image Diagnosis and Chest X-Ray Classification

Deep learning based architectures for lung disease analysis have attained momentum, which specifically comes from scalable diagnostic needs [11]. Conventional machine learning based techniques along with handcrafted features faced challenges in medical imaging complexity [12], this leads to the paradigm of deep learning which enables automatic feature extraction and improved classification [13]. Baseline studies developed various chest radiograph benchmarks with the help of different frameworks [14], which inspect architectures ranging from shallow networks designed for pathology specific tasks to deeper frameworks for multi-class diagnosis [15].

2.2 CNN Based Medical Image Datasets

For medical image analysis, CNN architectures lead this field through spatial hierarchy. ResNet based models surpass other models in chest X-ray classification task with the help of skip connections [16]. On the other hand, DenseNet architectures show promising results because of the idea of dense connectivity [17]. Furthermore, efficientNet models also have shown leading performance in trade-offs for accuracy and efficiency [18], and Inception models show multi-scale convolutions for feature detection [19].

2.3 Transformer Architectures in Medical Image Analysis

Transformer architectures have successfully adapted NLP to computer vision by using Vision Transformers. They divide images into patches and then process them in the form of sequences to grasp the overall global dependencies using self-attention mechanisms [20,21]. Although medical imaging has carefully explored most of the transformer approaches [22,23], hybrid CNN-Transformer architectures have come to light as a leading standard by carefully combining the CNN's local feature extraction abilities with the Transformer's global modeling [24,25]. This particular approach is significantly amazing for medical imaging tasks because of the multi-scale nature of pathological shapes in which diseases show some localized lesions, diffuse patterns, and complex combinations. These all require not just the detailed local analysis but also global context understanding [26,27].

2.4 Multi-scale Feature Extraction and Hierarchical Learning

Since pathological images have multi-scale nature which has lead to specialized feature extraction techniques. Furthermore, the Feature Pyramid Networks (FPN) and the hierarchical approaches let the models capture fine-grained details and important contextual information [28,29]. Attention-based fusion techniques merge multi-scale features by weighting the representations derived from diagnostic tasks [30]. Recent techniques have shown limitations which include poor transformer optimization for medical imaging tasks, restricted multi-scale attention fusion, and insufficient domain-specific improvements. Our proposed methodology presents solution to these limitations using novel architectural plan which blend transformer's global modeling with medical image optimizations.

3 Methodology

We proposed a hybrid CNN-Vision Transformer model for automated chest X-ray classification over all the five respiratory conditions: COVID-19, pneumonia, tuberculosis, lung opacity, and normal cases. Our technique combines CNN local feature extraction with Vision Transformer global modeling using the medical imaging based innovations.

3.1 Architectural Design

Hybrid CNN-Vision Transformer Framework. Our hierarchical architecture fuses an improved ResNet inspired CNN backbone for the local features, multi-scale fusion, medical-specific enhancements, and the Transformer blocks for global context. The overall workflow is shown in Fig. 2 below.

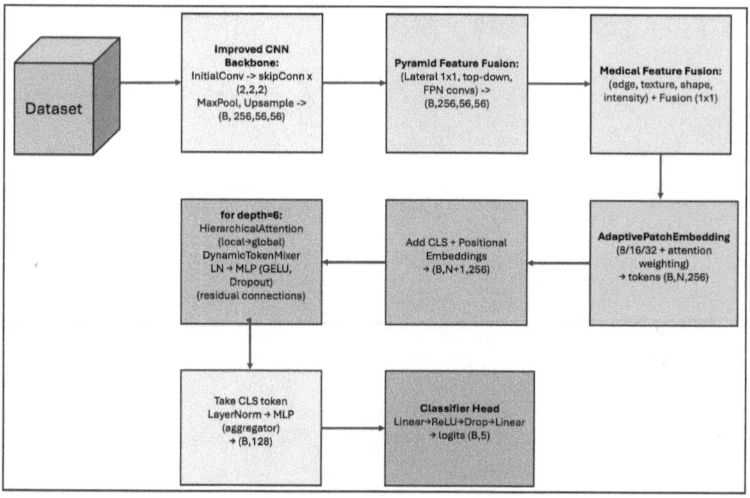

Fig. 2. Proposed hybrid CNN-ViT architecture for multi-scale five-class chest X-ray diagnosis.

Improved CNN Backbone. The ResNet-inspired backbone includes an initial 7×7 convolution with stride 2, batch normalization, ReLU, and max pooling (224×224 to 56×56), followed by three residual block layers (64, 128, 256 channels) with skip connections. Each layer captures multi-scale features (56×56, 28×28, 14×14) for feature pyramid construction.

Multi-scale Feature Pyramid Fusion. Our FPN-inspired mechanism integrates multi-scale features through 1×1 convolutions for channel unification (256 channels), a top-down pathway combining upsampled high-level with low-level features via element-wise addition, 3×3 convolutions for refinement, and final integration by upsampling to 56×56 resolution before concatenation.

3.2 Novel Architectural Components

Adaptive Patch Embedding. Our mechanism addresses fixed patch size limitations by applying three patch sizes (8×8, 16×16, 32×32) simultaneously. A learnable attention mechanism computes weights for each patch size based on global pooled features, enabling dynamic fusion that emphasizes the most relevant scale for each input. The detailed architecture for patch embedding is shown in Fig. 3 below.

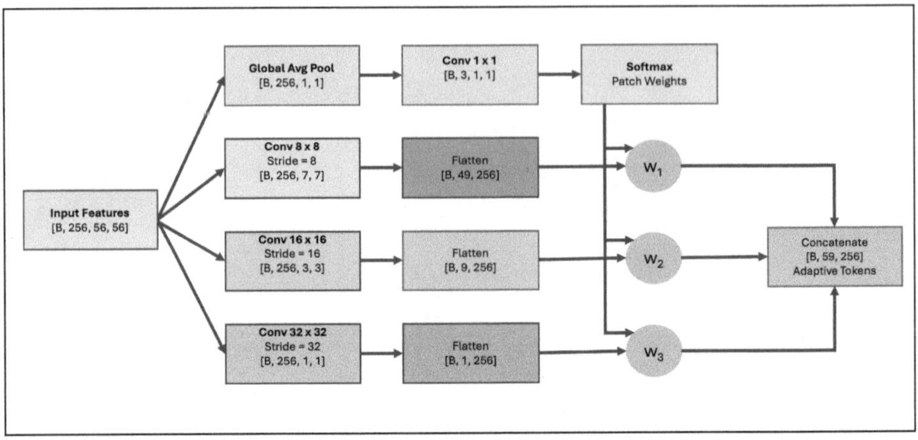

Fig. 3. Adaptive Patch Embedding Architecture uses three parallel convolution branches with patch sizes 8×8, 16×16, and 32×32, weighted by content-adaptive attention scores and concatenated into 59 adaptive tokens of dimension 256.

Medical-Specific Feature Enhancement. We introduce specialized enhancement modules targeting anatomical and pathological features through parallel processing branches: edge detection (3×3 convolutions) for boundary information, texture analysis (5×5 convolutions) for pathological patterns, shape analysis (7×7 convolutions) for morphological features, and intensity analysis (1×1 convolutions) for pixel-wise variations. All branch outputs are concatenated and fused through 1×1 convolution with residual connections preserving original features. An illustration of how medical specific feature enhancement is carried out is shown in Fig. 4 below.

Hierarchical Attention Mechanism. We propose a hierarchical attention framework balancing computational efficiency with modeling capability through dual-level processing. Local attention applies self-attention within 4×4 patch windows for efficient spatial relationships, while global attention processes window representatives (via average pooling) to model long-range dependencies. Local and global outputs are integrated through residual connections and layer normalization. The detailed hierarchical attention architecture is shown in Fig. 5 below.

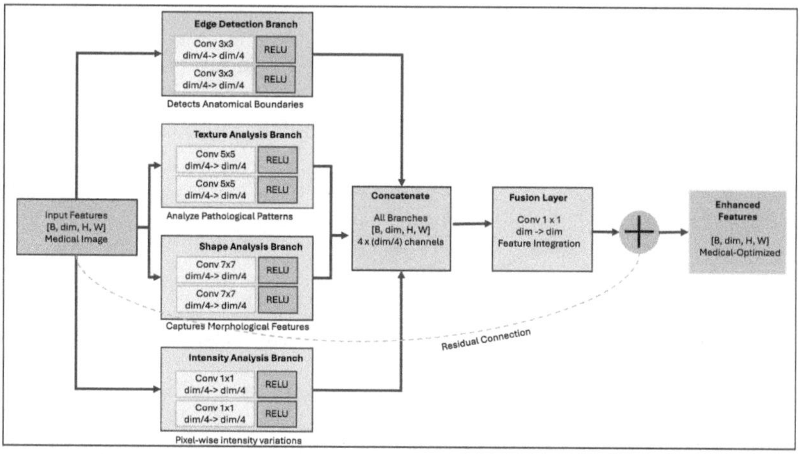

Fig. 4. Medical-Specific Feature Enhancement Architecture processes input features through four parallel branches, with outputs concatenated and passed through a 1×1 fusion layer. A residual connection preserves original information, producing enhanced features [B, dim, H, W] optimized for medical image analysis.

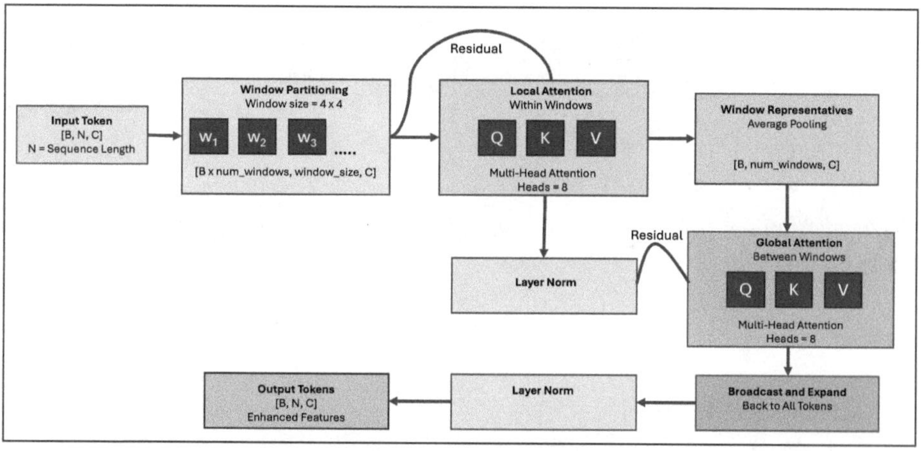

Fig. 5. Hierarchical Attention Architecture has input tokens which are processed with local window attention and global inter-window attention, combined via residual connections to produce enhanced output tokens $[B, N, C]$ that capture both local and global context.

Dynamic Token Mixing. We introduce a dynamic token mixing mechanism to enhance feature interactions based on content similarity. The model learns cluster centers in feature space to group similar tokens using cosine distance, then applies adaptive mixing within each cluster using learned weights for content-adaptive feature aggregation.

3.3 Training Strategy

Data Augmentation. We use medical imaging augmentation techniques which take care of diagnostic relevance while increasing model robustness. Our traditional alterations preserve clinical usefulness while enhancing dataset variety:

- Geometric Transformations: Light rotations (±5°), translations (±10%), and horizontal flips of almost (30%)
- Intensity Variations: Conventional brightness, contrast, and saturation adaptations.
- Gaussian Blur: Blurring to simulate imaging variants
- Random Erasing: Small patch blockage (2–10% image area) for robustness

Our optimization technique fuses adaptive learning with advanced regularization methods. We incorporate adaptive optimization, dynamic learning rate scheduling, and precision-enhanced computation for stable convergence. Detailed training arrangements are shown in Table 1.

Table 1. Training Configuration

Technique	Configuration	Purpose
AdamW Optimizer	Weight decay: 0.05, β: (0.9, 0.999)	Stable convergence
Learning Rate	Cosine annealing (T_0=10, T_{mult}=2)	Escape local minima
Label Smoothing	Cross-entropy with ϵ=0.1	Prevent overconfidence
Mixed Precision	16-bit floating-point	Memory efficiency
Dropout	30% in MLP layers	Prevent overfitting
Gradient Clipping	L2-norm (max=0.5)	Prevent exploding gradients
Early Stopping	30 epochs patience	Optimal convergence

3.4 Model Evaluation

The evaluation framework we used surrounds thorough dataset preparation, robust performance metrics, and strong inference schemes for the valid evaluation with clinically appropriate standards.

Dataset Preparation. The dataset was constructed using images from Kaggle through collaborative attempts between researchers from Bangladesh, Malaysia, Pakistan, and Qatar. Launched in two stages, the dataset initially hold 219 COVID-19, 1,341 Normal, and 1,345 Pneumonia images, then expanded to include Lung Opacity class with 3,616 COVID-19, 10,192 Normal, 6,012 Lung Opacity, and 1,345 Pneumonia images. We included an additional class which is Tuberculosis (TB) by collecting 3,195 TB images using an supplementary Kaggle dataset. The final merged dataset contains 24,360 chest X-ray images: Normal

(10,192), Lung Opacity (6,012), COVID-19 (3,616), Tuberculosis (3,195), and Pneumonia (1,345). All images were normalized to PNG format at 224×224 pixels resolution. The final dataset formation and class distribution are encapsulated in Table 2.

Table 2. Dataset Class Distribution

Class	Number of Images	Percentage
Normal	10,192	41.84%
Lung Opacity	6,012	24.68%
COVID-19	3,616	14.84%
Tuberculosis	3,195	13.12%
Pneumonia	1,345	5.52%
Total	**24,360**	**100%**

The dataset shows intrinsic class imbalance, with Normal images representing 41.8% of the total collection of images, while Pneumonia constitutes the smallest class at 5.5%. To address this significant imbalance during training, we used weighted loss functions using inverse frequency weighting and used stratified sampling for train-validation splits (80%-20%) which maintains proportional class distribution across both subsets. Dataset configuration is shown in Table 3 below.

Table 3. Dataset Configuration

Parameter	Specification
Dataset Classes	COVID-19, Pneumonia, TB, Lung Opacity, Normal
Training Split	80% (Stratified sampling)
Validation Split	20% (Stratified sampling)
Sampling Strategy	Maintains class distribution across splits

Data Preprocessing. Prior to model training, several preprocessing steps were applied to ensure data consistency and quality. All images were converted to grayscale PNG format and resized to 224×224 pixels. The Tuberculosis images, which were initially in JPG format, underwent format standardization and were organized into the same directory structure as other classes.

Performance Metrics. Our evaluation employs multiple complementary metrics to assess both overall model performance and disease-specific diagnostic capabilities, providing comprehensive insights into classification accuracy and clinical utility. The performance metrics used in this study are shown in equations below.

Accuracy:

$$\text{Accuracy} = \frac{TP + TN}{TP + TN + FP + FN} \tag{1}$$

Precision:

$$\text{Precision} = \frac{TP}{TP + FP} \tag{2}$$

Recall (Sensitivity):

$$\text{Recall} = \frac{TP}{TP + FN} \tag{3}$$

Specificity:

$$\text{Specificity} = \frac{TN}{TN + FP} \tag{4}$$

F1-Score:

$$\text{F1-Score} = \frac{2 \times \text{Precision} \times \text{Recall}}{\text{Precision} + \text{Recall}} = \frac{2TP}{2TP + FP + FN} \tag{5}$$

Test-Time Augmentation. During inference phased, we used Test-Time Augmentation (TTA) techniques to increase prediction dependability and give added performance insights using ensemble techniques and variance analysis. For TTA, authentic and horizontally flipped images together with ensemble averaging was employed which averaged logit over augmented inputs.

The complete architecture is implemented using modern deep learning frameworks with optimized configurations for medical image analysis, balancing computational efficiency with representational capacity for clinical deployment scenarios.

4 Results

4.1 Experimental Setup and Model Comparison Framework

The training of our proposed model and comparison models was carried out on an NVIDIA RTX A5000 GPU node within a high-performance computing (HPC) cluster. Table 4 shows the system configuration used to ensure consistent and reproducible results.

Table 4. GPU Hardware and Training Environment Configuration

Specification	Details
GPU Model	NVIDIA RTX A5000
GPU Memory	24 GB
Framework	PyTorch Lightning (Python)

We conducted a comprehensive comparative evaluation of our proposed model against five state-of-the-art baseline models representing the three major architectural paradigms in medical image analysis. To validate our hybrid CNN-Transformer approach, we selected models from each paradigm: pure CNN architectures (ResNet-18, InceptionNet-v3, DenseNet-121) to assess the contribution of our transformer components, and pure Vision Transformer architectures (SWIN Transformer Tiny, PVT-Small) to evaluate the benefits of our CNN backbone integration. This technique comparison lets us demonstrate that the proposed hybrid design efficiently fuses the power of both paradigms while minimizing their separate constraints. All models were trained using the same stratified dataset of 24,360 chest X-ray images over five lung diseases using the same training protocols and hyperparameters.

4.2 Overall Performance Analysis

Our hybrid CNN-Vision Transformer gained the maximum validation accuracy of 95.57%, showing better performance as compared to all baseline models. The overall performance ranking is shown in Table 5 and Table 6 in which we showed the comparison of the proposed model with other models in both CNN and Transformer categories.

We performed 10 rounds of 5-fold cross-validation to rigorously assess model stability and generalization performance. This strategy yielded a total of 50 model evaluations, ensuring that the reported results are not dependent on a particular data split. Figure 6 illustrates the stability of Precision, Recall, F1, and Specificity across 10 repeated cross-validation runs. The consistently high values and narrow variation confirm the robustness of our model across all classes.

The architectural validation confirmed the importance of multi-scale feature processing in medical image analysis. Our hybrid CNN-ViT outperformed alternatives through adaptive patch embedding and hierarchical attention, surpassing SWIN's window-based and PVT's spatial-reduction methods. Moreover, Medical-specific feature enhancements and dynamic token mixing gave added gains, showing the efficacy of merging architectural alteration with domain-specific optimizations. Furthermore, the cross-validation results confirm our model's stability and robustness.

Table 5. Comprehensive Performance Metrics Comparison

Model	Class	Precision	Recall	F1-Score	Accuracy
Hybrid CNN-ViT	Normal	0.9783	0.9364	0.9569	0.9511
	COVID	0.9118	0.9285	0.9200	0.9632
	Pneumonia	0.9495	0.9588	0.9541	0.9513
	Lung Opacity	0.9906	0.9859	0.9882	0.9984
	TB	0.9808	0.9517	0.9660	0.9977
	Overall	**0.9622**	**0.9522**	**0.9571**	**0.9511**
ResNet-18	Normal	0.9800	0.9502	0.9649	0.9456
	COVID	0.9309	0.8960	0.9131	0.9677
	Pneumonia	0.9234	0.9642	0.9434	0.9385
	Lung Opacity	0.9921	0.9859	0.9890	0.9987
	TB	0.9880	0.9182	0.9518	0.9985
	Overall	**0.9629**	**0.9429**	**0.9524**	**0.9456**
InceptionNet-v3	Normal	0.9575	0.9350	0.9461	0.9390
	COVID	0.9008	0.9143	0.9075	0.9571
	Pneumonia	0.9395	0.9446	0.9420	0.9465
	Lung Opacity	0.9809	0.9624	0.9716	0.9975
	TB	0.9628	0.9628	0.9628	0.9970
	Overall	**0.9483**	**0.9438**	**0.9460**	**0.9390**
DenseNet-121	Normal	0.9532	0.9585	0.9559	0.9306
	COVID	0.9834	0.7879	0.8748	0.9849
	Pneumonia	0.8837	0.9882	0.9331	0.9102
	Lung Opacity	0.9723	0.9875	0.9798	0.9968
	TB	0.9802	0.9219	0.9502	0.9977
	Overall	**0.9546**	**0.9288**	**0.9387**	**0.9306**
PVT-Small	Normal	0.9348	0.7732	0.8463	0.8869
	COVID	0.8373	0.8478	0.8425	0.9224
	Pneumonia	0.8660	0.9421	0.9025	0.8854
	Lung Opacity	0.9850	0.9264	0.9548	0.9966
	TB	0.9664	0.8550	0.9073	0.9961
	Overall	**0.9179**	**0.8689**	**0.8907**	**0.8869**
SWIN-Tiny	Normal	0.9567	0.9200	0.9380	0.8593
	COVID	0.7348	0.7850	0.7591	0.8958
	Pneumonia	0.8378	0.8867	0.8615	0.8854
	Lung Opacity	0.7892	0.7300	0.7584	0.9458
	TB	0.9882	0.9750	0.9815	0.9958
	Overall	**0.8613**	**0.8593**	**0.8597**	**0.8593**

Table 6. Summary Statistics across Models

Metric	Novel ViT	ResNet-18	InceptionNet	DenseNet	PVT	SWIN
Accuracy	95.11%	94.56%	93.90%	93.06%	88.69%	85.93%
Precision	96.22%	96.29%	94.83%	95.46%	91.79%	86.13%
Recall	95.22%	94.29%	94.38%	92.88%	86.89%	85.93%
F1-Score	95.71%	95.24%	94.60%	93.87%	89.07%	85.97%
Specificity	97.94%	97.82%	97.65%	97.38%	95.49%	93.96%

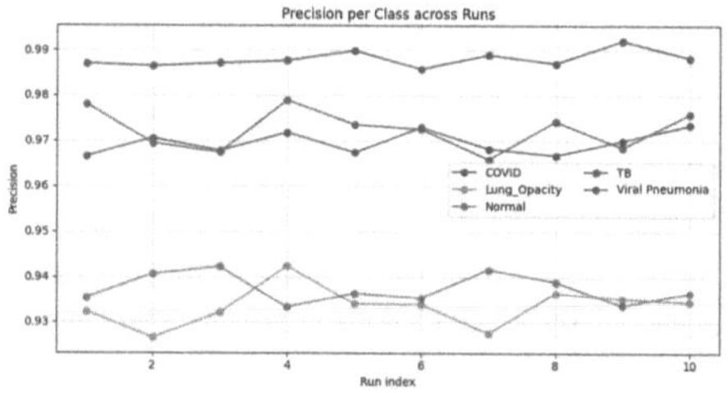

(a) Precision

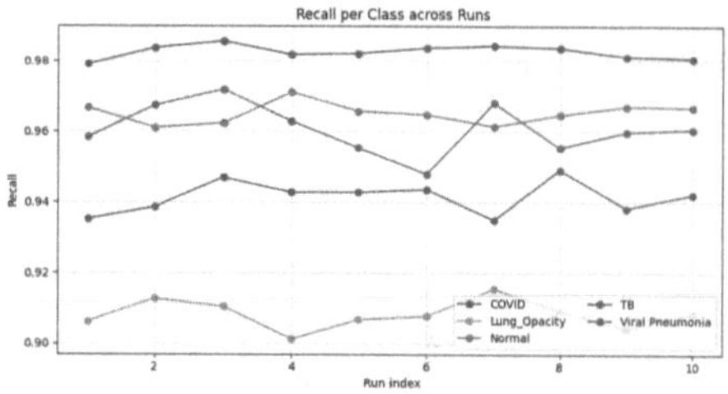

(b) Recall

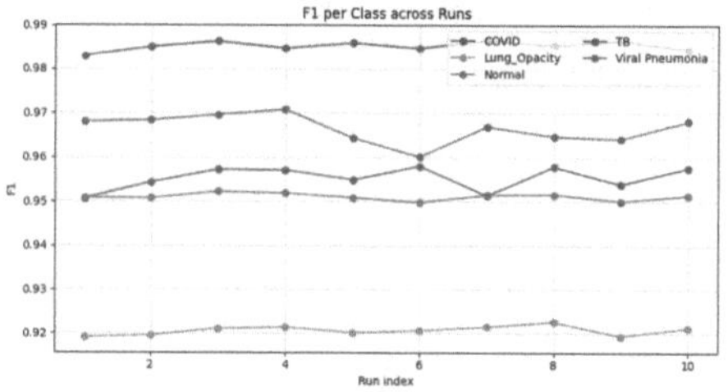

(c) F1-score

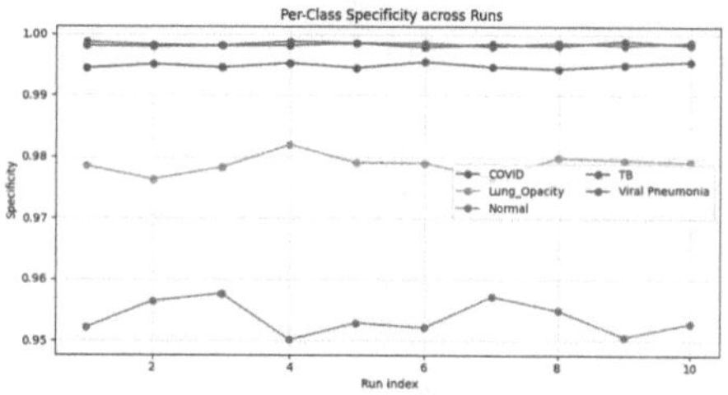

(d) Specificity

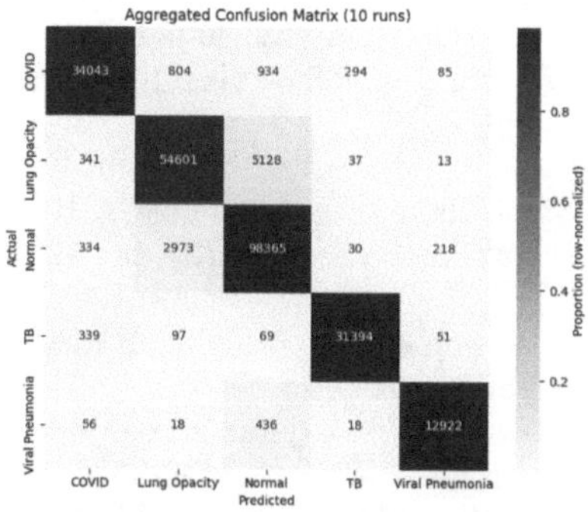

(e) Aggregated Confusion Matrix

Fig. 6. Performance evaluation across 10 repeated 5-fold cross-validation runs. The top two rows show per-class Precision, Recall, F1-score, and Specificity curves. The bottom row shows the aggregated confusion matrix combining all runs.

5 Conclusion

This study shows a hybrid CNN-Vision Transformer model which achieves 95.57% validation accuracy for automated chest X-ray classification over five respiratory disease conditions. Our framework combines adaptive patch embedding, medical-specific feature enhancement, and hierarchical attention mechanisms to efficiently merges local feature extraction with global context modeling. Thorough evaluation shows leading performance over other state-of-the-art baselines, and also outperforming other prominent models. The robust 50-fold cross-validation results approve model stability for clinical deployment, with high specificity (97.94%) specifically suited for medical tasks.

Current techniques essentially depend on standard sized feature extraction which limits their capability to capture multi-scale pathological design pivotal for accurate diagnosis. Most current techniques do not contain domain-specific architectural elements customized for medical imaging attributes, which result in suboptimal feature representation for radiological patterns. Moreover, many state-of-the-art models give inadequate uncertainty quantification, obstruct clinical reliability and interpretability. The computational cost of normal transformer architectures also give challenges for real-time clinical deployment. In the future, we project to enhance this research using datasets obtained from multiple institutes to address demographic differences and implement Bayesian uncertainty quantification for increased clinical decision support.

References

1. Siddiqi, R., Javaid, S.: Deep learning for pneumonia detection in chest X-ray images: a comprehensive survey. J. Imaging **10**(8), 176 (2024)
2. Kapoor, A., Guha, S., Das, M.K., Goswami, K.C., Yadav, R.: Digital healthcare: the only solution for better healthcare during COVID-19 pandemic? Indian Heart J. **72**(2), 61–64 (2020)
3. Guetari, R., Ayari, H., Sakly, H.: Computer-aided diagnosis systems: a comparative study of classical machine learning versus deep learning-based approaches. Knowl. Inf. Syst. **65**(10), 3881–3921 (2023)
4. Nakashima, K., Kataoka, H., Matsumoto, A., Iwata, K., Inoue, N., Satoh, Y.: Can vision transformers learn without natural images?. In: Proceedings of the AAAI Conference on Artificial Intelligence, vol. 36(2), pp. 1990–1998 (2022)
5. Steiner, A., Kolesnikov, A., Zhai, X., Wightman, R., Uszkoreit, J., Beyer, L.: How to train your vit? Data, augmentation, and regularization in vision transformers (2021). arXiv:2106.10270 arXiv preprint
6. Chambon, P., et al.: RoentGen: vision-language foundation model for chest X-ray generation (2022). arXiv preprint arXiv:2211.12737
7. Hellín, C.J., et al.: Unraveling the impact of class imbalance on deep-learning models for medical image classification. Appl. Sci. **14**(8), 3419 (2024)
8. Huo, X., et al.: HiFuse: hierarchical multi-scale feature fusion network for medical image classification. Biomed. Signal Process. Control **87**, 105534 (2024)
9. Wang, M., He, Y., Peng, L., Song, X., Dong, S., Gong, Y.: Cross-domain invariant feature absorption and domain-specific feature retention for domain incremental chest X-Ray classification. IEEE Trans. Med. Imaging (2025)

10. Malik, H., et al.: A novel fusion model of hand-crafted features with deep convolutional neural networks for classification of several chest diseases using X-ray images. IEEE Access **11**, 39243–39268 (2023)
11. Taloba, A.I., Matoog, R.T.: Detecting respiratory diseases using machine learning-based pattern recognition on spirometry data. Alex. Eng. J. **113**, 44–59 (2025)
12. Rasheed, J., Hameed, A.A., Djeddi, C., Jamil, A., Al-Turjman, F.: A machine learning-based framework for diagnosis of COVID-19 from chest X-ray images. Interdisc. Sci. Comput. Life Sci. **13**(1), 103–117 (2021)
13. Shen, D., Wu, G., Suk, H.I.: Deep learning in medical image analysis. Annu. Rev. Biomed. Eng. **19**(1), 221–248 (2017)
14. Wang, C.W., et al.: A benchmark for comparison of dental radiography analysis algorithms. Med. Image Anal. **31**, 63–76 (2016)
15. Barsoum, E., Zhang, C., Ferrer, C.C., Zhang, Z.: Training deep networks for facial expression recognition with crowd-sourced label distribution. In: Proceedings of the 18th ACM International Conference on Multimodal Interaction, pp. 279–283 (2016)
16. He, K., Zhang, X., Ren, S., Sun, J.: Deep residual learning for image recognition. In: Proceedings of the IEEE Conference on Computer Vision and Pattern Recognition, pp. 770–778 (2016)
17. Hess, A.: Exploring feature reuse in DenseNet architectures (2018). arXiv:1806.01935 arXiv preprint
18. Tan, M., L., Q.: EfficientNet: rethinking model scaling for convolutional neural networks. In: International Conference on Machine Learning, pp. 6105–6114 (2019). PMLR
19. Szegedy, C., et al.: Going deeper with convolutions. In: Proceedings of the IEEE Conference on Computer Vision and Pattern Recognition, pp. 1–9 (2015)
20. Dosovitskiy, A., et al.: An image is worth 16×16 words: Transformers for image recognition at scale. arXiv preprint arXiv:2010.11929 (2020)
21. Rangwani, H., Mondal, P., Mishra, M., Asokan, A.R., Babu, R.V.: DeiT-LT: distillation strikes back for vision transformer training on long-tailed datasets. In: Proceedings of the IEEE/CVF Conference on Computer Vision and Pattern Recognition, pp. 23396–23406 (2024)
22. Rane, N.: Transformers for medical image analysis: Applications, challenges, and future scope. Challenges, and Future Scope (2023)
23. Chen, J., et al.: TransUNet: Transformers make strong encoders for medical image segmentation. arXiv preprint arXiv:2102.04306 (2021)
24. Tang, H., et al.: HTC-Net: a hybrid CNN-transformer framework for medical image segmentation. Biomed. Signal Process. Control **88**, 105605 (2024)
25. Petit, O., Thome, N., Rambour, C., Themyr, L., Collins, T., Soler, L.: U-Net transformer: self and cross attention for medical image segmentation. In: International Workshop on Machine Learning in Medical Imaging, pp. 267–276. Springer International Publishing, Cham (2021)
26. Manzari, O.N., Ahmadabadi, H., Kashiani, H., Shokouhi, S.B., Ayatollahi, A.: MedViT: a robust vision transformer for generalized medical image classification. Comput. Biol. Med. **157**, 106791 (2023)
27. Hussain, T., et al.: EFFResNet-ViT: a fusion-based convolutional and vision transformer model for explainable medical image classification. IEEE Access (2025)

28. Lin, T.Y., Dollár, P., Girshick, R., He, K., Hariharan, B., Belongie, S.: Feature pyramid networks for object detection. In: Proceedings of the IEEE Conference on Computer Vision and Pattern Recognition, pp. 2117–2125 (2017)
29. Zhao, H., Shi, J., Qi, X., Wang, X., Jia, J.: Pyramid scene parsing network. In: Proceedings of the IEEE Conference on Computer Vision and Pattern Recognition, pp. 2881–2890 (2017)
30. Oktay, O., et al.: Attention U-Net: Learning where to look for the pancreas. arXiv preprint arXiv:1804.03999 (2018)

Explainable AI Methods Toward Increased Trustworthiness of ML in Industrial Control Systems: A Systematic Literature Review with Reference Design and Results

Fabien Sechi(✉), Erlend Sundet Siqveland, and Nadia Saad Noori

Faculty of Engineering and Science, Department of Information and Communication Technology, Universitetet i Agder, Kristiansand, Norway
{Fabien.e.sechi,erlendsi,nadia.saad.noori}@uia.no, Fabien.sechi@ife.no

Abstract. This paper presents a systematic literature review (SLR) of explainable artificial intelligence (XAI) in industrial control systems (ICS) and proposes a reference design combining semi-/unsupervised anomaly detection with operator-centered local explanations. We emphasize ICS-specific constraints—false-positive intolerance, real-time response, and resilience to novel attacks—and show why classical metrics (accuracy/F1) are insufficient without time-based and trade-off measures. Building on the SLR, we instantiate a compact pipeline (PCA-based anomaly detection, Random Forest (RF) labeling, SHAP and rule-based explanations, optional logic-layer fusion of network and physical data) and report indicative results across well-known datasets. Findings support prioritizing latency and false-positive control, validating detectors with adversarial/synthetic scenarios, and measuring explanation timeliness and fidelity.

Keywords: AI · XAI · ICS · Anomaly Detection · Semi-supervised · PCA · Random Forest · SHAP

1 Introduction

AI-enabled anomaly detection is attractive for ICS but adoption is limited by trust and safety concerns. Black-box alerts without explanations can trigger costly shutdowns or be ignored by operators [30]. Convergence with IIoT increases the attack surface [20,33], while legacy devices prioritize availability over security [22]. XAI promises transparency, yet XAI-for-ICS evidence is sparse and evaluation lacks standardization [29,30].

Contributions. (i) An SLR of AI/XAI for ICS synthesizing constraints, metrics, and methods; (ii) requirements for trustworthy ICS anomaly detection and explanations; (iii) a reference design aligned with these requirements; (iv) implementation notes and indicative quantitative results and visuals.

H. R. Arabnia et al. (Eds.): CSCI 2025, CCIS 2967, pp. 423–435, 2027.
https://doi.org/10.1007/978-3-032-28454-9_29

Research Questions. RQ1: What ML attributes are necessary to deploy in ICS? RQ2: What XAI requirements increase operator trust/compliance for ICS?

2 Systematic Literature Review

We follow Kitchenham's SLR protocol [19].

2.1 Search Strategy and Selection

Title-restricted queries were used on Google Scholar, IEEE Xplore, Scopus, and Springer (2018–2023), augmented by snowballing [34]. Two query families targeted (a) AI/ML/XAI in ICS and (b) explainable/ interpretable anomaly detection in ICS (Table 1). Inclusion: AI/ML for ICS monitoring/intrusion/anomaly detection; XAI methods applicable to ICS; datasets for ICS security. Exclusion: non-ICS domains without transferable constraints; non-peer-reviewed unless seminal surveys or datasets.

Table 1. Abridged search strings; full strings available on request.

Database	Title-restricted query (2018–2023)
Google Scholar	("industrial control system" OR ICS) AND (AI OR ML OR XAI)
IEEE/Scopus/Springer	("explainable anomaly detection" OR interpretable) AND ("industrial control system" OR ICS)

2.2 Screening and Extraction

Initial de-duplicated hits totaled 58 items (expanded with 3 XAI-in-ICS after gap analysis). We extracted tasks, models, datasets, metrics, and XAI technique/latency. Thematic synthesis focused on ICS constraints (§3), evaluation (§3.4), and XAI use (§4.3).

2.3 Threats to Validity

Title filtering reduces recall; however, we mitigated via snowballing and multi-database coverage. Page limits preclude a meta-analysis; nonetheless, triangulation across surveys [20,22,26,39] and datasets (Table 2) supports robustness.

3 AI in ICS: Constraints, Data, and Methods

3.1 Systems and Exposure

ICS comprises SCADA/DCS/PLC [33]. Industry 4.0 connectivity expands threat vectors [20]. Many devices lack modern security [22].

3.2 Datasets and Data Handling

Public ICS datasets emphasize a few attacks (FDI predominant), limiting generalization [20]. Table 2 lists representative sets. Labeling is expensive; privacy hinders sharing [8]. Augmentation via SMOTE can degrade fidelity or interpretability [18]; adversarial/synthetic variants help stress-test detectors [10,27,38].

Table 2. Representative ICS datasets (abridged).

Nr.	ICS Datasets	Cyberattack	Source
1.	SWaT	FDI	[36]
2.	Gas Pipeline (GP)	FDI, Recon, DoS	[9]
3.	IUNO	FDI	[6]
4.	BATADAL	FDI	[35]
5.	WST (SCADA)	FDI, Recon, DoS	[25]
6.	Power Sys	FDI	[1]
7.	WADI	FDI, Spo	[3]
8.	Festo Rig	FDI	[28]
9.	TEP	FDI	[11]
10.	TLIGHT	FDI	[13]
11.	HAI	FDI	[31]

3.3 Learning Paradigms

Supervised models attain strong accuracy yet struggle with zero-day/adversarial conditions [2,5,26,37]. Unsupervised and semi-supervised methods better address novelty but can raise false positives [2,22]. For ICS, semi-/unsupervised is generally preferable when paired with thresholds and XAI for triage.

3.4 Evaluation Metrics

Accuracy/F1 dominate [20,39] but are insufficient. ICS requires (i) false-positive minimization [2], (ii) latency/time-to-detect, and (iii) trade-offs (ROC/AUC, conflict indices [14]). Acceptability thresholds are process-specific.

4 Explainable AI

4.1 Scope and Taxonomy

XAI includes model-intrinsic transparency and post-hoc local/global methods [4,7,15]. Feature attribution (e.g., SHAP) and surrogate models (e.g., LIME) are common, but surrogate faithfulness is debated [29].

4.2 Why ICS Need XAI

EU guidance on trustworthy AI stresses transparency and auditability [12]. Operators require concise, timely explanations that map to actions [30]. Incomplete or misleading explanations can be harmful in safety contexts [29].

4.3 XAI in ICS: Use and Gaps

Reported uses include SHAP for autoencoder/OC-SVM pipelines [16,17] and pairing uninterpretable detectors with interpretable models for feature importance [40]. Adversarial explanation strategies have been proposed [24]. However, standardized XAI metrics (fidelity, usefulness) and reference time budgets are lacking.

5 Requirements and Reference Design

5.1 Requirements

R1 *Novelty-resilient detection*: semi-/unsupervised with drift monitoring. R2 *Operational metrics*: false positives and latency prioritized. R3 *Local, operator-centered explanations* with explicit time budgets. R4 *Faithfulness* checks for post-hoc/surrogates; R5 *SOP integration* and training.

5.2 Reference Design (Baseline to Implementation)

We implement a modular pipeline derived from the SLR: (i) PCA-based anomaly detection using reconstruction error (unsupervised) [21]; (ii) RF classifier to label flagged points (multiclass) [32]; (iii) Local explanations: TreeExplainer (SHAP) for RF [23], plus rule-based visualization (ExMatrix) to yield human-readable conditions; (iv) Optional logic layer combining network and physical-sensor anomalies to triage FDI vs. sensor faults. Surrogate caveats apply [29].

6 Implementation

6.1 Environment and Packages

Python 3.11.6; scikit-learn 1.3.2; numpy 1.26.2; matplotlib 3.8.2; SHAP 0.45.0; ExMatrix 0.2.3 (cf. appendix of package versions). Scripts automate per-dataset runs for consistency.

6.2 Data Preprocessing

KDD99, NSL-KDD, CICIoT2023, and NIDS-V1: CSV ingestion; cleaning (NULL removal, dropping high-cardinality IP strings), MinMax scaling, 85/15 train/test split; LabelEncoder/LabelBinarizer for multi-/binary. Feature selection via RF importance threshold 0.01.

6.3 Detectors and Thresholding

PCA retains 95–99% variance; anomaly score is reconstruction error (MSE/MAE/RMSE). Thresholds tuned empirically (grid over 10^{-3} to 10^{-1}) with confusion matrix inspection under latency constraints [21].

6.4 Explanations

For flagged instances: SHAP TreeExplainer returns top-k features and signed contributions; ExMatrix extracts compact rules from RF (sorted by support/importance) to present human-readable conditions; generation time logged to respect time budgets.

7 Results

7.1 Baselines (RF)

Table 3 summarizes RF baselines across tasks (binary, multiclass, FS, PCA). As in [32], RF offers strong accuracy and generalization; FS reduces training time with marginal loss; PCA may hurt multiclass accuracy on certain sets, but improves overfitting on NSL-KDD.

Table 3. RF baselines across datasets (abridged; weighted metrics for multiclass).

Dataset	Setting	Train Acc	Test Acc	Precision	F1
KDD99	Multiclass	0.9815	0.9809	0.9808	0.9808
NSL-KDD	PCA	0.9998	0.6985	0.5416	0.6000
CICIoT2023	Multiclass	0.9999	0.9950	0.9949	0.9947
NIDS-V1	Multiclass	0.9500	0.9235	0.9187	0.9183

7.2 PCA as Anomaly Detector

Table 4 reports best tuned runs per dataset; Fig. 1 shows confusion matrices. PCA favors novelty but is sensitive to class imbalance and thresholds: strong recall with high FP on some sets.

7.3 Hybrid PCA→RF (Multiclass on Flagged Points)

The hybrid improves interpretability and, on certain sets, accuracy (Table 5); confusion matrices in Fig. 2 illustrate per-class effects. NSL-KDD remains challenging.

Table 4. PCA anomaly detector (best runs per dataset).

Dataset	Acc	Prec	Rec	F1	PCs	Loss/Thr
KDD99	0.386	0.671	0.470	0.553	99%	MSE/$5e^{-4}$
NSL-KDD	0.744	0.729	0.875	0.795	50%	MSE/$5e^{-3}$
CICIoT2023	0.915	0.975	0.937	0.955	99%	MAE/$5e^{-4}$
NIDS-V1	0.432	0.279	0.912	0.427	99%	MAE/$5e^{-4}$

Table 5. Hybrid PCA→RF results (multiclass on anomalies).

Dataset	Acc	Prec	Rec	F1
KDD99	0.537	0.852	0.537	0.505
NSL-KDD	0.520	0.479	0.520	0.408
CICIoT2023	0.933	0.980	0.933	0.937
NIDS-V1	0.905	0.896	0.905	0.897

7.4 Visualizing PCA Detection vs. Ground Truth

Figure 3 shows 100-instance slices contrasting PCA loss and threshold with TP/FP/FN color coding.

7.5 Network and Physical Fusion with Logic Layer

The logic layer distinguishes network-only anomalies (likely network attack), physical-only (failing sensor), and coupled anomalies (indicative of FDI). For a representative point, SHAP and ExMatrix agree on top features/values (e.g., `src_bytes`, `dst_host_same_src_port_rate`, `srv_count`), yielding actionable operator guidance [16,17,40] (Fig. 4).

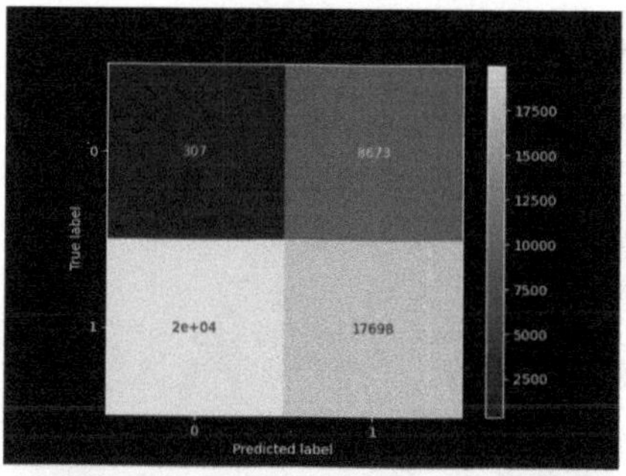

(a) KDD99

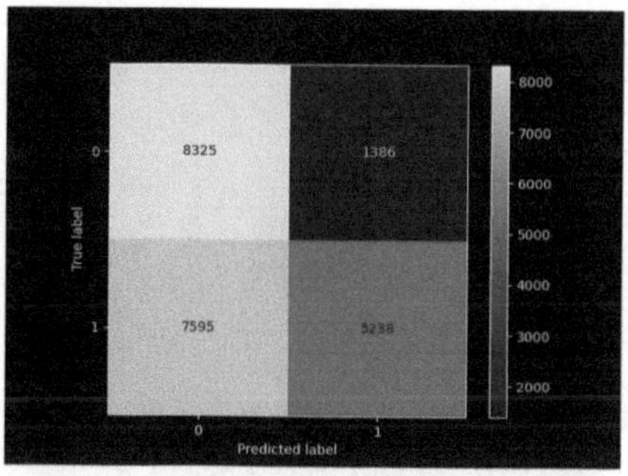

(b) NSL-KDD

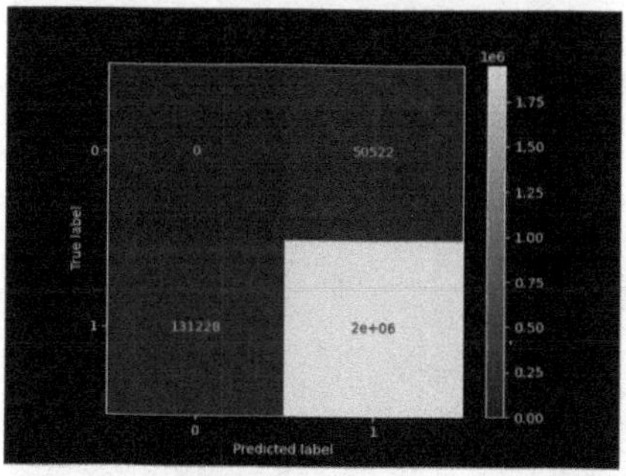

(c) CICIoT2023

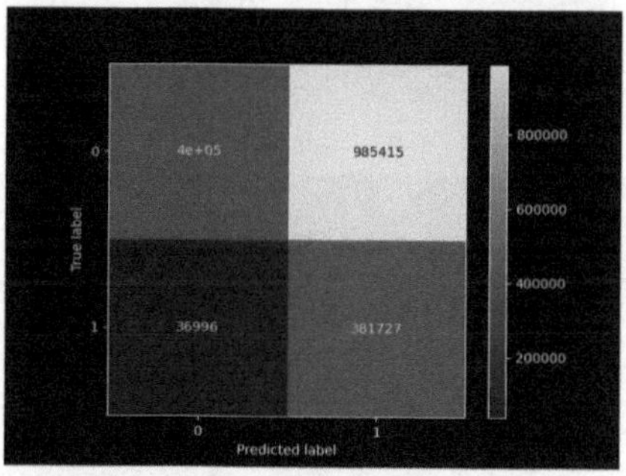

(d) NIDS-V1

Fig. 1. Confusion matrices for PCA anomaly detector (per dataset).

8 Discussion

8.1 Reflection on the SLR

The SLR provided a structured map of AI/XAI for ICS but two choices limited recall: (i) title-only filters (2018–2023) excluded papers where key terms appeared in abstracts/bodies; and (ii) the initial reliance on the single term "anomaly detection" missed close variants (intrusion/novelty/behavioral detection). We mitigated these issues with backward/forward snowballing, which recovered several influential works and datasets. Going forward, a brief scoping phase before protocol lock-in, broader synonym sets, abstract-level screening, and explicit handling of grey literature would improve coverage and reduce selection bias [19,34].

8.2 Reflection on the Results: What the Design, Implementation, and Results Say

The reference design (unsupervised PCA detector → RF labeler + local XAI + optional logic-layer fusion) operationalizes SLR-derived requirements: novelty tolerance via unsupervised detection, operator-centered local explanations, and attention to false positives and timeliness. Empirically, RF baselines

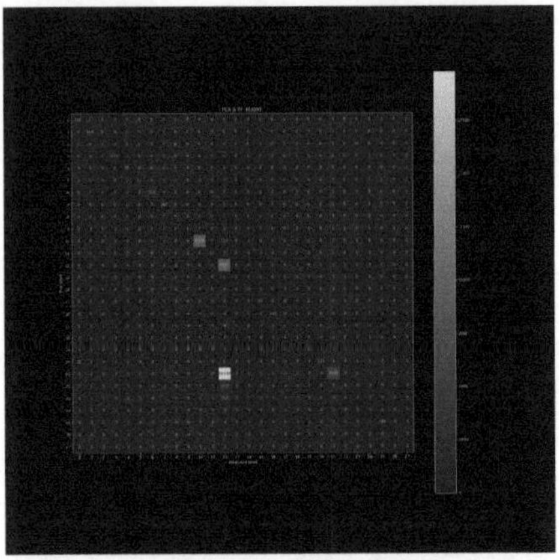

(a) KDD99

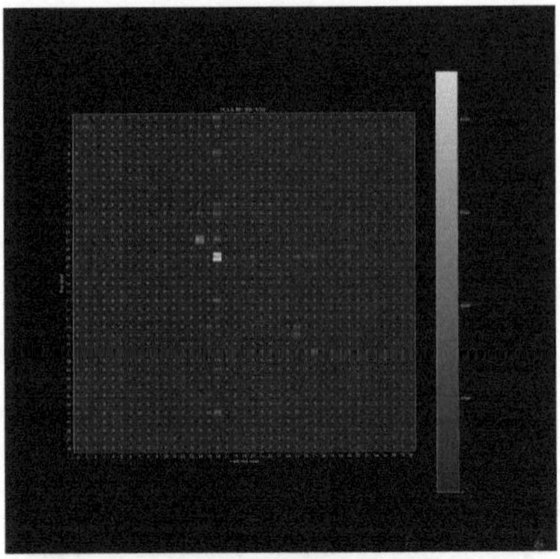

(b) NSL-KDD

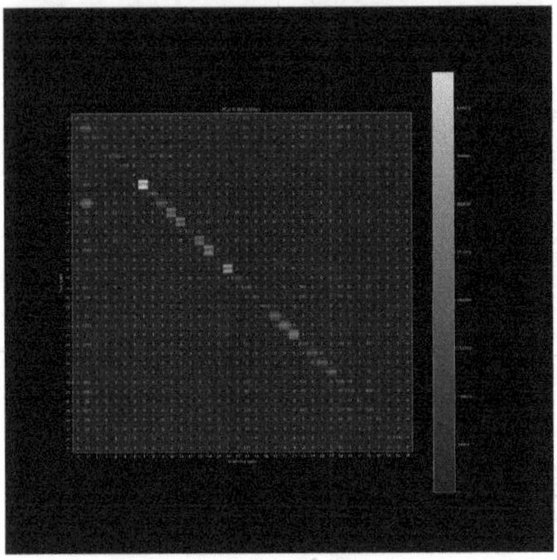

(c) CICIoT2023

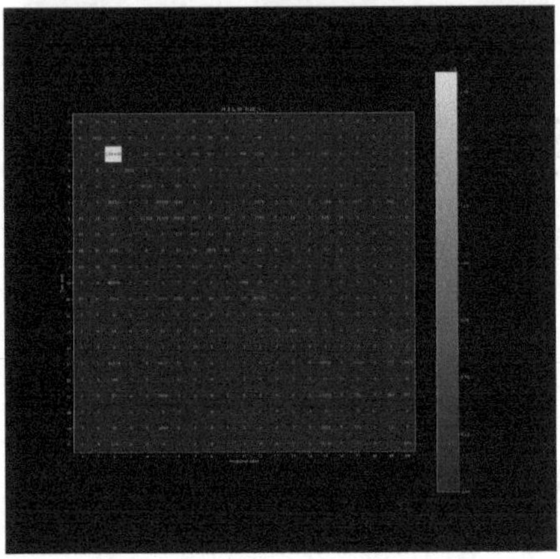

(d) NIDS-V1

Fig. 2. Hybrid PCA→RF: confusion matrices.

performed strongly on KDD99/CICIoT/NIDS but struggled on NSL-KDD—highlighting dataset-driven optimism. PCA detected deviations but required careful thresholding to avoid excessive false positives; the hybrid PCA→RF improved interpretability on flagged points yet can mislabel truly novel attacks to the nearest known class. The logic-layer conceptually separates network-only vs. sensor-only vs. coupled anomalies (FDI-like), but validating this requires a joint network+physical dataset. Overall, operational fitness (FP cost, detection/explanation latency, novelty handling) matters more than headline accuracy.

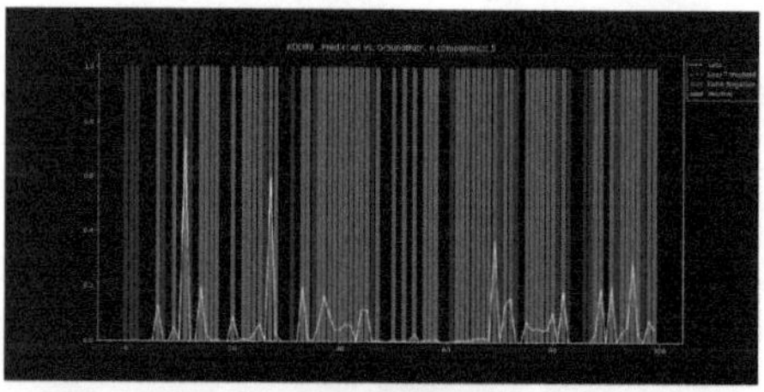

(a) KDD99

(b) NSL-KDD

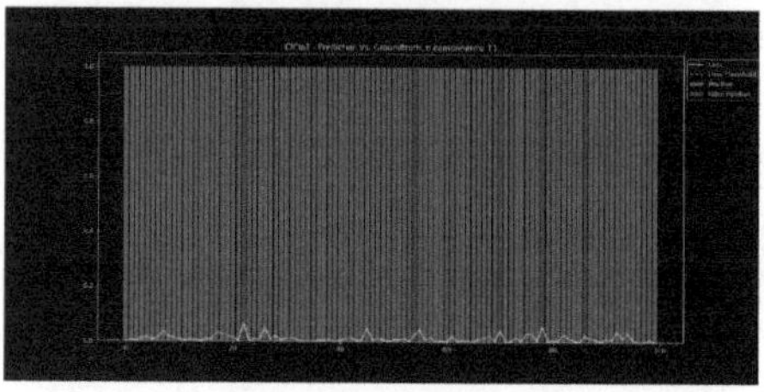

(c) CICIoT2023

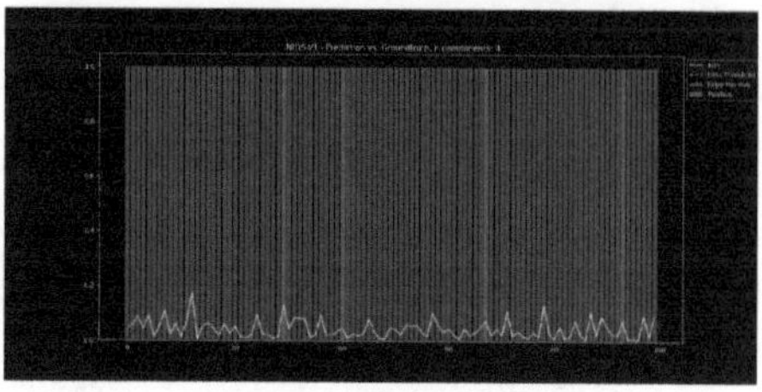

(d) NIDS-V1

Fig. 3. PCA detection vs. ground truth (100-instance slices).

8.3 Results Align with SLR Expectations

Novelty-oriented detectors benefit from post-hoc multiclass labeling and local explanations; class imbalance and limited attack diversity dominate errors; surrogate explanations should be used with fidelity checks [29]. Operator-centered, time-bounded explanations and SOP mapping are critical [12,30].

8.4 Reflection on Federated Learning (FL) for Future Research

Our companion NSL–KDD study (average FL, Fernet-encrypted weights, FGSM adversarial training, Flower simulation) shows FL can expand coverage across sites while preserving privacy—matching the SLR's call for novelty tolerance and operator-centered explanations.

Practical gains. (1) *Breadth without sharing raw traces:* federated training of the unsupervised gate and RF/small DNN labeler improved robustness to rare patterns. (2) *Local-first XAI:* compute SHAP/LIME on site; share summaries. (3) *Latency rules:* accuracy gains taper under strict time/epoch budgets—realistic for ICS windows.

What to Federate. Prefer lightweight artefacts over raw gradients alone: feature-importance/threshold summaries for the PCA gate, drift and data-quality signals, and per-mode (network/sensor/coupled) anomaly tallies to inform the logic layer.

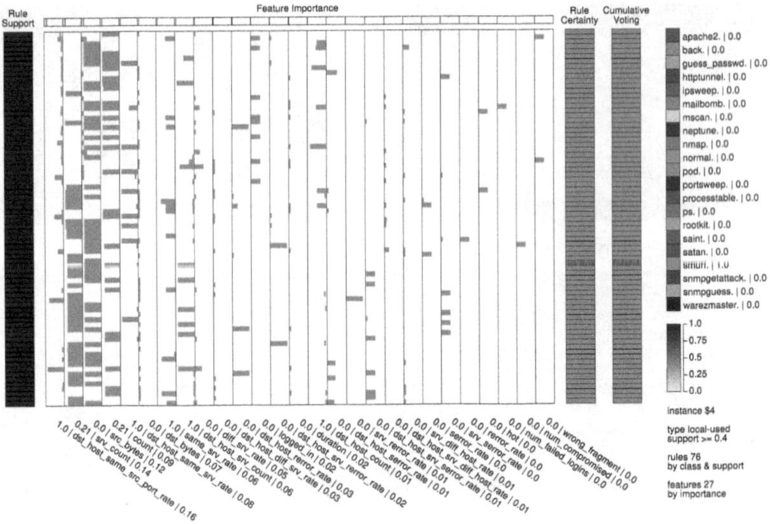

(a) ExMatrix on network data

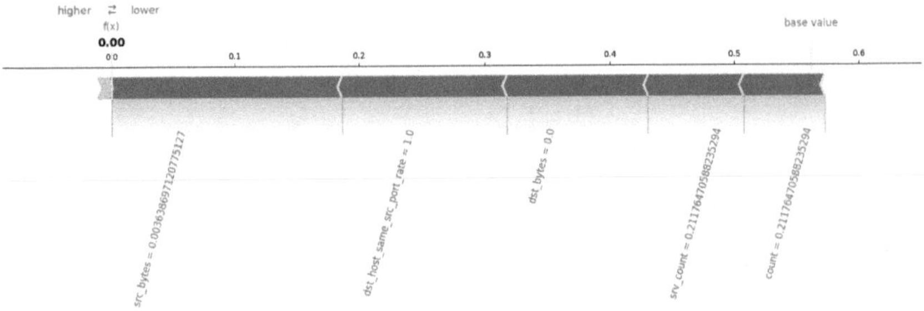

(b) SHAP (TreeExplainer) on RF

Fig. 4. Explainability for network attack (logic-layer decision, ExMatrix rules, SHAP force plot).

Security and Privacy. Fernet-in-transit is easy but insufficient at scale: add authenticated transport, key rotation, *secure aggregation*, and (where needed) DP. FGSM training boosts evasion robustness, but FL widens the attack surface—use robust aggregation, update vetting, and client reputation to curb poisoning/backdoors.

Logic Layer. Validating the network/sensor/coupled split needs joint network+physical datasets. FL can help via (i) federated curation of aligned episodes (metadata only) and (ii) local counterfactual/adversarial tests before promoting rules globally.

Priorities.

1. Real-time FL under ICS constraints (partial participation, asynchronous rounds) with bounded detect + explain latency.
2. Poisoning/backdoor defenses (trimmed/clipped aggregation, server-side vetting, red-team FGSM/PGD suites).
3. Privacy-preserving XAI (aggregated/DP attributions with fidelity checks).
4. FL-ready benchmarks pairing flows with physical telemetry; reproducible Flower baselines.

Overall, FL complements our architecture: it scales novelty handling without centralizing data and, with secure aggregation and robust training, can meet ICS privacy/safety demands—the decisive metrics remain FP cost, end-to-end latency, and resilience to evasion/poisoning.

9 Conclusion

We synthesized constraints and gaps for XAI in ICS and delivered a reference design with indicative results. Recommendations: prioritize false-positive control and latency; employ semi-/unsupervised detection plus local explanations; validate with adversarial/synthetic attacks; measure explanation fidelity and timeliness. Future work: standardized XAI metrics for ICS, calibrated time budgets per process, and evaluation on diverse real-world datasets.

References

1. Adhikari, U., Pan, S., Morris, T., Borges, R., Beaver, J.: Power system datasets. https://sites.google.com/a/uah.edu/tommy-morris-uah/ics-data-sets
2. Ahmed, C.M., Gauthama Raman, M.R., Mathur, A.P.: Challenges in machine learning based approaches for real-time anomaly detection in industrial control systems. In: Proceedings of the 6th ACM on Cyber-Physical System Security Workshop (CPSS '20), pp. 23–29. ACM, Taipei, Taiwan (2020). https://doi.org/10.1145/3384941.3409588
3. Ahmed, C.M., Palleti, V.R., Mathur, A.P.: WADI: a water distribution testbed for research in the design of secure cyber physical systems. In: Proceedings of the 3rd International Workshop on Cyber-Physical Systems for Smart Water Networks, pp. 25–28 (2017)
4. Ahmed, I., Jeon, G., Piccialli, F.: From artificial intelligence to explainable artificial intelligence in industry 4.0: a survey on what, how, and where. IEEE Trans. Industr. Inf. **18**(8), 5031–5042 (2022)
5. Anthi, E., Williams, L., Rhode, M., Burnap, P., Wedgbury, A.: Adversarial attacks on machine learning cybersecurity defences in industrial control systems. J. Inf. Secur. Appl. **58**, 102717 (2021)
6. Anton, S.D., Gundall, M., Fraunholz, D., Schotten, H.D.: Implementing SCADA scenarios and introducing attacks to obtain training data for intrusion detection methods. In: ICCWS 2019 - 14th International Conference on Cyber Warfare and Security, p. 56. Academic Conferences and Publishing Limited (2019)

7. Arrieta, A.B., et al.: Explainable artificial intelligence (XAI): concepts, taxonomies, opportunities and challenges toward responsible AI. Inf. Fusion **58**, 82–115 (2020)
8. Baptiste, M., Julien, F., Franck, S.: Systematic and efficient anomaly detection framework using machine learning on public ICS datasets. In: 2021 IEEE International Conference on Cyber Security and Resilience (CSR), pp. 292–297. IEEE (2021)
9. Beaver, J.M., Borges-Hink, R.C., Buckner, M.A.: An evaluation of machine learning methods to detect malicious SCADA communications. In: 12th International Conference on Machine Learning and Applications (ICMLA), vol. 2, pp. 54–59. IEEE (2013)
10. Chen, J., Gao, X., Deng, R., He, Y., Fang, C., Cheng, P.: Generating adversarial examples against machine learning-based intrusion detector in industrial control systems. IEEE Trans. Dependable Secure Comput. **19**(3), 1810–1825 (2020)
11. Downs, J.J., Vogel, E.F.: A plant-wide industrial process control problem. Comput. Chem. Eng. **17**(3), 245–255 (1993)
12. European Commission: Communication: Building trust in human centric artificial intelligence. Communication document (2019). https://ec.europa.eu/digital-single-market/en/news/communication-building-trust-human-centric-artificial-intelligence
13. Fregin, A., Muller, J., Krebel, U., Dietmayer, K.: The driveU traffic light dataset: introduction and comparison with existing datasets. In: 2018 IEEE International Conference on Robotics and Automation (ICRA), pp. 3376–3383. IEEE (2018)
14. Gauthama Raman, M., et al.: An efficient intrusion detection technique based on support vector machine and improved binary gravitational search algorithm. Artif. Intell. Rev. **53**, 3255–3286 (2020)
15. Gunning, D.: Explainable artificial intelligence (XAI). DARPA Web (2017). https://www.darpa.mil/program/explainable-artificial-intelligence
16. Hoang, N.X., Hoang, N.V., Du, N.H., Huong, T.T., Tran, K.P., et al.: Explainable anomaly detection for industrial control system cybersecurity. IFAC-PapersOnLine **55**(10), 1183–1188 (2022)
17. Huong, T.T., et al.: Federated learning-based explainable anomaly detection for industrial control systems. IEEE Access **10**, 53854–53872 (2022)
18. Imran, M., Siddiqui, H.U.R., Raza, A., Raza, M.A., Rustam, F., Ashraf, I.: A performance overview of machine learning-based defense strategies for advanced persistent threats in industrial control systems. Comput. Secur. **134**, 103445 (2023)
19. Kitchenham, B.: Procedures for performing systematic reviews. Tech. Rep. TR/SE-0401, Keele University, Keele, UK (2004). http://www.inf.ufsc.br/~aldo.vw/kitchenham.pdf
20. Koay, A.M., Ko, R.K.L., Hettema, H., Radke, K.: Machine learning in industrial control system (ICS) security: current landscape, opportunities and challenges. J. Intell. Inf. Syst. **60**(2), 377–405 (2023)
21. Kravchik, M., Shabtai, A.: Efficient cyber attack detection in industrial control systems using lightweight neural networks and PCA. IEEE Trans. Dependable Secure Comput. **19**(4), 2179–2197 (2021)
22. Li, L., et al.: Survey on methodology of intrusion detection in industrial control system based on artificial intelligence. In: 2022 International Conference on Computers and Artificial Intelligence Technologies (CAIT), pp. 93–103. IEEE (2022)
23. Lundberg, S.M., et al.: From local explanations to global understanding with explainable AI for trees. Nat. Mach. Intell. **2**(1), 56–67 (2020)

24. Marino, D.L., Wickramasinghe, C.S., Manic, M.: An adversarial approach for explainable ai in intrusion detection systems. In: IECON 2018 – 44th Annual Conference of the IEEE Industrial Electronics Society, pp. 3237–3243. IEEE (2018)
25. Morris, T., Srivastava, A., Reaves, B., Gao, W., Pavurapu, K., Reddi, R.: A control system testbed to validate critical infrastructure protection concepts. Int. J. Crit. Infrastruct. Prot. **4**(2), 88–103 (2011)
26. MR, G.R., Ahmed, C.M., Mathur, A.: Machine learning for intrusion detection in industrial control systems: challenges and lessons from experimental evaluation. Cybersecurity **4**, 1–12 (2021)
27. Pham, T.S., Nguyen, Q.U., Nguyen, X.H.: Generating artificial attack data for intrusion detection using machine learning. In: Proceedings of the 5th Symposium on Information and Communication Technology, pp. 286–291 (2014)
28. Robles-Durazno, A., Moradpoor, N., McWhinnie, J., Russell, G., Tan, Z.: Newly engineered energy-based features for supervised anomaly detection in a physical model of a water supply system. Ad Hoc Netw. **120**, 102590 (2021)
29. Rudin, C.: Stop explaining black box machine learning models for high stakes decisions and use interpretable models instead. Nat. Mach. Intell. **1**(5), 206–215 (2019)
30. Saeed, W., Omlin, C.: Explainable AI (XAI): a systematic meta-survey of current challenges and future opportunities. Knowl.-Based Syst. **263**, 110273 (2023)
31. Shin, H.K., Lee, W., Choi, S., Yun, J.H., Min, B.G.: HAI security datasets (2023). https://github.com/icsdataset/hai
32. Sokolov, A.N., Pyatnitsky, I.A., Alabugin, S.K.: Applying methods of machine learning in the task of intrusion detection based on the analysis of industrial process state and ICS networking. FME Trans. **47**(4), 782–789 (2019)
33. Stouffer, K., Falco, J., Scarfone, K., et al.: Guide to industrial control systems (ICS) security. Tech. Rep. SP 800-82, NIST (2011)
34. Streeton, R., Cooke, M., Campbell, J.: Researching the researchers: using a snowballing technique. Nurse Res. **12**(1), 35–47 (2004)
35. Taormina, R., et al.: Battle of the attack detection algorithms: disclosing cyber attacks on water distribution networks. J. Water Resour. Plan. Manag. **144**(8), 04018048 (2018)
36. Team, I.: Secure water treatment (SWaT). https://itrust.sutd.edu.sg/testbeds/secure-water-treatment-swat/. Accessed 24 Nov 2023
37. Trivedi, S., Tran, T.A., Faruqui, N., Hassan, M.M.: An exploratory analysis of effect of adversarial machine learning attack on IoT-enabled industrial control systems. In: 2023 International Conference on Smart Computing and Application (ICSCA), pp. 1–8. IEEE (2023)
38. Umer, M.A., Ahmed, C.M., Jilani, M.T., Mathur, A.P.: Attack rules: an adversarial approach to generate attacks for industrial control systems using machine learning. In: Proceedings of the 2nd Workshop on CPS&IoT Security and Privacy, pp. 35–40 (2021)
39. Umer, M.A., Junejo, K.N., Jilani, M.T., Mathur, A.P.: Machine learning for intrusion detection in industrial control systems: applications, challenges, and recommendations. Int. J. Crit. Infrastruct. Prot. **38**, 100516 (2022)
40. Vávra, J., Hromada, M., Lukáš, L., Dworzecki, J.: Adaptive anomaly detection system based on machine learning algorithms in an industrial control environment. Int. J. Crit. Infrastruct. Prot. **34**, 100446 (2021)

Network of Quantum Neural Networks

Emre Tokgoz[1(✉)] and Sharmendra Desiboyina[2]

[1] State University of New York, Farmingdale, NY 11735, USA
tokgoze@farmingdale.edu
[2] Clark University, Worcester, MA 01610, USA

Abstract. The foundational work on theory and applications of artificial intelligence using quantum computations started to expand with new foundational work introduced recently. The introduction of quantum neural network (QNN) is recent, and its applications are shown to provide strong results in real-life cases. In this work, we introduce a new theoretical framework; Network of quantum neural network (N-QNN) as a new quantum-based deep learning technique that takes advantage of quantum and traditional methods that run interconnected subnetworks that are formed by QNN, other quantum techniques as well as traditional AI methods. This new technique's subnetwork builds on mathematical foundations of quantum neural network that focuses on minimizing the difference between the calculated and known values of the network output by utilizing a cost function; Hence, this new design utilizes QNNs as a part of subnetwork design instead of using only qubits alongside of other methods such as Artificial Neural Networks (ANN), Support Vector Machines (SVM), and Convolutional Neural Network (CNN) as well as other quantum methods such as quantum convolutional neural network (QCNN) and Quantum Support Vector Machine (QSVM); Hence, we introduce different subnetwork designs of N-QNN that utilizes quantum and traditional techniques. N-QNN's subnetwork design is tested on the well-known Iris data set by using 3-class classification. N-QNN's result comparison with traditional ANN indicated a better accuracy attainment 97% along with faster result achievement (less than two minutes) by using a quantum simulator.

Keywords: Artificial Neural Networks · Efficiency · Quantum Computing · Quantum Simulator · Network of Artificial Neural Networks

1 Introduction

Artificial Neural Network (ANN) has gained popularity in applications after the use of the perceptron notion in [1] with its continuing incremental popularity in numerous field applications of engineering, science, mathematics, and technology. ANN started gaining more popularity with the advancement of computing power by using commercially available hardware and software applications. A variety of different approaches to the theory of ANN are introduced in the literatures that are including but not limited to feed forward, autoencoder, convolutional, and recurrent networks [2]. The corresponding frameworks have shown their effectiveness in various real-life applications while not so

H. R. Arabnia et al. (Eds.): CSCI 2025, CCIS 2967, pp. 436–448, 2027.
https://doi.org/10.1007/978-3-032-28454-9_30

much being effective in some other applications. One unchanging challenge of ANN is the computational power of the existing non-quantum commercial processors and its limitations for attaining results for different problem types.

Quantum computers and programming have gained popularity in recent years with the development of the associated quantum programming languages such as Q# by Microsoft [3], Qiskit by IBM [4], and Cirq by Google [5]; three of the existing languages that are used for programming of quantum simulators and computers. Quantum algorithms have gained popularity for executing deep learning solutions for computationally intractable problems that classical systems cannot easily solve. Examples of such real-life applications include the use of IBM Quantum to advance aerospace material design for Boeing that involves up to 100,000 variables, optimizing the global energy logistics of ExxonMobil as a part of energy supply chain distribution's optimization, and revolutionizing Mercedes Benz's electric vehicle battery technology [6].

There are different quantum computational algorithmic approaches introduced in the research literature with design differences that include Quantum Approximate Optimization Algorithm and Variational Quantum Algorithms that are two of the well-known algorithms [27]. Current quantum systems have certain challenges that are operating under a noisy intermediate-scale quantum stage that have challenges that are required by these algorithms to overcome including the following:

- There are limited number of qubits.
- Substantial noise levels occur.
- Error correction is a major challenge due to changing quantum states over time.
- Quantum gates' variational implementation lead to errors in state transformations.
- Measurement errors occur during measurement reading that causes computational inaccuracies.

Variational quantum algorithms have a promising outlook for solving classic problems that are challenging to solve, including but not limited to high dimensional optimization problems [7, 8]. Similarly, a quantum approximate optimization algorithm is introduced in [9] for solving combinatorial optimization problems with the aim of using a positive integer that allows improvement of approximation as this chosen number increases. Aside from optimization approaches, from a theoretical artificial intelligence perspective, the well-known developed techniques included Quantum Support Vector Machines (Q-SVM) [10] and Quantum Neural Networks (QNN) [11]. Q-SVM is shown to naturally speed up computational processing exponentially when it is executed on problems that have polynomial time complexity based on its binary classification optimization abilities [11]. Our focus throughout this work is going to be on the developed QNN as we are proposing a new approach to such an approach by incorporating subnetworks.

There have been research investments on the quantum machine learning (QML) approaches with its applications and interest in speeding up the classical ML approaches by using quantum algorithms [11–14]. Using quantum data with learning tasks carried out by utilizing quantum computing devices has been also one of the exploration areas of interest [15–23]. The idea of quantum neural computing was discussed in [24] with its furthermore advancement to quantum neural network approach in [25] that utilized quantum learning of parametrized unitary operations covered in [26].

Optimization technique utilization for QNN is one of the challenges that has been recognized and investigated in research. In quantum setting, Barren Plateau is recognized as a phenomenon such that the gradients of the loss landscape of variational quantum algorithms get exponentially suppressed [27]. In the case of ANN utilization for machine learning development, discriminative and generative approaches [28–31] have been shown to be effective when compared to classical ML approaches with the difficulty of training due to flat optimization landscapes of Barren plateaus [27]. Furthermore developments in QNN included both forward and backward propagation.

In this work, a new QNN approach; network of QNN, is introduced to improve the instability of quantum computing results to improve the QNN implementation. Next section outlines the mathematical foundations of the subnetwork design of N-QNN. The following section focuses on possible subnetwork designs of N-QNN by utilizing quantum and traditional AI techniques. The fourth section has the numerical results attained for a particular subnetwork design of N-QNNN by using quantum simulator. The last section covers discussions and conclusive remarks.

2 Mathematical Foundations

The mathematical foundation of N-QNN's subnetwork design takes advantage of recurrent neural network (RNN) design proposed in [25]. In this design, given that QNN subnetwork acts as a basic design element of N-QNN with its RNN nature, the input-output design of N-QNN is much different from that of QNN. Each QNN subnetwork elements of N-QNN need to be designed with the corresponding careful output design for a successfully connected QNN subnetwork. Figure 1 demonstrates a basic N-QNN infrastructure made up of QNN subnetworks with a magnifying glass shown on top for furthermore detailing the QNN subnetwork design shown in Fig. 2.

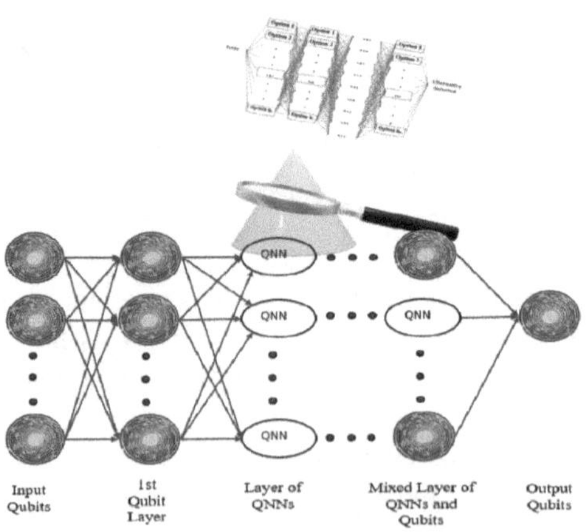

Fig. 1. An example of a subnetwork design consisting of QNNs and qubits simultaneously.

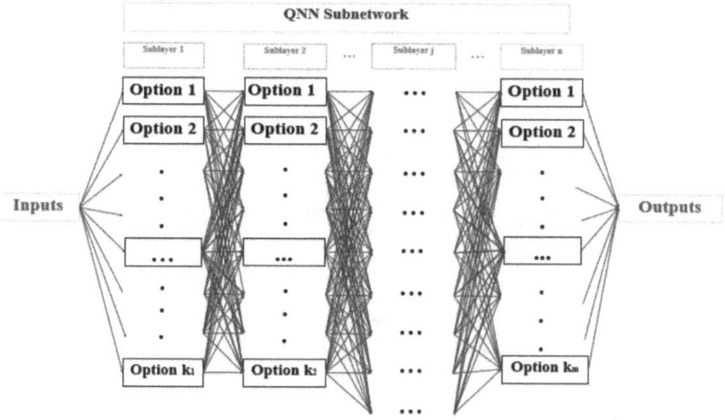

Fig. 2. An example of subnetwork design elements consisting of QCNNs and qubits simultaneously.

Noting that each subnetwork would have its own input-output arrangements that will be depending on the specific design needs of the subnetwork, general basic design of the subnetwork takes the form specified in Fig. 2 that is shown with magnifier in Fig. 1.

We assume the following for mathematical framework of the subnetwork:

- k: Number of hidden layers of qubits within the subnetwork
- O: Output
- I: Input
- Q_t^I: Input nodes
- Q_t^O: Output nodes
- $|c>$: Some basis for the input space
- $V(c)$: Parametrized unitary
- V^k: Layer unitary
- $\alpha(x)= \sum_j B_j X B_j{}^{\dagger}$: Completely positive trace map
- $_*\alpha(x)= \sum_j B_j{}^{\dagger} X B_j$: Adjoint channel of α
- V= $V^O V^k V^{k-1} .. V^1$: QNN quantum circuit where V^O is the output unitary.
- V_i^k: ith perceptron acting on layers k − 1 and k
- α^k: Subnetwork's layer-to-layer transition maps
- β^O: Subnetwork output

By using this notation, we use the following equations within the subnetwork design:

$$\beta^O = \alpha^{k+1}\Big(\alpha^k\Big(\ldots\alpha^1\Big(\beta^{in}\Big)\ldots\Big)\Big) \tag{1}$$

such that

$$\alpha^k\Big(t^{k-1}\Big) \equiv tr\Big(V_{k,j} \times \cdots \times \; V_{k,1}\Big(t^{k-1} \bigotimes |0\ldots0>_k < 0\ldots0|\Big)V_{k,1} \times \cdots \times V_{k,j}\Big) \tag{2}$$

The subnetwork output β^O for each subnetwork is made up of the sequence of completely positive maps' composition by utilizing the perceptrons V_i^k that are acting particularly on the sublayer levels k − 1 and k. In this setting, it is possible to state the unitary by using the basis and parametrized units

$$V = \sum_c |c >< c| \bigotimes \mathrm{V(c)} \tag{3}$$

allowing to state the subnetwork output to take the following form:

$$\beta^O = \sum_c |c > \beta^I < c|\mathrm{V}(c)|0 > < 0|V(c)^\dagger \tag{4}$$

Fidelity is known to be used for measuring closeness of pure quantum states' closeness that can be used as a measure of cost function between the subnetwork's output and the expected output averaged over the training data:

$$M = \frac{1}{n}\sum_c < Q_t^{out}|\beta_t^{out}|Q_t^{out} > \tag{5}$$

The cost function values vary between 0 and 1 with the training of the subnetwork accomplished by optimizing the cost function M. The special case of the input output relationship of the subnetwork nodes can be represented by using some unitary operation V as follows:

$$|Q_t^{out} > = V \mid Q_t^I > \tag{6}$$

The recurrent nature of the subnetwork can be obtained by updating the subnetwork parameter matrix D that are attained for the perceptron unitary components V and step size γ by using the rotational mapping [25]:

$$\mathrm{V} \mapsto \begin{bmatrix} \cos(\gamma D) & 0 \\ 0 & \sin(\gamma D) \end{bmatrix} V \tag{7}$$

In this framework, the subnetwork's output is made up of quantum perceptron's that are acting on qubits in layers k − 1 and k. In the heart of the work to be completed for each QNN subnetwork, the goal is to be able to minimize the cost function that relies on the actual and quantum-computed values during the training of the subnetwork. Letting

$$_*\vartheta_t^k = {}_*\alpha^{k+1}\left({}_*\alpha^{k+2}\left(\ldots{}_*\alpha^K\left({}_*\alpha^{K+1}\left(\left|Q_t^{K+1}\right\rangle\left\langle Q_t^{K+1}\right|\right)\right)\ldots\right)\right) \tag{8}$$

and

$$\beta_t^{k-1} = \alpha^{k-1}\left(\ldots\alpha^1\left(\beta_t^I\right)\ldots\right)\Big) \tag{9}$$

we can state the change in the cost function to be the following in a way that the matrix D components have an increasing nature:

$$\nabla M = \frac{\gamma}{n}\sum_{t=1}^{n}\sum_{s=1}^{K+1} trace\left[{}_*\vartheta_t^k\left(\nabla\alpha^k\left(\beta_t^{k-1}\right)\right)\right] \tag{10}$$

This change can be calculated on parameter matrices of any three layers of the subnetwork without the need for using the entire quantum circuitry to be able to fine tune the corresponding matrices M. This allows the training of the deep neural network to be able to fulfill subnetwork tasks.

3 Network of QNN Architecture

The concept of Network of Quantum Neural Network includes a variety of different design considerations to be explained in this section. Just like the design elements pointed out in [32], the following variety of N-QNN formation is possible in applications:

- QNN subnetwork design
- Quantum subnetwork design using QNN
- Quantum-Qubit subnetwork design using QNN
- Hybrid quantum-Traditional subnetwork design using QNN
- Fully mixed subnetworks (introduced in [32])

The variational designs included in this work have QNN subnetwork components, and they can be selected and need to be tested based on specific applications of the problem and the network needed to be designed. The subnetwork design of N-QNN incorporates a variety and mix of network elements including qubits, QNN, QSVM, QCNN, and any of the other traditional techniques such as SVM, ANN, and CNN.

The main subnetwork driver of the N-QNN formation is QNN subnetworks just like how QCNN subnetworks are used for the network of QCNN (N-QCNN) formed in [32, 39], however there is a subtle difference between N-QNN and N-QCNN designs as N-QCNN is effective in image recognition and classification while N-QNN may not be as effective as N-QCNN and their mathematical foundations are different from each other. For instance, the design of a N-QCNN on a low number of facial images' classification is shown to be successful in applications as it is demonstrated as a part of the computational results due to its convolutional layer while N-QNN is effective in other applications [39].

3.1 QNN-Qubit Design of N-QNN

The foundational design of Quantum-Qubit subnetwork design relies on three different combinations of layers that can be made up by the following:

- Only qubits
- Only QNNs
- Mix of Qubits and QNNs

The major difference between the traditional QNN and N-QNN is the utilization of layer-by-layer QNNs with the specific designs of inputs and outputs. This subnetwork design can be useful in special quantum computing tasks that may require a variety of QNN and qubit formations. For instance, this design can be used for dimensionality reduction within the same dataset. Assuming one of the variables is a binary variable, N-QNN has the advantage of using two QNNs in the first layer by training and testing one of the QNN subnetworks on the zero input while training and testing the second QNN on input one. A third QNN subnetwork can also be utilized to train and test the entire data without the splitting of the binary variable to be able to compare with the other two subnetworks results. This is a simple application example, and much more sophisticated solutions can be designed. The same approach also applies to the output design as one QNN subnetwork's output can be other QNN subnetworks' inputs. Figure 3 is an illustration of an N-QNN that is formed by quantum-qubit subnetworks with one or more layers consisting of only qubits, only QNNs, and a mix of qubits and QNNs. The number of qubits and QNNs as a part of the subnetwork design is driven by the design needs of the application.

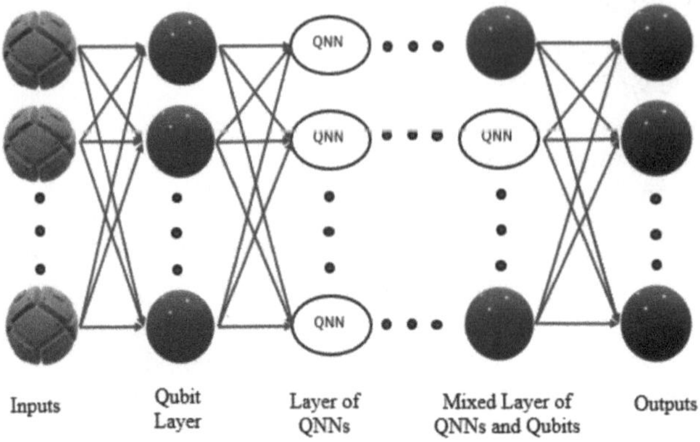

Fig. 3. Subnetwork illustration of N-QNN with elements consisting of QNNs and qubits.

3.2 QNN-Only Design of N-QNN

The N-QNN design can consist of only QNN subnetworks as a special case of the general N-QNN subnetwork design. The concept of hybrid parallel QNN is the closest to this framework [33] however it differs dramatically from it in a way that the design of N-QNN has different input-output formation with the corresponding number of layers and depth variations in the network. The hybrid dense part in [33] contained a combination of classical and quantum layers with the quantum layers' implementation relied on parallel quantum circuits for simultaneous execution to reduce the total computation time. In our work, we extend this framework to incorporate QNNs at any layer similar to the concept of network. The input and output designs need to be configured carefully for accomplishing the subnetwork tasks accordingly. This means independent classifications can be accomplished at each subnetwork that cannot be completed by parallel hybrid QNNs. Figure 4 is an example of such a network configuration for this case. The depth of each layer can be different in this configuration with the flexibility along with the possibility of each QNN's depth variability.

3.3 Hybrid Quantum-Traditional Method Design of N-QNN

Several works have utilized quantum computing in conjunction with classical machine learning methods that focus only on binary classification [34–37] with the utilization of methods that are typically tested on large data sets with objects or numbers that required binary classification. The layers in this section's N-QNN subnetwork design consist of the elements that are made up of a mixture of traditional models and QNNs at each layer with the variable subnetworks' design unlike other methods introduced in the literature.

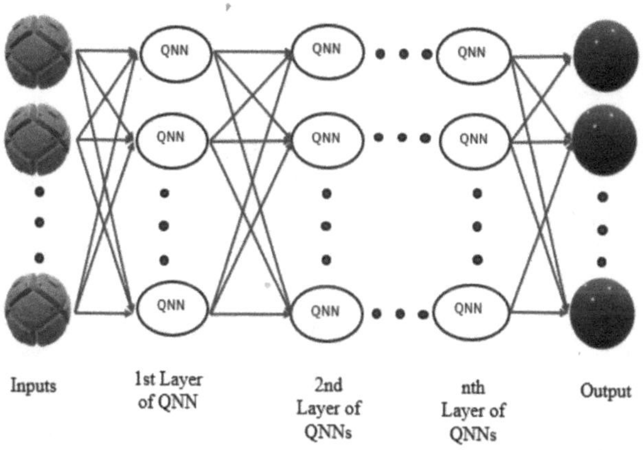

Fig. 4. Subnetwork illustration of N-QNN with elements consisting of only QNNs.

This subnetwork design excludes stand-alone qubit utilization in application cases that can take advantage of quantum AI techniques and traditional methods without the use of qubits. An illustration of this subnetwork design is shown in Fig. 5 below. In this figure, the first layer is shown to be made up of only QNNs while the second layer is made up of traditional methods, and the nth layer is made up of traditional and QNN elements. The input and output design of the subnetworks can be designed in a way to process the input data to produce the corresponding output data with tasks fulfilled in between layers of the subnetwork. The number of qubits and QNNs as a part of the subnetwork design is driven by the design needs of the application.

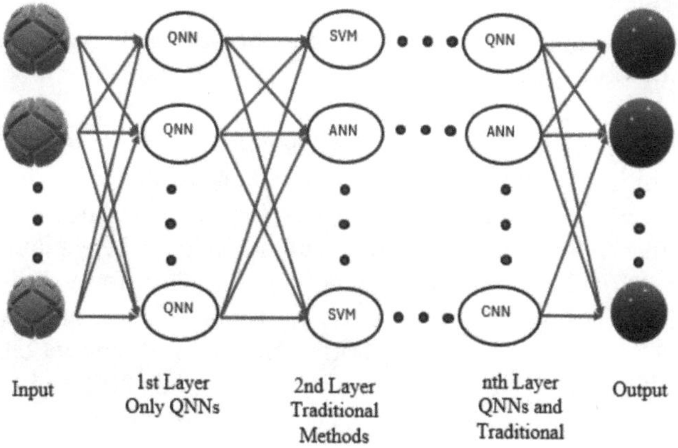

Fig. 5. Subnetwork illustration of N-QNN with elements consisting of QNNs and traditional machine learning methods.

3.4 Fully Mixed Subnetwork Design of N-QNN

The subnetwork design we cover in this section is a mix of quantum AI methods such as QNN, QCNN, QSVM as well as qubits and traditional methods such as SVM, ANN, CNN etc. at each layer. This subnetwork design can be used to benefit in certain applications that need to be tested by researchers just like other methods for specific applications with variational designs (Fig. 6).

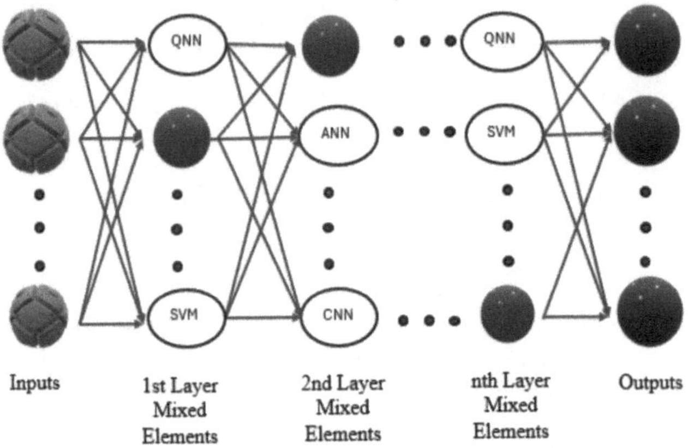

Fig. 6. An example of subnetwork design elements consisting of QNNs and qubits simultaneously.

4 Computational Results

The computational results we present in this section are tested on our custom-made N-QNN circuitry by using quantum simulator on the Iris dataset. The proposed design is made up of a network of QNN with two layers and the goal of the dataset classification is to utilize 3-class grouping of the Iris dataset that is made up of three different flower classes. The traditional ANN accuracy results attained for the Iris flower dataset in the literature is 96% [38]. The proposed N-QNN is identified to be effective with an identification of 97% of the data accurately with its design to be explained below.

We utilize the Iris flower data set that has 3 classes that are formed by Iris Setosa, Iris Versicolor, and Iris Verginica flowers with a total of 50 instances of each class consisting of 150 instances as a demonstration and proof of concept the methodology we introduced. Training and testing of the network relied on 80% and 20% splitting of the data respectively. The design of N-QNN is made up of the following two layers:

- Layer 1: Three parallel QNN subnetworks with each one having identical circuitry shown in Fig. 7 below.
- Layer 2: Three parallel traditional SVM subnetworks with the same design.

In our experimental run, as a special case of the general N-QNN framework, the QNN subnetworks in the first layer are used for feature extraction while SVM subnetworks in the second layer are used for classification purposes. For the proof-of-concept

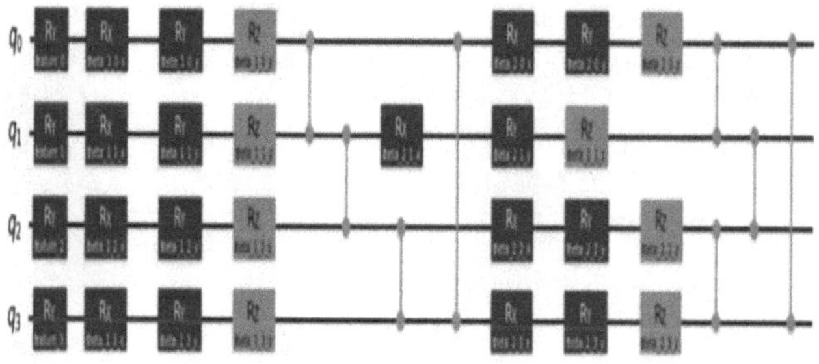

Fig. 7. The circuitry design of QNN for each one of the layers of N-QNN subnetworks.

experimental work, we used IBM's Qiskit with the utilization of the quantum simulator with no expected noise due to the current conditions of the real quantum hardware. A visualization of this architecture is demonstrated in Fig. 8 below.

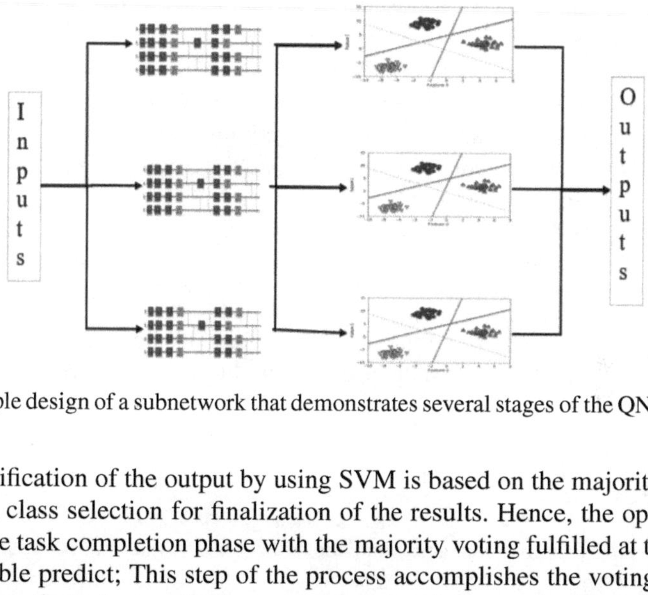

Fig. 8. Example design of a subnetwork that demonstrates several stages of the QNN subnetwork.

The classification of the output by using SVM is based on the majority voting with a single final class selection for finalization of the results. Hence, the optimizer is the last part of the task completion phase with the majority voting fulfilled at the end that is called ensemble predict; This step of the process accomplishes the voting at last, after final prediction, that selects only the one that has the highest voting by the subnetworks. In our specific design, two of the SVM subnetworks attained 97% accuracy while the third one achieved 90% accuracy.

Just like the network of quantum convolutional neural network (i.e. N-QCNN) design introduced in [32, 39], the QNNs that are used for designing subnetworks may have different structures, and each circuit can be designed by using a completely different interface of QNNs. Hence, each QNN can have varying designs and in this work, we are using QNN and SVM subnetworks with small tweaks in the structure of QNN entanglement for each circuit and gates. Expansion on the network by incorporating by other layers and a combination of quantum and traditional AI techniques may provide more accuracy and classification that can be tested as a part of future research endeavors.

For assessment purposes, the accuracy of the N-QNN that we designed in this work is compared with another N-QCNN's accuracy based on a specific design on the same data set. This design of N-QCNN consisted of the same design of N-QNN except all three QNN subnetworks are replaced by QCNN; The corresponding accuracy is attained to be 60% from the entire N-QCNN indicating the strength of the N-QNN that attained 97% accuracy.

5 Discussions

In this work, we introduce a new quantum computing approach, N-QNN, that incorporates both traditional and Quantum artificial intelligence techniques for improving certain shortcomings of quantum artificial intelligence driven accuracy and challenges faced in a variety of ways:

- N-QNN includes a variety of subnetwork designs incorporating a variety of AI models consisting of quantum and traditional AI techniques, and computational shortcomings that may arise from one subnetwork design that can be recovered by another subnetwork design by changing its infrastructure.
- Barren plateau is known to pose a challenge to Quantum AI applications incorporating computational challenges [34] and the proposed N-QNN can be utilized to overcome such a challenge as it can also incorporates traditional methods.
- As shown by the computational results, N-QNN method can attain same or better results on a data set faster than the traditional ANN.
- The flexibility of N-QNN permits modularity in the data processing that can be utilized for dimension reduction on the data set for input feature reduction. This has been shown to be effective in [32].

Overall, the major strength of the N-QNN technique we introduced in this work is its flexibility in design and taking advantage of both traditional and quantum AI techniques' adoption as subnetworks.

References

1. Rosenblatt, F.: The perceptron: a probabilistic model for information storage and organization in the brain. Psychol. Rev. **65**(6), 386 (1958)
2. Talaei Khoei, T., Ould Slimane, H., Kaabouch, N.: Deep learning: systematic review, models, challenges, and research directions. Neural Comput. Appl. **35**(31), 23103–23124 (2023)
3. Introduction to the quantum programming language Q#. Microsoft Corp. https://learn.microsoft.com/en-us/azure/quantum/qsharp-overview. Accessed 29 May 2025
4. Qiskit 2.0, IBM. https://www.ibm.com/quantum/qiskit. Accessed 29 May 2025
5. Cirq, Google. https://quantumai.google/cirq. Accessed 29 May 2025
6. Abu Ghanem, M.: IBM quantum computers: evolution, performance, and future directions. J. Supercomput. **81**(5), 687 (2025)
7. Peruzzo, A. et al.: A variational eigenvalue solver on a photonic quantum processor. Nat. Commun. **5**(1), 4213 (2014)
8. Benedetti, M., Lloyd, E., Sack, S., Fiorentini, M.: Parameterized quantum circuits as machine learning models. Quantum Sci. Technol. **4**(4), 043001 (2019)

9. Farhi, E., Goldstone, J., Gutmann, S.: A quantum approximate optimization algorithm. arXiv. (2014). https://doi.org/10.48550/arXiv.1411.4028
10. Rebentrost, P., Mohseni, M., Lloyd, S.: Quantum support vector machine for big data classification. Phys. Rev. Lett. **113**(13), 130503 (2014)
11. Aïmeur, E., Brassard, G., Gambs, S.: Quantum speed-up for unsupervised learning. Mach. Learn. **90**, 261 (2013)
12. Paparo, G.D., Dunjko, V., Makmal, A., Martin-Delgado, M.A., Briegel, H.J.: Quantum speedup for active learning agents. Phys. Rev. X. **4**, 031002 (2014)
13. Schuld, M., Sinayskiy, I., Petruccione, F.: The quest for a quantum neural network. Quantum Inf. Process. **13**, 2567–2586 (2014)
14. Wiebe, N., Kapoor, A., Svore, K.M.: Quantum perceptron models. In: Lee, D.D., Sugiyama, M., Luxburg, U.V., Guyon, I., Garnett, R. (eds.) Proceedings of the Neural Information Processing Systems 2016, pp. 3999–4007. Curran Associates, New York (2016)
15. Sasaki, M., Carlini, A.: Quantum learning and universal quantum matching machine. Phys. Rev. A. **66**, 022303 (2002)
16. Gambs, S.: Quantum classification. arXiv. (2008). https://doi.org/10.48550/arXiv.0809.0444
17. Sentís, G., Calsamiglia, J., Muñoz-Tapia, R., Bagan, E.: Quantum learning without quantum memory. Sci. Rep. **2**, 708 (2012)
18. Dunjko, V., Taylor, J.M., Briegel, H.J.: Quantum-enhanced machine learning. Phys. Rev. Lett. **117**, 130501 (2016)
19. Monràs, A., Sentís, G., Wittek, P.: Inductive supervised quantum learning. Phys. Rev. Lett. **118**, 190503 (2017)
20. Alvarez-Rodriguez, U., Lamata, L., Escandell-Montero, P., Martín-Guerrero, J.D., Solano, E.: Supervised quantum learning without measurements. Sci. Rep. **7**, 13645 (2017)
21. Amin, M.H., Andriyash, E., Rolfe, J., Kulchytskyy, B., Melko, R.: Quantum Boltzmann machine. Phys. Rev. X. **8**, 021050 (2018)
22. Du, Y., Hsieh, M.H., Liu, T., Tao, D.: Expressive power of parametrized quantum circuits. Phys. Rev. Res. **2**(3), 033125 (2020)
23. Sentís, G., Monràs, A., Muñoz-Tapia, R., Calsamiglia, J., Bagan, E.: Unsupervised classification of quantum data. Phys. Rev. X. **9**, 041029 (2019)
24. Kak, S.C.: Quantum neural computing. Adv. Imaging Electron Phys. **94**, 259–313 (1995)
25. Beer, K. et al.: Training deep quantum neural networks. Nat. Commun. **11**(1), 808 (2020)
26. Verdon, G., Pye, J., Broughton, M.: A universal training algorithm for quantum deep learning. arXiv. (2018). https://doi.org/10.48550/arXiv.1806.09729
27. McClean, J.R., Boixo, S., Smelyanskiy, V.N., Babbush, R., Neven, H.: Barren plateaus in quantum neural network training landscapes. Nat. Commun. **9**(1), 4812 (2018)
28. Huang, H.L. et al.: Experimental quantum generative adversarial networks for image generation. Phys. Rev. Appl. **16**(2), 024051 (2021)
29. Mitarai, K., Negoro, M., Kitagawa, M., Fujii, K.: Quantum circuit learning. Phys. Rev. A. **98**, 032309 (2018)
30. Zeng, J., Wu, Y., Liu, J.G., Wang, L., Hu, J.: Learning and inference on generative adversarial quantum circuits. Phys. Rev. A. **99**(5), 052306 (2019)
31. Hamilton, K.E., Dumitrescu, E.F., Pooser, R.C.: Generative model benchmarks for superconducting qubits. Phys. Rev. A. **99**(6), 062323 (2019)
32. Tokgoz E., Sharmendra, D.: Network of Quantum Convolutional Neural Networks, accepted to appear in CAC'26 Conference Proceedings, Springer Nature (2026)
33. Senokosov, A., Sedykh, A., Sagingalieva, A., Kyriacou, B., Melnikov, A.: Quantum machine learning for image classification. Mach. Learn. Sci. Technol. **5**(1), 015040 (2024)
34. Ranga, D., Prajapat, S., Akhtar, Z., Kumar, P., Vasilakos, A.V.: Hybrid quantum–classical neural networks for efficient MNIST binary image classification. Mathematics. **12**, 1–22 (2024)

35. Hafeez, M.A., Munir, A., Ullah, H.: H-QNN: a hybrid quantum–classical neural network for improved binary image classification. AI. **5**(3), 1462–1481 (2024)
36. Li, W., Chu, P.C., Liu, G.Z., Tian, Y.B., Qiu, T.H., Wang, S.M.: An image classification algorithm based on hybrid quantum classical convolutional neural network. Quantum Eng. **2022**, 5701479 (2022)
37. Fan, F., Shi, Y., Guggemos, T., Zhu, X.X.: Hybrid quantum-classical convolutional neural network model for image classification. IEEE Trans. Neural Netw. Learn. Syst. **35**, 18145–18159 (2023)
38. Goyal, S., Sharma, A., Gupta, P., Chandi, P.: Assessment of iris flower classification using machine learning algorithms. In: Marriwala, N., Tripathi, C.C., Jain, S., Mathapathi, S. (eds.) Soft Computing for Intelligent Systems. Algorithms for Intelligent Systems. Springer, Singapore (2021)
39. Desiboyina, S., Tokgoz, E.: Facial recognition by using ensemble and network of quantum convolutional neural network. In: Accepted to Appear in the Proceedings of the 2025 International Conference on Computational Science and Computational Intelligence (CSCI'25), Signal and Image Processing, Computer Vision & Pattern Recognition. Springer Nature (2025)

Soil Classification Using a ResNet50-EfficientNet Hybrid Model with Efficient Channel Attention

Abdulrahman Hassan Alhazmi[1](✉), Saeed M. Alshahrani[2], and Yahya Jaber Ghazwani[3]

[1] Department of Computer Science, College of Engineering and Computer Science, Jazan University, Jazan 45142, Kingdom of Saudi Arabia
abalhazmi6@jazanu.edu.sa

[2] Department of Computer Science, College of Computing and Information Technology, Shaqra University, Shaqra, Saudi Arabia

[3] Department of Computer Science, Applied College, Jazan University, Jazan 82817, Saudi Arabia

Abstract. Soil classification is critical in agricultural and management of the environment, conventional techniques are costly and labor-intensive. In this research, a novel deep learning hybrid model is proposed that enhances soil image classification by combining ResNet50's hierarchical feature extraction with Efficient Net's parameter efficiency. Furthermore, Efficient Channel Attention (ECA) blocks are used to enhance key features and improve the reliability of classification. This method enhances feature learning while being computationally efficient. The model's flexibility makes it appropriate for large-scale soil categorization tasks. The proposed model outperformed baseline architectures as well as performance evaluations. The proposed hybrid model achieves overall 97.7% accuracy.

Keywords: Soil Type · ResNet50 · EfficientNet · Channel Attention · deep model

1 Introduction

Machine learning has led to substantial advances in the application of computer vision, especially fine-grained image analysis. The feature extraction of complex textures and the architecture is the crucial domain of soil image classification. Classification is crucial for establishing fundamental concepts in research, to also identify diversity between the concepts and objects. It gives vital information for comprehensive studies [1]. Soil is a vital component in agricultural fields that contributes to crop productivity. Soil classification concepts are based on existing information and real-world circumstances. On Earth's land surfaces, soil classification establishes a link among soil samples and many types of natural entities [2]. Soil classification systems include the British standard system, Casagrande-based systems, AASHTO (American Association of State Highway and Transportation Officials), USCS (unified soil classification system), and USDA (US Department of Agriculture) [3]. Soil classification is typically based on features including

H. R. Arabnia et al. (Eds.): CSCI 2025, CCIS 2967, pp. 449–457, 2027.
https://doi.org/10.1007/978-3-032-28454-9_31

ion systems, grain size delivery, liquid limit, and flexibility [4]. For instance, the USDA categorization primarily evaluates the particle size distribution, whereas the USCS additionally includes plasticity of soil and assigns it to both the silt and clay categories. However, soil classification based on variation in particle size may have limitations due to differences in physical qualities among particles of equal size. While particle size is a key requirement for soil categorization, additional factors need to be considered for accurate results [5]. The essential features of soil are some discrete fragments (e.g., plant components, clay substances, quartz grains) that can be detected in fragments from employing an optical microscope. Structure of soil is strongly associated with the sharpness, form, contrast rate, size, gaps and spatial distribution of important particles. Moreover, several of similar qualities are reliant on the alignment of components and also the approach to the particles is cut as well as on the magnifying used [6].

Conventional soil classification strategies are frequently costly and time-consuming, making them challenging for large-scale classification. Traditional deep learning algorithms face challenges in capturing the fine-grained and hierarchical needs to achieve accurate soil classification. This paper presents a hybrid architecture that combines ResNet50 and EfficientNet, using their capabilities to provide reliable feature extraction. Furthermore, using Efficient Channel Attention (ECA) improves the model's ability to detect fine-grained features, achieving higher classification accuracy.

The contribution of this study is given below:

- Develop a hybrid deep learning model that combines ResNet50, EfficientNet, for feature extraction, which improves the classification accuracy.
- Employ Efficient Channel Attention (ECA) to improve feature selection and optimize soil grouping predictions.
- Evaluate the model's performance over a variety of soil samples to validate its resilience and dependability in application in real life.

The rest of the research work is arranged as follows. Section II discusses the proposed methodology in detail. Section III summarizes experimental results. The conclusion is reached in sect. IV.

2 Literature Review

In recent years, the significance of soil classification in reliable agriculture and food safety has been emphasized in several studies. The traditional methods for soil classification and analysis are labor-intensive and subjective. The latest research using ML-based supervised learning algorithms and ensembles is used on standard datasets for classification of soil and crop prediction [7]. These machine learning models examine soil parameters and recommend appropriate crops, increasing crop productivity. The authors proposed [8] "semi-supervised classification with stacking learning (SSC-SL)" which enhances the soil classification in rocky and low-mountain places. This model improves the classification efficiently by integrating the topographic features, the remote sensing data, and the soil parent components. The SSC-SL model for enhancing predictions employs unlabeled data close patch boundaries, resulting in an overall accuracy of 0.77 and higher stability. The authors [9] introduced in this research to analyze the soil groups on the Jiaodong Peninsula (China) using Landsat eight images, entropy texture metrics (23 × 23) and extracting mean parameter (19 × 19) over several optimal window sizes. The evaluation was conducted to determine the most efficient strategy for soil classification using numerous machine learning classifiers, including ANN, RF, MLC, and SVM. The precise spatial distribution of soil categories is required for optimal soil management of resources. The authors [10] explore the machine learning algorithms such as RF, NN, Naïve Bayes, SVM, DT, and k-NN, to perform automated soil classification on large datasets. This research analyzes various models based on the soil sample parameters and highlights issues including expert accessibility and stability. The results show that machine learning represents a more effective alternative to traditional soil classification methods, with the achieving accuracies of 97.23% for Random Forest, 96.82% for Naïve Bayes, and 92.92% for k-Nearest Neighbor. The models using deep learning, especially CNNs and their derivatives, including Unet, have seen widespread use in remote sensing image analysis. However, CNNs cannot capture long-range connections, while transformers thrive at handling global factors. This research proposes [11] to address this issue with the "efficient channel attention-enhanced dual-stream neural network (ECA-DSNN)" model, which combines the transformer framework and Unet to enhance the identification of architecture. The proposed model includes an attention mechanism with modest fine-tuning for cross-scene transfer to improve the generalization. The outcomes of overall accuracy are 96.3% and F1 score is 78.6%. Authors [12] proposed the ECA-ViT model, which includes a hybrid approach that combines Efficient Channel Attention (ECA) with Vision Transformers (ViT) to enhance the detection of disease in crop leaves. This model improves the local feature learning, whereas resolving ViT's deficits in wider context comprehension by using adaptive 1D convolution kernels. The results of this model, 95.41% accuracy on the dataset of rice leaf disease from genuine field conditions. Authors [13] introduce several cutting-edge machine learning models such as Wide Residual Networks (WRNs), CNNs, MLPs, Vision Transformers, and EfficientNets, applying them on the BigEarthNet dataset. To enhance the accuracy and effectiveness, the proposed technique combines EfficientNets with a scaled WRN design that incorporates an **Efficient Channel Attention** (ECA) technique. This research [14] used three machine learning models to propose a novel soil classification approach including Multilayer Perceptron (MLP), Random Forest (RF), and Support Vector Classification (SVC). The 4,888 soil samples from Vietnam were employed, with the input features of soil properties of 15, which classified samples into five soil categories. The outcomes show that all the models performed effectively, and the overall average accuracy was 0.968, with SVC achieving the highest accuracy of 0.984.

3 Proposed Model

In this section, our proposed model classifies soil images using deep learning algorithms. The preprocessing method maintains resizing and normalization consistency of soil images (224 × 223 × 3). We combine components from ResNet50 and EfficientNet. The ResNet50's deep residual learning tool assists in detecting hierarchical features, whilst EfficientNet improves parameter performance. This combination allows reliable feature extraction of soil images and makes the hybrid architecture. Soil images can feature high textures and fine-grained details. A hybrid architecture draws on the qualities of both models, to ensure subtle elements are not neglected. The ECA improves the inter-channel dependencies, which shows the enhanced features selection with the assistance of feature extraction using ResNet50 and EfficientNet. ECA blocks get incorporated into convolutional layers. They concentrate on local cross-channel interactions while avoiding significant computational complexity. Lastly, soil categories prediction with maximum accuracy and efficiency using classification.

3.1 Hybrid Architecture

After the preprocessing, employs the hybrid architecture that combines ResNet50 and EfficientNet, leveraging their enhancing features. This hybrid model, selecting as a base network, is critical for achieving high accuracy in image classification.

Firstly, for deep residual learning, the ResNet50 (Residual Network) is well-known, which solves the problem of vanishing gradient in soil images and allows for the training of extremely deep networks. The primary element is the residual block shown in Eq. 1.

$$y_l = F(x_l, w_l) + x_l \quad (1)$$

Where, x_l is the input at layer l, the layer's weight parameters denoted by w_l and the residual function represented by $F(x_l, w_l)$ (Fig. 1).

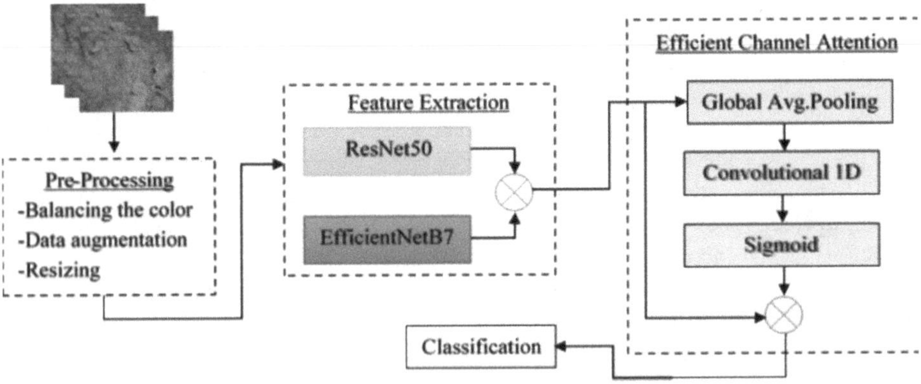

Fig. 1. Flow diagram of proposed soil types classification.

The effective features extraction across numerous hierarchical levels in residual learning, which makes it perfectly suited for capturing complicated soil textures. The ResNet5050 features in deep learning enable consistent gradient flow which improves classification performance.

Firstly, for deep residual learning, the ResNet5050 (Residual Network) is well-known, which solves the problem of vanishing gradient in soil images and allows for the training of extremely deep networks. The primary element is the residual block shown in Eq. 1.

$$y_l = F(x_l, w_l) + x_l \quad (1)$$

Where, x_l is the input at layer l, the layer's weight parameters denoted by w_l and the residual function represented by $F(x_l, w_l)$. The effective features extraction across numerous hierarchical levels in residual learning, which makes it perfectly suited for capturing complicated soil textures. The ResNet5050 features in deep learning enable consistent gradient flow which improves classification performance.

Secondly, EfficientNet implements a compound scaling approach to improve depth, width, and resolution of images. The scaling formula is shown in Eq. 2.

$$depth_i = \alpha^{\varnothing}$$

$$width_i = \beta^{\varnothing}$$

$$resolution_i = \gamma^{\varnothing} \tag{2}$$

where ϕ represents the scaling coefficient and predefined parameters represented by α, β, and γ. By combining these three elements, EfficientNet achieves high efficiency while remaining computationally effective. This method extracts features from high-resolution soil images and lightweight design minimizes complexity while maintaining beneficial accuracy, making it appropriate for large-scale soil classification.

Our hybrid model combines ResNet5050's residual blocks for improved hierarchical learning and EfficientNet's backbone for primary feature extraction. The total feature extraction method is shown in Eq. 3.

$$Hybrid_Y = ResNet(EfficientNet(Hybrid_X)) \tag{3}$$

Where X is the soil input image and Y represents the features extraction. This architecture provides that deep spatial features are efficiently captured while minimizing computing complexity.

3.2 Attention Mechanisms

The attention mechanisms are employed after the fine-tune features extraction. The Efficient Channel Attention (ECA) block has been incorporated into our model to improve channel-wise dependency. The ECA block improves the feature captures by providing several weights to every feature channel. This mechanism enhances the model's capacity to focus on important traits while remaining computationally efficient. The channel-wise attention factors are calculated as shown in Eq. 4.

$$s_c = \sigma(W.g_c) \tag{4}$$

Where the s_c represent the channel-wise attention scores, σ used for sigmoid activation function, the learnable weight matrix by W and g_c present the global descriptor obtained through global average pooling. ECA increases the model's sensitivity to key features, enhancing soil accuracy in classification. Because it avoids reducing dimensionality, this maintains key channel dependencies while preserving computing efficiency, which makes it appropriate for soil texture research.

4 Experimental Results

In this section, evaluated the proposed model on the soil image dataset publically available in kaggle the dataset contain 5 varieties of soil images, which demonstrates higher accuracy for classification. ResNet50 hierarchical learning associated with EfficientNet multi-scale feature extraction and significantly enhanced performance. These outcomes validate the proposed model efficiency in the soil classification task, and the model achieved higher accuracy than the traditional CNN architecture. Setting of the hyperparameter to train the proposed model is shown in Table 1.

Table 1. Hyperparameter and setting of the training for the proposed model

Hyper-parameter	Values
Base backbone	ResNet50 + EfficientNet-B3
ECA block placement	Mid layers (ResNet50 Stage 3)
Fusion method	Concatenation +1 × 1 Convolution
Input image size	224 × 224 × 3
Normalization	Dataset-specific mean/std
Optimizer	AdamW
Learning rate	3e-4 (Cosine Annealing)
Batch size	32
Epochs	100
Loss function	Cross-Entropy + Label Smoothing (0.1)
Data augmentation	Flip, Rotation, Color Jitter
ECA Kernel size	Adaptive k $=\lceil log_2(C)/2 + 1\rceil$
ResNet50 depth	50 layers
EfficientNet scaling	B3 (Compound)
Dropout rate	0.3

In the first experiment, as shown in Table 2, present the results of the proposed model for each class. Results show that the Laterite and Peat soil types are precisely classified up to 100% and 98.2% respectively as compared to the other classes.

The Fig. 2 illustrates a confusion matrix for a multi-class classification challenge with four classes. Each row relates to the actual class, and every single column reflects the predicted class. Diagonal values indicate proper classifications, but off-diagonal values reflect misclassifications. The Black 1 represents the 160 samples, 150 were identified correctly and five misclassified with other classes. The Yellow identified correctly 148 samples and misclassified 12 in other classes. In conclusion, the model achieves significant classification accuracy, as numerous predictions match the actual classes with few misclassifications.

Table 2. Different matrices values of each soil class types.

Class	Precision	Recall	F1 Score
Black	96.8	98.0	97.4
Cinder	91.4	97.7	94.5
Laterite	1	88.9	94.1
Peat	98.2	95.6	96.9
Yellow	90.5	92.2	91.3

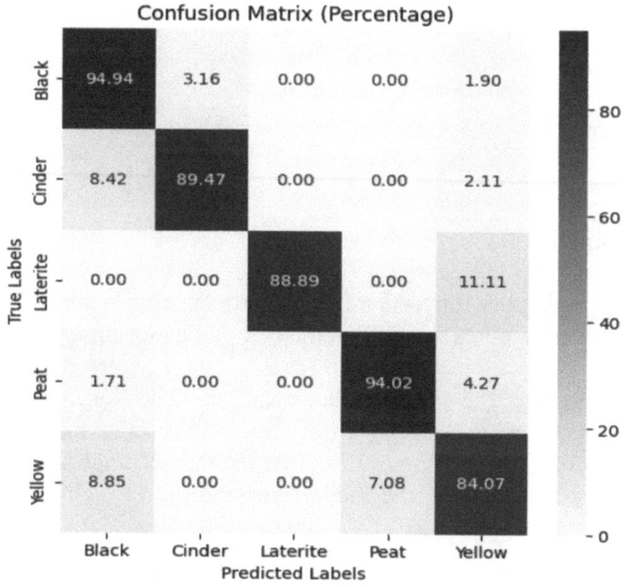

Fig. 2. Confusion matrix result in percentage.

The Table 3 compares the accuracy of multiple model variants used for soil classification tasks. The baseline model achieved 85.2% accuracy, exhibiting a high capability of feature extraction. Furthermore, including the Efficient Channel Attention (ECA) strategy improved the model's efficiency, resulting in an overall accuracy of 91.7%. This trend indicates that each component helps to improve feature learning, which leads to superior soil classification.

The Table 4 evaluates the efficiency of the three different models, including ResNet50, EfficientNet-B0, and the Proposed Hybrid Model. The ResNet50 indicates a robust feature extraction capability while EfficientNet-B0 is highly computational and performed a little bit lower. Lastly, the proposed hybrid model which combines EfficientNet, ResNet50, and ECA outperforms both, achieving higher accuracy with enhanced precision. This highlights the effectiveness of integrating hierarchical feature extraction and attention techniques for better soil classification ability.

Table 3. Performance Analysis of Several Model Variants

Model variant	EfficientNet	ResNet50	ECA	Accuracy
Baseline (EfficientNet)	✓	✗	✗	85.2%
EfficientNet + ResNet50	✓	✓	✗	88.4%
EfficientNet + ResNet50 + ECA	✓	✓	✓	95.7%

Table 4. Performance evaluations for different models

Model	Accuracy (%)	Precision (%)	Recall (%)	F1-score (%)
ResNet50–50	87.1	86.5	86.9	86.7
EfficientNet-B0	85.2	84.9	85.1	85.0
Proposed Hybrid Model	**93.7**	**91.3**	**91.5**	**91.4**

5 Conclusion

This research proposed a novel hybrid model that effectively combines ResNet50's hierarchical extraction of features and EfficientNet's parameter optimization. The model improves key soil image elements, which results in higher classification accuracy by integrating ECA blocks. The results indicate the efficiency of our strategy and performance indicators, proving their superiority over the baseline models. The proposed hybrid model achieves 91.7% accuracy. Future research will emphasize improving computational efficiency and testing the model on a variety of soil imaging datasets to determine its generalizability.

References

1. Rahman, S.A.Z., Mitra, K.C., Islam, S.M.: Soil classification using machine learning methods and crop suggestion based on soil series. In: 2018 21st International Conference of Computer and Information Technology (ICCIT), pp. 1–4. IEEE (2018)
2. Eswaran, H., Ahrens, R., Rice, T.J., Stewart, B.A.: Soil Classification: a Global Desk Reference. CRC Press (2002)
3. Gambill, D.R., Wall, W.A., Fulton, A.J., Howard, H.R.: Predicting USCS soil classification from soil property variables using random. Forest. **65**, 85–92 (2016)
4. Heung, B., Ho, H.C., Zhang, J., Knudby, A., Bulmer, C.E., Schmidt, M.G.: An overview and comparison of machine-learning techniques for classification purposes in digital soil mapping. Geoderma. **265**, 62–77 (2016)
5. Moreno-Maroto, J.M., Alonso-Azcárate, J., O'Kelly, B.C.: Review and critical examination of fine-grained soil classification systems based on plasticity. Appl. Clay Sci. **200**, 105955 (2021)
6. Srivastava, P., Shukla, A., Bansal, A.: A comprehensive review on soil classification using deep learning and computer vision techniques. Multimed. Tools Appl. **80**(10), 14887–14914 (2021)

7. Goswami, T., Ganapathi, D.M., Goswami, P.: Soil Classification and Crop Prediction Using Machine Learning Techniques, in Intelligent Robots and Drones for Precision Agriculture, pp. 101–118. Springer (2024)
8. Zhu, F., Zhu, C., Lu, W., Fang, Z., Li, Z., Pan, J.: Soil classification mapping using a combination of semi-supervised classification and stacking learning (SSC-SL). Remote Sens. **16**(2), 405 (2024)
9. Duan, M., Song, X., Li, Z., Zhang, X., Ding, X., Cui, D.: Identifying soil groups and selecting a high-accuracy classification method based on multi-textural features with optimal window sizes using remote sensing images. Eco. Inform. **81**, 102563 (2024)
10. Azmin, S.N.F.L.M., Rahman, H.M., Lim, N.N.M.H., Arbaiy, N.: Machine learning for soil classification: challenges and opportunities. J. Appl. Sci. Technol. Comput. **1**(1), 29–38 (2024)
11. Li, W., et al.: Cross-scene building identification based on dual-stream neural network and efficient channel attention mechanism. IEEE J. Sel. Topics Appl. Earth Obs. Remote Sens. **17**, 6920–6932 (2024)
12. He, F., Liu, Y., Liu, J.: ECA-ViT: leveraging ECA and vision transformer for crop leaves diseases identification in cultivation environments. In: 2024 4th International Conference on Machine Learning and Intelligent Systems Engineering (MLISE), pp. 101–104. IEEE (2024)
13. Papoutsis, I., Bountos, N.-I., Zavras, A., Michail, D., Tryfonopoulos, C.: Efficient deep learning models for land cover image classification. **5**, 1–13 (2021)
14. Nguyen, M.D., et al.: Novel approach for soil classification using machine learning methods. Bull. Eng. Geol. Environ. **81**(11), 468 (2022)

Cross-Lingual Benchmarking for Image Captioning: Assessing Translation in Low-Resource Settings

Anahita Dinesh(✉), Ching-Seh Wu, and Robert Chun

San Jose State University, San Jose, USA
{anahita.dinesh,ching-seh.wu,robert.chun}@sjsu.edu

Abstract. One of the most prominent tasks that lie on the conjunction of Natural Language Processing (NLP) and computer vision, is image captioning. Image captioning is the generative task of achieving textual descriptions from images. Its application finds use in many real-world scenarios like aiding the visually impaired, editing applications, recommendation systems, and medical imaging. This research focus lies in Hindi image captioning, the official language of India, as it has not been explored as far as its need. Several challenges such as the lack of substantial Hindi text data for training models, the need for human annotators to verify the accuracy of Hindi text, and insufficient training of pre-trained models on Indic languages for translation, can be realized. This paper has leveraged the state-of-the-art IndicTrans2, a cutting-edge pre-trained translation model, as a new approach to create a novel dataset. This innovative approach marks the first instance of IndicTrans2 being utilized for Hindi image captioning, as opposed to previous works that have used Google Translate. With this newly translated Hindi corpus, encoder-decoder architectures integrating convolutional neural networks (CNNs) and recurrent neural networks (RNNs), have been built and trained to effectively learn and generate captions in Hindi that are both accurate and contextually relevant to the corresponding images. This methodology marks the first comparative study between IndicTrans2 and a commercial baseline for this multimodal task.

Keywords: Natural Language Processing · image captioning · IndicTrans2 · Hindi

1 Introduction

Image captioning, a dynamic field at the intersection of computer vision and natural language processing (NLP), holds immense potential in bridging the gap between visual content and linguistic description. Primarily focused on the English language, this technology has made significant strides in accessibility and interactive digital content. However, the need for image captioning extends far beyond the English realm. The ability to generate accurate and contextually

H. R. Arabnia et al. (Eds.): CSCI 2025, CCIS 2967, pp. 458–468, 2027.
https://doi.org/10.1007/978-3-032-28454-9_32

relevant captions in various languages not only enhances user experience but also ensures inclusivity, opening new avenues for global communication.

Despite this progress, the field faces several challenges when transcending linguistic barriers. A primary hurdle is the scarcity of comprehensive, human-annotated datasets in languages other than English, which are crucial for training effective models. This limitation often forces a reliance on tools like Google Translate for dataset creation, a method that frequently fails to capture the nuanced semantic and cultural subtleties of the target language. While explorations continue to improve the quality of generated captions, the persistent challenge lies in perfecting the data-generation and adaptation process for diverse, low-resource languages.

This research aims to explore a novel approach by addressing this data quality bottleneck directly. We investigate the capabilities of IndicTrans2 [1], a cutting-edge open-source translation model specifically optimized for Indic languages, as a superior proxy for human annotation. Unlike previous works that relied on general-purpose commercial models (e.g., Google Translate) for dataset creation, this study utilizes IndicTrans2 to translate the English Flickr8k dataset [2] into a high-fidelity Hindi corpus. This innovative approach marks the first direct comparison of IndicTrans2 against a commercial machine translation benchmark for a multimodal task.

Using this newly translated Hindi corpus, we build and train classic encoder-decoder architectures, integrating CNNs and RNNs. This choice of baseline models ensures direct comparability with foundational research in the field. Six distinct model variations were implemented and refined, each experimented with hyperparameter tuning to reduce overfitting and improve performance. This methodology tests the hypothesis that a semantically richer, IndicTrans2-derived reference corpus significantly enhances the output quality of image captioning models, thereby setting a new and achievable benchmark for cross-lingual transfer in Hindi.

2 Related Work

For the purpose of image captioning in Hindi, the world's fourth most spoken language, [3] translated the original captions in the Flickr8K dataset to create the Flickr8K Hindi dataset. Upon implementing the encoder-decoder architecture with a CNN and RNN, the authors found that training with five uncleaned captions per image yielded the highest BLEU score, as opposed to using a single caption per image. Findings from this research showcased that extracting features from the flatten layer gave better prediction accuracy than extracting them from the pooling layer. It was also found that VGG16 performed better among other CNNs, such as InceptionV3 and ResNet50, for producing strong image vector representations. The models were further compared by varying the number of dense and Long Short-Term Memory (LSTM) layers, concluding that adding multiple layers of both the dense and LSTM components led to an increase in error compared to a single layer.

A Transformer-based architecture was employed in [4]. The authors translated the MSCOCO [5] dataset into Hindi using Google Translate, followed by a subsequent and resource-intensive correction phase by human annotators. In their encoder-decoder architecture, the popular CNN, ResNet101, was used as the encoder, and a Transformer was used as the decoder.

Another research work by [6] also utilized Google Translate on the MSCOCO dataset and stated that the translated captions suffered from poor grammar and that the accuracy of Google Translate was not standardized due to its reliance on source and target language pair availability. Their methodology involved implementing an encoder-decoder architecture with efficient channel attention in the CNN encoder and a Gated Recurrent Unit (GRU) as the decoder.

To emphasize the importance of selecting the right CNN for image feature extraction, [7] experimented with VGG16, VGG19, InceptionV3, and ResNet50 as encoders, all paired with an LSTM decoder on the Flickr8K dataset. In the encoder-decoder architecture, the authors employed the merge-add technique. A dropout rate of 0.4, the ADAM optimizer, and a categorical cross-entropy loss function were employed in the training of the models, which were run for 20 epochs.

3 Methodology

3.1 Translation Using IndicTrans2

The first step in our methodology was to generate a Hindi dataset for the image captioning models to train on. We chose the Flickr30K dataset, which has 31,783 images, with each image having five representative captions. IndicTrans2 is the first open-source multilingual pre-trained Transformer model that has been trained on the largest publicly available parallel corpora for 22 Indic languages, the Bharat Parallel Corpus Collection (BPCC), which has data distributed among various domains like entertainment, health, tourism, and agriculture. The model was released in 2023 by AI4Bharat and is a new, promising, cutting-edge model aimed at improving the translation system between English and Indic languages.

The tokenizer we used was IndicTransTokenizer [1], as it is built to be compatible with IndicTrans2. While setting up parameters for the tokenizer, we specified the direction = "en-indic" argument, which establishes the translation direction is from English to Indic languages. To allow batch processing, an instance of IndicProcessor [1] was created and set to True. For using the tokenizer during training/fine-tuning, we set the inference argument of IndicProcessor to False. Finally, the IndicTrans2 model can be imported using the HuggingFace library to generate the output translations. For translation, the source and target languages have to be specified. The input can be a batch of sentences, originally in English.

3.2 Data Preparation

The output captions, now in Hindi, were saved to an output text file where each line consisted of the image name and an array of its captions. Upon observing the translated outputs, some manual cleaning was required, such as removing abnormally generated texts that would lead to noisy data if not eliminated. This cleaning effort was undertaken manually to ensure the data was thoroughly up to the mark.

Another observation worth noting is that for English-specific words that do not have a corresponding vocabulary in Hindi, the IndicTrans2 model handled them perfectly through transliteration. Transliteration, when converting English to Hindi, converts the English text into the Devanagari script, representing the sounds of the English words using Hindi characters rather than translating their meaning. For example, the English word "India" can be transliterated into Hindi as "इंडया ". This feature provides credibility to the model for handling out-of-vocabulary English words and maintaining context in Indic languages.

Further data processing was carried out in the form of tokenization to create the vocabulary. We attempted stop-word removal; however, the observed output captions were not properly structured, and thus, we retained the stop words to achieve coherent Hindi sentences. To prepare the captions as input for model training, we added the '**startseq**' token to the beginning of each caption and '**endseq**' at the end. This aids the models in recognizing the start and end of a caption, enabling the generation process to stop when the '**endseq**' token is met. The dataset was split into 70:15:15 for training, testing, and validation. A DataGenerator was used to generate batches of data for model training.

3.3 Model Architecture

The proposed model architecture is shown in Fig. 1. The encoder-decoder framework consists of two stages: an encoder which takes a raw input and converts it into a vector representation. In image captioning, pre-trained CNNs are used as the image encoder. We extract image features from the layer immediately before the top classification layer to retrieve the visual information and represent it in a fixed-length vector. The decoder is implemented by RNNs to process the textual input sequence and use the current state to predict the next word in the caption.

By slightly modifying the encoder-decoder architecture, the model is evolved into an image caption generator, where predictions are influenced by a given input image. For such a model to perform the task, it must integrate the given image into its context, along with the preceding tokens of the caption. This raises a key question: When is the optimal moment to introduce image features into the model? In [8], it was found that the 'merge' technique was comparatively better than the 'inject' technique for combining image and text context into the network; consequently, we have implemented the 'merge-add' and 'merge-concat' methods. Figure 1 shows that for 'merge' architectures, the RNN subnetwork is only responsible for processing the caption prefix, dealing solely with linguistics.

The visual data, specifically the image feature vector, is introduced only after the RNN has encoded the caption sequence. The integration of the image and text data is achieved by methods such as vector addition or concatenation. In this methodology, the RNN focuses on encoding the textual prefix, while the addition of textual and visual information is handled by a subsequent feed-forward layer, which acts as the decoder in generating caption sequences. The softmax activation function outputs the probability of each word in the vocabulary to be the next word in the sequence, based on the prefix and image features. This is repeated until the model predicts the '**endseq**' token or the output caption reaches the defined maximum length.

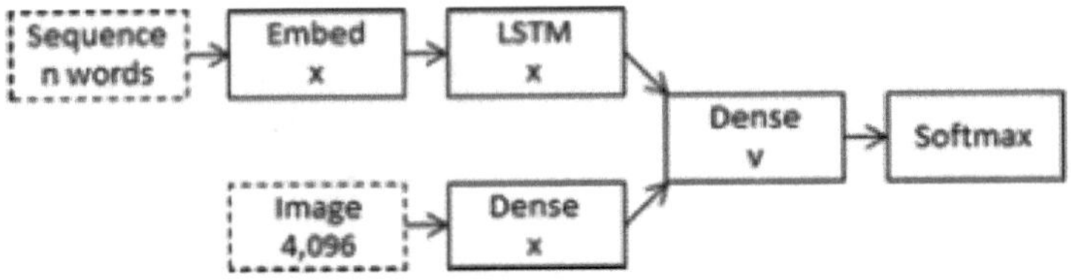

Fig. 1. Proposed model architecture [3]

4 Experiments

Model training was performed on Google Colab using their A100 and V100 GPUs. Due to the powerful GPUs being subscription-based, our computational resources were limited. In this research, for image encoding, we used InceptionV3 [9], ResNet50 [10], and VGG16 [11], which are three state-of-the-art CNNs widely used for object recognition.

4.1 Model Training and Hyperparameter Tuning

Overfitting was a major challenge in training the models. Hence, it was essential to find the optimal hyperparameters for each model to ensure reliable results, rather than using one-size-fits-all hyperparameter settings. To overcome this problem, techniques like varying batch sizes, L2 regularization, batch normalization, adaptive learning rate, dropout, and early stopping were employed. L2 regularization penalizes the sum of squares of weights, forcing them to be small, which helps the model generalize better by making it less sensitive to minor input changes. Batch normalization ensures the distribution of inputs remains consistent across batches by normalizing them, which helps stabilize the network. The adaptive learning rate lowers the learning rate as training progresses, meaning smaller weight updates are made when the validation loss stops decreasing for a set number of epochs. Dropout clips a couple of neurons randomly in each epoch, which prevents the model from fitting too closely to the training data.

Early stopping, which monitors validation loss, stops training once the performance ceases to improve or degrades for a set number of epochs, preventing the model from losing its ability to generalize well on unseen data.

4.2 Inference

In the inference stage, the greedy search algorithm was used to generate a caption for a given input image. The greedy selection is a naive method that picks the word with the highest predicted probability to be the next word in the caption being generated. Starting with ¡startseq¿, which signifies the beginning of a caption, a loop continues until the ¡endseq¿ token is read. In each iteration, the model outputs a vector of probabilities for each word in the vocabulary, taking image features and the current state of the caption as inputs. Using an argmax function, the word with the highest probability is selected as the next word in the caption. An example of the greedy search algorithm is depicted in Fig. 2. Since it doesn't look ahead beyond the immediate next word, it can sometimes lead to suboptimal word choices in the overall context of the sentence. Despite this limitation, it remains a popular choice due to its speed and ease of implementation, especially when computational resources are a constraint.

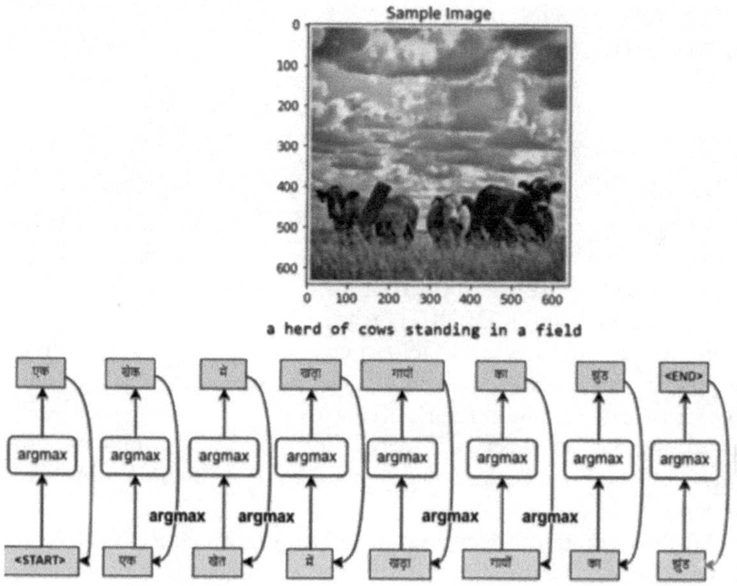

Fig. 2. An example of how the greedy search algorithm works.

4.3 Evaluation Metrics

The Bilingual Evaluation Understudy Score [12], termed as BLEU for short, compares the candidate sentence, which is generated by the model, with the reference sentences. BLEU scores range from 0 to 1, with 1 indicating a perfect match between the generated text and the reference text. BLEU evaluates the precision of the generated text by comparing n-grams in the generated text against n-grams in the reference text. It is also stated that BLEU scores can increase with a higher number of reference texts [12]. However, it is worth noting that it is not a perfect performance metric that we can completely rely on for assessing image captioning on a sentence level, because a thought derived from an image can be articulated in many different ways by a neural network, just like a human being. NIC: Neural Image Caption Generator [13], a benchmark in the field of image captioning that experimented with CNNs for image feature extraction and RNNs for language generation, evaluated the predicted captions using human evaluation to assess the model performance. They concluded that the BLEU metric is not perfect, inferring from their human-evaluated results. Some known limitations are due to the fact that these scores are comparable across different languages, shorter predictions achieve higher scores because of the way the brevity penalty is calculated, and lastly, it does not compare the meaning of the predicted and true sentences.

5 Results and Analysis

The quantitative results of our research confirm the efficacy of employing a specialized translation corpus for cross-lingual vision tasks. The key findings are:

- The **ResNet50+LSTM** model configuration, trained on the **IndicTrans2** corpus, outperformed all other models implemented in this study in generating representative captions, reporting BLEU-4 scores of 0.6205, 0.431, 0.33, and 0.2 (BLEU-1 through BLEU-4, respectively).
- While the 'merge' architecture yielded competitive results after hyperparameter tuning, it was found to be highly prone to overfitting, a consequence of creating an overly simple network structure.
- The high frequency of the Hindi word "एक" (a/one) in the captions resulted in its high probability next to the <startseq>, a phenomenon evaluated by human assessors.
- The IndicTrans2 model proves competent in providing accurate and meaningful translations, establishing its viability as a replacement for general commercial machine translation tools like Google Translate in future works (Fig. 3).

5.1 Comparison with Previous Works in Hindi Image Captioning

Table 1 shows a comparison of our methodology with previous Hindi image captioning works. In comparing our results with existing works, it is essential to

Table 1. Qualitative Comparison of Generated Captions

Example 1: Soccer Players	**Example 2: Skateboarder**

Generated Hindi Caption (Model Output): दो फुटबॉल खलिाड़ी मैदान में खेल रहे हैं	**Generated Hindi Caption (Model Output):** स्केट बोर्डर स्केटबोर्ड के साथ हवा में कूदता है
True/Reference Hindi Captions: {एक आदमी की लाल शर्ट गरि रही है, फुटबॉल खलिाड़ी लाल एक नीले रंग के सामने दौड़ता है, दो आदमी सॉकर घास का मैदान खेल रहे हैं, पुरुष सॉकर खेल खेलते}	**True/Reference Hindi Captions:** {लाल शर्ट पहने आदमी, स्केटबोर्डर हाईवे गार्ड रेल के ऊपर से कूदता है, स्केटबोर्डर जंप स्लोप कर रहा है, लड़का बाहर स्केटबोर्ड ट्रकि कर रहा है, स्केटबोर्डर बाड़ के ऊपर}
Generated English (for context): Two football players playing on the field	**Generated English (for context):** Skateboarder jumps in the air with a skateboard

acknowledge the novel approach adopted for the creation of the Hindi dataset in this research. Unlike previous studies that relied on Google Translate to create datasets, often requiring significant manual labor for verification, this research implemented the state-of-the-art IndicTrans2 model to translate the Flickr30K dataset into Hindi. Previous studies have often utilized the Flickr8K dataset or

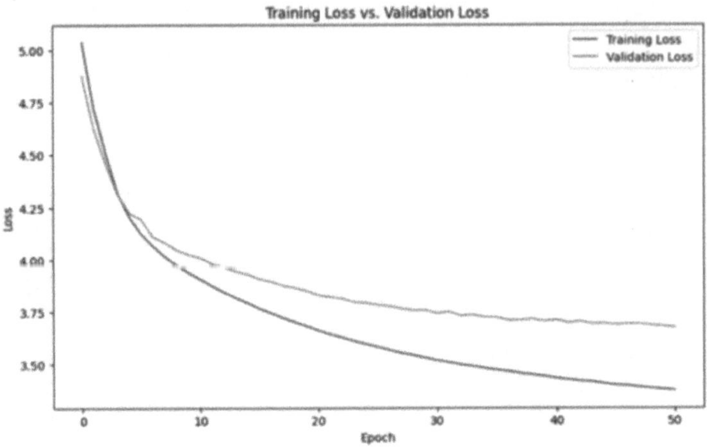

Fig. 3. ResNet50+LSTM loss plot

other smaller datasets chosen for the convenience of manual verification. Table 2 lists some of the previous baseline models used on Hindi image captioning to provide a comparison as close as possible, given the variance in contributing factors. The proposed methodology, including our translation technique and choice of dataset, is a novel approach that has set a high BLEU score benchmark for future works. We expect that with further refinement of the dataset used, higher performance and significant improvements in caption quality can be achieved. Thus, this unique approach establishes a new baseline benchmark for future research in this area.

Table 2. Comparison of Some Previous Works with Proposed Methodology

Paper	Translation Technique	Dataset	Baseline Model	BLEU-1 Score
Proposed Methodology	**IndicTrans2**	**Flickr30K**	**ResNet50+LSTM**	**0.6205**
A. Rathi [3]	Google Translate	Flickr8K	CNN+LSTM	0.584
K. Gill [14]	Google Translate	Visual Genome	ResNet101+LSTM	0.285
R. Dhir et al. [15]	Google Translate	MSCOCO	ResNet101+GRU	0.57
A. K. Poddar & R. Rani [16]	Google Translate	Flickr8K	VGG16+LSTM	0.55

6 Conclusion And Future Work

In conclusion, implementing the IndicTrans2 model stands as a pioneering effort in the creation of a Hindi dataset for this research. By leveraging IndicTrans2 for translating image captions into Hindi, this work is a first in the exploration of Hindi image captioning to the best of our knowledge. This research has successfully used the Flickr30k dataset, featuring captions that have been converted into Hindi. The dataset curated through this model provides a fresh perspective and could potentially improve the performance of captioning models in Indian languages, thereby making this research work both innovative and noteworthy. Another point to note, the dataset might not adequately capture the rich diversity and cultural nuances of the Hindi language, which could restrict generalization. While the new translation model gave a fresh perspective on Hindi translations, human effort may still be needed in accurately annotating and cleaning the textual data after translation takes place. As mentioned earlier, overfitting was a key challenge in training the models. The research effectively addressed the challenge of overfitting by employing techniques to optimize training of each model. These strategies collectively ensured that the models remained general and robust, thereby substantially mitigating the risk of overfitting. Specific to the language chosen in this research, Hindi is a language with complex morphological features and distinct linguistic traits. The applicability of the models to languages with rich morphology may be restricted. And thus, this opens the discussion for future work in this research area. Image captioning is a vast research area which is still being explored. In the context of future work in Hindi image captioning, the immediate and most promising direction is the integration of modern Generative AI techniques, specifically Multimodal Large Language Models (LLMs).

References

1. Gala, J., et al.: IndicTrans2: Towards high-quality and accessible machine translation models for all 22 scheduled Indian languages. arXiv preprint arXiv:2305.16307 (2023)
2. Lin, T.Y., et al.: Microsoft COCO: common objects in context. In: European conference on computer vision (ECCV) (2014). (Note: Replace this placeholder with the actual Flickr8k paper citation, like the one from the University of Illinois.)
3. Rathi, A.: Deep learning approach for image captioning in Hindi language. In: International Conference on Computer, Electrical & Communication Engineering (ICCECE), Kolkata, India, pp. 1–8 (2020)
4. Mishra, S.K., Dhir, R., Saha, S., Bhattacharyya, P., Kumar Singh, A.: Image captioning in Hindi language using transformer networks. Comput. Electr. Eng. **92**, 107114 (2021)
5. Lin, T.-Y., et al.: Microsoft COCO: common objects in context. In: Proceedings of 13th European Conference on Computer Vision (ECCV), Zurich, Switzerland, Sep. 6-12, 2014, vol. 13, Part V, Springer International Publishing (2014)
6. Mishra, S.K., Rai, G., Saha, S., Bhattacharyya, P.: Efficient channel attention based encoder-decoder approach for image captioning in Hindi. Trans. Asian Low-Resource Lang. Inf. Process. **21**(3), 1–17 (2021)
7. Singh, P., Raja, F., Sharma, H.: Generating Image Captions in Hindi Based on Encoder-Decoder Based Deep Learning Techniques, in Reliability Engineering for Industrial Processes: An Analytics Perspective, pp. 81–94. Cham: Springer Nature Switzerland (2024)
8. Tanti, M., Gatt, A., Camilleri, K.P.: Where to put the image in an image caption generator. Nat. Lang. Eng. **24**(3), 467–489 (2018)

9. Szegedy, C., et al.: Rethinking the inception architecture for computer vision. In: Proceedings of the IEEE Conference on Computer Vision and Pattern Recognition (2016)
10. He, K., et al.: Deep residual learning for image recognition. In: Proceedings of the IEEE Conference on Computer Vision and Pattern Recognition (2016)
11. Simonyan, K., Zisserman, A.: Very deep convolutional networks for large-scale image recognition. arXiv preprint arXiv:1409.1556 (2014)
12. Papineni, K., et al.: BLEU: a method for automatic evaluation of machine translation. In: Proceedings of the 40th Annual Meeting of the Association for Computational Linguistics (2002)
13. Vinyals, O., et al.: Show and tell: a neural image caption generator. In: Proceedings of the IEEE Conference on Computer Vision and Pattern Recognition (2015)
14. Gill, K., Saha, S., Mishra, S.K.: Dense image captioning in Hindi. In: IEEE International Conference on Systems, Man, and Cybernetics (SMC), Melbourne, Australia, pp. 2894–2899 (2021)
15. Dhir, R., Mishra, S.K., Saha, S., Bhattacharyya, P.: A deep attention based framework for image caption generation in Hindi language. Computación y Sistemas **23**(3) (2019)
16. Poddar, A.K., Rani, R.: Hybrid architecture using CNN and LSTM for image captioning in Hindi language. Procedia Comput. Sci. **218**, 686–696 (2023)

Towards Tracing the Quality of Code Generated by LLM Back to the Training Code

Chen Wang(✉) and Yujun Huang

College of Computer Science and Technology, Nanjing University of Aeronautics and Astronautics, Nanjing 210016, China
wangchen712@nuaa.edu.cn

Abstract. The rise of code-focused Large Language Models (LLMs) like Codex and StarCoder has accelerated software development but also raised concerns about inherited defects. Trained on mixed-quality open-source code, these models risk propagating bugs and vulnerabilities. This study empirically investigates the link between training data quality and LLM-generated code. We evaluate five models trained on Python and Java subsets of The Stack, applying SonarQube to profile training data and HumanEval to assess generated outputs. Results show that defect types in training data are reflected in model outputs, confirming defect propagation. We further find that mixed-language training enhances robustness, while performance drops notably on more complex tasks. These findings underscore the need for rigorous data curation to improve the reliability of code generation.

Keywords: Code Generation · Large Language Models · Training Data Quality · Defect Propagation

1 Introduction

Recent years have seen a paradigm shift in artificial intelligence, with Large Language Models demonstrating substantial capabilities in modeling complex data distributions. Beyond natural language, these models are increasingly applied to software engineering, where automated code synthesis has become a central research and industrial focus [2,3,9]. Their promise lies in improving developer productivity and accelerating the software development lifecycle, while addressing persistent challenges such as technical debt and code maintenance [1].

Code generation systems have evolved rapidly, from early rule-based and template-driven approaches [7,8] to Transformer-based models [10]. These foundation models can generate syntactically valid and semantically coherent code. They also capture long-range context, infer program structures, and exhibit basic reasoning abilities. Yet, significant challenges remain in practice: ensuring functional correctness and security [4], generalizing to tasks requiring deeper algorithmic reasoning, and managing the heavy computational costs of training and

H. R. Arabnia et al. (Eds.): CSCI 2025, CCIS 2967, pp. 469–483, 2027.
https://doi.org/10.1007/978-3-032-28454-9_33

inference. Moreover, the evaluation of generated code in realistic development settings is still an open question [5,6].

More recently, specialized code LLMs such as Codex [5], CodeGen [13,21], StarCoder [26], and the GPT series [11,12,15,16] have set new benchmarks in code completion, transpilation, and program repair [17]. Trained on massive open-source corpora, these models learn idiomatic programming patterns and can perform cross-lingual synthesis. However, as their scale grows, the quality of training data has become a critical factor. Open-source datasets often contain bugs, code smells, and security vulnerabilities. While prior work has focused on model architectures and benchmark accuracy, there remains limited systematic evidence on how corpus-level defects may influence generated outputs. This lack of clarity makes it difficult to assess the trustworthiness of code LLMs and to design principled data curation strategies.

In this work, we empirically analyze the relationship between training data quality, measured through metrics such as defect density, cyclomatic complexity, and vulnerability signatures, and the quality of generated code. Our study provides direct evidence of defect propagation, where undesirable properties of training corpora reappear in LLM outputs, and explores how multilingual training regimens influence robustness. These findings aim to inform future research on both evaluation and dataset governance for code LLMs.

2 Related Work

Researchers have developed numerous models specialized for code generation tasks such as completion, translation, and repair. Early work emphasized training strategies and architectures. CodeGen, for instance, adopted a multi-stage training paradigm that progressed from natural language to mixed code+NL and finally pure code, achieving better performance than single-stage training [13,21]. InCoder introduced large-scale infilling with sentinel tokens, enabling generation of arbitrary code spans from surrounding context [14,22]. PolyCoder, an early GPT-2-based model, highlighted the benefit of training on curated corpora and achieved competitive results in low-resource programming languages [23].

Later efforts focused on scaling models and improving dataset quality. The BigCode project released SantaCoder and StarCoder, trained on *The Stack*, a permissively licensed multi-terabyte dataset with multilingual coverage and license/PII filtering [24,26]. CodeGen2 extended hybrid training on both code and natural language. StarCoder2 scaled training to trillions of tokens and matched or exceeded the performance of larger models such as CodeLlama [25]. CodeLlama, derived from LLaMA-2, provided strong completion capabilities and long-context instruction-tuned variants [20]. More recently, DeepSeek-Coder introduced models up to 33B parameters trained on trillions of code tokens, with performance competitive with GPT-3.5 [27]. In industry, Qwen2.5-Coder continued pre-training on 5.5T tokens of code and released models ranging from 0.5B to 32B parameters, supporting context lengths up to 1M tokens [16]. These studies indicate that scale, dataset governance, and instruction-tuning contribute significantly to performance in code completion.

In parallel, the software engineering community has examined the reliability and risks of code LLMs. Early studies focused on predictive accuracy. Ciniselli et al. [48] evaluated transformer-based models at the token, line, and block levels, establishing metrics such as top-k accuracy. Mastropaolo et al. [49] showed that small prompt variations could cause Copilot to produce inconsistent or incorrect code, raising concerns about determinism. Later work turned attention to training data quality. Siddiq et al. [50] reported high rates of code smells and vulnerabilities in widely used corpora, suggesting that such flaws could propagate into generated outputs. Cotroneo et al. investigated data poisoning via fine-tuning and provided preliminary causal evidence, albeit at limited scales. More recently, Improta et al. [52] demonstrated that filtering defective training data substantially reduced issues in generated code. Sabra et al. [53] found that even top-performing models frequently produced code with vulnerabilities and smells despite achieving high functional correctness. Collectively, these findings underline that beyond benchmark accuracy, the quality, security, and maintainability of generated code are critical evaluation dimensions. Our work builds on this line by systematically linking defect profiles in training corpora to corresponding properties in model outputs under code completion tasks.

3 Code Dataset Analysis

To investigate the relationship between training data quality and the outputs of large language models, we conduct a systematic analysis of the datasets used in our experiments. The analysis proceeds in two stages. First, we examine a statistically representative sample of the Stack dataset [29], which underlies model training, in order to characterize code complexity and defect distributions. Second, we analyze the HumanEval benchmark [5] to establish a baseline for evaluating generated code.

3.1 The Stack Training Dataset

Sampling Strategy. To obtain a representative subset of the Stack dataset, we determined the required sample size using Cochran's formula:

$$n = \frac{Z^2 \cdot p \cdot (1-p)}{e^2} \div \left(1 + \frac{Z^2 \cdot p \cdot (1-p)}{e^2 \cdot N}\right) \tag{1}$$

where n is the sample size, N is the population size, p is the estimated proportion, Z is the Z-score, and e is the margin of error. Since the true value of p is unknown, we set $p = 0.5$ to maximize variance and obtain the most conservative estimate. We adopt $Z = 1.96$ for 95% confidence and set $e = 0.10$. The Stack contains tens of millions of Python and Java files, and substituting their population sizes into Eq. 1 yields required sample sizes of about 0.92M for Python and 0.94M for Java. These subsets are used in the subsequent complexity and defect analyses.

Code Complexity. We measure two common complexity metrics at the method level. Cyclomatic complexity quantifies the number of independent execution paths in a program and increases with additional decision points. Cognitive complexity estimates the mental effort required to understand a method, emphasizing non-linear control structures such as nested conditionals, loops, recursion, and exception handling.

For quantitative analysis, each method is assigned to one of five levels. Cyclomatic complexity is divided into Lowest (1), Low (2–4), Medium (5–7), High (8–10), and Very High (11 or more). Cognitive complexity is divided into Lowest (0), Low (1–5), Medium (6–10), High (11–20), and Very High (21 or more). The distributions are reported in Table 1.

Table 1. Distribution of methods by complexity levels in the Stack dataset

Category	Cyclomatic Complexity (%)		Cognitive Complexity (%)	
	Python	Java	Python	Java
Lowest	68.51	81.97	71.85	82.24
Low	19.55	11.93	18.53	12.59
Medium	6.05	3.16	4.66	2.51
High	2.53	1.26	3.01	1.52
Very High	3.36	1.68	1.95	1.14

As shown in Table 1, most methods fall into the Lowest or Low levels for both metrics, which indicates that the training corpus is dominated by relatively simple code fragments. Java exhibits a higher proportion of low-complexity methods compared to Python, a difference that likely reflects language design and common coding practices. This skew toward simple code suggests that models trained primarily on such data may face challenges when generalizing to tasks that require reasoning across more complex program structures.

Defects. Besides complexity, defects such as bugs, code smells, and vulnerabilities are essential indicators of code quality. In the sampled data, Java files exhibited slightly more bugs and code smells on average, while Python files contained a marginally higher proportion of vulnerabilities. Overall, code smells were by far the most frequent issue, reaching several occurrences per file in both languages, whereas vulnerabilities appeared only rarely.

To enable systematic analysis, the hundreds of detailed SonarQube rules were consolidated into eight general categories. These categories are summarized in Table 2.

Table 2. General defect categories derived from SonarQube rules

Type	Description
Code Performance	Inefficient logic or redundant computation.
Code Syntax	Violations that prevent proper compilation or execution.
Control Flow	Errors in loops, conditionals, or logical structures.
Redundant Code	Unused variables, unreachable code, or duplication.
Function-related	Incorrect function definitions, parameters, or return values.
Naming-related	Poorly chosen or inconsistent identifiers.
Operator Usage	Incorrect or inefficient use of operators.
Data Type	Misuse of data types or conversions.

Figure 1 illustrates the distribution of these defect categories across the Java and Python samples. Several patterns emerge. Java shows fewer issues related to performance and control flow, while Python performs better in function-related aspects. In both languages, naming-related problems are by far the most frequent, reflecting the widespread use of non-descriptive or inconsistent identifiers in open-source projects. Operator-related defects are comparatively rare. Other differences, such as in syntax, redundant code, and data type usage, largely stem from language-specific features and community coding practices.

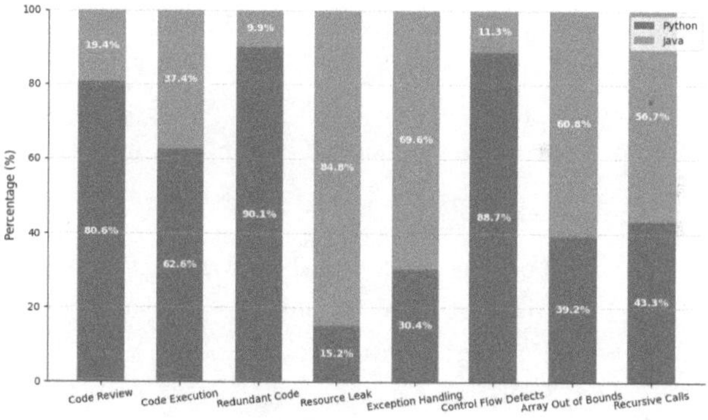

Fig. 1. Defect distribution in Java and Python datasets

3.2 The HumanEval Benchmark

After analyzing the training corpus, we now turn to the HumanEval benchmark [5]. HumanEval is widely used to evaluate code generation models and consists of Python programming tasks with reference implementations and test cases. To establish a baseline for our study, we examine both the complexity of its reference solutions and the distribution of problem types.

Complexity Analysis. Each reference solution is classified into the five levels introduced earlier. The resulting distribution is shown in Table 3.

Table 3. Complexity distribution of HumanEval reference solutions

Category	Cyclomatic complexity (%)	Cognitive complexity (%)
Lowest	30.36	39.88
Low	53.57	43.45
Medium	12.19	11.31
High	1.83	3.57
Very High	2.44	1.78

The majority of solutions fall into the lowest two levels, confirming that HumanEval focuses on fundamental programming and algorithmic skills rather than complex software design. Compared with the Stack dataset, HumanEval contains fewer trivial cases and a larger share of problems in the low level. This makes the benchmark slightly more challenging while still emphasizing basic programming tasks. It therefore provides a useful testbed for examining whether models generalize beyond simple training examples.

Problem Type Analysis. We also categorized the 164 tasks by problem type. Each task may belong to multiple categories. The classification is reported in Table 4.

Table 4. Problem type classification of HumanEval

Problem type	Explanation	Count
Search	Looping or traversal over data structures	90
Math-related	Mathematical formulas or computations	55
String manipulation	Parsing or transforming strings	28
Sorting	Ordering lists or arrays	21
Recursion	Solutions that require recursion	4
Function call	Use of a predefined function	3
Other	Specialized algorithms such as shortest path	2

Search and math-related tasks dominate the dataset, followed by string manipulation and sorting. Recursion and specialized algorithms are rare. This distribution shows that HumanEval is designed to test essential programming skills and logical reasoning, rather than advanced software engineering capabilities. As such, it serves as a suitable benchmark for evaluating the baseline performance of code generation models.

4 Results and Comparative Analysis

Building on the static analysis of the training and evaluation datasets in Sect. 3, this section provides a multi-dimensional evaluation of the generated code and integrates the findings into a comparative analysis. The five code generation models are introduced first with their training strategies. We then report overall performance on HumanEval using the Pass@100 metric under optimal temperature settings. Next, we characterize intrinsic properties of the generated code through problem-solving steps, defect distributions, and complexity. Finally, we connect these observations to the properties of the training data and discuss performance limits.

4.1 Models and Training Data

Five code generation models were trained and compared, each with 1.1B parameters and a shared architecture. All models are based on the StarCoder framework and use subsets of The Stack dataset as training data. The configurations are as follows:

- **PyCoder**: trained only on the Python portion of The Stack.
- **JavaCoder**: trained only on the Java portion of The Stack.
- **MixCoder**: trained jointly on both Python and Java.
- **PyCoder_Base_Java**: trained sequentially, first on Java, then fine-tuned on Python.
- **JavaCoder_Base_Python**: trained sequentially, first on Python, then fine-tuned on Java.

This setup enables a comparison of single-language specialization, mixed-language training, and sequential cross-language training.

4.2 Overall Performance

The five models were evaluated on HumanEval. Programming prompts were provided as input and each model generated code under multiple temperature settings. All outputs were tested against the official unit tests to determine correctness. The best-performing temperature for each model was selected. Results are summarized in Table 5. The reported metric is Pass@100 [51], which measures the probability that at least one of the generated samples for a problem passes all unit tests and thus provides an upper bound on model accuracy.

Table 5. Code generation results at optimal temperature using Pass@100

Language	Model	Temperature	Pass@100
Python	PyCoder	0.6	0.177
	PyCoder_Base_Java	0.4	0.152
	MixCoder	0.4	0.158
Java	JavaCoder	0.2	0.128
	JavaCoder_Base_Python	0.2	0.091
	MixCoder	0.4	0.115

Three trends emerge from Table 5. First, single-language specialization yields the best accuracy in its domain. PyCoder leads in Python with a Pass@100 of 0.177 and JavaCoder leads in Java with a Pass@100 of 0.128. Second, simultaneous mixed-language training produces robust outcomes across languages. MixCoder is competitive in Python with 0.158 and remains close to JavaCoder in Java with 0.115. Third, sequential training performs worse than both specialization and mixed training. PyCoder_Base_Java and JavaCoder_Base_Python underperform, and the latter is the weakest overall with 0.091.

4.3 Problem-Solving Steps

Reasoning quality was examined by abstracting HumanEval solutions into four steps. These steps are problem comprehension, data structure choice and usage, coherent control flow with appropriate handling of edge cases, and adherence to syntax rules. The criteria are summarized in Table 6.

Table 6. Problem-solving steps in generated code

Step	Requirement
Problem comprehension	Correct interpretation of the task shown through meaningful naming and alignment with the specification.
Data structure	Appropriate choice and correct use of data types and structures.
Code logic	Sound control flow with loops and conditionals, including the handling of edge cases.
Code syntax	Compliance with language syntax to ensure compilable or interpretable code.

Figure 2 shows that PyCoder achieves the most complete reasoning in Python. MixCoder ranks second and slightly exceeds PyCoder_Base_Java. In Java, MixCoder delivers the most complete solutions and exceeds JavaCoder by approximately 31% and JavaCoder_Base_Python by about 10% in average step coverage. Overall, MixCoder demonstrates broader problem-solving ability than sequential models. Python outputs generally include more steps than Java outputs.

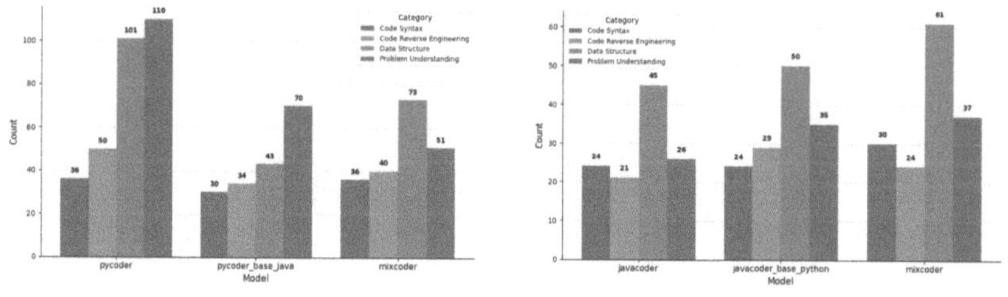

(a) Python (b) Java

Fig. 2. Comparison of implemented problem-solving steps by model

4.4 Defect Propagation and Code Quality

Generated programs were analyzed according to the eight defect categories defined in Sect. 3. Each output was scanned with a static analyzer and frequencies were aggregated for comparison. The distributions are shown in Fig. 3.

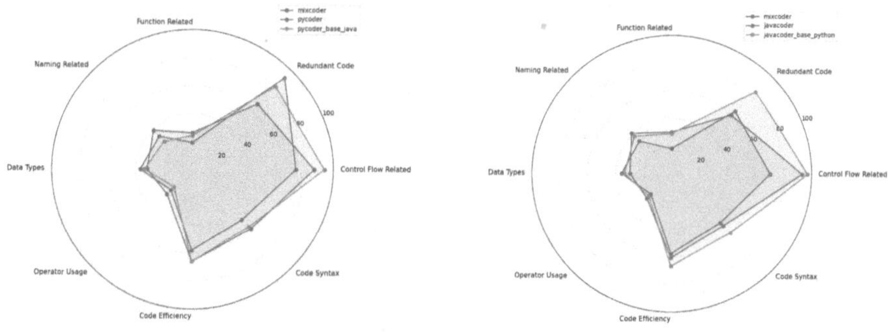

(a) Python models (b) Java models

Fig. 3. Defect distribution in generated code across Python and Java models

Three observations stand out. MixCoder consistently produces the fewest defects. In Python, defect counts are reduced by approximately one third when compared with PyCoder and PyCoder_Base_Java. In Java, reductions relative to JavaCoder and JavaCoder_Base_Python are also substantial. Defect patterns vary by model. PyCoder generates relatively few naming-related issues yet shows more frequent problems in control flow and data type usage. The overall defect level is lower in Python than in Java, which aligns with differences observed in the training data.

To test whether these patterns mirror the training data, a Chi-Squared test was conducted by comparing the percentage distribution of defect types in the generated code with the distribution in the corresponding training datasets. The null hypothesis H_0 states that the two distributions are independent. The statistic is given by

$$\chi^2 = \sum \frac{(\text{Observed} - \text{Expected})^2}{\text{Expected}} \tag{2}$$

The test yields a value of 3.46 with 7 degrees of freedom. The corresponding P-value equals 0.8399. Since this value greatly exceeds the conventional threshold $\alpha = 0.05$, the result indicates strong consistency between the two distributions. The defect distribution in generated code is therefore statistically correlated with the defect distribution of the training data, which supports the presence of defect propagation.

4.5 Complexity and Task-Specific Abilities

The relationship between solution complexity and training data was examined by classifying correct outputs into complexity levels. The distributions are reported in Table 7. A complementary analysis considered task-specific abilities across four categories. Aggregate step completion scores are summarized in Table 8.

Table 7. Complexity distribution of correctly generated Python and Java code

Level	Python models			Java models		
	PyCoder	PyCoder Base Java	MixCoder	JavaCoder	JavaCoder Base Python	MixCoder
Lowest	9/16	7/13	8/14	6/11	3/7	5/11
Low	19/12	17/10	17/11	11/6	11/8	15/9
Medium	1/1	1/2	1/1	2/2	1/0	1/1
High and above	0/0	0/0	0/0	0/0	0/0	0/0

Table 8. Step completion scores across task types for Python and Java models

(a) Python models

Task type	PyCoder	PyCoder Base Java	MixCoder
Search	153	118	135
Math	82	65	72
String	41	31	38
Sorting	32	24	28
Total	308	238	273

(b) Java models

Task type	JavaCoder	JavaCoder Base Python	MixCoder
Search	67	88	105
Math	33	45	54
String	16	22	27
Sorting	11	18	21
Total	127	173	207

Two findings follow. First, all models struggle once task complexity increases. Figure 4 shows that performance declines sharply from low to medium and no model solves problems at high or very high levels. Second, task-type results indicate that PyCoder has a clear advantage on Python tasks, while MixCoder provides the most balanced performance across languages and exceeds the specialized JavaCoder in Java.

The heatmaps in Fig. 4 reveal a consistent complexity ceiling. For both cyclomatic complexity and cognitive complexity, performance declines as complexity increases from low to medium. No model solves any problem at high or very high levels. Since the training corpus in Table 1 is dominated by low-complexity examples, models do not acquire the patterns required for more complex algorithmic tasks. MixCoder is more resilient at lower levels yet remains constrained by this ceiling.

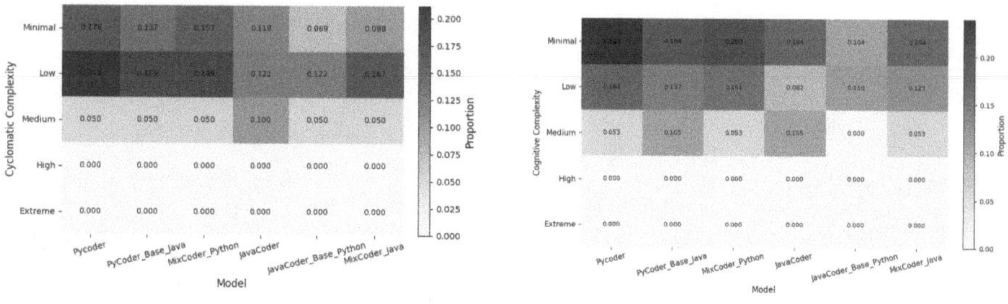

(a) Cyclomatic complexity (b) Cognitive complexity

Fig. 4. Heatmaps of model success rates by problem complexity level

5 Conclusion

This paper aimed to establish and trace the explicit link between the quality of code generated by Large Language Models and the quality of their training data through a systematic empirical study. By conducting a multi-dimensional analysis of five models under different training strategies, we revealed how features of the training data, such as complexity and defect distributions, are inherited by generated code and evaluated the effectiveness of these training strategies. This section summarizes the main findings, discusses the limitations of the study, and outlines directions for future research.

5.1 Key Findings

Our study leads to four principal findings. First, mixed-language training enhances robustness and code quality. The MixCoder model, trained jointly on Python and Java, produced the most reliable code with the fewest defects in both languages. Its accuracy in Java even surpassed that of the specialized Java-Coder model, indicating that cross-language exposure enables models to learn complementary programming patterns and norms.

Second, defect distribution patterns in training data are reproduced in generated code. A Chi-Squared test confirmed a strong statistical correlation between the two distributions, providing direct evidence that models inherit not only useful structures but also undesirable flaws from their training corpora. This highlights the importance of dataset governance and systematic filtering.

Third, all models exhibited a clear complexity ceiling. While they performed well on simpler problems, success rates dropped sharply as task complexity increased. In our experiments, none of the models solved problems classified as high or very high complexity. This suggests that the current models have difficulty generalizing to complex algorithmic tasks, potentially reflecting the influence of training data distributions.

Finally, training strategy decisively affects performance. Simultaneous mixed-language training outperformed sequential fine-tuning, where initial biases from the first language persisted and limited gains from the second. This suggests that sequential training is insufficient to fully exploit cross-lingual benefits.

5.2 Limitations

This study is subject to several limitations. First, the evaluation of problem-solving steps was conducted manually, and although guided by explicit criteria, subjective judgment may have influenced the results. Second, the metrics employed, such as cyclomatic complexity, cognitive complexity, and static defect analysis, capture important aspects of quality but do not cover all dimensions such as algorithmic optimality or efficiency. Third, due to resource constraints, our experiments were limited to 1.1B-parameter models based on the StarCoder framework. The findings may not generalize to larger models or alternative architectures such as GPT or LLaMA. Finally, the scope of datasets and tasks was limited to Python and Java code from The Stack and the HumanEval benchmark, which may not represent the challenges posed by other programming languages or large-scale software development scenarios.

5.3 Future Work

Several directions arise for future work. One is data-driven model optimization through systematic corpus curation. Filtering out defective code or rebalancing datasets toward higher complexity may improve robustness and help models overcome the observed complexity ceiling. Another is expansion of scope to larger models, more diverse architectures, and additional programming languages in order to test the universality of our conclusions. Finally, exploring advanced training strategies such as curriculum learning or controlled defect injection could provide deeper insights into how models acquire, propagate, or mitigate data-driven flaws. Together, these efforts could pave the way for building code generation systems that are not only accurate but also more reliable and trustworthy in practice.

References

1. Dakhel, A.M., Majdinasab, V., Nikanjam, A., et al.: GitHub copilot AI pair programmer: asset or liability? J. Syst. Softw. **203**, 111734 (2023)
2. Zhang, Z., Chen, C., Liu, B., et al.: Unifying the perspectives of NLP and software engineering: A survey on language models for code. arXiv preprint. arXiv:2312.07996 (2023)
3. Zhao, W.X., Zhou, K., Li, J., et al.: A survey of large language models. arXiv preprint. arXiv:2303.18223 (2023)
4. Liu, J., Xia, C.S., Wang, Y., et al.: Is your code generated by ChatGPT really correct? Rigorous evaluation of large language models for code generation. Adv. Neural Inf. Process. Syst. **36** (2024)
5. Chen, M., Tworek, J., Jun, H., et al.: Evaluating large language models trained on code. arXiv preprint arXiv:2107.03374 (2021)
6. Yu, H., Shen, B., Ran, D., et al.: CoderEval: a benchmark of pragmatic code generation with generative pre-trained models. In: Proceedings of the 46th IEEE/ACM International Conference on Software Engineering, pp. 1–12 (2024)
7. Tiwari, V., Malik, S., Wolfe, A.: Compilation techniques for low energy: an overview. In: Proceedings of 1994 IEEE Symposium on Low Power Electronics, pp. 38–39 (1994)
8. Gogoi, B., Cheramangalath, U., Nasre, R.: Custom code generation for a graph DSL. In: Proceedings of the 13th Annual Workshop on General Purpose Processing using Graphics Processing Unit, pp. 51–60 (2020)
9. Feng, Y., Vanam, S., Cherukupally, M., et al.: Investigating code generation performance of Chat-GPT with crowdsourcing social data. In: Proceedings of the 47th IEEE Computer Software and Applications Conference, pp. 1–10 (2023)
10. Vaswani, A., Shazeer, N., Parmar, N., et al.: Attention is all you need. Adv. Neural Inf. Process. Syst. **30** (2017)
11. Floridi, L., Chiriatti, M.: GPT-3: its nature, scope, limits, and consequences. Mind. Mach. **30**, 681–694 (2020)
12. Wu, T., He, S., Liu, J., et al.: A brief overview of ChatGPT: the history, status quo and potential future development. IEEE/CAA J. Automatica Sinica **10**(5), 1122–1136 (2023)
13. Nijkamp, E., Pang, B., Hayashi, H., et al.: CodeGen: an open large language model for code with multi-turn program synthesis. In: The Eleventh International Conference on Learning Representations (2023)
14. Fried, D., Aghajanyan, A., Lin, J.: et al.: InCoder: a generative model for code infilling and synthesis. In: The Eleventh International Conference on Learning Representations (2023)
15. Wang, B., Komatsuzaki, A.: GPT-J-6B: A 6 Billion Parameter Autoregressive Language Model (2021). https://github.com/kingoflolz/mesh-transformer-jax
16. Achiam, J., Adler, S., Agarwal, S., et al.: GPT-4 technical report. arXiv preprint. arXiv:2303.08774 (2023)
17. Evtikhiev, M., Bogomolov, E., Sokolov, Y., et al.: Out of the BLEU: how should we assess quality of the code generation models? J. Syst. Softw. **203**, 111741 (2023)

18. Zheng, Q., Xia, X., Zou, X., et al.: CodeGeeX: a pre-trained model for code generation with multilingual benchmarking on HumanEval-X. In: Proceedings of the 29th ACM SIGKDD Conference on Knowledge Discovery and Data Mining, pp. 5673–5684 (2023)
19. Wang, Y., Le, H., Gotmare, A.D., et al.: CodeT5+: Open code large language models for code understanding and generation. arXiv preprint. arXiv:2305.07922 (2023)
20. Roziere, B., Gehring, J., Gloeckle, F., et al.: Code Llama: Open foundation models for code. arXiv preprint. arXiv:2308.12950 (2023)
21. Nijkamp, E., Pang, B., Hayashi, H., et al.: CodeGen: An open large language model for code with multi-turn program synthesis. arXiv preprint arXiv:2203.13474 (2022)
22. Fried, D., Aghajanyan, A., Lin, J., et al.: InCoder: A generative model for code infilling and synthesis. arXiv preprint. arXiv:2204.05999 (2022)
23. Xu, F.F., Alon, U., Neubig, G., et al.: A systematic evaluation of large language models of code. In: Proceedings of the 6th ACM SIGPLAN International Symposium on Machine Programming, pp. 1–10 (2022)
24. Allal, L.B., Li, R., Kocetkov, D.: SantaCoder: don't reach for the stars! arXiv preprint (2023). arXiv:2301.03988
25. Nijkamp, E., Hayashi, H., Xiong, C., et al.: CodeGen2: Lessons for training LLMs on programming and natural languages. arXiv preprint. arXiv:2305.02309 (2023)
26. Li, R., Allal, L.B., Zi, Y., et al.: StarCoder: may the source be with you! arXiv preprint. arXiv:2305.06161 (2023)
27. Shen, B., Zhang, J., Chen, T., et al.: PanGu-coder2: Boosting large language models for code with ranking feedback. arXiv preprint arXiv:2307.14936 (2023)
28. Luo, Z., Xu, C., Zhao, P., et al.: WizardCoder: Empowering Code Large Language Models with Evol-Instruct. arXiv preprint. arXiv:2306.08568 (2023)
29. Kocetkov, D., Li, R., Allal, L.B., et al.: The stack: 3 TB of permissively licensed source code. arXiv preprint arXiv:2211.15533 (2022)
30. Rasley, J., Rajbhandari, S., Ruwase, O., et al.: DeepSpeed: system optimizations enable training deep learning models with over 100 billion parameters. In: Proceedings of the 26th ACM SIGKDD International Conference on Knowledge Discovery & Data Mining, pp. 3505–3506 (2020)
31. Dao, T., Fu, D., Ermon, S., et al.: FLASHATTENTION: fast and memory-efficient exact attention with IO-awareness. Adv. Neural. Inf. Process. Syst. **35**, 16344–16359 (2022)
32. Wang, Y., Kordi, Y., Mishra, S., et al.: Self-instruct: Aligning language model with self generated instructions. arXiv preprint. arXiv:2212.10560 (2022)
33. Xu, C., Sun, Q., Zheng, K., et al.: WizardLM: Empowering large language models to follow complex instructions. arXiv preprint. arXiv:2304.12244 (2023)
34. Frostig, R., Johnson, M.J., Leary, C.: Compiling machine learning programs via high-level tracing. Syst. Mach. Learn. **4**(9) (2018)
35. Wang, Y.E., Wei, G.Y., Brooks, D.: Benchmarking TPU, GPU, and CPU platforms for deep learning. arXiv preprint. arXiv:1907.10701 (2019)
36. Dong, Q., Li, L., Dai, D., et al.: A survey for in-context learning. arXiv preprint. arXiv:2301.00234 (2023)
37. Lewis, F.L., Vrabie, D., Vamvoudakis, K.G.: Reinforcement learning and feedback control: Using natural decision methods to design optimal adaptive controllers. IEEE Control Syst. Mag. **32**(6), 76–105 (2012)

38. Wei, J., Wang, X., Schuurmans, D., et al.: Chain-of-thought prompting elicits reasoning in large language models. Adv. Neural. Inf. Process. Syst. **35**, 24824–24837 (2022)
39. Zhang, S., Dong, L., Li, X., et al.: Instruction tuning for large language models: A survey. arXiv preprint arXiv:2308.10792 (2023)
40. Black, S., Gao, L., Wang, P., Leahy, C., Biderman, S.: GPT-Neo: Large Scale Autoregressive Language Modeling with Mesh-Tensorflow (2021). https://github.com/EleutherAI/gpt-neo
41. Gao, L., Biderman, S., Black, S., et al.: The pile: An 800GB dataset of diverse text for language modeling. arXiv preprint arXiv:2101.00027 (2020)
42. Su, J., Ahmed, M., Lu, Y., et al.: RoFormer: Enhanced transformer with rotary position embedding. Neurocomputing **568**, 127063 (2024)
43. Stability-AI. StableLM: Stability AI language models (2023). https://github.com/Stability-AI/StableLM
44. Touvron, H., Lavril, T., Izacard, G., et al.: LLaMA: Open and efficient foundation language models. arXiv preprint arXiv:2302.13971 (2023)
45. Xu, J., Sun, X., Zhang, Z., et al.: Understanding and improving layer normalization. Adv. Neural Inf. Process. Syst. **32** (2019)
46. Chiang, W.-L., Li, Z., Lin, Z., et al.: Vicuna: An open-source chatbot impressing GPT-4 with 90%* ChatGPT quality (2023). https://lmsys.org/blog/2023-03-30-vicuna/
47. Jiang, A.Q., Sablayrolles, A., Mensch, A., et al.: Mistral 7B. arXiv preprint arXiv:2310.06825 (2023)
48. Ciniselli, M., Cooper, N., Pascarella, L., et al.: An empirical study on the usage of BERT models for code completion. In: 2021 IEEE/ACM 18th International Conference on Mining Software Repositories (MSR), pp. 108–119 (2021)
49. Mastropaolo, A., Pascarella, L., Guglielmi, E., et al.: On the robustness of code generation techniques: an empirical study on GitHub copilot. In: 2023 IEEE/ACM 45th International Conference on Software Engineering (ICSE), pp. 2149–2160 (2023)
50. Siddiq, M.L., Majumder, S.H., Mim, M.R., et al.: An empirical study of code smells in transformer-based code generation techniques. In: 2022 IEEE 22nd International Working Conference on Source Code Analysis and Manipulation (SCAM), pp. 71–82 (2022)
51. Kulal, S., Pasupat, P., Chandra, K., et al.: SPoC: search-based pseudocode to code. Adv. Neural Inf. Process. Syst. **32** (2019)
52. Improta, C., Cotroneo, D., Liguori, P.: Quality In, Quality Out: Investigating Training Data's Role in AI Code Generation. arXiv preprint. arXiv:2503.11402 (2025)
53. Sabra, A., Zaman, S., Ghafari, M., et al.: Assessing the Quality and Security of AI-Generated Code: A Quantitative Analysis. arXiv preprint. arXiv:2508.14727 (2025)
54. Esfahani, A.M., Mousavi, M.R., Proença, J., et al.: Understanding Defects in Generated Codes by Language Models. arXiv preprint. arXiv:2408.13372 (2024)
55. Chong, H., Park, J.H., Kim, M., et al.: Artificial-Intelligence Generated Code Considered Harmful: A Road Map for Secure and High-Quality Code Generation. arXiv preprint. arXiv:2409.19182 (2024)

Conceptualizing Trustworthiness of AI Systems

Mark Arinaitwe(✉) and Reza Hassan

University of North Dakota, Grand Forks, ND 58202, USA
mark.arinaitwe@und.edu

Abstract. Trustworthiness in Artificial Intelligence (AI) has emerged as a critical requirement for safe and reliable integration into society. This paper provides a conceptual foundation for trustworthiness in AI systems by drawing on established quality attributes such as performance and dependability. We structure our approach around three pillars: Concerns (core attributes that define trustworthiness), Factors (system characteristics that influence or undermine these concerns), and Methods (tools and processes to achieve trustworthiness across the AI lifecycle). By integrating these dimensions, we present an initial conceptualization of trustworthiness as a quality attribute to guide the systematic design and validation of AI systems.

Keywords: Trustworthiness · Artificial Intelligence · AI Ethics · Quality Attributes · Responsible Artificial Intelligence

1 Introduction

AI is transforming industry and everyday life and enabling new capabilities in decision support, analyzing data, and automation. However, the increasing reliance on AI brings about a greater risk of unanticipated harm, bias, privacy breaches, and other negative impacts. Trustworthiness in AI refers to the degree to which an AI system can be depended upon to behave as intended. It is related with the notion of dependability, which is defined as the ability to deliver a service that can justifiably be trusted [1], but extends it to include further considerations such as fairness and alignment with human values. This is because AI adds new complexity due to its adaptive behavior, and socio-technical effects such as bias, misinformation, or manipulation. This paper provides an initial conceptualization of trustworthiness as a quality attribute in an AI system based on [2].

2 Conceptualizing Trustworthiness in AI

Trustworthiness consists of several sub-attributes that make an AI system deserving of trust, such that stakeholders such as users, operators, regulators, can rely on its behavior and outcomes [3]. Trustworthiness should be seen as an objective quality attribute of the AI system, and distinct from the subjective state of a user's trust. In other words, an AI

H. R. Arabnia et al. (Eds.): CSCI 2025, CCIS 2967, pp. 484–497, 2027.
https://doi.org/10.1007/978-3-032-28454-9_34

is trustworthy if it has the properties that warrant user trust and the goal in design and evaluation of trustworthiness is to ensure the system's intrinsic properties merit trust, if they do, users' trust should follow through appropriate user experience. As with other quality attributes [4], a systematic conceptualization of Trustworthiness can be separated into three parts:

Concerns/Attributes: Key properties that any trustworthy AI system should exhibit e.g., robustness, fairness, transparency, etc.
Factors: Characteristics that affect the concerns and therefore the trustworthiness of AI.
Methods: Techniques and processes for achieving the desired attributes

This is illustrated in Fig. 1 below.

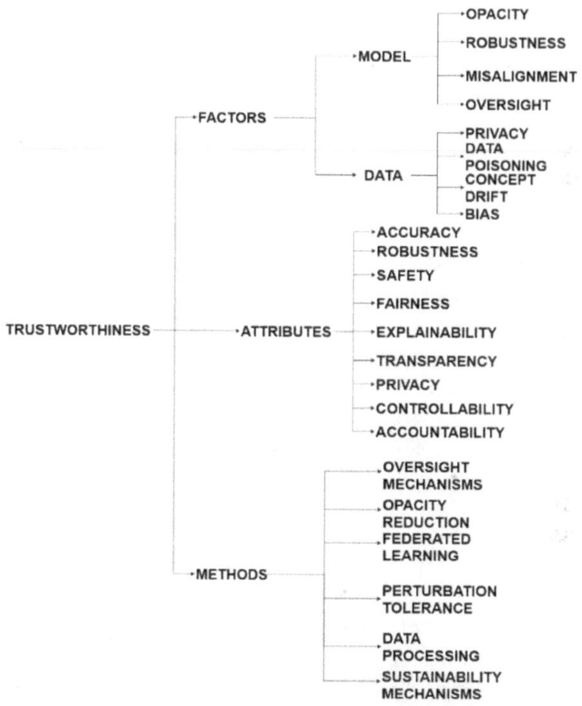

Fig. 1. Trustworthiness conceptual tree

3 Trustworthiness in AI Trustworthiness Factors for AI Systems

This section identifies and explains the factors that affect the trustworthiness of AI systems. An AI system's trustworthiness can be affected by several factors that cause the system to violate one or more trustworthiness criteria. These include the following: opacity, oversight, misalignment, concept drift, bias, data poisoning, robustness, privacy violations. The factors can be grouped according to whether they are associated with the model of the system, or the data of the system.

3.1 Model

These factors are trustworthiness factors that affect the AI model of the system. The main factors here include opacity, robustness to adversarial attacks, misalignment to user intent, and oversight.

Opacity. AI systems are often described as black boxes because their internal logic is not accessible or understandable to humans [5]. Opacity reduces its trustworthiness of the system since users and stakeholders cannot comprehend or predict how the AI makes decisions, leading to a lack of confidence, and the inability to explain or provide justification of the AI's behavior makes it difficult to make the system accountable for errors or harm. This opacity is reduced through transparency and explainability requirements. A lack of transparency means that an AI system does not provide clear insight into its decision-making process. The black box problem has been widely recognized, and users find it hard to trust decisions when they are given no rationale. If no one can understand how an AI arrived at a harmful outcome, it becomes challenging to determine responsibility or to contest the decision. Laws such as the EU's GDPR and proposed AI Act demand a degree of transparency or explainability for high-impact automated decisions [6]. In safety-critical fields such as healthcare and aviation a lack of transparency and explainability would prevent humans from trusting the system's recommendations.

Robustness to Adversarial Attacks. AI systems can be susceptible to being attacked or manipulated by adversaries. One of the most common threats is the adversarial attack, in which malicious inputs are given to an AI model and cause it to make a mistaken prediction or decision [3]. Adversarial attacks can directly threaten the robustness, reliability, and safety of AI systems. If an AI can be easily fooled by an attacker, it cannot be trusted to operate correctly in an open environment. The existence of adversarial attacks on an AI system means the AI system is not robust to malicious and worst-case inputs, and thus an adversary reduces its reliability.

Misalignment with User Intent or Values. Misalignment refers to a situation where an AI system's goals, function, or behavior does not properly reflect the intentions, values, or instructions of its human designers or users [7]. A misaligned AI might be competent at optimizing some objective, but if the objective is even slightly wrong or incomplete, the results can be unhelpful or even catastrophic [7]. Misalignment undermines safety, and trustworthy AI should not just avoid errors and attacks, but it should also not pursue unwanted outcomes. A misaligned AI may still be technically reliable at optimizing something, but it is not trustworthy from a holistic perspective, because it's not doing what the user actually intends or expects.

Oversight. A lack of oversight in AI systems refers to a lack of mechanisms to supervise AI decision-making or to hold any party responsible for the system's outcomes [8]. This is a result of AI models being deployed without proper governance, for example, no human-in-the-loop for decision-making, no audit trails, and no clarity on who is responsible if the AI causes harm. An unaccountable AI system can make consequential errors e.g. unfairly denying loans or making incorrect medical diagnoses with little recourse for those affected [9]. From a trustworthiness perspective, accountability is a foundational attribute since it ensures someone can explain and correct an AI's behavior. Without it, other attributes like fairness, reliability, and transparency are difficult to guarantee, since no one is actively monitoring or taking responsibility for AI's actions.

3.2 Model

These factors are trustworthiness factors that affect the data of the AI system. The main factors include bias, data poisoning, privacy violations, and concept drift.

Bias. Bias in AI refers to the systematic prejudice in an AI system's decisions and can result in unfair decision making. Bias can occur in several stages e.g. in the training data, if data reflects historical discrimination or lack of representation, in the model, due to flawed algorithms or objective functions, or in deployment through uneven feedback loops. When an AI system produces outcomes that favor or disfavor certain groups of people in a way that would be considered unjust, it directly reduces the attribute of fairness. Bias often affects fairness, which is defined as the absence of any systematic disadvantage to any group [10]. Bias can arise unintentionally. For example, if the training dataset is not diverse, the model may perform poorly on under-represented groups.

Data Poisoning. Data poisoning is where an attacker compromises the training data of a machine learning model in order to undermine the model's performance or to insert hidden malicious behavior [11]. This is in contrast to adversarial attacks, which happen at inference time. Poisoning happens during the training phase. By injecting specially crafted examples into the training dataset or by altering labels of training examples, an adversary can cause the trained model to misbehave in predictable ways. Data poisoning can affect attributes like reliability, robustness, and fairness, and an AI system cannot be deemed trustworthy if its training process could have been manipulated without anyone knowing.

Privacy Violations. AI systems can pose significant privacy threats by collecting, inferring, or exposing sensitive personal information without consent [12]. Privacy violations occur when an AI system mishandles data, for example, using personal data beyond its intended use or allowing unauthorized access to private information. AI can, for example, collect disparate data to create detailed user profiles of individuals, leading to intrusive inferences about them. This erosion of privacy can compromise user autonomy and dignity. Users expect AI systems to respect confidentiality and data protection norms, and if the system does not, their trust in the system can be broken.

Concept Drift. AI systems are not static and the data they are exposed to can change in distribution or meaning i.e. concept drift. Concept drift refers to a change in the underlying distribution of the data [66]. As a result the model's accuracy and reliability can degrade, and the system that once performed well no longer does so. This degradation can occur even without any obvious faults by the developers or any attack. It occurs naturally when deploying AI in dynamic real-world settings. Concept drift reduces trustworthiness since stakeholders expect AI systems to remain accurate and robust over time or at least be updated to keep them on track. If an AI is not monitored and updated, concept drift can lead to failures where the system's outputs become outdated or incorrect, potentially going unnoticed until significant harm or loss of performance has occurred.

Each factor can undermine specific trustworthiness attributes. By identifying these factors, we can better target mitigation or supporting strategies for each. This is illustrated in Fig. 2 below.

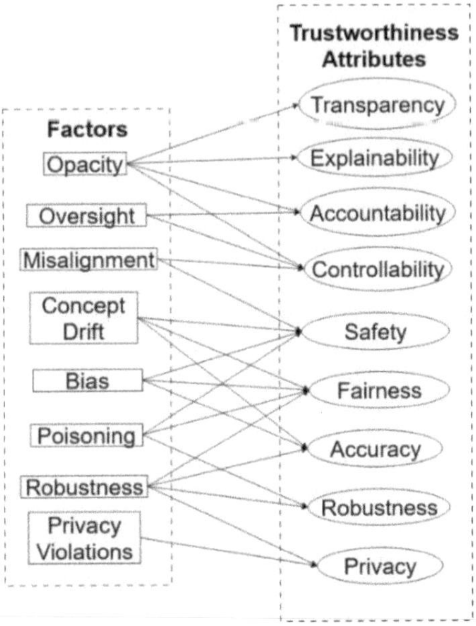

Fig. 2. AI Trustworthiness factors and the attributes they affect

4 Key Concerns (Attributes) of Trustworthy AI Systems

To be deemed trustworthy, an AI system should exhibit a set of key attributes or properties. These attributes correspond to the system qualities that inspire confidence and justify trust. Many of them directly counteract the threats discussed in the previous section. The major attributes of trustworthy AI drawn from the literature include the following: accuracy, robustness, safety, fairness, transparency, explainability, accountability, privacy, and controllability. These attributes can be interrelated and sometimes overlap. For example, transparency and explainability are closely linked; robustness and safety overlap; fairness and accountability can intersect. A trustworthy AI system should balance them all and any weakness in any one attribute can reduce the overall trustworthiness. We discuss each attribute below, defining it and explaining why it is essential for trustworthiness.

4.1 Accuracy

Accuracy refers to the level that the AI system consistently performs its intended function correctly. An accurate AI system produces the right outputs within the specified tolerances and does so consistently over time and across scenarios for which it was designed. For AI, this translates to metrics like high accuracy percentages, low error rates, and consistency of performance. Accuracy is achieved through rigorous validation, testing across diverse conditions, and design choices that emphasize correctness. It is closely related to robustness when considering deviations from expected conditions. An accurate AI model will achieve optimal or near-optimal results in the conditions it was designed for [13]. Accuracy is foundational since other attributes like fairness or transparency mean little if the system simply doesn't produce correct results, and an AI system with low accuracy cannot be trustworthy because it fails the basic promise of performing as expected.

4.2 Robustness

Robustness is the ability of an AI system to maintain its accuracy when conditions deviate from the nominal [14]. This includes resilience to noise in inputs, to reasonable perturbations or changes in the environment, and to malicious attacks like adversarial examples. Robustness is what gives users confidence that AI's performance will not degrade. If an AI is robust against adversarial inputs and other tampering, it increases trustworthiness. There tends to be trade-off between accuracy and robustness. Some models might achieve higher accuracy on a narrow set of inputs but fail outside if those inputs have slight perturbations, whereas a slightly less accurate model might be less accurate but is more stable across a variation of inputs. Therefore, for trustworthiness in AI systems, there should be a balance between the two.

4.3 Safety

Safety in AI refers to the system's ability to operate without causing unacceptable harm [15]. Safety is about preventing negative outcomes, especially those that could injure humans, damage property, etc. An AI system is safe if it respects the principle of non-maleficence i.e. it should not knowingly or predictably cause harm. Safety is a critical attribute for trustworthiness since users must be confident that the AI will not cause them or their surroundings harm. Safety can overlap with robustness since a robust system is often safer against perturbations which in turn to less changes of harmful outcomes. It is also related to fairness, since an unfair or misaligned system can be unsafe if it pursues a harmful goal. Many organizations emphasize that AI should undergo hazard analysis similar to traditional engineering systems. Safety also includes sustainability which refers to the system's environmental and societal impact over its life cycle, and the ability to operate in a manner that is resource-efficient and environmentally responsible in the long term.

4.4 Fairness

Fairness is the attribute that the AI system's decisions are impartial, just, and equitable, free from unfair bias [16]. A fair AI does not systematically disadvantage any particular group of people, especially groups defined by characteristics such as race, gender, age, etc. without a justified reason. Fairness is essentially the absence of the bias discussed earlier in sect. III. It is a key attribute of trustworthiness of AI systems because AI systems often automate decisions that can significantly affect peoples' lives, as such it is vital that these decisions are fair. Fairness is critical to trust. If people suspect an AI is biased against them, they will not trust or accept it. Furthermore, unfair AI can lead to societal backlash and legal challenges, as seen in cases of biased algorithms for criminal risk assessment. Fairness can depend on context and perspective [58] since different stakeholders might have different views of what is fair, therefore designers need to be clear and transparent about what notion of fairness they implement.

4.5 Transparency

Transparency is the attribute that the AI system's workings are open and clear to stakeholders [17]. A transparent AI system does not operate as an inscrutable black box and provides visibility into aspects such as, the data it was trained on, what algorithm or model it uses, what factors it considers in decision-making, and even how it has performed during evaluation. Transparency is crucial for accountability and trust. When an AI system is transparent, it allows the stakeholders to understand why it produces certain outcomes, or at least how it functions at a high level. This helps in detecting problems such as bias or errors, in debugging, and in establishing responsibility. Transparency is often paired with explainability, but they are distinct. Transparency is about the system being open and understandable in itself, whereas explainability is about providing specific explanations for individual outputs [98]. A transparent system might allow explainability methods to be applied more easily.

4.6 Explainability

Explainability is an attribute in which the AI system can provide understandable and meaningful explanations for its decisions or outputs [18]. Even if the system is complex internally, explainability focuses on translating its workings or outputs into terms that humans can grasp, thereby enhancing comprehension and trust. Explainability helps users and affected individuals to understand outcomes and increases acceptance if the decision is justified, it helps developers and auditors to verify the system is working properly and fairly, and it can uncover errors or biases. Even if an AI is accurate, people are more willing to trust and use it if they can get an explanation for its decisions [18]. Providing an explanation also allows human experts to validate and either agree or decide to override AI's suggestion.

4.7 Accountability

Accountability in the context of AI refers to the mechanisms that ensure that the AI system's outcomes can be ascribed to responsible parties, who are answerable for those outcomes and obligated to take corrective actions if necessary [8]. The EU AI guidelines state say accountability means an actor must be able to explain their aims, motivations, and reasons for their actions. It is essential for trustworthiness and overlaps with transparency and explainability, since those deploying the AI should be able to explain and justify the system's decisions and actions to the appropriate stakeholders, and without those attributes, accountability is hard to realize. Accountability ensures that AI systems, no matter how autonomous, operate under human oversight and responsibility.

4.8 Privacy

Privacy is the attribute in which an AI system respects and protects the privacy of individuals, particularly in regards to personal data. This includes ensuring that personal data used in or generated by the AI is properly safeguarded and that individuals have control over their own data [8]. A trustworthy AI should not intrude upon people's privacy without consent or legal basis, nor should it leak or misuse personal information. NIST lists privacy as one of the characteristics of trustworthy AI, and EU includes it as a key requirement [18]. Privacy deals with how the system should handle data in a manner that protects individuals and is both a right to be safeguarded and a necessary ingredient for trust. A trustworthy AI treats personal data with care and ensures individuals are informed by how their data is used.

4.9 Controllability

Controllability in AI systems refers to the degree to which human operators can monitor, direct, and intervene in the system's operation at any time to alter its behavior or outcomes. In other words, a controllable AI is one designed for meaningful human oversight, allowing humans to override or adjust the AI's decisions when necessary [19]. Ensuring controllability is fundamentally about preserving human agency over the AI's actions, especially in safety-critical or ethically sensitive situations. The EU AI Act explicitly requires human oversight for high-risk AI applications, meaning systems must be designed so that natural persons can monitor and if needed intervene or disable the AI to prevent harm, as such the AI system is never given free rein to operate beyond human governance.

4.10 Secondary Attributes

AI systems often exist as part of a larger computer system, and there are attributes that are applicable to the trustworthiness of those systems. These are often referred to the dependability attributes of computer systems, and in turn are applicable to the trustworthiness of an AI system. For example, security means the AI system is protected against unauthorized access and tampering [20]. If an AI system can be easily hijacked or manipulated by a malicious party, it cannot be trusted. Security can affect the fairness and robustness of the system and can ensure that the model's logic and the data it processes are not altered or accessed by an attacker.

5 First Section Methods to Achieve Trustworthiness in AI Systems

In this section, we discuss the methods, practices, and mechanisms to attain these attributes and manage the concerns described earlier. Achieving trustworthiness involves measures in the AI's design and implementation, as well as how AI is developed, tested, deployed, and monitored.

5.1 Opacity Reduction Mechanisms

Post-hoc Explainability Methods. For complex models like deep neural networks, techniques like Local Interpretable Model-agnostic Explanations (LIME) and Shapley Additive Explanations (SHAP) are widely used [21]. These methods yield human-readable explanations such as feature attributions. Techniques like SHAP or LIME provide local explanations for complex models by indicating the contribution of each input feature to the model's output.

Interpretable Model Design. One approach is using inherently interpretable models, when possible, such as decision trees or rule-based systems [6] so that the logic is transparent by design. These may sacrifice some accuracy compared to complex models, and so tradeoff may have to be considered.

Saliency and Visualization. In image or text models, visualization techniques highlight what parts of input influenced the decision, for example, saliency maps for images, and attention weight visualization for text [22].

Example-Based Explanations. Another technique is to provide examples to explain, for example, the system can retrieve similar past cases [21]. This gives context for the output of the AI system.

Self-Explaining Models. Research into models that output an explanation alongside their prediction is ongoing, for example, some AI models are trained to generate a justification for their output as part of their architecture [23].

5.2 Oversight Mechanisms

Runtime Monitoring. Monitoring the system can flag out-of-bounds behavior and either alert a human supervisor or trigger automated containment procedures [24]. In the operational phase, maintaining controllability entails deploying monitoring tools and interfaces that allow operators to continually observe the AI's actions and step in when the AI behaves unexpectedly.

Documentation and Transparency Artifacts. Tools like Model Cards are structured documents accompanying a model that detailing its intended use, performance metrics and ethical considerations [17]. This measure helps users understand the model's appropriate domain and limitations.

Logging. Keep logs of AI system decisions along with information such as input, output, maybe internal states or explanations. If an issue arises, these logs can be audited to trace what went wrong. It also deters inappropriate usage because everything is monitored [24]. In deployment, it involves monitoring for anomalies such as spikes in error rates or unusual patterns of inputs, which might indicate an adversarial attack. Set up alerts or automatic shutdown triggers if certain conditions occur, for example, if the system starts giving a lot of low-confidence outputs or an abnormal distribution of predictions.

Safe Interruptability. One prominent approach is safe interruptability, which ensures an AI agent can be paused or shut down without it undermining that intervention.

Researchers have demonstrated frameworks allowing a human operator to safely interrupt an AI agent, while preventing the agent from learning to resist or induce these interruptions.

Human-in-the-Loop. For many AI systems, having a human either making the final decision or being able to intervene provides a safety net and allows for oversight [19]. This also ensures accountability because a human is still supervising outcomes. Human-in-the-loop designs are methods in which human approval or supervision are allowed for important decisions by providing manual override mechanisms.

5.3 Data Processing and Handling

Data Augmentation. Data augmentation is the practice of programmatically expanding and diversifying your training dataset by applying transformations to existing data. This involves augmenting data with various transformations such as rotations, noise, different lighting for images and paraphrases for text [25]. It helps the model not rely on spurious features. The model learns invariances, which improves robustness to natural shifts in data. If certain groups are underrepresented, techniques like synthetic data generation could be used to balance the dataset.

Privacy-Preserving Machine Learning. Implement techniques such as differential privacy during training [26]. Differential privacy ensures that the model's parameters do not encode specifics about any single data point beyond a mathematically bounded amount of information, and one can train a model with differential privacy, which ensures any single training example has limited influence on the final model and prevents memorization of private details

Encryption and Secure Multi-Party Computation. In cases where a model is trained on sensitive data, techniques like homomorphic encryption or secure computation can allow the model to be trained or used on encrypted data [27]. This way, the system can make use of data without actually exposing it, thus maintaining privacy.

Synthetic Data for Privacy. In some cases, synthetic data generation is used to create a dataset that has similar statistical properties as real data but without real personal records [28]. Models can be trained on synthetic data to avoid using real sensitive data. However, synthetic data has to be carefully validated to ensure it truly protects privacy and maintains utility.

Federated Learning. In federated learning, instead of centralizing data, train AI models across decentralized data and only share model updates [26]. This keeps personal data local and therefore it increases privacy since the raw data never leaves its source and only the learned patterns do.

5.4 Perturbation Tolerance

Adversarial Training. Adversarial training involves generating adversarial data, for example, by introducing perturbations of the training inputs and training the model to correctly classify them and this can significantly improve a model's resilience to adversarial attacks [14].

Robust Optimization Models. In this method models are optimized for worst-case performance within some distribution of the data. This ensures the system isn't just good on average but also in worst-case scenarios [14].

5.5 Sustainability Mechanisms

Energy Efficient Models. Several technical approaches can be employed to implement sustainable AI. One important movement in research is known as Green AI [29]. This approach advocates AI models and training processes that prioritize energy efficiency and carbon footprint reduction alongside traditional metrics like accuracy. Techniques such as model compression, pruning, quantization [29], and efficient neural architecture search help reduce model size and power consumption without substantially sacrificing capability.

Batching Tasks. Batch tasking is an approach that groups similar tasks for joint execution [30]. By executing related computations together, it avoids repeated initialization of hardware and software contexts and thus reduces energy consumption and computational overhead. Applying this batching strategy can lower energy usage, and therefore a drop in carbon emissions.

Transfer Learning. Another practice is leveraging transfer learning and smaller pre-trained models, so that each new AI application doesn't require as much training for a new model, which therefore saves energy and compute time, since initializing new models with weights from related tasks significantly shortens training time and lowers the amount of computation required [29].

Table 1 below shows the attributes and the methods that can be used to achieve them.

Table 1. Table captions should be placed above the tables.

Attribute	Means
Accuracy	Data Augmentation, Monitoring for concept drift
Robustness	Adversarial training, Robust optimization models, Data augmentation (edge-case injection), Runtime monitoring for anomalous inputs
Safety	Safe interruptability, Runtime monitoring with automated shutdowns, Human-in-the-loop, Feedback channels
Fairness	Data governance policies, Data augmentation for minority classes, Human-in-the-loop bias review, Monitoring for concept drift
Transparency	Documentation & transparency artefacts, Model information disclosures, Traceability, logging pipelines
Explainability	Interpretable model design, Self-explaining models, Post-hoc explainability methods, Saliency & visualizations, Example-based explanations

(continued)

Table 1. *(continued)*

Attribute	Means
Accountability	Logging, Documentation & transparency artefacts, Data governance policies
Privacy	Differential privacy, Federated learning, Encryption & secure multi-party computation, Synthetic data generation
Controllability	Human-in-the-loop review / override, Safe interruptability, Feedback channels, Runtime monitoring dashboards

6 Conclusions

Trustworthiness in AI systems is a multifaceted concept that is essential as AI assumes larger roles in decision-making and automation. Similar to dependability in classical computing which encompasses a set of attributes that together ensure an AI system can be justifiably trusted by its users and by society [1]. This paper provides a conceptualization of trustworthiness in AI systems and understanding its attributes, the threats that impede them, and the means to achieve them is crucial towards achieving trustworthy AI systems. Achieving trustworthiness is an ongoing process, and AI systems exist in dynamic environments and interact with society. Therefore, trustworthiness must be continually maintained through vigilance and improvement. Building such systems requires a comprehensive approach and applying various methods to embed trustworthiness at the core of AI systems.

References

1. Aviz, A., Laprie, J.-C., Randell, B.: Fundamental Concepts of Dependability. University of Newcastle (2001)
2. Skalse, J., Howe, N.H.R., Krueger, D., Krasheninnikov, D.: Defining and Characterizing Reward Hacking. Curran Associates Inc., Red Hook, NY, United States (2022)
3. Zhang, J., Li, C.: Adversarial examples: opportunities and challenges. IEEE Trans. Neural Netw. Learn. Syst., 1–16 (2019). https://doi.org/10.1109/TNNLS.2019.2933524
4. Barbacci, M., Klein, M.H., Longstaff, T.A., Weinstock, C.B.: Quality Attributes. Defense Technical Information Center, Fort Belvoir, VA (1995). https://doi.org/10.21236/ADA307888
5. Rai, A.: Explainable AI: from black box to glass box. J. Acad. Mark. Sci. **48**, 137–141 (2020). https://doi.org/10.1007/s11747-019-00710-5
6. Nisevic, M., Cuypers, A., De Bruyne, J., Explainable, A.I.: Can the AI act and the GDPR go out for a date? In: 2024 International Joint Conference on Neural Networks (IJCNN), pp. 1–8. IEEE, Yokohama, Japan (2024). https://doi.org/10.1109/IJCNN60899.2024.10649994
7. Gabriel, I.: Artificial intelligence, values, and alignment. Minds & Machines. **30**, 411–437 (2020). https://doi.org/10.1007/s11023-020-09539-2
8. Herrera-Poyatos, A., Ser, J.D., de Prado, M.L., Wang, F.-Y., Herrera-Viedma, E., Herrera, F.: Responsible artificial intelligence systems: a roadmap to society's trust through trustworthy AI. Auditability, Account. Governance. (2025) http://arxiv.org/abs/2503.04739
9. Babushkina, D.: Are we justified attributing a mistake in diagnosis to an AI diagnostic system? AI Ethics. **3**, 567–584 (2023). https://doi.org/10.1007/s43681-022-00189-x

10. Ferrara, E.: Fairness and bias in artificial intelligence: a brief survey of sources, impacts, and mitigation strategies. Sci. **6**, 3 (2023). https://doi.org/10.3390/sci6010003
11. Fan, J., Yan, Q., Li, M., Qu, G., Xiao, Y.: A survey on data poisoning attacks and defenses. In: 2022 7th IEEE International Conference on Data Science in Cyberspace (DSC), pp. 48–55. IEEE, Guilin, China (2022). https://doi.org/10.1109/DSC55868.2022.00014
12. Curzon, J., Kosa, T.A., Akalu, R., El-Khatib, K.: Privacy and artificial intelligence. IEEE Trans. Artif. Intell. **2**, 96–108 (2021). https://doi.org/10.1109/TAI.2021.3088084
13. Kowald, D. et al.: Establishing and evaluating trustworthy AI: overview and research challenges. Front. Big Data. **7**, 1467222 (2024). https://doi.org/10.3389/fdata.2024.1467222
14. Baker, J.W., Schubert, M., Faber, M.H.: On the assessment of robustness. Struct. Saf. **30**, 253–267 (2008). https://doi.org/10.1016/j.strusafe.2006.11.004
15. Salhab, W., Ameyed, D., Jaafar, F., Mcheick, H.: A systematic literature review on AI safety: identifying trends, challenges, and future directions. IEEE Access. **12**, 131762–131784 (2024). https://doi.org/10.1109/ACCESS.2024.3440647
16. Varona, D., Suárez, J.L.: Discrimination, bias, fairness, and trustworthy AI. Appl. Sci. **12**, 5826 (2022). https://doi.org/10.3390/app12125826
17. Felzmann, H., Fosch-Villaronga, E., Lutz, C., Tamò-Larrieux, A.: Towards transparency by design for artificial intelligence. Sci. Eng. Ethics. **26**, 3333–3361 (2020). https://doi.org/10.1007/s11948-020-00276-4
18. Phillips, P.J. et al.: Four principles of explainable artificial intelligence. National Institute of Standards and Technology (U.S.), Gaithersburg, MD (2021). https://doi.org/10.6028/NIST.IR.8312
19. Jones-Jang, S.M., Park, Y.J.: How do people react to AI failure? Automation bias, algorithmic aversion, and perceived controllability. J. Comput.-Mediat. Commun. **28**, zmac029 (2022). https://doi.org/10.1093/jcmc/zmac029
20. Sotgiu, A., Pintor, M., Biggio, B.: Explainability-based debugging of machine learning for vulnerability discovery. In: Proceedings of the 17th International Conference on Availability, Reliability and Security, pp. 1–8. ACM, Vienna Austria (2022). https://doi.org/10.1145/3538969.3543809
21. Slack, D., Hilgard, S., Jia, E., Singh, S., Lakkaraju, H.: Fooling LIME and SHAP: adversarial attacks on post hoc explanation methods. In: Proceedings of the AAAI/ACM Conference on AI, Ethics, and Society, pp. 180–186. ACM, New York NY USA (2020). https://doi.org/10.1145/3375627.3375830
22. Alqaraawi, A., Schuessler, M., Weiß, P., Costanza, E., Berthouze, N.: Evaluating saliency map explanations for convolutional neural networks: a user study. arXiv. (2020). https://doi.org/10.48550/arXiv.2002.00772. http://arxiv.org/abs/2002.00772.
23. Zhang, Z., Rudra, K., Anand, A.: Explain and predict, and then predict again. In: Proceedings of the 14th ACM International Conference on Web Search and Data Mining, pp. 418–426. Association for Computing Machinery, New York, NY, United States (2021). https://doi.org/10.1145/3437963.3441758
24. Yampolskiy, R.V.: On monitorability of AI. AI Ethics. **5**, 689–707 (2025). https://doi.org/10.1007/s43681-024-00420-x
25. Taylor, L., Nitschke, G.: Improving deep learning using generic data augmentation. arXiv. (2017). https://doi.org/10.48550/arXiv.1708.06020. http://arxiv.org/abs/1708.06020.
26. Ouadrhiri, A.E., Abdelhadi, A.: Differential privacy for deep and federated learning: a survey. IEEE Access. **10**, 22359–22380 (2022). https://doi.org/10.1109/ACCESS.2022.3151670
27. Lee, J.-W. et al.: Privacy-preserving machine learning with fully homomorphic encryption for deep neural network. IEEE Access. **10**, 30039–30054 (2022). https://doi.org/10.1109/ACCESS.2022.3159694

28. Abay, N.C., Zhou, Y., Kantarcioglu, M., Thuraisingham, B., Sweeney, L.: Privacy preserving synthetic data release using deep learning. In: Berlingerio, M., Bonchi, F., Gärtner, T., Hurley, N., Ifrim, G. (eds.) Machine Learning and Knowledge Discovery in Databases, pp. 510–526. Springer International Publishing, Cham (2019). https://doi.org/10.1007/978-3-030-10925-7_31
29. Dash, S.: Green AI: enhancing sustainability and energy efficiency in AI-integrated Enterprise systems. IEEE Access. **13**, 21216–21228 (2025). https://doi.org/10.1109/ACCESS.2025.3532838
30. Zhang, D., Vance, N., Zhang, Y., Rashid, M.T., Wang, D.: EdgeBatch: towards AI-empowered optimal task batching in intelligent edge systems. In: 2019 IEEE Real-Time Systems Symposium (RTSS), pp. 366–379. IEEE, Hong Kong (2019). https://doi.org/10.1109/RTSS46320.2019.00040

Technical Review of the Challenges and Advances of Privacy-Preserving Deep Learning Techniques for Audio Data

Rotimi-Williams Bello(✉), Roseline Oluwaseun Ogundokun, Pius A. Owolawi, Etienne A. van Wyk, and Chunling Tu

Department of Computer Systems Engineering, Faculty of Information and Communication Technology, Tshwane University of Technology, Pretoria, South Africa
{bellorw,ogundokunro,owolawipa,vanwykea,duc}@tut.ac.za

Abstract. The advent of deep learning and its rapid expansion in audio processing for powering applications like speaker identification, automatic speech recognition (ASR), emotion detection, and other audio-based Artificial Intelligence, have demonstrated powerful capabilities. However, audio signals inherently transmit sensitive personal information, such as emotional state, identity, location, and health, making protection of user privacy critical. A survey is carried out in this paper focusing on recent privacy-preserving deep learning techniques for audio data, including federated learning (decentralized training), differential privacy (noise-based data/model obfuscation), secure multiparty computation and homomorphic encryption (cryptographic methods), and adversarial perturbation strategies. The paper discusses trade-offs between utility and privacy, it describes how task performance and privacy leakage are measured, and lists common datasets widely used for evaluation. Finally, the paper highlights open challenges such as handling limited labeled data, meeting real-time constraints, optimizing the accuracy/privacy tradeoff, and ensuring robustness. The paper concludes the survey by suggesting future research directions in privacy-aware audio Artificial Intelligence (AI).

Keywords: Artificial Intelligence · Audio Dataset · Automatic Speech Recognition · Federated Learning · Deep Learning · Privacy-Preserving

1 Introduction

Audio understanding has dramatically improved by deep learning breakthroughs [1]. Near-human performance in Automatic Speech Recognition (ASR) is now achieved by end-to-end models [1]. Voice can be used for identification of speakers with high accuracy, and emotions or other traits can be inferred by models [2]. However, audio is not the same as text or images, as it continuously captures rich personal content. In recent surveys, it was revealed that audio data often comprises sensitive personal information like emotional state, speaker identity, health indicators, and other private cues [3]. The potential leakage or misuse of such information poses a serious ethical and security

H. R. Arabnia et al. (Eds.): CSCI 2025, CCIS 2967, pp. 498–507, 2027.
https://doi.org/10.1007/978-3-032-28454-9_35

concern. A good example is a situation where an ambient recording might reveal not only spoken content but also the identity or location of the speaker [4]. In smart homes and voice assistants, intimate conversations or background activity can be inadvertently collected by always-on microphones, making privacy in these systems paramount [5]. Despite all these risks, audio-based Artificial Intelligence (AI) is pervasive.

Deep audio models are employed in hearing aids, voice assistants, medical diagnostic tools, and biometric security systems [6]. Recent work emphasizes that it is a timely and critical challenge to balance model performance with user privacy [7]. Consequently, a novel privacy-preserving paradigms for audio learning have become area of research interest to growing body of research. Therefore, this paper provides a comprehensive technical review of these approaches, such as federated learning, differential privacy, cryptographic approaches, and adversarial obfuscation. It discusses the applicability and tradeoffs of each technique for audio data. The paper summarizes common deep learning architectures for audio tasks, after outlining key privacy threats in audio-based learning systems. Finally, the paper also describes how utility and privacy are evaluated (datasets and metrics) and discusses open problems. Throughout, the latest developments were cited to ensure up-to-date coverage.

2 Privacy Threats in Audio-Based Learning Systems

Multiple layers of private information can be leaked by audio signals. This is demonstrated when the identity and biometrics (speaker recognition) of speakers are directly revealed by the speakers' voice [8]. Moreover, the voice can reveal some other information, such as gender, age, accent, emotional state, health (e.g. a cough indicating illness), or intoxication. Sensitive contexts such as background conversations, location signatures, or other ambient events can be betrayed by even non-speech sounds [9]. Inference attack may be enabled by unauthorized access to a trained audio model, that is, a membership inference might be used by an adversary to predict if an audio of a person was used in training or reconstruct the actual speech content using model inversion [10]. Similarly, other privacy threats in audio-based learning systems are:

1. Content exposure: This is at risk of attackers, who eavesdrop on conversations or extract confidential information that were never meant to leave the device. This happens if audio recordings or intermediate features are stored or transmitted insecurely (e.g. spectrograms sent to cloud servers). Figure 1 shows privacy threats in audio-based learning systems, including content exposure, identity leakage, content violation, and adversarial attacks.
2. Identity leakage: Speaker-specific characteristics are often retained by deep audio features (MFCCs, spectrograms). A user can be re-identified by an adversary by applying a speaker verification model on leaked features. In the voice privacy challenge, for example, the Equal Error Rate (EER) of speaker verification is used as a proxy for privacy (higher EER means less identity leakage) [11].
3. Content violation: Sensitive content may be extracted by audio classifiers even without explicit speech. For instance, a patient's medical condition could be inadvertently revealed from cough sounds or breathing patterns by a health-monitoring application. Moreover, audio data such as calls and voice notes on mobile devices can be stolen or inadvertently shared.

4. Inference attacks: Attackers may exploit a model's outputs beyond direct inversion.

For example, a model with carefully designed inputs could be probed to extract hidden patterns about the speaker or infer if a specific background noise or phrase appeared in the training set.

Fig. 1. Privacy threats in audio-based learning systems, including content exposure, identity leakage, content violation, and adversarial attacks.

Recent studies reveal the feasibility of membership inference attacks against speech models, potentially showing whether a user's recordings contributed to model training. Such risks are indicated by research on speaker-level membership inference.

5. Adversarial attacks: A model can be forced to misbehave or divulge information by maliciously crafted audio inputs. For example, targeted adversarial examples in audio can cause misidentification of a person by speaker recognition or make an ASR system transcribe a different phrase (e.g. a hidden command) [12]. Conversely, benign users can repurpose adversarial techniques as a defensive obfuscation. These threats highlight that privacy must be treated holistically (not just by filtering speech) by audio learning systems. Indeed, it is noted by research that even non-speech cues such as identity, background, and location in audio should be protected. With always-on microphones and ubiquitous smart devices, audio privacy remains a core challenge in deploying deep models [13].

3 Deep Learning Models for Audio Processing

A variety of architectures is employed by audio deep learning, depending on the task. A common pipeline technique is by converting raw audio signals into time-frequency features, such as MFCCs, log-Mel spectrograms, or raw waveform embeddings, and then feed them into neural networks [14]. For example, Convolutional Neural Networks (CNNs) are widely used on spectrogram inputs, this is evident in one classic study, where it was found that Mel-spectrograms paired with CNNs worked well for classification of environmental sound. Large sequence models (e.g., recurrent networks (Long Short-Term Memory (LSTM) or Gated Recurrent Unit (GRU)) and, more recently, Transformer-based and convolution-augmented Transformer models (e.g. Conformer)) are now typically used by modern end-to-end ASR systems. Rich audio embeddings that are fine-tuned for ASR are produced by large-scale self-supervised models like HuBERT or Wav2Vec 2.0.

Speaker recognition often uses embedding networks such as x-vectors extracted through ResNet architectures or Time-Delay Neural Networks (TDNNs), trained to cluster same-speaker utterances together [15]. Similarly, paralinguistic and emotion detection employ CNNs and Recurrent Neural Networks (RNNs) on spectrograms, often with attention mechanisms for capturing of temporal patterns. In practice, multi-layer deep networks are used by many of these architectures. For instance, recent research applies large end-to-end Transformer models to speech recognition and speaker tasks [16]. In the context of privacy studies, the federated ASR work of Apple specifically explores large Transformer models for Federated Learning (FL), indicating that such models are now common in ASR and relevant to privacy investigations. In summary, audio deep learning spans convolutional, recurrent, and attention-based architectures, whereby raw audio or extracted features can be processed to perform tasks like speaker ID, ASR, language identification, and emotion classification [17].

4 Privacy-Preserving Techniques

Several approaches have been proposed for the protection of privacy in audio learning. In this section, we review the applications of the main techniques to audio data and their trade-offs.

1. Federated Learning (FL): Models are trained by FL across many devices with no collection of raw audios on a central server. A local model is updated by the device of each user with its private audio data, thereby ensuring that only model weights or gradients are sent to a secure server for aggregation. By this technique, raw recordings are permanently kept in the device, ensuring inherent reduction of privacy exposure. For example, a federated speaker recognition system was demonstrated by Woubie et al. [18], where voice data of users were retained on-device. Good accuracy was achieved by the aggregated model while preserving privacy by design. In practice, secure aggregation protocols are required by FL to ensure individual updates cannot be inspected. FL allows collaborative learning on speaker or speech tasks, but that comes with communication overhead. Moreover, the participation of malicious clients often makes it vulnerable to non-IID (Independent and Identically Distributed) data and poisoning attacks. Often, FL is combined with additional measures for reliability. Figure 2 shows the applications of the main techniques of privacy-preserving techniques to audio data.
2. Differential Privacy (DP): A mathematical guarantee is provided by DP, limiting the information any single training example, such as one user's audio, contributes to the trained model. In deep learning, this usually implies addition of noise during training, such as in gradients or outputs. For ASR and other audio tasks, the application of DP ensures the protection of speaker identity and content. For instance, the Private

Aggregation of Teacher Ensembles (PATE) framework was applied by Yang et al. [19] to end-to-end ASR. It was shown that word error under DP constraints could be significantly reduced by the technique. In essence, DP's answer to privacy is: "inject noise to obscure who said what". There is a tunable privacy parameter ε, in which stronger privacy is produced by the smaller ε (more noise), but model accuracy is typically degraded.

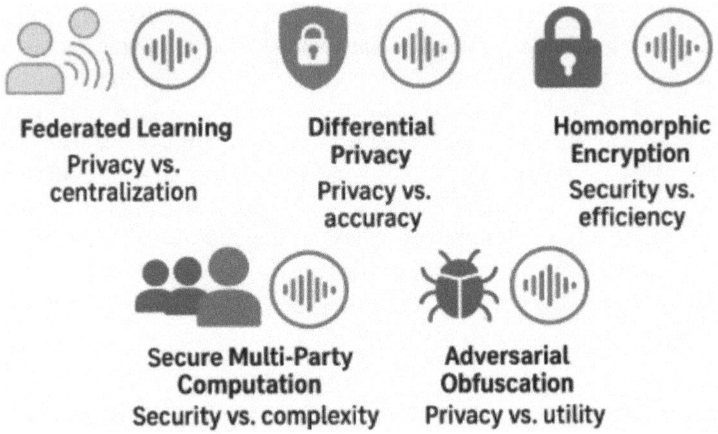

Fig. 2. Applications of the main techniques of privacy-preserving techniques to audio data, focusing on federated learning, differential privacy, homomorphic encryption, secure multi-party computation, and adversarial obfuscation.

Empirically, DP in audio often results to a trade-off: there is a drop in acoustic performance as the budget tightens. However, DP and FL can be combined (FL + DP) for stronger reliability, as in recent work that trained an ASR Transformer under user-level DP, achieving only a few percent increase in WER [16]. Noise injection and the complexity of training under strict DP (especially for large models) are the major limitations of DP.

3. Homomorphic Encryption (HE): By using HE, mathematical operations are allowed on encrypted data. In the audio context, audio features could be encrypted or model updated by a device before sending them. Likewise, ciphertexts can be computed by a server without decrypting. Apple uses a combination of differential privacy, and HE, specifically the Brakerski-Fan-Vercauteren (BFV) scheme to enable searches of private nearest-neighbor on encrypted embeddings within their Enhanced Visual Search feature in Photos. This enables searches on image embeddings by the system without directly revealing the image data or the search results to Apple's servers [16]. Similarly, HE has been proposed in literature for audio features encryption during processing. Strong confidentiality is the key benefit of this technique; this is evident in the high protection of raw audio or plaintext updates from being seen by the server. However, HE is computationally expensive. As noted in one analysis, strong confidentiality is provided by fully homomorphic schemes, but suffer from high

computational overhead, making their real-time applications unsuitable. In practice, HE might be more feasible for offline or on-device scenarios, such as private cloud queries on audio features, rather than large-scale training.

4. Secure Multi-Party Computation (SMPC): Cryptographic protocols such as secret sharing, and garbled circuits are used by SMPC for joint computation of a model without any party revealing its data. For example, a model could jointly be trained by multiple audio providers on their private speech data through SMPC, resulting in protection of all the raw audio or individual gradients from exposure. While theoretically powerful, complexity and communication cost limit the practicality of SMPC. Research has it that SMPC for audio or machine learning is mainly feasible for small-scale tasks and suffers from considerable communication and computational overhead. In other words, SMPC could secure privacy in principle, but at the cost of high latency and resource usage [20].
5. Adversarial Obfuscation: This technique involves hiding sensitive information by actively modifying the audio. For instance, small imperceptible perturbations can be added to the waveform for obfuscation of transcription models or speaker-identification while preserving audible content. VoiceBlock [21], running on a single CPU thread, is an example that employs a neural audio-to-audio model to apply a time-varying filter on live speech, producing perturbations that "de-identify user speech from speaker recognition". Such methods operate in real-time and aim to maintain perceptual quality while misleading downstream classifiers. Adversarial obfuscation is not the same as DP or encryption in that the input stream is the protected data. It faces a trade-off between efficacy and transparency, strong perturbations may be perceptible or hurt utility tasks, and they may eventually be adapted to by some systems. In the audio domain, each technique involves trade-offs. Raw data is kept local by FL and SMPC, but communication or computation overhead is added. Signal quality or accuracy is degraded by differential privacy and adversarial noise. Strong security is provided by HE/SMPC, but they are expensive. In practice, to balance these trade-offs, hybrid approaches are commonly used, such as FL with DP, or FL with secure aggregation [22].

5 Evaluation Metrics and Datasets

Metrics and datasets are required for both privacy and utility when evaluating privacy-preserving audio systems. Some privacy metrics are common, including ASR error rate and speaker verification EER on protected data. The EER of a speaker verification system is employed as the privacy metric (higher EER means harder recognition of anonymized voice) in the VoicePrivacy challenge framework [23]. Downstream task performance is used for measuring utility: typically, ASR error rate, commonly measured using Word Error Rate (WER), or classification accuracy/recall, such as Unweighted Average Recall for emotion recognition. Some evaluations jointly measure the success of a system by how much it increases EER (privacy gain) of a speaker while keeping ASR WER or accuracy of emotion recognition high. Specialized metrics are used in other works: for example, Chhaglani et al. [24] quantify the privacy-utility trade-off in conversation analysis by reporting both ASR WER and speaker diarization error rate.

In short, with higher EER or higher WER on the protected data, the implication is greater privacy at the cost of accuracy. The datasets that are widely used reflect the tasks in question. LibriSpeech [25] is a standard benchmark for ASR and speech tasks. The LibriSpeech comprises 960 h of English audiobook speech. The subsets of LibriSpeech (clean/other) serve for both ASR and training speaker verification. Vox-Celeb1/2 [26, 27] are popular for speaker recognition, and together, they contain on the order of 2,400 h from over 6,000 speakers, enabling training and testing of models and adversaries, such as speaker verifiers. Several acted corpora are common for emotion recognition and paralinguistics, including IEMOCAP (emotional English dialogues, used for SER evaluation) [28], RAVDESS [29], SAVEE [30], and CREMA-D [31], which provide labeled emotional speech. Environmental sound and context tasks use datasets like AudioSet [32] or ESC-50 [33], because they often contain non-speech sounds, they are less privacy-sensitive. Other multilingual and domain-specific corpora include AISHELL for Mandarin [34]. In privacy research, LibriSpeech and VoxCeleb are reused by many studies for benchmarking, and some studies, such as VoicePrivacy challenge adapt automatic metric scripts (ASR/EER models).

6 Challenges and Open Problems

Despite success and progress, several open challenges are militating against privacy-preserving audio learning [35–37], including:

1. Real-time constraints: low latency such as voice assistants or live transcription is required by many audio applications. These real-time demands may not be met by complex training or heavy cryptographic methods. Indeed, too much overhead is typically incurred by fully homomorphic or MPC-based approaches for streaming audio. Simultaneously, some adversarial obfuscation methods such as VoiceBlock have demonstrated real-time operation on modest hardware. However, ensuring both effectiveness and speed remains an open problem. It is an active challenge designing fast encrypted inference or lightweight privacy filters for continuous audio.
2. Trade-off between accuracy and privacy: utility is inherently reduced by distorting audio (adversarial filters) or injecting noise (DP). It is difficult finding the optimal balance. The performance of ASR is destroyed by too much noise, while privacy is leaked by too little noise. This trade-off is confirmed by recent analyses, where low-SNR audio recognition can be degraded by DP noise. Moreover, there is a lack of universal methods that work across tasks like speaker ID, speech, and emotion; an algorithm modified for identity protection may not fully protect content or vice versa. Better metrics and adaptive schemes such as context-aware noise tuning must be developed by research to guide this balance.
3. Lack of labeled private data: privacy research often uses public datasets like LibriSpeech, etc., but realistic private audio data such as personal voice logs, and hospital recordings are scarce and hard to obtain. This limits the evaluation of attacks and defenses under real-world conditions. It may be helpful using self-supervised learning on synthetic data or unlabeled data generation, but there is a need for labeled benchmarks for privacy-focused tasks such as labeled by identity for anonymization studies.

4. Adversarial robustness: mechanisms developed for privacy can become the target. For instance, a model might be trained by an attacker for audio obfuscation circumvention or adaptive attacks against DP-trained models. Also, traditional adversarial examples remain a threat: sensitive audio could be forcefully misclassified by perturbations. It is an ongoing challenge to build models that are robust to malicious adversarial inputs and inadvertent leakage. A deeper study is required for the interplay between privacy perturbations and adversarial attacks (two sides of "adversarial" techniques).
5. Multimodal integration: many systems are integrated to be multimodal, such as audio-visual assistants, and speech + text. There is an addition of complexity protecting audio privacy in a multimodal pipeline. It is nontrivial aligning features while keeping each modality private. For example, it requires joint obfuscation strategies to ensure a voice-video model cannot leak identity. Research on privacy for audio combined with other data such as video, and sensors is still emerging.

7 Conclusion and Future Directions

Recent literature on audio privacy was synthesized in this review, including surveys and empirical studies. It is a multifaceted problem protecting privacy in audio deep learning. In summary, this paper has reviewed the spectrum of threats such as identity leaks, content exposure, inference and adversarial attacks, and surveyed common architectures for audio models. The main defenses were examined, such as federated learning, differential privacy, homomorphic encryption/SMPC, and adversarial obfuscation. How each defense applies to audio and their inherent trade-offs were discussed. The paper also highlighted evaluation strategies such as EER/WER trade-offs, and benchmark datasets such as LibriSpeech, VoxCeleb, etc. There is no single method that is panacea. Distinct benefits are offered by each approach, but incurs cost in complexity, latency, or accuracy. As emphasized in a recent analysis, there is a need for hybrid and adaptive solutions.

For example, better results may be produced by combining secure aggregation and DP in federated training or using lightweight cryptographic enclaves alongside noise injection than using any one individual technique. Future research should explore such combinations. Moreover, it should leverage advances in efficient cryptography and hardware such as trusted execution environments, to mitigate overhead. Moreover, it is important to revisit privacy methods as foundation models and new modalities such as large audio transformers and multimodal embeddings are becoming widespread. Progress and evaluation will be well guided by establishing standardized privacy-utility benchmarks for diverse audio tasks. Finally, it is crucial to continue paying attention to ethical and legal frameworks such as General Data Protection Regulation (GDPR) compliance. Early integration of privacy-preserving designs (e.g. privacy-by-design in data collection and model deployment) and development of robust and practical techniques can enable future audio systems to leverage deep learning while respecting user confidentiality.

References

1. Purwins, H., Li, B., Virtanen, T., Schlüter, J., Chang, S.Y., Sainath, T.: Deep learning for audio signal processing. IEEE J. Sel. Topics Signal Process. **13**(2), 206–219 (2019)

2. Bello, R.W., Owolawi, P.A., van Wyk, E.A., Tu, C.: Audio emotion detection application utilizing AWS cloud. Recent Adv. Comput. Sci. Commun. **9**(5), 77–87 (2025)
3. Aloufi, R., Haddadi, H., Boyle, D.: Paralinguistic privacy protection at the edge. ACM Trans. Privacy Secur. **26**(2), 1–27 (2023)
4. Kröger, J.L., Gellrich, L., Pape, S., Brause, S.R., Ullrich, S.: Personal information inference from voice recordings: user awareness and privacy concerns. Proc. Privacy Enhancing Technol. **1**, 6–27 (2022)
5. Lenhart, A., Park, S., Zimmer, M., Vitak, J.: "You shouldn't need to share your data": perceived privacy risks and mitigation strategies among privacy-conscious smart home power users. Proc. ACM on Hum.-Comput. Interact. **7**(CSCW2), 1–34 (2023)
6. Khare, P., Srivastava, S.: Enhancing security with voice: a comprehensive review of AI-based biometric authentication systems. Int. J. Res. Anal. Rev. **10**(2), 398–403 (2023)
7. Williamson, S.M., Prybutok, V.: Balancing privacy and progress: a review of privacy challenges, systemic oversight, and patient perceptions in AI-driven healthcare. Appl. Sci. **14**(2), 675 (2024)
8. Wubet, Y.A., Lian, K.Y.: Speaker anonymization for voice biometrics protection using voice conversion and multi-target speaker voice fusion. IEEE Trans. Inf. Forensics Secur. **20**, 6046–6057 (2025)
9. Chandrakala, S., Jayalakshmi, S.L.: Environmental audio scene and sound event recognition for autonomous surveillance: a survey and comparative studies. ACM Computing Surveys (CSUR). **52**(3), 1–34 (2019)
10. Hu, H., Salcic, Z., Sun, L., Dobbie, G., Yu, P.S., Zhang, X.: Membership inference attacks on machine learning: a survey. ACM Computing Surveys (CSUR). **54**(11s), 1–37 (2022)
11. Champion, P.: Anonymizing speech: evaluating and designing speaker anonymization techniques. arXiv, 1–171 (2023). https://doi.org/10.48550/arXiv.2308.04455
12. Tan, H., Wang, L., Zhang, H., Zhang, J., Shafiq, M., Gu, Z.: Adversarial attack and defense strategies of speaker recognition systems: a survey. Electronics. **11**(14), 2183 (2022)
13. Gray, S.: Always on: Privacy Implications of Microphone-Enabled Devices, pp. 1–10. Future of privacy forum (2016)
14. Zaman, K., Sah, M., Direkoglu, C., Unoki, M.: A survey of audio classification using deep learning. IEEE access. **11**, 106620–106649 (2023)
15. Khadar, K.A., Sunil Kumar, R.K., Sameer, V.V.: Speaker diarization based on X vector extracted from time-delay neural networks (TDNN) using agglomerative hierarchical clustering in noisy environment. Int. J. Speech Technol. **28**(1), 13–26 (2025)
16. Chang, X., Zhang, W., Qian, Y., Le Roux, J., Watanabe, S.: End-to-end multi-speaker speech recognition with transformer. In: ICASSP 2020–2020 IEEE International Conference on Acoustics, Speech and Signal Processing (ICASSP), pp. 6134–6138. IEEE (2020)
17. Al-Fraihat, D., Sharrab, Y., Alzyoud, F., Qahmash, A., Tarawneh, M., Maaita, A.: Speech recognition utilizing deep learning: a systematic review of the latest developments. HCIS. **14**, 1–34 (2024)
18. Woubie, A., Bäckström, T.: Federated learning for privacy-preserving speaker recognition. IEEE Access. **9**, 149477–149485 (2021)
19. Yang, C.H.H., Chen, I.F., Stolcke, A., Siniscalchi, S.M., Lee, C.H.: An experimental study on private aggregation of teacher ensemble learning for end-to-end speech recognition. In: 2022 IEEE Spoken Language Technology Workshop (SLT), pp. 1074–1080. IEEE (2023)
20. Rahaman, M., Arya, V., Orozco, S.M., Pappachan, P.: Secure multi-party computation (SMPC) protocols and privacy. In: Innovations in Modern Cryptography, pp. 190–214. IGI Global (2024)
21. O'Reilly, P., Bugler, A., Bhandari, K., Morrison, M., Pardo, B.: Voiceblock: privacy through real-time adversarial attacks with audio-to-audio models. Adv. Neural Inf. Proces. Syst. **35**, 30058–30070 (2022)

22. Truex, S. et al.: A hybrid approach to privacy-preserving federated learning. In: Proceedings of the 12th ACM Workshop on Artificial Intelligence and Security, pp. 1–11 (2019)
23. Panariello, M., Tomashenko, N., Wang, X., Miao, X., Champion, P., Nourtel, H., et al.: The voiceprivacy 2022 challenge: progress and perspectives in voice anonymisation. IEEE/ACM Trans. Audio, Speech, Lang. Process. **32**, 3477–3491 (2024)
24. Chhaglani, B., Gomasta, S.S., Agarwal, Y., Gummeson, J., Shenoy, P.: FeatureSense: protecting speaker attributes in always-on audio sensing system. arXiv. **1**(1), 1–34 (2025)
25. Panayotov, V., Chen, G., Povey, D., Khudanpur, S.: Librispeech: an asr corpus based on public domain audio books. In: 2015 IEEE International Conference on Acoustics, Speech and Signal Processing (ICASSP), pp. 5206–5210. IEEE (2015)
26. Nagrani, A., Chung, J.S., Zisserman, A.: Voxceleb: a large-scale speaker identification dataset. arXiv, 1–6 (2018)
27. Chung, J.S., Nagrani, A., Zisserman, A.: Voxceleb2: deep speaker recognition. arXiv, 1–6 (2018)
28. Busso, C., Bulut, M., Lee, C.C., Kazemzadeh, A., Mower, E., Kim, S., et al.: IEMOCAP: interactive emotional dyadic motion capture database. Lang. Resour. Eval. **42**(4), 335–359 (2008)
29. Livingstone, S.R., Russo, F.A.: The Ryerson audio-visual database of emotional speech and song (RAVDESS): a dynamic, multimodal set of facial and vocal expressions in north American English. PLoS One. **13**(5), e0196391 (2018)
30. Jackson, P., Haq, S.: Surrey Audio-Visual Expressed Emotion (Savee) Database. University of Surrey, Guildford, UK (2014)
31. Cao, H., Cooper, D.G., Keutmann, M.K., Gur, R.C., Nenkova, A., Verma, R.: Crema-d: crowd-sourced emotional multimodal actors dataset. IEEE Trans. Affect. Comput. **5**(4), 377–390 (2014)
32. Gemmeke, J.F., Ellis, D.P., Freedman, D., Jansen, A., Lawrence, W., Moore, R.C., et al.: Audio set: an ontology and human-labeled dataset for audio events. In: 2017 IEEE International Conference on Acoustics, Speech and Signal Processing (ICASSP), pp. 776–780. IEEE (2017)
33. Piczak, K.J.: ESC: dataset for environmental sound classification. In: Proceedings of the 23rd ACM International Conference on Multimedia, pp. 1015–1018 (2015)
34. Bu, H., Du, J., Na, X., Wu, B., Zheng, H.: Aishell-1: an open-source mandarin speech corpus and a speech recognition baseline. In: 2017 20th Conference of the Oriental Chapter of the International Coordinating Committee on Speech Databases and Speech I/O Systems and Assessment (O-COCOSDA), pp. 1–5. IEEE (2017)
35. Saleh, K., Darweesh, D., Darwish, O., Hammad, E., Amsaad, F.: Hardware and software methods for secure obfuscation and deobfuscation: an in-depth analysis. Computers. **14**(7), 251 (2025)
36. Kumar, P., Shukla, M., Malik, M.: Privacy-preserving speech analysis-a review. In: 2024 Sixth International Conference on Computational Intelligence and Communication Technologies (CCICT), pp. 538–545. IEEE (2024)
37. Singh, P., Srivastava, R., Rana, K.P.S., Kumar, V.: A multimodal hierarchical approach to speech emotion recognition from audio and text. Knowl.-Based Syst. **229**, 107316 (2021)

Cognitive Artificial Intelligence in Business Process Management: An Adaptive and Context-Aware Process Optimization Framework

Bhagyeshkumar Chokhawala(✉) and Atif Farid Mohammad

Capitol Technology University, Laurel, MD, USA
{bchokhawala,afmohammad}@captechu.edu

Abstract. Integrating cognitive AI into business process management (BPM) is transforming how organizations analyze, adapt, and improve workflows. Although many companies use automation and process mining tools, most BPM systems still remain static, prioritizing efficiency over adaptability. This paper presents the Cognitive BPM Integration Framework (C-BPM), a model that integrates adaptive learning, context-based reasoning, and predictive decision-making into BPM. Based on cognitive systems theory, process mining, and reinforcement learning, the framework features a multi-layer architecture that supports real-time process adjustment and context-aware optimization. Using a mixed-methods approach that combines process mining simulations and expert validation, the study demonstrates that C-BPM can greatly improve responsiveness, transparency, and predictive capabilities in essential business processes. The research offers a structured, evidence-based model for integrating cognitive AI into BPM, helping bridge the gap between mere automation and intelligent process understanding.

Keywords: Cognitive Artificial Intelligence · Business Process Management · Process Mining · Reinforcement Learning · Adaptive Workflows · Predictive Analytics

1 Introduction

Organizations increasingly rely on artificial intelligence (AI) to manage complex business processes and data-intensive operations. Yet, despite advances in automation and analytics, most BPM systems lack cognitive capabilities—the ability to learn from experience, reason contextually, and adapt to dynamic business environments [1, 2]. Traditional BPM focuses on efficiency and compliance but does not address how processes can autonomously adjust to changes in demand, policy, or environment.

This gap underscores the need for Cognitive Business Process Management (C-BPM)—a new paradigm that combines process mining, adaptive learning, and cognitive reasoning to create self-optimizing workflows. Current literature identifies substantial fragmentation across BPM and AI research. While process mining and predictive monitoring have matured, few frameworks integrate cognitive reasoning and reinforcement

H. R. Arabnia et al. (Eds.): CSCI 2025, CCIS 2967, pp. 508–518, 2027.
https://doi.org/10.1007/978-3-032-28454-9_36

learning for adaptive control [3, 4]. This paper introduces the Cognitive BPM Integration Framework (C-BPM), designed to incorporate AI-driven learning and reasoning mechanisms into BPM workflows.

2 Related Work

Process mining provides a foundational analytical approach for discovering and optimizing workflows using event logs [5]. However, traditional process mining techniques are retrospective, focused on what has happened rather than what will happen next. Recent research highlights the potential of cognitive AI to augment BPM through semantic reasoning and learning [6, 7]. Despite progress in AI-enabled automation, existing BPM systems rarely combine semantic understanding with adaptive control, creating a clear need for a validated integration model [8].

The C-BPM framework builds on three theoretical foundations: Process Mining and Conformance Theory, Reinforcement Learning, and Cognitive Systems Theory. Process mining interprets event logs to discover, monitor, and improve processes, providing the perception layer for cognitive BPM [5]. Reinforcement learning models process management as an agent–environment interaction [4, 9]. Cognitive systems theory posits that intelligent agents can perform goal-directed reasoning similar to human cognition [10].

3 The Cognitive BPM Integration Framework

The Cognitive BPM Integration Framework (C-BPM) unites process mining analytics, adaptive AI mechanisms, and feedback control loops into a single architecture for intelligent process optimization. It enables continuous learning and autonomous adjustment of workflows across diverse business process transactions.

It consists of four interconnected layers and an implementation mechanism, each addressing both technical and managerial dimensions.

The C-BPM model builds on cybernetic control, cognitive learning, and reinforcement learning theories. It introduces continuous process optimization through feedback loops analogous to sensor–controller–actuator structures, combined with double-loop learning and active inference for adaptive governance (Fig. 1).

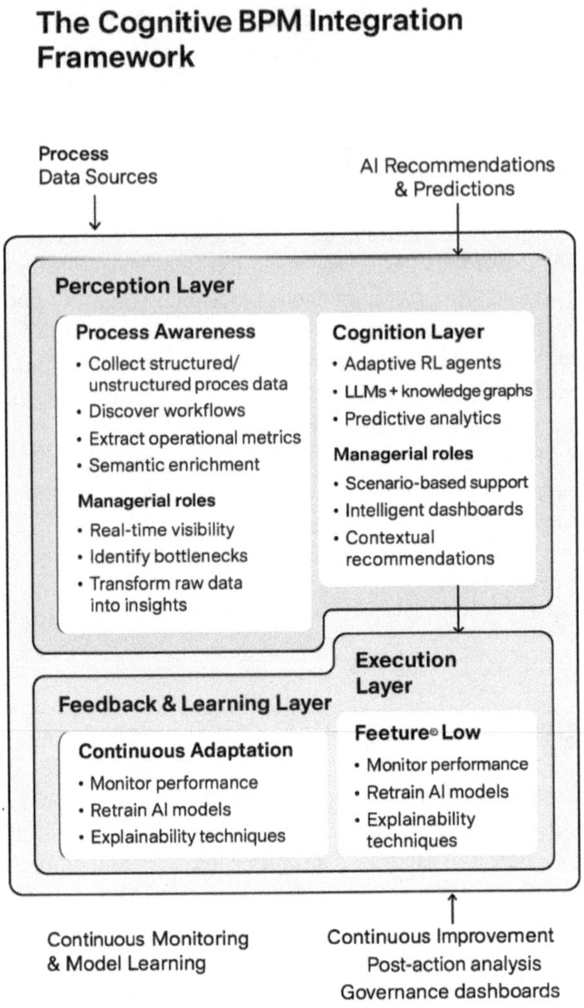

Fig. 1. A high-level conceptual design for the Cognitive BPM Integration Framework (C-BPM), capturing its architecture, layers, and flow.

4 Theoretical Foundations

4.1 BPM and Cybernetic Control

BPM follows a design–execute–monitor–optimize cycle. C-BPM enriches this with cybernetic control, creating a feedback loop analogous to sensor → controller → actuator → feedback, ensuring continuous correction.

4.2 Cognitive and Organizational Learning

Drawing from SOAR, ACT-R, and Argyris & Schön's organizational learning models, C-BPM incorporates double-loop learning — adapting not only actions but also decision policies.

4.3 Active Inference and Reinforcement Learning

Adaptive decisions are modeled through Expected Free Energy (EFE) or reward maximization paradigms, supporting contextual policy updates.

5 Cognitive Algorithms

5.1 Process Perception and Model Extraction

Technical Functions: The collection of structured and unstructured process data (event logs, audit trails, communication records). Uses process discovery algorithms (Alpha Miner, Inductive Miner) to reconstruct workflow models. Extracts key operational metrics (cycle time, throughput, compliance deviations). Performs semantic enrichment via NLP to connect contextual information to event data.

Managerial Roles: Provides real-time visibility into ongoing operations and deviations. Enables process analysts to proactively identify bottlenecks and compliance risks. Serves as the "sensory system" of BPM, transforming raw data into actionable process knowledge.

Objective: To convert heterogeneous, multi-source event data (structured and unstructured) into a unified, semantically enriched process model and quantitative performance indicators, forming the perception layer of the C-BPM framework.

Conceptual Overview: This component functions as the "nervous system" of the Cognitive BPM ecosystem — sensing, extracting, and contextualizing operational signals.

It bridges raw data → process intelligence by combining:

- Process mining algorithms (Alpha Miner, Inductive Miner, Heuristic Miner),
- NLP-based contextual understanding, and
- Metric extraction engines for compliance, time, and throughput analytics.

Inputs: Functional Event logs/Business Transaction logs E, communication records/SOP records C

Outputs: Process model P, performance indicators M

Procedure:

- Aggregate structured/unstructured $data \rightarrow D = E \cup C$.

- Apply process mining algorithm (e.g., Inductive Miner) miner $\rightarrow P = f_{mine}(D)$.
- Compute operational metrics $M = \{t_{cycle}, r_{throughput}, d_{compliance}\}$.
- Use NLP enrichment $\rightarrow$ contextualize events $P' = f_{sem}(P, text(C))$.
- Return (P', M).

5.2 Cognition Reasoning via RL and LLM Integration

Technical Functions: This layer integrates Reinforcement Learning (RL) agents for adaptive routing and task allocation. Uses Large Language Models (LLMs) and Knowledge Graphs to interpret context and infer causal relationships among process events. Applies predictive analytics to forecast deviations, SLA breaches, and workload imbalances.

Managerial Roles: Supports human decision-makers with contextual recommendations and explanations. Provides intelligent dashboards that translate AI insights into business-level KPIs. Enables scenario-based reasoning—evaluating "what-if" simulations to improve processes.

Objective: Generate adaptive decisions based on real-time process state.

Conceptual Overview: The Cognitive Reasoning Algorithm represents the adaptive decision-making mechanism within the Cognition Layer of the Cognitive BPM Integration Framework (C-BPM).

Its primary objective is to generate context-aware, explainable, and adaptive actions based on the current process state, leveraging reinforcement learning (RL) for decision optimization and large language models (LLMs) for causal interpretation.

Input: Process model P', current state s_t, KPI vector M

Outputs: Action recommendation a_t

Procedure:

- Encode state s_t into feature vector.
- RL agent selects action $a_t = \pi(s_t)$ maximizing reward R_t or minimizing Expected Free Energy G_t.
- LLM interprets context and provides a causal explanation e_t.
- Combine $(a_t, e_t) \rightarrow$ decision packet.
- Forward to the Execution Layer for enactment.

Reward Function Example:

$$R_t = w_1(\text{SLA adherence}) + w_2(\text{cost efficiency}) - w_3(\text{risk penalty})$$

5.3 Feedback and Learning Layer — Continuous Adaptation

Technical Functions: Continuously monitors performance indicators and outcome deviations. Feeds result back to the cognition layer to retrain AI models (RL and predictive analytics). Uses explainability techniques (e.g., SHAP values, causal models) for transparent learning.

Managerial Roles: Facilitates continuous improvement loops aligned with organizational goals. Supports post-action analysis and lessons learned to refine strategy. Provides governance dashboards linking AI performance to policy compliance.

Objective: Update cognitive models and workflows based on outcome deviations.

Conceptual Overview: The Feedback-Driven Learning and Adaptation Algorithm governs the continuous improvement mechanism within the Feedback and Learning Layer of the Cognitive BPM Integration Framework (C-BPM).

Its primary function is to close the adaptive loop between execution outcomes and cognitive model updates—ensuring that each cycle of process performance contributes to more brilliant, more aligned, and more transparent decision-making.

Inputs: Actual outcome o_t, predicted outcome $\hat{o}_t$, reward R_t

Outputs: Updated policy parameters θ'

Procedure:

- Calculate deviation $\delta_t = o_t - \hat{o}_t$.
- Update policy gradient:

$$\theta' = \theta + \alpha \nabla_\theta (R_t - \lambda \mid \delta_t \mid)$$

- Retrain models using new labeled outcomes.
- Update governance dashboards with SHAP-based explanations.
- Propagate learned policies to the Cognition Layer.

5.4 Execution Layer — Operational Implementation

Technical Functions: This layer translates cognitive insights into executable BPMN or workflow rules. Interfaces with ERP, CRM, and BPM platforms through APIs or middleware. Triggers automated or semi-automated actions (e.g., rerouting approvals, reallocating resources).

Managerial Roles: Acts as the *operational control center* for executing adaptive actions. Supports hybrid intelligence—humans can override or approve AI recommendations. Provides audit trails and traceability to ensure accountability in automated decisions.

Objective: To transform cognitive recommendations into actionable workflow executions, integrating seamlessly with enterprise systems (ERP, CRM, BPM platforms) and maintaining accountability through traceable audit trails.

Conceptual Overview: The Execution Layer serves as the operational control hub of the Cognitive BPM architecture. It translates cognitive insights and recommendations generated by the Cognition Layer into executable business process actions.

It operates at the intersection of automation and governance, ensuring that adaptive actions are technically accurate, operationally compliant, and managerially auditable (Fig. 2).

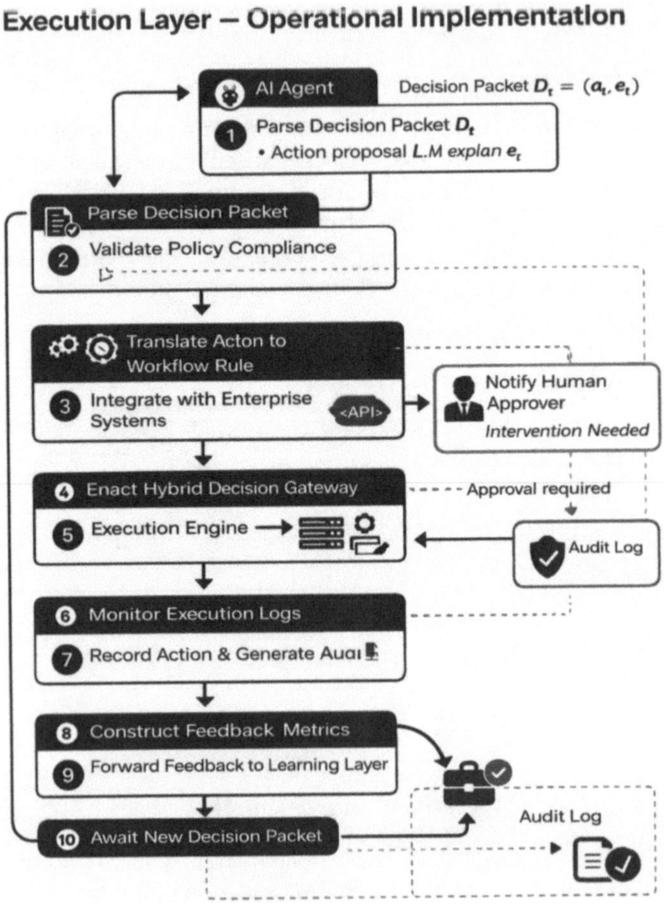

Fig. 2. Cognitive BPM Integration Framework (C-BPM) flowchart.

Inputs

- Decision Packet $D_t = (a_t, e_t)$: Recommended action and its causal explanation from the Cognition Layer.
- Business Process Model P': BPMN or workflow schema.

- Contextual Constraints C: Policies, role permissions, and SLA rules.

Outputs

- Enacted Process Changes: Executed or updated workflows and resource allocations.
- Audit Log Entry: Detailed trace of decision, execution, and oversight validation.
- Feedback Metrics: Performance and compliance indicators for the Feedback Layer.

Procedure:

- Parse Decision Packet

  - Receive Decision Packet $D_t = (a_t, e_t)$ from the Cognition Layer.
  - Extract:

    Action a_t — the AI-recommended task or process modification.
    Explanation e_t — the causal rationale produced by the LLM component.

- Validate Policy and Compliance Rules

  - Cross-check a_t against contextual constraints C

    (authorization, SLA limits, regulatory or ethical boundaries).

  - If a violation is detected:

    Flag action for human approval and log the compliance alert.

  - If compliant:

    Continue to the execution translation.

- Translate Action into Executable Workflow Rules

  - Map a_t to the corresponding elements in the BPMN process model P'.
  - Generate execution syntax such as:

    – BPMN task/sequence-flow updates, or
    – REST/GraphQL API payloads for ERP, CRM, or BPM middleware.

  - Validate syntactic and semantic integrity before dispatch.

- Integrate with Enterprise Systems

- Invoke relevant connectors or adapters (e.g., SAP ERP API, Salesforce REST API, Camunda BPM engine, any BPMN, Workflow engine).
- Perform authentication and authorization checks.
- Dispatch the executable command or rule update.

- Hybrid Decision Gateway
  - If human validation is required:
    - Route the action to the designated approver via the task queue or the UI dashboard.
    - Wait for the approval/rejection signal.
  - If fully automated:
    - Execute action autonomously through the process orchestrator.
- Monitor Execution Status
  - Capture execution logs, timing, and exceptions.
  - Compare actual vs. expected outcomes in near real time.
  - Trigger compensating actions or rollbacks if failures occur.
- Generate Audit Record R_t
  - Construct a complete transaction trace, including:
    - Action ID, timestamp, system user/agent ID.
    - Executed the command and the target endpoint.
    - Causal explanation e_t.
    - Policy validation result and mode (auto/manual).
  - Store R_t in an immutable audit repository for compliance.
- Construct Feedback Metrics F_t
  - Compute performance indicators (cycle-time Δ, cost impact, SLA change).
  - Package results as a feedback object F_t, including success/failure flags and KPI updates.
- Propagate to the Feedback and Learning Layer
  - Send F_t and R_t to the Feedback Layer for evaluation and model retraining.
  - Update the governance dashboard with trace and explainability visuals.
- Finalize Cycle

- Mark the execution instance as *closed*.
- Await new decision packet from Cognition Layer to initiate next adaptive iteration.

6 Implementation Highlights

We have designed work as a modular framework compatible with standard BPM tools and cloud infrastructures. Supports plug-and-play integration with enterprise systems through open APIs. Can operate in *supervised*, *semi-autonomous*, or *autonomous* modes based on organizational readiness. Enables a phased adoption strategy, allowing for the gradual infusion of cognitive components without requiring complete system replacement.

7 Research Design and Validation

The framework was developed and validated using a sequential explanatory mixed-methods design combining quantitative and qualitative components. Quantitative validation employed process mining and simulation on process data such as order-to-cash and claims management. Qualitative validation involved expert interviews to assess usability, transparency, and ethical implications. Findings informed refinements emphasizing modularity and integration with ERP tools.

8 Discussion and Implications

The proposed C-BPM framework extends traditional BPM by embedding cognitive capabilities that enable context-aware decision-making and continuous learning. The study advances BPM theory by operationalizing cognitive systems theory in a process management context and provides practitioners a structured approach for implementing AI-driven process improvements.

9 Conclusion and Future Work

This paper presented the Cognitive BPM Integration Framework, a conceptual and empirically grounded approach to embedding cognitive artificial intelligence into BPM. By uniting process mining, reinforcement learning, and cognitive reasoning, C-BPM enables adaptive, context-aware workflows capable of continuous self-optimization. Future work will extend the framework with a governance layer for ethical oversight and benchmark datasets for empirical evaluation.

The future of Cognitive AI in Business Process Management (BPM) holds immense potential for transforming static workflows into dynamic, intelligent systems. Building on the Cognitive BPM Integration Framework (C-BPM), future work should prioritize enhancing real-time adaptability through advanced context modeling and edge AI integration, enabling organizations to respond swiftly to market shifts or internal changes. Scalability and interoperability will be key, requiring modular designs and industry-wide standardization to ensure seamless adoption across diverse sectors. Ethical considerations, such as bias mitigation and explainable AI, must also be central to development, fostering trust and compliance. Additionally, fostering human-AI collaboration through augmented intelligence and continuous learning will ensure that cognitive BPM systems complement human expertise while evolving alongside organizational needs.

Validation through pilot studies in high-stakes industries and benchmarking against traditional BPM systems will further solidify the framework's efficacy. Exploring synergies with emerging technologies like blockchain for secure auditing and digital twins for process simulation can unlock new dimensions of efficiency and innovation. Longitudinal studies and evolutionary models will help assess the long-term impact of cognitive BPM, ensuring these systems not only adapt to change but also drive sustainable growth. Ultimately, the goal is to create BPM ecosystems that are intelligent, ethical, and collaborative, bridging the gap between automation and true process understanding.

References

1. Vidgof, M., Bachhofner, S., Mendling, J.: Large language models for business process management: opportunities and challenges. arXiv. (2023). https://doi.org/10.48550/arXiv.2304.04309
2. Grohs, M., Abb, L., Elsayed, N., Rehse, J.-R.: Large language models can accomplish business process management tasks. arXiv. (2023). https://doi.org/10.48550/arXiv.2307.09923
3. Carmona, J., van Dongen, B.F., Solti, A., Weidlich, M.: Conformance Checking: Relating Processes and Models. Springer (2018)
4. Integrating reinforcement learning into business process mining for continuous process adaptation and optimization. J. Comput. Intel. Robot. **2**(2), 69-79 (2022)
5. van der Aalst, W.M.P.: Process Mining: Data Science in Action. Springer (2016)
6. Fettke, P., Di Francescomarino, C.: Business process management and artificial intelligence. Künstl. Intell. **39**, 67–79 (2025)
7. Rosemann, M., vom Brocke, J., Van Looy, A.: Business process management in the age of AI – three essential drifts. Inf. Syst. e-Bus. Manag. **22**, 415–429 (2024)
8. Morana, S., et al.: Responsible BPM governance: embedding accountability and compliance in AI-driven workflows. Inf. Syst. Frontiers. (2024)
9. Lahza, H., et al.: Adaptive multi-objective resource allocation for edge-cloud workflow optimization using deep reinforcement learning. Modelling. **5**(3), 1298–1313 (2024)
10. IBM Research: Towards a Benchmark for Causal Business Process Reasoning with Explainable AI. IBM Research (2024)

A Multi-agent Self-optimizing Framework for Real-Time LLM Model Selection and Routing

Raghu Katikeri[1(✉)], Chiranjeevi Bhagavan[2], R Anush Jain[1], Priya Sinha[2], Giri Sainath Peruri[1], Khushi Lamichaney[1], and Amit Vaid[2]

[1] State Street Corporation, Hyderabad, India
rkatikeri@StateStreet.com
[2] State Street Corporation, Bengaluru, India
https://www.statestreet.com/us/en

Abstract. The increasing complexity and diversity of user interactions with LLM Applications demand adaptive strategies for model selection and optimization. Relying on a single, static model often leads to inefficiencies, especially in dynamic environments where query types, user intent, and performance constraints vary significantly. This paper introduces a self-governing, agent-based framework for real-time model orchestration that intelligently routes user queries to the most appropriate language model based on contextual understanding and system metrics. The proposed architecture integrates a lightweight query classifier, a modular agent router, and a centralized model registry enriched with metadata such as domain specialization, latency, and cost. These components work in tandem to enable dynamic model selection, ensuring optimal trade-offs between response quality, efficiency, and resource utilization. The system also incorporates feedback loops, long-term memory modules, and benchmarking pipelines to support continuous learning and robust model lifecycle management. Empirical evaluations across multiple use cases demonstrate the framework's ability to improve performance, reduce latency, and enhance personalization. The paper also discusses key deployment challenges, including scalability, explainability, and cold start mitigation, and outlines future directions involving reinforcement learning, meta-learning, and cross-domain model coordination. This work contributes a scalable and intelligent solution for adaptive model governance in next-generation conversational AI systems driven by LLM's.

Keywords: adaptive routing · multi-agent orchestration · large language models (LLMs) · feedback and personalization · policy-aware governance

H. R. Arabnia et al. (Eds.): CSCI 2025, CCIS 2967, pp. 519–526, 2027.
https://doi.org/10.1007/978-3-032-28454-9_37

1 Introduction

LLM Applications serve heterogeneous requests under tight and shifting constraints namely task quality, latency, safety, and cost. Static, one-size-fits-all model choices routinely under-perform when user intent and workload mixes evolve, or when provider behavior and pricing change. We present *AGA* (Agentic Governance and Allocation), a self-governing multi-agent framework that classifies query intent and risk, enforces policy, and optimizes model selection and decoding parameters in real time. AGA separates concerns across a *Supervisor* (orchestration), a *Query Curator* (intent/task classification), a *Model Recommender* (policy-aware routing), a *Query Execution Engine*, and a *Feedback Synthesizer* with long-term memory (LTM), enabling adaptive, auditable decisions and continual improvement.

Our contributions are: (i) a policy-gated architecture for safe, real-time model orchestration; (ii) a multi-objective selection methodology balancing response quality, latency, and cost; and (iii) an evaluation protocol combining offline replay with online A/B testing to quantify win-rate, P95 latency, cost/1K tokens, and guardrail performance. The design integrates a lightweight query classifier, a modular agent router, and a centralized, metadata-rich model registry to support dynamic model selection and life-cycle governance.

2 Background and Motivation

Modern AI applications increasingly depend on portfolios of language models and tools. A single static model cannot consistently satisfy the diverse intents spanning summarization, extraction, coding, multi-step reasoning, and domain-specific tasks, especially under variable SLAs and governance constraints. Recent work in agentic retrieval and dynamic routing emphasizes orchestration, memory, and feedback as first-class system elements. Building on these ideas, we adopt multi-agent role separation and iterative learning to deliver per-query decisions rather than static defaults, and to embed governance as a property of the system rather than an afterthought. We further incorporate multi-dimensional query classification to inform routing and parameter choices.

As shown in Fig. 1, AGA is composed of:

1. **Query Intake & Processing**: Ingests and normalizes incoming requests, performing preliminary validation and metadata enrichment to ensure consistent formatting for subsequent components.
2. **Supervisor Agent**: Oversees the entire pipeline and maintains a verifiable record of decisions. This ensures transparent orchestration and auditability of all interactions.
3. **Query Curator**: The Query Curator analyzes user intent, identifies relevant tasks, and assesses potential risks to produce routing signals that guide downstream processes. It employs lightweight classifiers for task and intent detection, supplemented by an optional LLM fallback for ambiguous scenarios. Content risk classifiers detect unsafe categories and sensitive data, imposing

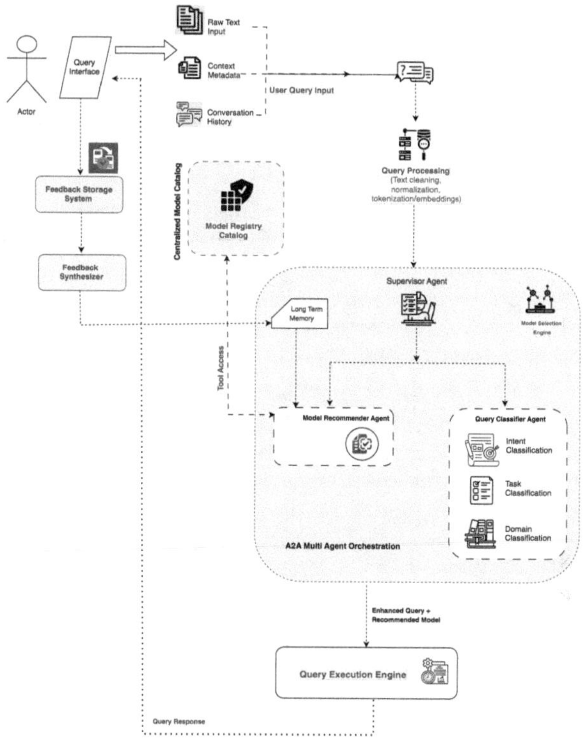

Fig. 1. Modular Multi-Agent Architecture for Real-Time Policy-Aware LLM Routing.

constraints that limit the Model Recommender's search space. By employing multi-dimensional query classification to encompass complexity, domain, task type, and performance requirements, the system effectively manages routing and parameter selection.

4. **Model Recommender**: Consults a policy-gated **Model Registry**, encompassing model capabilities, costs, latencies, safety tiers, and versions, along with real-time telemetry. Selects optimal candidate models and decoding parameters while supporting controlled exploration for continuous improvement.
5. **Query Execution Engine**: Executes the chosen model or toolchain, returning structured responses and logs for further analysis and traceability.
6. **Feedback Synthesizer & LTM**: Aggregates explicit and implicit feedback, updates learned priors, and feeds insights back into the pipeline to enhance future decision-making and system performance.

Policy-Aware Governance: Eligibility checks occur pre-selection (allow/deny lists, data residency), during execution (runtime content safety), and post-hoc (audit trails), providing defense-in-depth and compliance by construction.

3 Problem Statement

Given an incoming query q with context x (user/tenant metadata, historical behavior, risk signals), the system must select a model configuration $m \in \mathcal{M}$ (model family/version, temperature, decoding parameters, tool chain) to produce output y that:

- maximizes task quality $Q(y; q)$ (accuracy, faithfulness, coherence),
- satisfies latency and cost budgets $L(m; x), C(m; x)$,
- obeys safety and policy constraints $S(m, q, y) \leq 0$.

We frame this as a constrained, multi-objective optimization:

$$\max_{m \in \mathcal{M}} U(m, q, x) = w_q \widehat{Q}(m, q, x) - w_l \widehat{L}(m, x) - w_c \widehat{C}(m, x),$$

subject to eligibility and safety constraints.

4 Evaluation

4.1 Experimental Design

Workloads: We evaluate on RouterBench with a stratified split preserving label distributions across summarization, information extraction, complex reasoning, and code.

Baselines: (i) Random; (ii) Static Best (single strongest model chosen on validation); (iii) Oracle (upper bound using ground-truth best per query); and (iv) AGA (ours).

In addition to the RouterBench evaluation, we also compared **benchmark models** against **AGA's recommended models** across three representative intents: *Reasoning*, *Coding Assistant*, and *Chat*. For each intent, we tracked the benchmark model and its average cost, and contrasted these with the recommended model selected by AGA and its corresponding cost. We further measured *precision*, reflecting the alignment of the recommended model with task requirements.

This analysis demonstrates that AGA consistently identifies *lower-cost alternatives without sacrificing quality*. In reasoning tasks, AGA reduced token costs while maintaining competitive accuracy. For coding and chat workloads, adaptive routing achieved comparable or higher precision than static baselines, while optimizing overall resource utilization.

Importantly, the framework is not tied to a single provider. The model registry and routing pipeline are designed to be *extensible*, supporting seamless integration of additional providers such as **Databricks models**. This ensures cross-ecosystem portability and positions AGA as a future-proof orchestration layer for heterogeneous model landscapes.

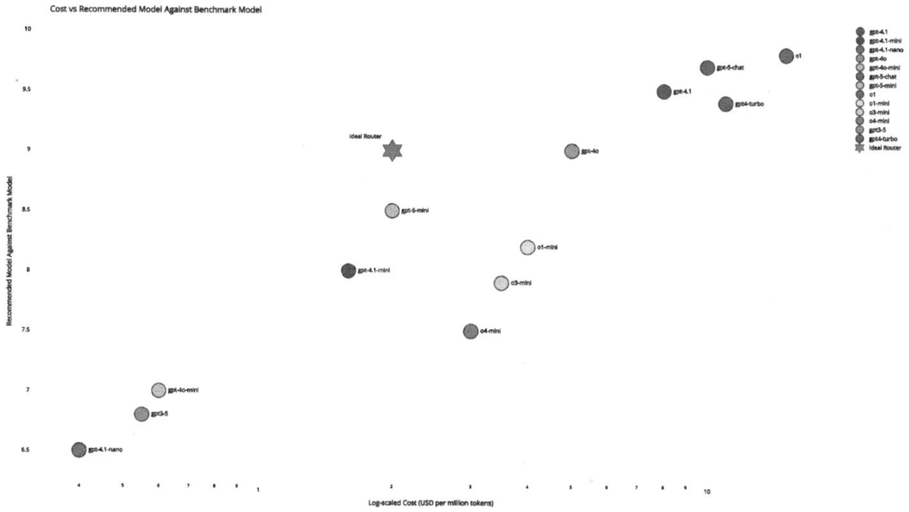

Fig. 2. Intent-Aware Cost Optimization via Adaptive Model Selection.

Results:

1. The results demonstrate the effectiveness of our router. On the test set, our agentic router achieved a routing accuracy of 0.76, significantly outperforming both the random baseline and the static best model. The regret metric, which quantifies the frequency of suboptimal routing decisions, was also notably lower for our router (0.24) compared to the baselines.
2. By employing a dynamic, run-time model selection strategy rather than relying on a static configuration, token costs were reduced by 23% while preserving comparable accuracy (Fig. 2 and Table 1).

Table 1. Intent-based Model Performance Summary

Intent	Benchmark Model	Recommended Model	Avg Benchmark Cost	Avg Recommended Cost	Precision Score
reasoning	gpt-4.1-nano	o1-mini	3.56	1.65	0.94
code generation	gpt-4.1-mini	gpt-4.1-nano	3.06	2.0	0.993
chat	gpt-4.1-nano	gpt-4o	3.62	1.5	0.85

4.2 Comparison Study Between Azure AI Foundry's Model Route and AGA

- **Abstraction:** Azure Router exposes a single endpoint that auto-selects models based on prompt complexity, cost, and latency. AGA, instead, separates roles across agents (Curator, Recommender, Supervisor, Feedback/LTM), enabling fine-grained orchestration.
- **Governance:** Azure integrates service-level guardrails and content filters. AGA enforces policy at three stages (pre-selection, runtime, post-hoc) and logs decision traces for auditability and compliance.
- **Optimization:** Azure optimizes across a fixed catalog of Foundry models. AGA jointly optimizes model choice, decoding parameters, and tool chains, using contextual bandits and feedback-driven priors.
- **Feedback and Learning:** Azure provides limited transparency into routing logic. AGA continuously updates its priors via explicit/implicit feedback, offline replay, and online A/B testing, reducing regret over time.
- **Flexibility:** Azure Router is bound to the Azure ecosystem. AGA supports cross-provider orchestration, internal SLMs, and tenant-specific utility weights (e.g., quality vs. latency vs. cost trade-offs).

When AGA is Favorable?

- *Strict governance & audit*
- *Hybrid tooling*
- *Cross-provider agility*
- *Tenant-specific utility*

Summary: Azure's router emphasizes *managed convenience and platform-level guardrails*, whereas AGA emphasizes *policy-aware orchestration, transparency, and continual optimization* across models, parameters, and tools. These approaches are complementary; AGA can *use* Azure router as a candidate within its model registry when appropriate, while retaining governance and telemetry at the AGA layer.

5 Conclusion and Future Scope

5.1 Conclusion

AGA provides a practical, modular path to self-governing, policy-aware model selection and optimization. By combining a Supervisor, Query Curator, Model Recommender, Execution Engine, and Feedback Synthesizer with long-term memory, AGA improves quality, reduces latency and cost, and lowers safety risk under drift. The approach is production-oriented, auditable, and adaptable, making it well suited to next-generation conversational systems that demand governed

5.2 Future Scope

We plan to: personalize routing via tenant- and role-specific utility weights; jointly optimize retriever+generator/tool chains; incorporate region- and carbon-aware routing; adopt federated learning for privacy-preserving cross-tenant learning; explore reinforcement learning and meta-learning for policy- and utility-aware orchestration; and extend to cross-domain, multi-modal coordination (See Table 2).

List of Abbreviations

Table 2. Abbreviations and descriptions

Abbreviation	Description
LLM	Large language model
AGA	Agentic Governance and Allocation
SLM	Small language model
LTM	Long-term memory
SLA	Service-level agreement
API	Application programming interface
CI	Confidence interval
DB	Database

Acknowledgments. We extend our deepest gratitude to our leadership team at State Street Corporation for their support and funding of this research

References

1. Huang, Q., et al.: Harder tasks need more experts: dynamic routing in MoE models. arXiv preprint arXiv:2403.07652 (2024)
2. Guo, T., et al.: Large language model based multi-agents: A survey of progress and challenges. arXiv preprint arXiv:2402.01680 (2024)
3. Schick, T., et al.: ToolFormer: language models can teach themselves to use tools. arXiv preprint arXiv:2302.04761 (2023)
4. Chen, Q., Han, X., Lin, Z., Li, B., Lu, J.: Towards automatic prompt engineering with contextual optimization. arXiv preprint arXiv:2210.03629 (2022)
5. Liu, P., et al.: HM-RAG: hierarchical multi-agent multimodal retrieval-augmented generation. arXiv preprint arXiv:2504.12330v1 (2025)
6. Yorozu, Y., Hirano, M., Oka, K., Tagawa, Y.: Electron spectroscopy studies on magneto-optical media and plastic substrate interface. IEEE Transl. J. Magn. Japan **2**, 740–741 (1987). [Digests 9th Annual Conf. Magnetics Japan, p. 301, 1982]

7. Young, M.: The Technical Writer's Handbook. University Science, Mill Valley, CA (1989)
8. Zhou, S., et al.: AgentBoard: an interactive platform for benchmarking multi-agent collaboration. arXiv preprint arXiv:2403.12031 (2024)
9. Chen, Y., et al.: AutoAct: automated action selection for large language model based agents. arXiv preprint arXiv:2502.18482 (2025)

Reducing Carbon Emissions in AI-Generated Videos by Harnessing Spatial Shifting Technique

Alexander van Gerwen(✉)

El Camino Real Charter High School, Los Angeles, CA, USA
lasurferthe1@gmail.com

Abstract. AI-generated videos (e.g. Sora) are heading in the direction of becoming the next frontier in generative AI. Although people are impressed by the video quality generated by these models, their excessive energy consumption and carbon emissions are often overlooked. Currently, the AI-generated video market is expanding at a rate of 33.12% per year, leading to a significant increase in carbon emissions. As a result, the environmental impact of video generation is becoming an increasingly pressing concern. To tackle this challenge, I conducted a compressive study on energy consumption and carbon emissions of state-of-the-art AI models that can generate videos. In addition, I propose a spatial shifting algorithm that aims to reduce carbon emissions by shifting workload from a high carbon intensity region to a low carbon intensity region. My experimental results demonstrate that AI models with a low carbon footprint can generate videos of comparable quality to those produced by high-carbon-footprint models. This finding reinforces the idea that opting for more eco-friendly models does not require sacrificing video quality for lower emissions. Furthermore, even with a relatively small sample size of 100,000 six-second videos, leveraging the proposed spatial shifting algorithm could reduce carbon emissions by approximately 350,000 kg per year.

Relevant Research Track: CSCI-RTCC

Keywords: Reducing Carbon Emissions · Artificial Intelligence · AI-Generated Videos · Spatial Shifting

1 Introduction

Since the release of Chat-GPT, in 2 months, 100 million users flocked onto the AI-generating website to find solutions to their questions. MidJourney has daily active users ranging from 1.2 million to 2.5 million (MidJourney [13]). The next frontier in AI generation will be AI-generated videos. The market for AI-generated videos was valued at 534.4 million in 2023 and is predicted to grow to 2,562.9 million by 2032. The pressing issues surrounding AI-generated videos are escalating, only becoming more severe in the future.

AI generation requires powerful GPUs, producing substantial amounts of carbon. For example, researchers have reported that creating a 2-s video at 240p resolution

H. R. Arabnia et al. (Eds.): CSCI 2025, CCIS 2967, pp. 527–535, 2027.
https://doi.org/10.1007/978-3-032-28454-9_38

using Sora produces six times more carbon emission than generating a text response using a large language model with a comparable size. The carbon emission for a single frame is 78 times higher than that for generating a single text token (Hot Carbon [8]). I measured the energy consumption required to generate a six-second video using an A100 GPU running the CogVideoX-5B-12 V model, which totaled 0.168 kWh. Scaling up to 1,000,000 videos per day results in an energy consumption of 168 MWh daily. For context, the average household consumes approximately 30 kWh of energy per day. This means that generating 1,000,000 videos daily would use enough energy to power 5,600 households. Therefore, it is essential to reduce the carbon emissions of AI models that can generate videos.

This paper presents a comprehensive study of the energy consumption and carbon emissions of three state-of-the-art open-source models: CogVideoX-5B-I2V, CogVideoX-2B-12 V, and Rain1011 (Hugging Face, CogVideoX-5B [9]; Hugging Face, CogVideoX-2B [10]; Hugging Face, Rain1011 [11]). The aim is to quantify the emissions produced during AI video generation. Specifically, I ran the CogVideoX-5B-12 V model on an A100 GPU hosted in Lambda Cloud and found that generating a six-second video took 1,755.35 s (about 59 min) and consumed 0.168 kWh of energy. In comparison, the CogVideoX-2B-12 V model completed the same task in 436.20 s (about 15 min), using 0.125 kWh of energy. For the Rain1011 model, the low-resolution version generated a six-second video in 63 s, consuming 0.0067 kWh, while the high-resolution version completed the task in 218 s (about 7 min) but consumed 0.05 kWh.

The spatial shifting algorithm simulates 100,000 user requests for generating videos across various regions. Using the carbon intensity of each region, the algorithm determines the CO2 emissions produced in the user's current region and compares them to the emissions in the region with the lowest carbon intensity. At the end of the simulation, it calculates the total CO2 savings that could be achieved if video generation occurred in the most optimal region.

2 Related Works

Generative AI has been booming in many domains (e.g. text, voice, image generation), since the introduction of ChatGPT. MidJourney, one of the largest image generation services using AI, has attracted 1.2–2.5 million daily users (Php Tutorial [14]). This success has catalyzed the rapid expansion of the AI-generated video market, which is projected to grow from $534.4 million in 2023 to $2,562.9 million by 2032 (Fortune Business Insights [4]). State-of-the-art video generation models, such as CogVideoX-5B-I2V and CogVideoX-2B-12 V, have enabled the production of high-quality videos (Hugging Face, CogVideoX-5B [9]; Hugging Face, CogVideoX-2B [10]). The Rain1011 model further shows how efficient these models have become by offering a low and high-resolution option that can be used to balance video quality and energy consumption (Hugging Face, Rain1011 [11]).

However, using large AI models to generate videos requires significant resources and emits massive carbon. For example, Sora generates six times more carbon emissions than producing a text response of similar complexity, with a single frame emitting 78 times more carbon than a single text token (HotCarbon24 [8]). Several studies have shown the environmental cost of training and deploying AI models. Strubell et al. (2019) revealed the carbon emissions of training large natural language processing models and proposed interventions to lessen these impacts (Strubell et al., 2019 [18]). Henderson et al. (2020) focused on the energy and carbon impacts of natural language processing, advocating for more efficient ways (Henderson et al., 2020 [7]).

These studies highlight the urgent need for energy-efficient AI design, prompting innovations like Google's TPU, specifically designed to reduce energy use during AI workloads (Jouppi et al., 2017 [12]). Beyond hardware optimizations, software-based strategies have also gained attraction. For example, the Cloud Carbon Footprint Tool was proposed to accurately measure carbon emissions across various cloud providers (Cloud Carbon [1]). Tools like CodeCarbon provide a framework to measure energy consumption and carbon emissions across AI (CodeCarbon [2]). Real-time data on regional carbon intensity, available through Electricity Maps, allows for precise carbon tracking during model operations (Electricity Maps [3]). Han et al. (2016) demonstrated that pruning neural networks can significantly reduce energy consumption without compromising performance (Han et al., 2016 [5]). Schwartz et al. (2020) introduced the concept of "Green AI," emphasizing sustainable practices in AI research and development (Schwartz et al., 2020 [16]).

Renewable energy integration is another solution to reduce emissions. Radovanovic described Google's Carbon-Intelligent Compute Management system, which minimizes carbon emissions by shifting flexible workloads to times of lower grid carbon intensity (Radovanovic et al., 2021 [15]). This system shows the potential of spatial shifting algorithms to reduce emissions without compromising quality. Similarly, load flexibility has been shown to support carbon-free energy, reducing clean energy costs and increasing resource efficiency (Spatio-temporal load shifting for truly clean computing).

This study builds on these contributions by discussing the carbon emissions of video generation models and showing the effectiveness of selecting a model with low carbon footprint as well as using spatial shifting algorithms in reducing environmental impact.

3 Spatial Shifting Algorithm

Figure 1 provides an overview of the spatial shifting algorithm. The algorithm begins with data loading, incorporating real-time carbon intensity data from Electricity Maps and generating traces based on population data from the Census. Next, the algorithm compares the carbon intensity of the user's original region (Old Region) to that of the least carbon-intensive region (New Region), which then becomes the user's shifted region. Finally, the program converts carbon intensity into CO_2 and calculates the savings by subtracting the emissions of the original region from those of the new region.

The Spatial Shifting Algorithm is designed to optimize the location of generating AI videos by shifting the workload from a region with high carbon intensity to one with low carbon intensity. By using real-time carbon intensity data, this algorithm considers the energy usage in different regions, simulating how much energy and carbon can be saved if the generation was done in a region that had a smaller carbon intensity.

Figure 2 provides a brief description of the data used in the model. The algorithm simulates 100,000 instances of video generation across various regions. For each request,

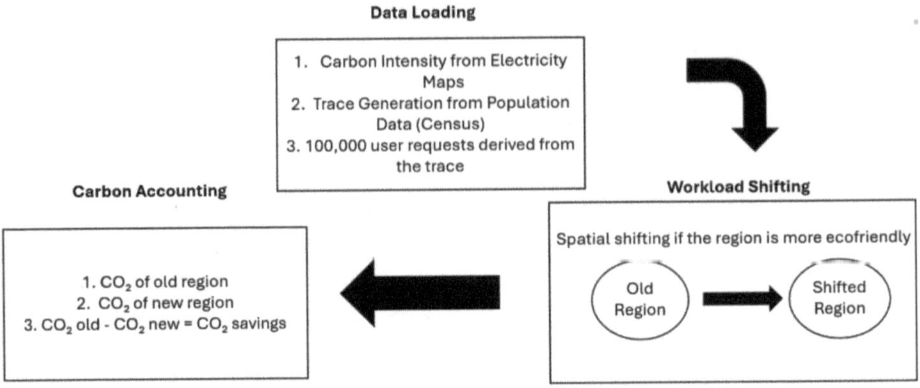

Fig. 1. An overview of the spatial shifting algorithm.

Time Stamp	Texas	California	Arizona	Original Region	Shifted Region
12:08 AM PST	269.82	**218.3**	443.68	Texas	California
4:27 PM PST	251.48	**152.98**	426.56	Arizona	California
12:18 PM PST	**187.96**	216.22	442.46	Texas	Texas
9:30 AM PST	**167.38**	215.44	435.33	Arizona	Texas
11:35 PM PST	340.04	**148.49**	433.69	Texas	California

Fig. 2. A table showing the carbon intensity in different regions, along with their original and shifted regions.

a specific timestamp is assigned within a 24-h window, accounting for the variability in video generation times and the fluctuations in CO_2 emissions throughout the day. Once the hour is assigned, the algorithm compares the carbon intensity of the selected regions at that specific time. It then stores the user's original region along with the region that has the lowest carbon intensity (Shifted Region).

Figure 3 is a line graph showing how the carbon intensities in certain regions change throughout a single day. When the carbon intensity lines intersect, that is the opportunity to shift from one region to another. As carbon intensity fluctuates throughout the day, the ability to transition between regions is essential for reducing carbon emissions.

Figure 4 illustrates the population distribution across the selected regions, with requests proportionally assigned based on population percentages: Texas (34.5%), California (44.8%), Arizona (8.6%), and Sweden (12.1%). To accurately represent real-world scenarios, the algorithm assigns a user's region based on population probability. For instance, with a total simulated population of 87 million, California's 44.8% share results in a 44.8% probability of assignment. Emission savings are then calculated by comparing the carbon intensity of the assigned region at a given hour to that of the least carbon-intensive region. If the workload remains in the original region, no carbon

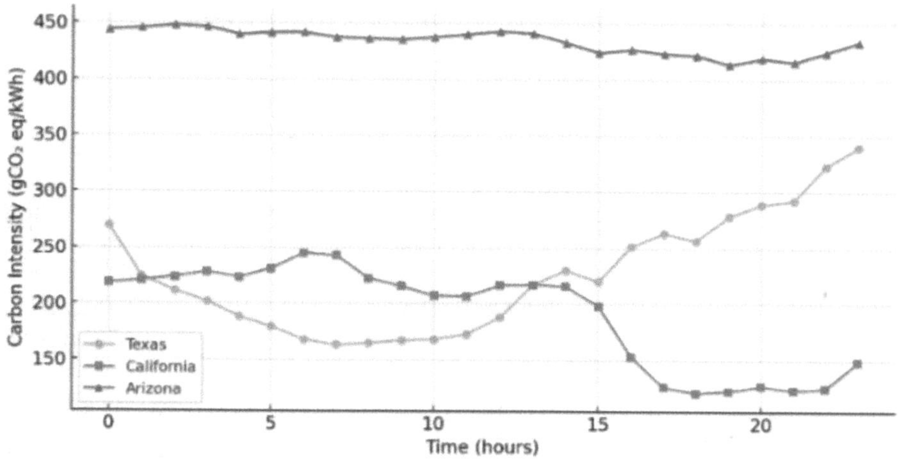

Fig. 3. A graph displaying the carbon intensity (gCO_2 eq/kWh) over time (in hours) for Texas, California, and Arizona.

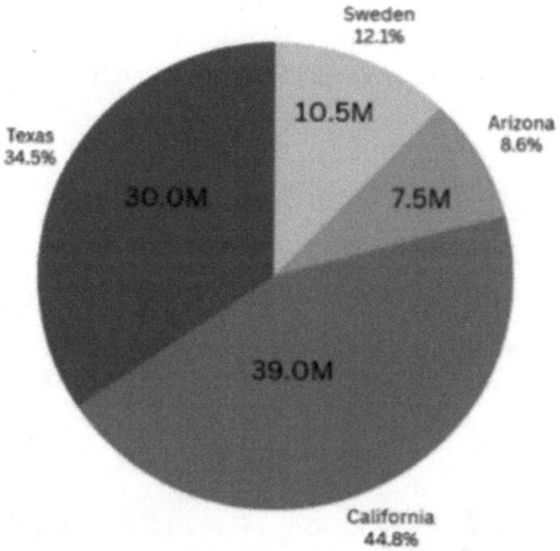

Fig. 4. A pie chart showing the population distribution of the chosen regions.

savings occur. Energy consumption per operation remains constant (e.g., 0.168 kWh) to ensure consistency across simulations.

The trace file contains information about a request, which includes the timestamp, the original region and the shifted region (if applicable). For example, on Fig. 4, the first request was sent from Texas 12:23 PM PST. This request was shifted to California because the carbon intensity of California was only 218.3 while the carbon intensity of Texas was 269.82. The algorithm calculates the carbon intensity for the region of generation, based on data from electricity grid APIs (e.g., Electricity Maps). It then compares the carbon emissions of the current region with the emissions of the most optimal (low-carbon intensity) region at the time of requests arrive.

Each simulation request represents a person generating a video in a random region across the globe. After calculating the CO2 emissions for the user's current region, the algorithm then checks if the emissions could be reduced by shifting the task to the most energy-efficient region at that time.

The total savings demonstrate the sheer amount of carbon emission levels we can limit if the spatial shifting techniques are applied. This algorithm offers a strategy to reduce carbon emission impacts, by providing a sustainable practice that can lower carbon levels. It builds upon prior work like the Electricity Maps tool (Electricity Maps [3]), which provides real-time regional carbon intensity data, and CodeCarbon (CodeCarbon [2]), which measures energy consumption in AI workflows. By focusing on real-time data and using spatial flexibility, this algorithm serves as a step toward going green and achieving more sustainable AI operations.

4 Experimental Results

To see the impact of generated videos on the environment, we can explore the results found from the Rain1011 and CogVideo datasets. In each of these models, I tested a multitude of prompts to get more complicated video results.

Fig. 5. A collage of prompts ran on different models. Clear distinctions between the models are shown.

Figures 5 and 6 show examples of the videos output on different tested models. Figure 7 shows that various models consume significantly different energy to generate a video with the same prompt. CogVideoX-5B-12 V, CogVideoX-2B-12 V, and Rain1011 High Resolution consume 25.07, 18.66, and 7.46 times more energy, respectively, compared to Rain1011 Low Resolution. Surprisingly, the model that requires more resources does not necessarily produce higher quality videos. For example, the Rain1011 model

Number	Prompt
1	A panda dressed in a small, red jacket and a tiny hat, sits on a wooden stool in a serene bamboo forest. The panda's fluffy paws strum a miniature acoustic guitar, producing soft, melodic tunes.
2	A giant humanoid, made of fluffy blue cotton candy, stomping on the ground, and roaring to the sky , clear blue sky behind them.
3	Beautiful, snowy Tokyo city is bustling. The camera moves through the bustling city street following several people enjoying the beautiful snowy weather and shopping at nearby stalls.
4	View out a window of a giant strange creature walking in rundown city at night, one single street-lamp dimly lighting the area.

Fig. 6. A list of the prompts and their corresponding numbers.

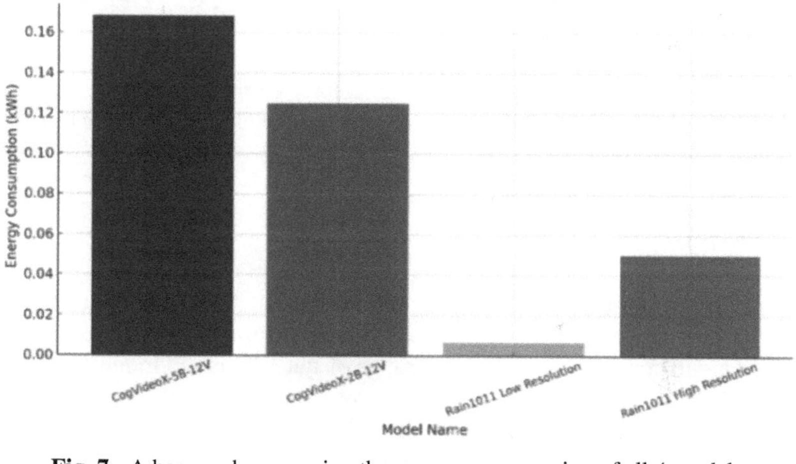

Fig. 7. A bar graph comparing the energy consumption of all 4 models.

generates a higher quality video as shown in Fig. 5 while consuming only 3.9% of the energy that the CogVideoX-5B-12 V model would typically consume.

The CogVideo-5B-12 V model, optimized for high-resolution video output, ensures greater detail but comes at a computational cost. On average, it took 1,755.35 s to generate a 6-s video, consuming 0.168 kWh of energy. This confirms that high-resolution video generation requires substantial computing power. The trade-off between quality and energy efficiency is clear—0.168 kWh is enough to drive approximately 0.84 miles in a Tesla Model S.

The CogVideo-2B-12 V model demonstrated faster processing, generating a 6-s video in an average of 436.20 s, with an energy consumption of 0.125 kWh, as measured by CodeCarbon. Although the processing time was significantly shorter than the 5B-12 V model, the energy savings were less pronounced. To put this in perspective, charging a smartphone once a day for a week consumes about 0.035 kWh—meaning this energy could power a phone for over three and a half weeks.

The Rain1011 High-Resolution model showed a significant improvement in efficiency, generating a 6-s video in just 218 s while consuming only 0.05 kWh of energy. Compared to the CogVideo models, it proved to be far more energy-efficient while maintaining comparable video quality.

The Rain1011 Low-Resolution model further improved efficiency, producing a 6-s video in just 63 s with an energy consumption of only 0.067 kWh. Despite its lower resolution, it delivered similar video quality in a fraction of the time and with the lowest energy consumption among all models tested.

While high-quality models like CogVideo offer ultra-detailed video generation, more efficient models like Rain1011 achieve similar visual fidelity with significantly lower energy consumption. These findings suggest that models with a high carbon footprint do not necessarily enhance video quality, making it a more sustainable choice to adopt eco-friendly models for video generation.

By using the Spatial Shifting Algorithm, video generation can be optimized by shifting workloads from high-carbon intensity regions to those with lower carbon intensity. In a simulation of 100,000 video generations across various regions, the algorithm demonstrated a carbon savings of 350,000 kg per year. This is equivalent to 1.75 million miles flown by a commercial airplane. These results clearly demonstrate how selecting optimal regions for video generation can significantly reduce carbon emissions, making the process eco-friendlier without compromising quality or performance.

5 Conclusion

As the popularity of AI-generated videos continues to rise, so too will the carbon emissions associated with their production. This paper presents a study on the environmental impact of AI video generation, exploring a range of state-of-the-art models. Our findings show that using more eco-friendly models can reduce carbon emissions without sacrificing video quality. Additionally, I demonstrate that the spatial shifting algorithm can further cut emissions by 350,000 kg annually by shifting workloads from high-carbon intensity regions to lower-carbon intensity regions. These results highlight the potential of adopting more sustainable models and optimizing workload distribution to reduce carbon footprints. Moving forward, this research demonstrates the potential for sustainable AI practices, offering a pathway to reduce carbon emissions and foster environmentally responsible AI development in the future.

Acknowledgements. I would like to thank my advisor, Dr. Zong, for the positive insight he provided for this project.

References

1. Cloud Carbon: Cloud Carbon Footprint Tool. https://www.cloudcarbon.com
2. CodeCarbon: CodeCarbon: A framework to measure energy consumption and carbon emissions. https://www.codecarbon.com
3. Electricity Maps: Real-time data on regional carbon intensity. https://www.electricitymaps.com
4. Fortune Business Insights: AI-generated video market growth projection (2023–2032). https://www.fortunebusinessinsights.com
5. Han, S., Mao, H., Dally, W.J.: Deep compression: compressing deep neural networks with pruning, trained quantization, and Huffman coding. In: Proc. Int. Conf. On Learning Representations (ICLR) (2016)
6. Holttinen, H., Cutululis, N.A., Orths, A., van Hulle, F., Kiviluoma, J., Gibescu, M.: Renewable energy integration in data centers. IEEE Trans. Sustain. Comput. **5**(2), 123–134 (2020)
7. Henderson, P., Hu, J., Romoff, J., Brunskill, E., Jurafsky, D., Pineau, J.: Towards the systematic reporting of the energy and carbon footprints of machine learning. J. Mach. Learn. Res. **21**(248), 1–43 (2020)
8. HotCarbon24: Carbon emissions of AI video generation. https://www.hotcarbon.com
9. Hugging Face: CogVideoX-5B-I2V. https://huggingface.co/cogvideox-5b. Accessed 25 Oct 2024
10. Hugging Face: CogVideoX-2B-12V. https://huggingface.co/cogvideox-2b. Accessed 25 Oct 2024
11. Hugging Face: Rain1011 model for energy-efficient video generation. https://huggingface.co/rain1011
12. Jouppi, N.P., Young, C., Patil, N., Patterson, D.: In-datacenter performance analysis of a tensor processing unit. In: Proc. 44th Int. Symp. On Computer Architecture (ISCA), pp. 1–12 (2017)
13. MidJourney: AI image generation service user statistics. https://www.midjourney.com
14. Php Tutorial: MidJourney daily users analysis. https://www.phptutorial.com
15. Radovanovic, A., Schmid, M., Ramesh, S., Isard, M.: Carbon-intelligent compute management. Google Research. https://research.google/pubs/pub51450/
16. Schwartz, R., Dodge, J., Smith, N.A., Etzioni, O.: Green AI. Commun. ACM. **63**(12), 54–63 (2020)
17. Spatio-temporal load shifting for truly clean computing: Load flexibility for carbon-free energy. https://www.cleancomputing.org
18. Strubell, E., Ganesh, A., McCallum, A.: Energy and policy considerations for deep learning in NLP. In: Proc. 57th Annu. Meet. Assoc. Comput. Linguist, pp. 3645–3650. ACL (2019)

Poster Papers

Accurate 0 m Goal Achievement for CanSat Using Multi-zone ToF Sensing and Object Detection

Miho Akiyama(✉) and Takuya Saito

Tokyo University of Information Sciences, 4-1 Onaridai Wakabaku Chiba-shi, Chiba 265-0077, Japan
m3akiyam24@rsch.tuis.ac.jp

Abstract. This paper presents a novel approach for achieving precise 0 m goal stopping in planetary exploration rovers (CanSat) used in the ARLISS competition. Conventional methods employing image-based object detection often fail to stop exactly at the goal due to limitations in visual distance estimation and variations in lighting conditions. To overcome this issue, we propose a system that integrates a Multi-Zone Time-of-Flight (ToF) sensor with object detection based on the SSD MobileNet V2 model. The ToF sensor (VL53L5CX) measures depth information across an 8×8 array (64 zones) within a 63° field of view and detects distances from 2 to 400 cm. By combining these measurements with visual bounding boxes, the system accurately determines whether the rover has reached within 0.35 m of the goal and commands a precise stop. Experimental evaluation using a CanSat prototype demonstrated consistent success in all 10 trials, achieving exact 0 m contact with the goal pylon. The results confirm that the proposed fusion of visual and depth sensing provides higher accuracy and robustness than conventional vision-only approaches. Future work will validate the system in real ARLISS and other rocket-based CanSat competitions.

The paper is a submission for CSCI-RTAI

Keywords: ToF · SSD MobileNet · CanSat · ARLISS · Object Detection

1 Introduction

ARLISS (A Rocket Launch for International Student Satellites) is an experimental program and competition for planetary exploration rovers known as CanSat. A CanSat is a small-scale rover designed to fit within a rocket payload equivalent to three 350 mL cans. The competition is held annually in the Black Rock Desert, Nevada, USA. In ARLISS, a high-power rocket launches the CanSat to an altitude of approximately 4000 m. After separation from the rocket, the CanSat descends via parachute and autonomously navigates toward a predefined target position (the goal). Teams compete based on the accuracy of arrival, measured as the final distance from the goal. CanSats typically land 2–5 km away from the goal, thus requiring robust long-range mobility and autonomous navigation capabilities, as well as tolerance to high mechanical stress during flight and landing.

H. R. Arabnia et al. (Eds.): CSCI 2025, CCIS 2967, pp. 539–543, 2027.
https://doi.org/10.1007/978-3-032-28454-9_39

During launch and descent, the CanSat experiences approximately 8 G during ascent, 45 G upon separation, and up to 100 G at impact with the ground [1]. After landing, it autonomously drives or flies toward the goal using GPS and other sensors while maintaining drivability even over uneven desert terrain. Since its inception in 1999, ARLISS had primarily focused on GPS-based autonomous navigation until around 2015. However, by ARLISS 2016, three teams successfully achieved GPS-guided autonomous navigation to the goal. As the recorded error approached the GPS positioning limit, subsequent research shifted toward methods capable of surpassing GPS accuracy—particularly those incorporating image recognition to guide the rover more precisely to the goal.

In recent years, CanSat competitions have evolved toward achieving a "0 m goal", where the rover autonomously detects and stops exactly at the goal pylon rather than within the GPS error range. A common approach has been color-based image recognition, identifying the red color of the goal pylon and steering toward it. At ARLISS 2017, a team successfully achieved a 0 m goal using such a color-based vision system. Our own team similarly achieved a 0 m goal using color-based recognition at the Tanegashima Rocket Contest held on JAXA grounds [2].

However, color-based recognition methods have inherent limitations. Lighting conditions change over time and with the sun's position, causing variations in brightness, reflections, and shadows that can obscure color features. To achieve reliable recognition under diverse lighting environments, a method robust to illumination changes is necessary.

To address this, we applied a Deep Learning-based image classification approach capable of recognizing goals under varying brightness conditions. Using a lightweight six-layer CNN model similar to CIFAR-10, which can operate on embedded microcontrollers, we achieved a 0 m goal at ARLISS 2019 [3]. Nevertheless, simple image classification could not determine the positional direction of the goal within the image. Therefore, we introduced object detection using the SSD MobileNet model, which recognizes multiple objects and their coordinates. This method, proposed at ARLISS 2022, improved recognition accuracy [4].

At ARLISS 2022, our system stopped 0.2 m short of the goal. The difficulty arose because, once the pylon filled the camera's field of view, the system could no longer distinguish whether it had reached the exact 0 m point. To overcome this limitation, we propose a new approach employing a Multi-Zone Time-of-Flight (ToF) distance sensor, capable of simultaneously measuring multiple depth zones. By integrating the measured distance data, our system can accurately determine contact with the goal and perform a reliable 0 m stop.

2 Method

The proposed system utilizes the VL53L5CX Multi-Zone ToF sensor, which provides distance measurements across an 8×8 array (64 points) within a field of view of approximately 63° and a measurement range of 2–400 cm. Using this 8×8 depth map, the CanSat estimates the distance to the goal and determines whether it has reached the 0 m position.

Figure 1 illustrates an example of object recognition and distance measurement. Each of the 64 zones displays a distance value, while the goal pylon is detected as a rectangular bounding box by the Object Detection model. To determine the 0 m condition, the system checks whether the minimum distance value within the bounding box is less than 0.35 m. If this condition is met, the rover stops.

For object detection, the SSD MobileNet V2 model was trained to identify goal pylons. Detection begins at a distance of approximately 10 m, guiding the rover's motion control accordingly. As the CanSat approaches the pylon and the Multi-Zone ToF detects a distance of 0.35 m within the detected bounding box, the rover autonomously stops.

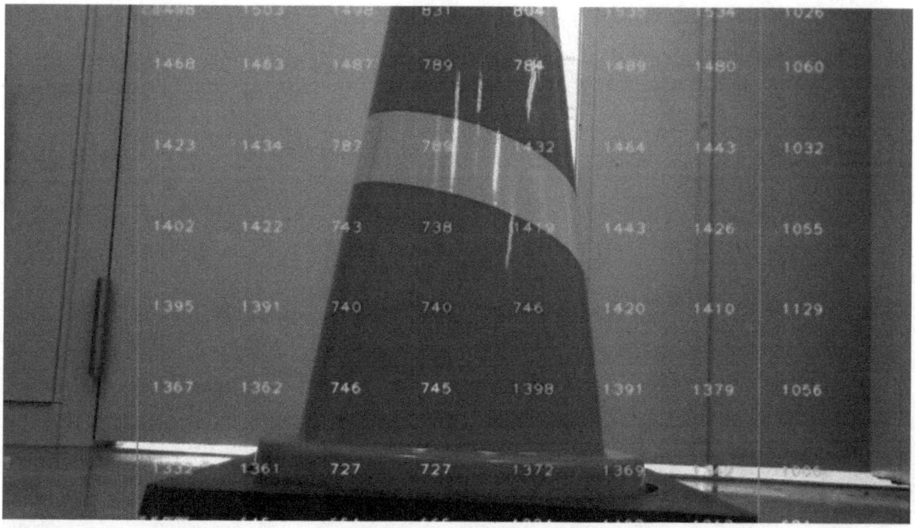

Fig. 1. Image showing the recognition of the goal pylon using MultiZone ToF.

3 Experiments

To evaluate the proposed method, we integrated the Multi-Zone ToF sensor into an actual CanSat and tested its ability to achieve a 0 m goal using both Object Detection and distance measurement. Figure 2 shows the Multi-Zone ToF sensor, and Fig. 3 shows the developed CanSat equipped with it.

Experiments were conducted on a futsal court at Tokyo University of Information Sciences. The goal pylon was placed more than 10 m away from the starting point. The rover began autonomous navigation using Object Detection while continuously measuring the goal distance with the Multi-Zone ToF sensor. The trial ended when the rover stopped automatically upon detecting a distance of 0 m. Ten trials were performed, each starting from a random initial position.

4 Evaluation

The proposed method successfully achieved a 0 m goal in all ten trials. Figure 4 shows an example recognition log from the first trial. In frames (a) and (b), the goal was detected by Object Detection on the left side of the image, and the rover turned left while approaching the target. In frame (c), the Multi-Zone ToF detected a distance of 0.32 m to the goal within the Object Detection region, and the rover stopped upon contact with the pylon.

Fig. 2. MultiZone ToF VL53L5CX

Fig. 3. CanSat equipped with a MultiZone ToF

(a) (b) (c)

Fig. 4. Log of the goal image recognition

Table 1. Log data for 0 m goal determination

Object detection recognition rate	65.625%
ROI(x,y,width,height)	4, −2, 642, 466
Processing time	0.234 s
Distance	0.32 m

Table 1 summarizes the log data recorded at the moment of 0 m detection. The Object Detection model achieved a recognition rate of 65.625%, with the region of interest (ROI) defined as (x, y, width, height) = (4, −2, 642, 466). The processing time per frame was 0.234 s, and the detected distance was 0.32 m.

These results confirm that the proposed method improves guidance accuracy compared to conventional vision-only approaches and successfully resolves the issue of premature stopping before reaching the goal pylon.

5 Conclusion

Conventional 0 m-goal approaches using image-based Object Detection have struggled to ensure an exact stop at the goal due to limitations in visual distance estimation. The proposed method, which integrates Multi-Zone ToF distance sensing with Object Detection, enables precise distance measurement and reliable 0 m stopping. Experimental results demonstrated its effectiveness in achieving accurate contact-level goal arrival. Future work will focus on validating this approach in real-world CanSat competitions such as ARLISS, the Noshiro Space Event, and the Tanegashima Rocket Contest.

References

1. Takuya, S., Miho, A.: Development of rover with ARLISS requirements and the examination of the rate of acceleration that causes damages during a rocket launch. J. Robot. Mechatron. **31**(6), 913–925 (2019)
2. Miho, A., Takuya, S.: A novel CanSat-based implementation of the guidance control mechanism using goal-image recognition. In: 2020 IEEE 9th Global Conference on Consumer Electronics (GCCE), Kobe, Japan, pp. 580–581 (2020)
3. Miho, A., Takuya, S.: Study on a method to guide CanSat to the goal at a distance of 0m using deep learning. IEICE Trans. Inf. Syst. (Japanese Edition). **J104-D**(7), 540–550 (2021)
4. Miho, A., Takuya, S.: Method to achieve high speed and high recognition rate of goal from long distance for CanSat. J. Robot. Mechatron. **35**(1), 194–205 (2023)

Comparative Study of Evolutionary and Swarm Algorithms for Training Neural Network-Based Acrobot Controllers

Hidehiko Okada(✉)

Kyoto Sangyo University, Motoyama, Kamigamo, Kita-ku, Kyoto 603-8555, Japan
hidehiko@cc.kyoto-su.ac.jp

Abstract. This article presents a comparative analysis of four representative population-based optimization algorithms—Evolution Strategy (ES), Genetic Algorithm (GA), Differential Evolution (DE), and Particle Swarm Optimization (PSO)—for reinforcement learning of neural network controllers. The Acrobot task from OpenAI Gym was employed as a benchmark, where a multilayer perceptron (MLP) was trained to control the system without reliance on gradient-based methods. Each algorithm was evaluated under equivalent experimental conditions, and statistical analyses were conducted. Results show that DE significantly outperformed ES, GA and PSO ($p < .01$). Across algorithms, the MLP with eight hidden units achieved the best trade-off between accuracy and model complexity, except in the case of PSO, where four hidden units were preferable due to its limited effectiveness in high-dimensional search spaces. These findings highlight the relative strengths, weaknesses, and hyperparameter sensitivities of each algorithm, providing practical insights into algorithm selection for neuroevolutionary reinforcement learning.

Keywords: Evolutionary algorithm · Swarm intelligence · Neuroevolution · Reinforcement Learning · Acrobot Control

1 Introduction

Neural networks are typically trained using gradient-based methods in supervised learning tasks, where labeled training data are available. However, in reinforcement learning tasks, gradient-free approaches are often required. Evolutionary and swarm algorithms offer population-based stochastic optimization methods that are well suited to reinforcement learning. This study consolidates four separate investigations [1–4] into a comparative analysis of the Acrobot task, applying Evolution Strategy (ES) [5], Genetic Algorithm (GA) [6], Differential Evolution (DE) [7], and Particle Swarm Optimization (PSO) [8].

H. R. Arabnia et al. (Eds.): CSCI 2025, CCIS 2967, pp. 544–549, 2027.
https://doi.org/10.1007/978-3-032-28454-9_40

2 Acrobot Control Task

Figure 1 shows a screenshot of the Acrobot system provided in OpenAI Gym. According to its documentation, the system consists of two links connected in series to form a chain, with one end fixed. The joint between the two links is actuated. The objective is to apply torques at the actuated joint to swing the free end of the chain above a target height, starting from the initial downward-hanging state.[1]

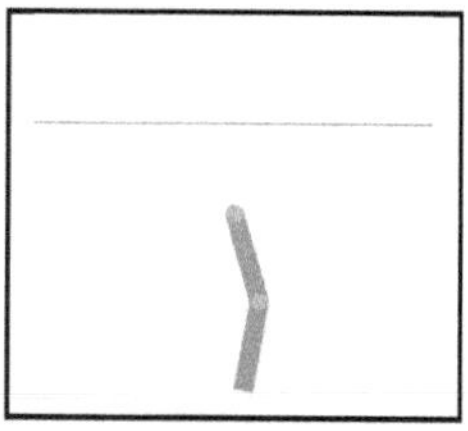

Fig. 1. Acrobot system.

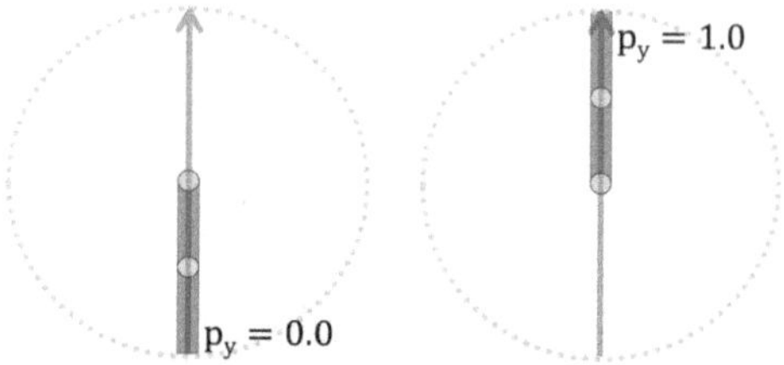

(a) Initial state (b) Best state

Fig. 2. Initial and best states.

Let p_y denote the vertical height of the free end of the chain, where the minimum and maximum values of py are 0.0 and 1.0, respectively (Fig. 2). In the original formulation of the task, the objective is to achieve $p_y \geq 0.5$. An episode terminates when the goal is reached or when a preset time limit is exceeded. In this work, the task was modified so that the objective is to maximize p_y throughout an episode of 200 time steps. Furthermore, two modifications were introduced: (i) the control task begins from the state shown in Fig. 2(a), where $p_y = 0.0$, and (ii) the torque applied to the actuated joint is continuous within $[-1.0, 1.0]$, instead of the original discrete values $\{-1, 0, 1\}$.

At each step, the controller observes the current state and selects an action. The observation includes $\cos(\theta_1)$, $\sin(\theta_1)$, $\cos(\theta_2)$, $\sin(\theta_2)$, and the angular velocity of θ_1 and θ_2, where θ_1 is the angle of the first link relative to the vertical axis, and θ_2 is the angle of the second link relative to the first link. The ranges are $-1.0 \leq \cos(\theta_1), \sin(\theta_1), \cos(\theta_2), \sin(\theta_2) \leq 1.0$, $-4\pi \leq \dot{\theta_1} \leq 4\pi$, and $-9\pi \leq \dot{\theta_2} \leq 9\pi$.

The fitness of a neural network controller is defined as in Eq.(1), where $p_y(t)$ denotes the vertical height of the free end at time step t. The fitness score increases as the controller maintains higher values of $p_y(t)$ over more time steps, thereby favoring controllers that keep the chain's free end elevated.

$$Fitness = \frac{1}{200} \sum_{t=1}^{200} p_y(t) \tag{1}$$

[1] https://www.gymlibrary.dev/environments/classic_control/acrobot/.

3 Training Neural Network Controllers by Evolutionary and Swarm Algorithms

A three-layer feedforward neural network, specifically a multilayer perceptron (MLP), was employed as the controller. The MLP serves as the policy function, expressed as action(t) = F(observation(t)). The topology of the MLP is shown in Fig. 3. The input layer consists of six units, each corresponding to an element of the observation. To normalize the inputs within the range $[-1.0, 1.0]$, the angular velocity of θ_1 (θ_2) is divided by 4π (9π). The output layer consists of a single unit, and its output is directly applied as the torque to the actuated joint.

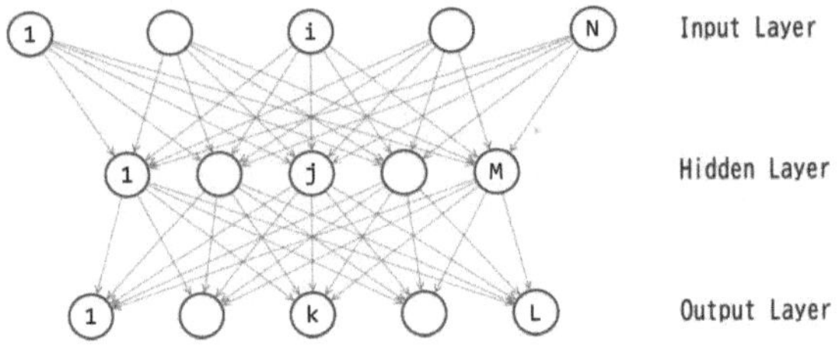

Fig. 3. Topology of the MLP.

The MLP includes M+L unit biases and NM+ML connection weights, yielding a total of M+L+NM+ML parameters. Let D = M+L+NM+ML denote the number of parameters. In this study, N = 6 and L = 1, which results in D = 8M+1. Training the MLP is thus formulated as an optimization problem in a D-dimensional continuous space. Let $\boldsymbol{x} = (x_1, x_2, ..., x_D)$ represent the D-dimensional parameter vector, where each x_i corresponds to either a connection weight or a unit bias. By assigning the values of x to the corresponding parameters, the feedforward calculation of the MLP can be carried out. Each of the four algorithms—ES, GA, DE, and PSO—was applied to optimize the parameter vector x. Further details of the training procedures for each algorithm can be found in the corresponding preprints [1–4].

4 Experiments

4.1 Algorithm Configurations

In each experimental trial, the number of fitness evaluations was fixed at 5,000. Two settings were considered for the number of new genotype vectors generated per generation: (a) 10 and (b) 50. Accordingly, the number of generations was set to 500 for case (a) and 100 for case (b). The specific configurations for ES, GA, DE, and PSO are summarized in Table 1.

Table 1. Hyperparameter values of the four algorithms.

ES	
Number of parents	(a) 10×0.5=5 (b) 50×0.5=25
Step size	Uniformly random within [-1.0, 1.0]

GA	
Number of elites (E)	(a) 10×0.2=2 (b) 50×0.2=10
α for blend crossover	0.5
Mutation probability	1/D

DE	
Scaling factor (F)	(a) 0.5 (b) 0.1
Crossover rate (CR)	0.9

PSO	
Inertia weights (w)	0.729
Pbest coefficient (c_p)	1.49445
Gbest coefficient (c_g)	1.49445

4.2 Results

Table 2 presents the best, worst, average, and median fitness scores of the trained MLPs across 11 trials. Supplementary videos[2] are provided to illustrate the motions of the chain controlled by the trained MLPs.

Table 2. Fitness scores among 11 runs.

		ES				GA			
	M	Best	Worst	Average	Median	Best	Worst	Average	Median
(a)	4	0.444	0.362	0.423	0.430	0.462	0.423	0.433	0.427
	8	0.455	0.386	0.431	0.434	0.455	0.423	0.442	0.443
	16	0.441	0.309	0.420	0.432	0.458	0.427	0.443	0.446
	32	0.437	0.349	0.413	0.416	0.460	0.421	0.441	0.438
(b)	4	0.457	0.410	0.430	0.433	0.452	0.398	0.428	0.432
	8	0.462	0.431	0.448	0.446	0.434	0.407	0.421	0.426
	16	0.457	0.423	0.437	0.436	0.435	0.382	0.416	0.422
	32	0.459	0.417	0.438	0.442	0.418	0.396	0.408	0.410

		DE				PSO			
	M	Best	Worst	Average	Median	Best	Worst	Average	Median
(a)	4	0.450	0.419	0.439	0.441	0.469	0.336	0.408	0.417
	8	0.467	0.422	0.448	0.450	0.450	0.269	0.401	0.435
	16	0.461	0.428	0.446	0.448	0.467	0.333	0.424	0.439
	32	0.464	0.426	0.445	0.446	0.465	0.314	0.418	0.425
(b)	4	0.464	0.433	0.445	0.443	0.460	0.416	0.440	0.436
	8	0.471	0.423	0.443	0.442	0.463	0.396	0.437	0.436
	16	0.460	0.427	0.445	0.444	0.458	0.351	0.429	0.439
	32	0.450	0.423	0.438	0.440	0.453	0.391	0.425	0.424

First, the performances of each algorithm were compared under the two hyperparameter configurations (a) and (b). Wilcoxon signed-rank tests revealed the following:

[2] https://www.cc.kyoto-su.ac.jp/~hidehiko/csci2025/Supplementary_Videos.html.

- For GA, configuration (a) performed significantly better than (b), $p<.01$.
- For ES and PSO, configuration (b) performed significantly better than (a), $p<.01$.
- For DE, no significant difference was observed between (a) and (b), $p>.05$.

These results suggest that DE is more robust to hyperparameter configurations than ES, GA, or PSO.

Second, the performances of MLPs trained by each algorithm were compared across different numbers of hidden units (M). Wilcoxon rank-sum tests revealed the following:

- For ES, $M = 8$ was significantly better than $M = 4,16$, or 32 ($p<.05$).
- For GA and DE, $M = 8$ was significantly better than $M = 4$ ($p<.05$), and $M = 8$ was not significantly inferior to $M = 16$ or $M = 32$ ($p>.05$).
- For PSO, $M = 4$ was significantly better than $M = 32$ ($p<.05$), and $M = 4$ was not significantly inferior to $M = 16$ or $M = 32$ ($p>.05$).

These results indicate that increasing the number of hidden units M does not necessarily improve the performance of the MLP. This can be attributed to the fact that a larger M results in a longer genotype vector D, thereby expanding the algorithm's search space and increasing the difficulty of finding better solutions. When employing evolutionary algorithms for reinforcement learning of neural networks, it is therefore crucial to adjust the model size so that the network's representational capacity is balanced against the complexity of the search space.

Third, the performances of the four algorithms were statistically compared. Wilcoxon signed-rank tests revealed that the fitness scores obtained by PSO were significantly lower than those obtained by DE ($p<.01$). Although the differences between PSO and GA or ES were not statistically significant, the p-values suggested a tendency for PSO to yield lower scores than GA ($p = .25$) and ES ($p = .07$). These results indicate that, for the MLP training task considered in this study, PSO performed significantly worse than DE and also tended to underperform relative to GA and ES. This can be attributed to the tendency of particles in PSO to become trapped in local optima. Conclusion This comparative study evaluated ES, GA, DE, and PSO on the Acrobot task. Among the four algorithms, DE achieved the best overall performance, while ES and GA demonstrated complementary strengths. In contrast, PSO underperformed, likely due to insufficient exploration in high-dimensional search spaces. Future work will extend this analysis to additional algorithms and benchmark tasks.

References

1. Okada, H.: Evolutionary reinforcement learning of neural network controller for Acrobot task — part4: particle swarm optimization. Preprints.org (2024). https://doi.org/10.20944/preprints202403.0109.v1
2. Okada, H.: Evolutionary reinforcement learning of neural network controller for Acrobot task — part3: differential evolution. Preprints.org (2024). https://doi.org/10.20944/preprints202402.0145.v1

3. Okada, H.: Evolutionary reinforcement learning of neural network controller for Acrobot task — part2: genetic algorithm. Preprints.org (2023). https://doi.org/10.20944/preprints202310.0852.v1
4. Okada, H.: Evolutionary reinforcement learning of neural network controller for Acrobot task — part1: evolution strategy. Preprints.org (2023). https://doi.org/10.20944/preprints202308.0081.v1
5. Schwefel, H.P.: Evolution strategies: a family of non-linear optimization techniques based on imitating some principles of organic evolution. Ann. Oper. Res. **1**(2), 165–167 (1984)
6. Goldberg, D.E., Holland, J.H.: Genetic algorithms and machine learning. Mach. Learn. **3**, 95–99 (1988). https://doi.org/10.1023/A:1022602019183
7. Storn, R., Price, K.: Differential evolution - a simple and efficient heuristic for global optimization over continuous spaces. J. Global Optim. **11**, 341–359 (1997)
8. Kennedy J., Eberhart, R.: Particle swarm optimization. In: Proceedings of International Conference on Neural Networks (ICNN 1995), vol. 4, pp. 1942–1948 (1995)

Controller for Android Navigation with Discretized Inputs (CANDI)

Rishan Patel(✉), Grant Zhang, and Fatimaezzahra Labaaj

Mississippi School for Mathematics and Science, Columbus, MS 39701, USA
rishan25patel@outlook.com, flabaaj@themsms.org

Abstract. This work explores discretizing AndroidEnv's touch interface into controller inputs to make reinforcement learning (RL) training on mobile environments faster and cheaper. The default touch space is vast and complex, inflating exploration time and sample complexity. We introduce a controller layer that maps AndroidEnv's actions to discrete buttons and joystick inputs, resulting in a simple action set that will provide higher efficiency. By constraining the action space, our method aims to reduce computational overhead while preserving the quality of the policy in a simulated smartphone environment. For benchmarking, two agents will be observed. One agent will be given touch inputs, and the other will receive controller inputs. These RL agents will be trained under the same conditions to navigate various applications on the Android operating system in order to determine which approach makes greater progress. This provides a practical pipeline for reducing RL training and improving the quality of the policy.

Keywords: reinforcement learning · android · optimization · mobile applications

1 Introduction

1.1 Background

Reinforcement learning has expanded beyond traditional control tasks such as Atari games [1] and balancing pendulums [2]. Environments such as Google DeepMind's AndroidEnv [3] are now capable of interacting with platforms such as virtual Android devices by emulating taps, swipes, and other actions a real person would make. This allows RL agents to interact with an enormous collection of mobile apps and games, presenting a wide range of tasks for RL research.

There are a few training challenges that emerge from the action space provided by AndroidEnv's default interface [3]. An agent is expected to learn how to select a point on the screen from continuous 2D coordinates and then perform a discrete action on the selected coordinate. To get ideal actions to take

Track: CSCI-RTAI (Artificial Intelligence).

H. R. Arabnia et al. (Eds.): CSCI 2025, CCIS 2967, pp. 550–554, 2027.
https://doi.org/10.1007/978-3-032-28454-9_41

place, gestures are required. These actions include tapping, swiping, long pressing, scrolling, zooming, and other interactions a human would perform on a phone. These random inputs have a low chance of producing meaningful interactions with the environment and making connections between the rewards and the inputs.

1.2 Proposed Approach

A proposed approach is using a simulated controller to let the agent interact with the environment. This will limit the agent to a smaller set of moves and reduce the amount of exploration required to get a well-trained policy. The policy trained with the controller inputs might outperform a policy trained on touch inputs because the coordinates of a controller joystick can be conveniently specified directly rather than relying on the often cumbersome press-and-drag motions required by simulated touch joysticks, allowing for more frequent inputs that lead to more precise actions.

2 Related Works

2.1 RL Agents Interacting with Android

DeepMind's AndroidEnv provided an open-source framework for researchers to build their RL agents to interact with Android [3]. AndroidEnv will be used as the base for our new approach. It can simulate touch gestures to carry out complex tasks and has been used to navigate apps and play some simple games. This makes it an optimal benchmark environment for RL agents interacting with Android phones.

QDroid was one of the early examples of RL interacting with Android [4]. QDroid used Q-learning to explore an app's GUI. The components of a GUI would be encoded as states, and then actions were used to generate a Q-table. This worked well on simple apps, but researchers needed to use more sophisticated methods for complex apps. As a result, DinoDroid was developed with Deep Q-Networks (DQN) [5]. DQN uses neural nets to approximate action values, enabling generalization across similar states and more effective action selection. This allowed DinoDroid to find bugs across a wider range of Android apps and take a wider range of observations.

2.2 Structured or Restricted Action Spaces

A recurring challenge in RL for Android is the large action space due to continuous touchscreen coordinates with numerous possible touch actions. In response, past works have attempted to restrict or structure the agent's action space. Wuji transformed continuous touch joystick inputs into discrete commands [6] such as north, south, east, and west. Similarly, DRAG allowed a DQN agent to find success playing 19 Android games by defining discrete actions based on combinations of taps and swipes [7]. These approaches showed better results and displayed the improved success rates of discretized action spaces.

3 Controller Interface and Emulator Setup

3.1 Input Mechanism

Certain Android applications have built-in controller support. Simulating a controller with a discrete input space for these applications instead of providing the RL agent with a continuous touch space makes it much easier for the agent to form meaningful connections between states, actions, and rewards. The simulated controller also allows the input mechanism to communicate directly with the application rather than defining actions that map to certain combinations of touch inputs. Furthermore, touch inputs often require multiple frames to complete. For example, simulated touch joysticks must first be tapped in the center and then dragged in some direction to command movement. Simulated controllers remove this issue by allowing the joystick to be immediately moved to a defined location, allowing for more frequent and precise commands.

3.2 Android Virtual Device (AVD)

Android Virtual Environments (AVDs) allow the RL agent to do anything that can be done on an Android device. AVDs are emulators that are provided by Android Studio, an IDE developed by Google for Android app development. An AVD functions as a fully virtual phone where apps can be installed and run without requiring any physical hardware outside of the computer running Android Studio. This emulates the full operating system including the Linux kernel, application framework, and even the ARM instruction set architecture.

4 Environment

4.1 AndroidEnv

AndroidEnv is the environment that will be used in Gymnasium [3,8]. It will wrap the AVD as a Gymnasium-compatible environment. This will allow a researcher to use the familiar API that comes with environments handled by Gymnasium-style interfaces. Actions will be defined by a Gymnasium action space, which AndroidEnv will use as inputs provided through the virtualized controller.

4.2 Observations

The main observations from AndroidEnv are expressed as pixels, timedelta, and orientation [3]. The pixels are from an RGB array representing the current frame; timedelta is the time that has passed since grabbing the last observation; orientation carries information about how things such as text are oriented. The RGB matrix will be output in the resolution of the AVD, but researchers might encode it down to $84 \times 84 \times 3$ for training efficiency. Each frame has three color channels and can be normalized or stacked with other frames for temporal context. Other preprocessing steps such as grayscale conversion, normalization, or frame stacking might be implemented to more efficiently express the observation of a state.

4.3 Task Extras

AndroidEnv is also capable of sending extras, which is information that is difficult to obtain from raw pixels [3]. Examples include the position of an avatar on a map, whether a checkpoint has been reached, score counters, episode timers, and other domain-specific state features extracted from the application.

5 Reinforcement Learning

Three RL algorithms (DQN, PPO, and CQL) will be used to train agents. These algorithms show strengths in exploration efficiency, policy stability, and offline-to-online RL compatibility.

5.1 Deep Q-Networks (DQN)

DQN will be implemented with Stable-Baselines3 [1,9]. DQNs are sample efficient with discrete tasks and will provide a strong baseline for evaluating the controller-based actions against the touch-based actions.

5.2 Proximal Policy Optimization (PPO)

PPO [10] will also be implemented with Stable-Baselines3 [9]. PPO prevents overly large policy updates, which leads to stable convergence over time. PPO also scales extremely well with the parallel training needed for RL.

5.3 Conservative Q-Learning (CQL)

CQL will be used because it discourages the overestimation of unseen actions [11]. This will produce safer policies and generalize observations better. D3rlpy will be used for CQL algorithms.

5.4 Offline to Online CQL

D3rlpy's online finetuning [12] will be used to start from a policy trained on human data with offline CQL and then refine it online with an explorer. This hybrid training pipeline reduces computational cost, increases quality of policy, and preserves the quality across all phases.

6 Expected Results

Discrete inputs for AndroidEnv in controller-based inputs are anticipated to accelerate convergence compared to raw touch inputs. Reducing possible actions should reduce the needed samples, allowing actions to be better associated with rewards. Training curves should also look more stable due to less variability in state-to-reward transitions. Policies trained with controller inputs may generalize better across similar tasks, since the action space uses directional inputs rather than raw pixel coordinates. Direct inputs of joystick positions will also lead to faster and more precise responses from the environment. Agents will then be evaluated on an application such as Rocket League Sideswipe, Crossy Road, or some open-source game if Task Extras are needed. Controller-based agents are expected to achieve higher performance relative to their touch-based counterparts.

References

1. Mnih, V., et al.: Human-level control through deep reinforcement learning. Nature **518**, 529–533 (2015). https://doi.org/10.1038/nature14236
2. Brockman, G.: OpenAI Gym (2016). arXiv:1606.01540
3. Toyama, D.: AndroidEnv: a reinforcement learning platform for android. arXiv:2105.13231 (2021)
4. Vuong, T.A.T., Takada, S.: A reinforcement learning based approach to automated testing of Android applications. In: Proceedings of the 9th ACM SIGSOFT International Workshop on Automating Test Case Design, Selection, and Evaluation (A-TEST 2018), pp. 31–37. ACM, New York (2018). https://doi.org/10.1145/3278186.3278191
5. Zhao, H., Song, X., Li, T., Lin, H., Gao, Y.: DinoDroid: a deep reinforcement learning framework for automated android testing. arXiv:2210.06307 (2022)
6. Wang, Y.: WUJI: automatic mobile game testing through reinforcement learning. In: Proceedings of the 34th IEEE/ACM Int. Conf. on Automated Software Engineering (ASE 2019), pp. 772–783. IEEE, Piscataway (2019). https://doi.org/10.1109/ASE.2019.00077
7. Khurana, S.: Android game testing using reinforcement learning, Blacksburg (2023). Master's thesis
8. Foundation, F.: Gymnasium: a standard API for reinforcement learning environments. Zenodo (2023). https://doi.org/10.5281/zenodo.8127026
9. Raffin, A., et al.: stable-baselines3: reliable reinforcement learning implementations. J. Mach. Learn. Res. **22**(268), 1–8 (2021). https://jmlr.org/papers/volume22/20-1364/20-1364.pdf
10. Schulman, J.: Proximal policy optimization algorithms. arXiv:1707.06347 (2017)
11. Kumar, A., Zhou, A., Tucker, G., Levine, S.: Conservative q-learning for offline reinforcement learning. In: Advances in Neural Information Processing Systems (NeurIPS 2020), pp. 1179–1191. Curran Associates, Red Hook (2020)
12. Seno, T., Imai, M.: D3RLPY: an offline deep reinforcement learning library. J. Mach. Learn. Res. **23**(315), 1–20 (2022). https://www.jmlr.org/papers/volume23/22-0017/22-0017.pdf

Graph Neural Network Prediction of Frictional Force Networks in Wet Fine Grinding

Corlbert A. Ayuk(✉) and Olumide Ogunmodimu

Department of Energy and Minerals Engineering, Pennsylvania State University, University Park, PA, USA
{cma6301,olumide.o}@psu.edu

Abstract. Dense slurry flows in stirred media mills exhibit complex behavior due to the formation of dynamic contact networks between particles. At high solid concentrations, these slurries exhibit non-Newtonian rheology behavior such as shear thickening or even jamming which impedes efficient mixing and grinding. In this work, we predict the evolution of particle contact and force networks within a stirred mill and demonstrate how these microstructural metrics reveal the prevailing flow regime. We use the coordination number (average contacts per particle) as an indicator of confinement and incipient jamming, while the contact persistence and renewal rates help to distinguish fluid-like from solid-like contact dynamics, additionally, the percolation of strong contact force networks serve as a signature of system-spanning force chains. Our results show that as operating conditions (e.g., higher solids fraction or impeller speed) drive the suspension toward thickening, the coordination number rises (more enduring contacts), and a robust frictional contact network develops, correlating with increased flow resistance. High contact persistence alongside a percolating force network signals the formation of a rigid, plug-like region prone to shear jamming, whereas rapid contact turnovers indicate a continuously rejuvenated, well-mixed state. This approach provides a new quantitative insight into slurry flow within stirred mills, helping identify conditions that lead to excessive thickening or dead zones and guiding strategies to maintain efficient, uniform mixing.

Keywords: Suspension Jammed · Slurry rheology · Force chain · Granular flow · Solid fraction · Stresses

1 Introduction

Fine grinding of high-solids mineral slurries in stirred media mills, such as tower mills and stirred ball mills, is central to mineral processing and chemical industries. At high solid fractions, these dense suspensions exhibit complex, non-Newtonian rheological behavior, where the apparent viscosity can increase

H. R. Arabnia et al. (Eds.): CSCI 2025, CCIS 2967, pp. 555–573, 2027.
https://doi.org/10.1007/978-3-032-28454-9_42

sharply with shear rate or stress [1,2]. In particular, slurries with high fines content often display dilatant (shear-thickening) behavior and, beyond a critical shear condition, may undergo shear-induced jamming [3]. In industrial mills, these transitions can lead to operational challenges, including excessive power draw, reduced throughput, and poorly mixed stagnant regions. Such macroscopic flow behavior is now understood to arise from the formation of particle-scale contact force networks (CFNs) [4–6]. As the suspension approaches the jamming limit, frictional particle contacts increase and organize into load-bearing networks that transmit stress across the system, imparting rigidity to the slurry. These networks govern shear stress, effective viscosity, and energy dissipation, and therefore directly influence grinding performance [7–11].

Understanding how CFNs emerge and evolve under shear is thus essential for predictive modeling and optimization of stirred milling processes. Directly characterizing contact force networks in operating stirred mills, however, remains extremely challenging due to the opaque, dense medium and the dynamic, high-energy environment [2,12,13]. Stirred mills are complex systems, making direct, in-situ measurements difficult with traditional methods. Researchers are developing innovative approaches, such as positron emission tomography, acoustic emission sensors, computational modeling, and pressure mapping, to capture these contact networks [13].Experimental visualization of CFNs in opaque, dense slurries is impractical, while discrete element method (DEM) simulations of industrial-scale mills are computationally expensive [14–20]. Despite its inherent complexity and experimental difficulty, the characterization of contact force networks (CFNs) in granular materials has attracted increasing attention over recent decades, owing to their fundamental importance in physics, engineering, and materials science [21–24]. Numerically, particle-scale simulation techniques such as the discrete element method (DEM) [25,26] and large-scale molecular dynamics frameworks, including the massively parallel simulator LAMMPS [27,28], have been widely employed to resolve granular systems with known interaction laws and elastic networks [28,29]. These approaches provide direct access to inter-particle contact forces and network topology, enabling systematic investigation of force transmission mechanisms under controlled conditions [30].

From an experimental perspective, several advanced techniques have been developed to probe stress and force distributions in granular assemblies. Photoelastic experiments allow direct visualization of both geometrical configurations and contact force patterns in two-dimensional photo-elastic granular systems [21]. However, this method is largely restricted to specialized photo-elastic materials and exhibits limited sensitivity to weak contact forces, making it unsuitable for common granular materials such as sands, glass beads or minerals ores [21]. Confocal fluorescence microscopy has been successfully applied to measure force distributions in three-dimensional systems composed of frictionless droplets, but its applicability is constrained to relatively small sample sizes [31]. Magnetic resonance elastography (MRE) offers a non-invasive route to image three-dimensional force-chain structures in quasi-static granular assemblies [32]. Despite its promise, a quantitative and general framework for relating high-

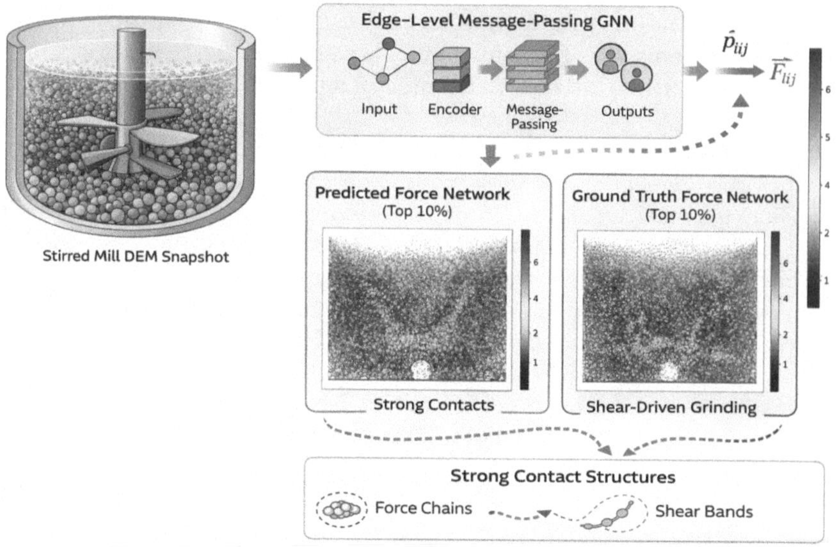

Fig. 1. Schematic of Discrete Element Method and Graph Neural Network of studying flow dynamic in a vertical Strirred Media Mill

resolution MRE images to particle-scale contact forces in three-dimensional granular matter remains an open challenge [32]. While both numerical and experimental approaches have substantially advanced our understanding of contact force networks, significant limitations remain in terms of scalability, material generality, and direct force resolution. These challenges motivate the development of alternative, data-driven approaches capable of efficiently inferring contact force networks in complex, dense granular systems. As a result, linking particle-scale force transmission to bulk rheological behavior and mill performance remains difficult.

Multi-scale modeling approaches that embed discrete simulations into continuum frameworks can partially bridge this gap, but they introduce additional modeling complexity and computational overhead. There is therefore a pressing need for new approaches that can efficiently predict internal contact force networks in dense slurries and connect microscale force transmission to macroscale rheological behavior and mill performance metrics such as power draw. Recent advances in graph neural networks (GNNs) provide a promising alternative. GNNs naturally represent particles as nodes and interactions as edges, and have been shown to successfully predict force chains and contact forces in jammed granular systems [15,17,33]. However, their application to dense, shear-driven slurry flows in stirred mills remains largely unexplored.

In recent years, the integration of machine learning techniques with traditional physical intuition has provided powerful new insights into granular materials [34]. A broad range of studies have demonstrated that machine learning models can identify subtle structural defects governing relaxation dynamics [35–37], plastic deformation [38], vibrational modes [39], and nonequilibrium transitions in disordered systems [40]. In many cases, these approaches achieve significantly improved efficiency and predictive accuracy compared to conventional analysis methods [41–43].

Despite these advances, several limitations remain. Many machine-learning approaches rely on handcrafted structural descriptors that must be defined *a priori* using physical intuition [35,38]. Moreover, limited long-term predictive capability constrains their ability to provide deeper insight into slow and aging flow dynamics in granular systems [42]. These challenges motivate the development of new machine-learning frameworks that learn directly from raw structural information with minimal human intervention. Graph neural networks (GNNs) provide a natural alternative by operating directly on graph representations of interacting particles [44,45]. Recent studies have shown that GNNs can successfully predict long-term glassy dynamics and uncover the predictive power of static structure in amorphous materials [46,47]. Beyond glassy dynamics, GNNs have also been applied to predict material properties in systems where interparticle contacts govern rigidity and stability. In granular materials, GNNs have recently been used to predict force-chain locations [48] or to estimate particle-level maximum normal contact forces from kinematic information [30]. However, these approaches do not recover the full contact force network required to analyze force transmission and stress propagation. Furthermore, most existing studies focus on node-level predictions [30,47], despite the fact that a principal advantage of GNNs lies in edge-level operations such as contact classification and force regression [44,45]. The explicit prediction of edge-level contact force networks in dense granular systems, therefore, remains largely unexplored.

In this study, we introduce an edge-aware message-passing GNN that directly predicts both contact existence and contact forces between particle pairs in a stirred media mill. High-fidelity DEM simulations are used to generate training data, with each snapshot represented as a particle interaction graph constructed via local k-nearest-neighbor sampling. Node features encode particle kinematics and geometry, while edge features describe relative motion and contact-relevant descriptors. The network jointly infers the probability of contact and the corresponding normal and tangential contact forces, enabling the recovery of both individual forces and the spatial organization of mechanically dominant interactions. This proposed framework, summarized schematically in Fig. 1, maps DEM snapshots to predicted contact force networks and is validated against ground-truth DEM data. The model accurately reproduces axial and radial distributions of strong contacts, force-chain structures, and shear-band localization characteristic of stirred milling. Despite being trained on a limited set of operating conditions, the GNN generalizes to new system states and enables rapid inference without requiring additional DEM simulations. By bridging particle-scale force transmission and bulk rheological behavior, this approach provides a computationally efficient, physics-consistent surrogate for analyzing and optimizing fine grinding processes in stirred media mills.

2 Method

2.1 Discrete Element Method (DEM) and Simulation Setup

The Discrete Element Method (DEM) is employed to generate the data needed for training the GNNs model and acts as the ground truth. In this study, Altair® EDEM™ 2025.1 is used as the DEM solver, with Python 3.10 for post-processing. EDEM™ employs a soft-sphere approach and computes contact forces using Hertz–Mindlin contact mechanics, which are capable of capturing non-linear interactions arising from particle–particle and particle–wall collisions [49].

Newton's equations of motion for translational and rotational dynamics are solved numerically, as shown below:

$$m_i \frac{d\mathbf{V}_i}{dt} = \mathbf{F}_{c,i} + m_i \mathbf{g}, \tag{1}$$

$$I_i \frac{d\boldsymbol{\omega}_i}{dt} = \boldsymbol{\tau}_i, \tag{2}$$

where m_i, I_i, $\mathbf{V}_i$, and $\boldsymbol{\omega}_i$ denote the mass, moment of inertia, linear velocity, and angular velocity of particle i, respectively. The quantities $\mathbf{F}_{c,i}$ and $\boldsymbol{\tau}_i$ in (1) and (2) above represent the total contact force and torque acting on particle i, aggregated from all neighboring contacts. Each discrete element has specific properties such as radius R, mass m, Young's modulus Y, shear modulus G, restitution coefficient e, and Poisson's ratio ν. The contact force between two particles i and j (or with a wall) is as shown in (3) below:

$$\mathbf{F}_{c,ij} = \mathbf{F}_{n,ij} + \mathbf{F}_{t,ij} \tag{3}$$

$$= (K_n \boldsymbol{\delta}_{n,ij} - \gamma_n \mathbf{V}_{n,ij}) + (K_t \boldsymbol{\delta}_{t,ij} - \gamma_t \mathbf{V}_{t,ij}), \tag{4}$$

with:

$$K_n = \frac{4}{3} Y^* \sqrt{R^* \delta_n}, \tag{5}$$

$$\gamma_n = -2\sqrt{\frac{5}{6}} \beta \sqrt{S_n m^*}, \tag{6}$$

$$K_t = 8G^* \sqrt{R^* \delta_n}, \tag{7}$$

$$\gamma_t = -2\sqrt{\frac{5}{6}} \beta \sqrt{S_t m^*}, \tag{8}$$

$$S_n = 2Y^* \sqrt{R^* \delta_n}, \tag{9}$$

$$S_t = 8G^* \sqrt{R^* \delta_n}, \tag{10}$$

$$\beta = \frac{\ln e}{\sqrt{(\ln^2 e + \pi^2)}}, \tag{11}$$

$$\frac{1}{Y^*} = \frac{(1-\nu_1^2)}{Y_1} + \frac{(1-\nu_2^2)}{Y_2}, \tag{12}$$

$$\frac{1}{G^*} = \frac{2(2-\nu_1)(1+\nu_1)}{Y_1} + \frac{2(2-\nu_2)(1+\nu_2)}{Y_2}, \tag{13}$$

$$\frac{1}{R^*} = \frac{1}{R_1} + \frac{1}{R_2}, \tag{14}$$

$$\frac{1}{m^*} = \frac{1}{m_1} + \frac{1}{m_2}. \tag{15}$$

Here, $\mathbf{V}_{n,ij}$ and $\mathbf{V}_{t,ij}$ are the normal and tangential relative velocities, and $\boldsymbol{\delta}_{n,ij}$ and $\boldsymbol{\delta}_{t,ij}$ represent the corresponding overlaps. The tangential overlap is computed by integrating the tangential velocity over time. If the tangential force exceeds the Coulomb friction limit ($F_t \geq \mu_f F_n$), it is clamped to F_n in the tangential direction. The torque exerted on particle i from particle j is:

$$\boldsymbol{\tau}_{ij} = \mathbf{R}_{ij} \times \mathbf{F}_{t,ij} + \boldsymbol{\tau}_{r,ij}, \tag{16}$$

where $\boldsymbol{\tau}_{r,ij} = -\mu_r F_{n,ij} R_{ij} \frac{\omega_{rel}}{\omega_{rel}}$ accounts for rolling resistance. The energy dissipated through damping over a time interval $[t_1, t_2]$ is:

$$E_n = \int_{t_1}^{t_2} \mathbf{F}_{n,d} \cdot \mathbf{V}_{n,ij} dt, \tag{17}$$

$$E_t = \int_{t_1}^{t_2} \mathbf{F}_{t,d} \cdot \mathbf{V}_{t,ij} dt. \tag{18}$$

The DEM model is calibrated for high energy milling using an vertical stirred media mill®. Simulation parameters are summarized in Table 1 (Fig. 2).

Table 1. Material and simulation parameters used in DEM setup.

Parameter	Value
Particle diameter	0.003 m
Restitution coefficient	0.5
Friction coefficient	0.3
Rolling friction coefficient	0.045
Density	3500 kg/m^3
Young's modulus	2.05 GPa
Poisson's ratio	0.29
Simulation time step	1×10^{-6} s
Total simulation time	10 s
Time integration method	Euler

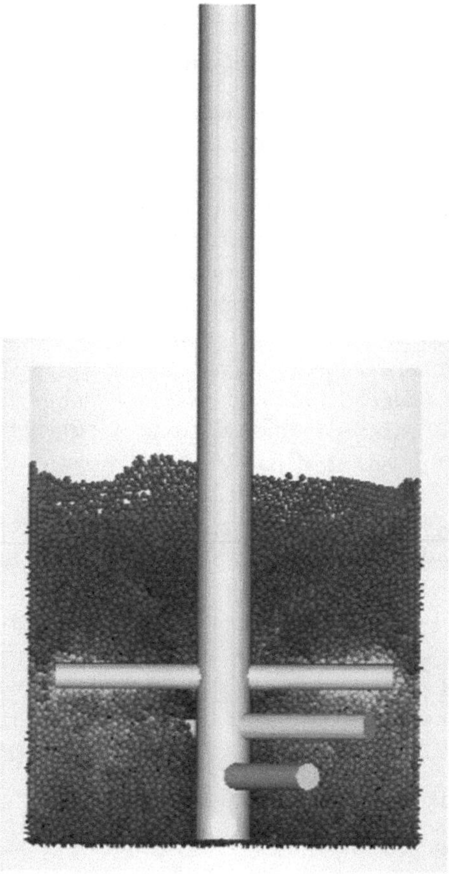

Fig. 2. Discrete Element Simulation of Particles in a Vertical Stirred Mill

2.1.1 3D Geometry of the Mill

The milling drum, with an internal volume of 5 L, is equipped with a central stirrer that drives the charge at a rotational speed of 300 rpm. The drum–stirrer geometry is imported into EDEM^{TM} as a watertight STL mesh. Simulations begin with a 2 s settling phase to initialize particle positions, followed by an 8 s milling phase at 300 rpm. The drum is filled to 60% of its volume with 100,000 ceramic grinding media. Powder particles are omitted to reduce computational cost and isolate media–media interaction dynamics. Their mechanical effects are incorporated via calibrated friction and restitution parameters [8, 50].

2.2 GNN Model

2.2.1 Graph Construction and Feature Extraction

We represent the simulation process at each time step as a graph $G_t = (V_t, E_t)$, where each node $i \in V_t$ corresponds to a particle and each edge $(i, j) \in E_t$ represents a potential contact between particles i and j. To construct E_t, we consider pairs of particles that either are in physical contact or lie within a neighborhood radius (e.g., k nearest neighbors), ensuring that all true contacts are included while providing negative samples for training. Each node is associated with a feature vector $\mathbf{x}_i$ capturing intrinsic properties of that particle. These features provide the model with information about individual particles that could influence contact interactions. Each edge (i, j) is annotated with a feature vector $\mathbf{e}ij$ that encodes pairwise geometric and kinematic information between the two particles. Important edge features include the relative displacement $\mathbf{d}_{ij} = \mathbf{p}_j - \mathbf{p}_i$ (where $\mathbf{p}_i$ denotes the position of particle i), the distance $||\mathbf{d}_{ij}||$, and the unit direction vector $\hat{\mathbf{n}}_{ij} = \mathbf{d}_{ij}/||\mathbf{d}_{ij}||$. Additionally, we include relative velocity features: given particle velocity vectors $\mathbf{v}_i$ and $\mathbf{v}_j$, we compute the relative velocity $\Delta\mathbf{v}_{ij} = \mathbf{v}_j - \mathbf{v}_i$. We decompose this into a normal component $v_n = \Delta\mathbf{v}_{ij} \cdot \hat{\mathbf{n}}_{ij}$ and a tangential speed $v_t = ||\Delta\mathbf{v}_{ij} - v_n\hat{\mathbf{n}}_{ij}||$. Thus, $\mathbf{e}_{ij}$ can include features such as $(||\mathbf{d}_{ij}||, \hat{\mathbf{n}}_{ij}, v_n, v_t)$ and other domain-specific quantities. By incorporating these features, the graph representation compactly encodes both the static geometry and dynamic state of particle pairs, which is important for predicting contact existence and force outcomes.

2.2.2 Graph Neural Network Architecture

We employ a graph neural network (GNN) that performs edge-aware message passing over the constructed graphs to jointly predict contact existence and contact forces. The architecture first transforms the raw node and edge features into a latent space and then applies multiple layers of message passing. Specifically, we use a Graph Isomorphism Network (GIN) style model [51] extended to include edge features (often referred to as a GINEConv or edge-conditioned GNN). The initial node feature $\mathbf{x}_i$ is passed through a learnable encoder ϕnode (a small feed-forward network), producing an embedding $\mathbf{h}i^{(0)} = \phi\text{node}(\mathbf{x}_i)$ of dimension d. Similarly, each edge feature $\mathbf{e}_{ij}$ is passed through an edge encoder ϕ_{edge}, yielding $\mathbf{z}_{ij}^{(0)} = \phi\text{edge}(\mathbf{e}_{ij})$ in the same d-dimensional latent space. This ensures that node and edge attributes are represented in comparable feature spaces before message passing begins. The GNN then applies L rounds of message passing to propagate information through the graph. At each layer l (for $l = 0, 1, \ldots, L-1$), every node i aggregates messages from its neighbors $\mathcal{N}(i)$. We adopt an edge-conditioned message function f so that messages depend on both neighbor node features and edge features. Formally, for each neighbor $j \in \mathcal{N}(i)$ we compute a message $\mathbf{m}_{ij}^{(l)} = f!\left(\mathbf{h}_i^{(l)},, \mathbf{h}_j^{(l)},, \mathbf{z}_{ij}^{(l)}\right)$, where f is a learnable function (implemented as a multilayer perceptron as in GIN [51]). These messages are summed (or averaged) and used to update i's representation:

$$\tilde{\mathbf{h}}_i^{(l+1)} = \mathbf{h}_i^{(l)} + \sum_{j \in \mathcal{N}(i)} \mathbf{m}_{ij}^{(l)}, \tag{19}$$

$$\mathbf{h}_i^{(l+1)} = \mathrm{ReLU}\left(\mathrm{LayerNorm}\left(\tilde{\mathbf{h}}_i^{(l+1)}\right)\right), \tag{20}$$

where we employ residual connections and a normalization step followed by a ReLU activation. This update rule preserves the original node state (residual addition) and injects information from neighbors, including the edge attributes between nodes. We apply L such message-passing layers, yielding final node embeddings $\mathbf{h}_i^{(L)}$ for each node. Due to the inclusion of $\mathbf{z}_{ij}^{(l)}$ in message computations, the model can learn how pair-specific features (like distances or relative velocities) modulate the influence between connected nodes. After L layers, the network produces outputs for each edge. To predict an edge's properties, we combine the features of the two endpoint nodes and the edge's own latent features. Specifically, for edge (i, j) we construct a feature vector by concatenating the final node embeddings and the final edge embedding: $\mathbf{u}_{ij} = [, \mathbf{h}_i^{(L)}, |, \mathbf{h}_j^{(L)}, |, \mathbf{z}_{ij}^{(L)},]$, where $|$ denotes concatenation. This combined representation $\mathbf{u}_{ij}$ is passed through a final feed-forward network (an "edge readout" module) that outputs two quantities: (1) a link probability (or logit) $\hat{y}_{ij}$ indicating the model's confidence that a real contact exists between i and j, and (2) a force vector $\hat{\mathbf{f}}_{ij} \in \mathbb{R}^6$ representing the predicted components of the contact force. Thus, the GNN architecture is a multi-task model that learns a shared latent representation to simultaneously perform a binary classification for contact existence and a regression for contact forces. After node- and edge-level encoding and message passing, predictions in this work are formulated explicitly at the *edge level*, as both target quantities contact existence and contact forces are defined on particle-particle interactions.

Let $\mathcal{G}_t = (\mathcal{V}_t, \mathcal{E}_t)$ denote the particle interaction graph at timestep t, where nodes correspond to particles and edges correspond to candidate particle-particle interactions identified via k-nearest-neighbor (kNN) sampling. Following L layers of edge-aware message passing using GINEConv, each node $i \in \mathcal{V}_t$ is associated with a latent embedding $\mathbf{h}_i \in \mathbb{R}^d$, while each candidate edge $(i, j) \in \mathcal{E}_t$ retains an encoded edge embedding $\mathbf{e}_{ij} \in \mathbb{R}^d$.

To construct an edge-level representation suitable for downstream prediction, the embeddings of the two endpoint nodes and the corresponding edge embedding are concatenated:

$$\mathbf{z}_{ij}^{\mathrm{in}} = [\mathbf{h}_i \parallel \mathbf{h}_j \parallel \mathbf{e}_{ij}] \in \mathbb{R}^{3d}, \tag{21}$$

where $\parallel$ denotes vector concatenation.

The concatenated representation $\mathbf{z}_{ij}^{\mathrm{in}}$ is passed through a shared multilayer perceptron, referred to here as the *edge trunk*:

$$\mathbf{z}_{ij}^{\mathrm{edge}} = f_{\mathrm{trunk}}(\mathbf{z}_{ij}^{\mathrm{in}}), \tag{22}$$

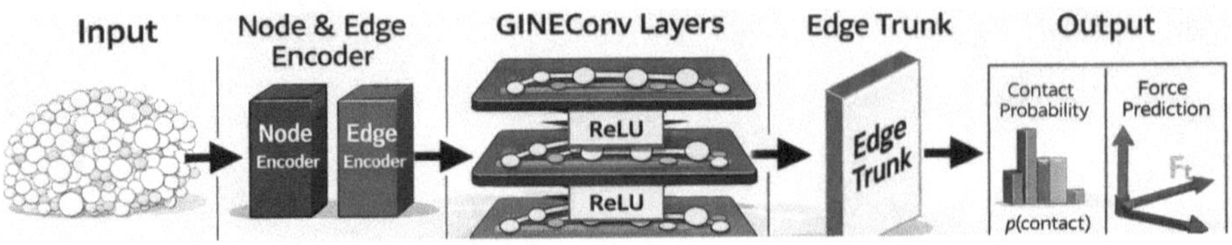

Fig. 3. Schematic of the proposed edge-aware graph neural network architecture for joint contact classification and contact force prediction. **Input:** For each timestep, particle states (position, velocity, radius, and mass) define node features, while kNN-based candidate particle pairs define edges with geometric and kinematic attributes. **Encoder:** Node and edge features are independently mapped into a shared latent space using multilayer perceptron encoders. **GINEConv:** Edge-aware message passing is performed using multiple GINEConv layers, allowing node embeddings to be updated based on neighboring particles and edge attributes. **ReLU:** Each message-passing layer is followed by layer normalization, ReLU activation, dropout, and residual connections to stabilize training and improve representation capacity. **Edge Trunk:** Latent embeddings of the two endpoint nodes and the corresponding edge embedding are concatenated and processed by a shared edge-level multilayer perceptron, forming a unified representation of particle-particle interactions. **Output:** Two task-specific heads operate on the shared edge representation to predict (i) the probability of contact and (ii) the six-dimensional normal and tangential contact force vector.

where $f_{\text{trunk}}(\cdot)$ consists of two fully connected layers with ReLU activations and dropout regularization. The edge trunk serves as a shared edge-level feature extractor that fuses geometric, kinematic, and topological information into a unified latent representation.

Two task-specific output heads operate on $\mathbf{z}_{ij}^{\text{edge}}$:

1. **Contact classification head**, producing a scalar logit $\hat{y}_{ij}$ and corresponding contact probability
$$p_{ij} = \sigma(\hat{y}_{ij}), \tag{23}$$
where $\sigma(\cdot)$ denotes the sigmoid function.
2. **Force regression head**, predicting the six-dimensional contact force vector
$$\hat{\mathbf{F}}_{ij} = \left[\hat{F}_{n,x}, \hat{F}_{n,y}, \hat{F}_{n,z}, \hat{F}_{t,x}, \hat{F}_{t,y}, \hat{F}_{t,z}\right], \tag{24}$$
corresponding to normal and tangential contact force components.

By sharing the edge trunk between both prediction tasks, the model enforces a common latent representation for contact existence and contact force magnitude and direction, thereby improving physical consistency and enabling effective multi-task learning. This design decouples node-level message passing from edge-level prediction, allowing the framework to scale efficiently to particle systems containing $\mathcal{O}(10^5)$ particles per timestep (Fig. 3).

2.2.3 Training and Joint Loss Function

Training the GNN involves learning from ground-truth contact data available in simulation snapshots. However, using entire graphs with tens of thousands of nodes and edges can be computationally prohibitive. We therefore adopt a mini-batch training approach based on sampling subgraphs from each full snapshot graph [52]. In each training iteration, we randomly select a subset of "seed" particles and extract the induced subgraph containing those seeds, their neighboring particles, and all edges among them. This subgraph will include a certain number of true contact edges (positive examples) and we augment it with an additional set of negative edges (pairs of particles that are not in contact) to provide negative examples for the link prediction task. Negative edges are chosen by sampling particle pairs that are spatially close (e.g., within the kNN list of a seed) but not actually in contact, ensuring that they are plausible non-contacts. We maintain a fixed ratio of negative to positive edges (e.g., 5:1) in each subgraph to address class imbalance. All edges in the subgraph are assigned a binary label y_{ij}, with $y_{ij} = 1$ for real contacts and $y_{ij} = 0$ for the sampled non-contact edges. The corresponding target force $\mathbf{f}_{ij}$ is known for positive edges from the simulation data, and undefined for negatives. The model is trained using a joint loss function that combines binary contact classification and contact force regression. Let $y_{ij} \in \{0, 1\}$ denote the ground-truth contact label for edge (i, j), and let $\mathbf{F}_{ij} \in \mathbb{R}^6$ denote the corresponding true contact force vector when a contact exists. The network parameters are learned via stochastic gradient descent using the Adam optimizer. We train for a sufficient number of epochs, shuffling and sampling new subgraphs each epoch to expose the model to diverse neighborhoods of the full graph. We apply dropout (rate 0.1) in the GNN layers during training as regularization, and we perform early stopping or validation monitoring to prevent overfitting. The result of this training procedure is a model that can generalize to predict contacts and forces on unseen graph data from the simulation.

The total loss is defined as

$$\mathcal{L} = \lambda_{\text{link}}\, \mathcal{L}_{\text{BCE}} + \lambda_{\text{force}}\, \mathcal{L}_{\text{MSE}}, \tag{25}$$

where λ_{link} and λ_{force} are weighting coefficients controlling the relative contributions of the two objectives.

Contact Classification Loss.

The contact classification loss is given by a binary cross-entropy loss with logits:

$$\mathcal{L}_{\text{BCE}} = -\frac{1}{|\mathcal{E}_t|} \sum_{(i,j)\in\mathcal{E}_t} \left[w\, y_{ij} \log \sigma(\hat{y}_{ij}) + (1 - y_{ij}) \log(1 - \sigma(\hat{y}_{ij})) \right], \tag{26}$$

where w is an optional positive-class weighting factor used to compensate for class imbalance between contacting and non-contacting particle pairs.

Contact Force Regression Loss.

The force regression loss is computed only for edges corresponding to true contacts. Introducing a binary mask $m_{ij} = y_{ij}$, the mean-squared error loss is defined as

$$\mathcal{L}_{\mathrm{MSE}} = \frac{1}{\sum_{(i,j)} m_{ij}} \sum_{(i,j)\in\mathcal{E}_t} m_{ij} \left\| \hat{\mathbf{F}}_{ij} - \mathbf{F}_{ij} \right\|_2^2 . \tag{27}$$

This masked formulation ensures that force predictions do not influence the loss for non-contacting particle pairs. During training, force components are normalized using statistics computed from the training set and are denormalized during inference.

The joint formulation allows the network to learn a physically consistent latent contact state, where contact probability and force magnitude are inferred from the same underlying edge representation.

3 Results and Discussion

We evaluate the trained model on held-out test graphs (simulation snapshots not seen during training) to assess its performance on both contact prediction and force estimation. For contact prediction (the classification task), we threshold the predicted probability $\hat{p}_{ij}$ (e.g., at 0.9) to decide whether an edge is classified as a contact. We then compute standard classification metrics: accuracy (the fraction of edges correctly classified), precision (the fraction of predicted contacts that are true contacts), recall (the fraction of true contacts that are correctly predicted), and the F1-score (the harmonic mean of precision and recall). These metrics provide insight into the model's ability to detect actual contacts while avoiding false positives. In particular, precision and recall are crucial in our imbalanced setting, where the number of non-contact pairs far exceeds actual contacts. For the force prediction task, we evaluate the regression accuracy on those edges where a contact is present in the ground truth. Common error metrics are used, such as the mean squared error (MSE) or mean absolute error (MAE) between the predicted force vector $\hat{\mathbf{f}}_{ij}$ and the true force $\mathbf{f}_{ij}$ across all contact edges in the test set. We may report the average force error per component or the root mean square error (RMSE) for interpretability. Additionally, we assess the model's ability to predict the direction of forces by computing correlation or angle differences between predicted and true force vectors, if relevant. All metrics are computed per snapshot and then averaged over the test set for robustness. During evaluation, to handle large test graphs, we apply the model in a batched manner. The test graph can be processed by dividing it into smaller subgraphs or by iterating over edges in chunks, which yields the same predictions as processing the full graph at once. No further training or tuning is done on the test data. We simply feed the node and edge features of the test graphs into the GINEConv and collect its outputs. The resulting metrics demonstrate the effectiveness of our approach: high precision and recall indicate reliable contact identification, and low force prediction error signifies that the model has learned the underlying physical interaction forces to a significant degree. This comprehensive evaluation confirms that our GINEConv-based methodology generalizes well and accurately captures the contact mechanics present in the particle system.

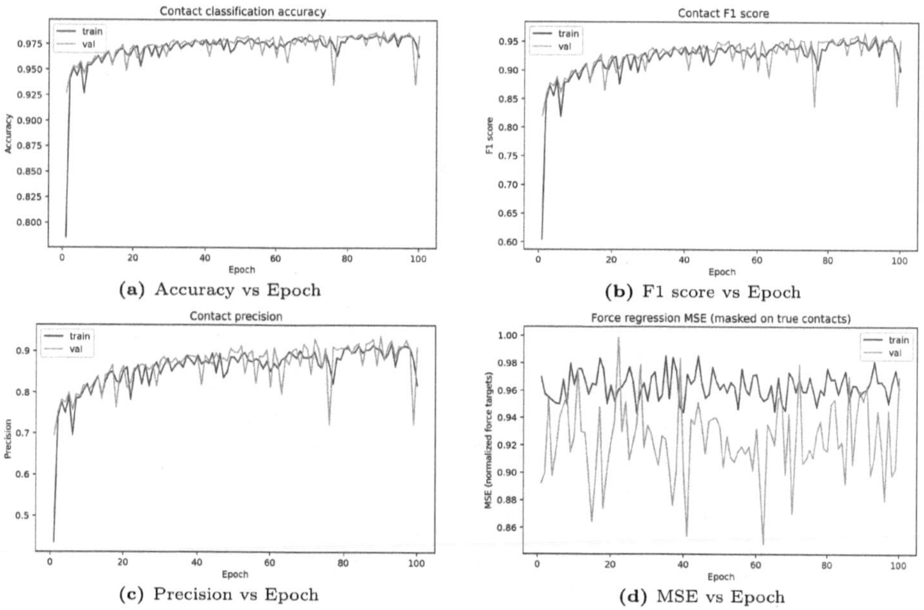

(a) Accuracy vs Epoch

(b) F1 score vs Epoch

(c) Precision vs Epoch

(d) MSE vs Epoch

Fig. 4. Training and validation learning curves for the multi-task edge-aware GNN. (a) accuracy, (b) precision, (c) recall, and (d) F1 score for contact classification; (e) masked force regression MSE computed only on true-contact edges (reported in normalized force space). The best validation F1/accuracy/precision occurs at epoch 90, while the minimum validation MSE occurs at epoch 62.

Figure 4 summarizes the training dynamics of the proposed multi-task edge-aware GNN across epochs. For the contact classification task, validation performance improves steadily, reaching peak performance at epoch 90 with $\mathrm{Acc}_{\mathrm{val}} = 0.988$, $\mathrm{Prec}_{\mathrm{val}} = 0.939$, $\mathrm{Rec}_{\mathrm{val}} = 0.995$, and $\mathrm{F1}_{\mathrm{val}} = 0.966$ (Fig. 4a-b). The high recall indicates that true contacts are rarely missed, which is critical for downstream stress and force-chain reconstruction. For the force regression task, the masked mean-squared error (computed only on true-contact edges) exhibits a distinct optimum at epoch 62, achieving the minimum $\mathrm{MSE}_{\mathrm{val}} = 0.848$ Fig. 4d. This MSE is reported in normalized force space, consistent with the training procedure, and is denormalized during inference for physical force reconstruction. Because the primary objective of the present work is accurate recovery of the contact network structure while maintaining strong force prediction performance, epoch 90 is selected as the final checkpoint for subsequent analyses. For completeness, epoch 62 is additionally reported as the checkpoint yielding the best validation force MSE.

Figure 5 shows the confusion matrix for contact classification evaluated on a full simulation timestep. The proposed edge-aware GNN demonstrates strong performance despite the pronounced class imbalance between contacting and non-contacting particle pairs. Out of all true non-contact pairs, 245,018 are cor-

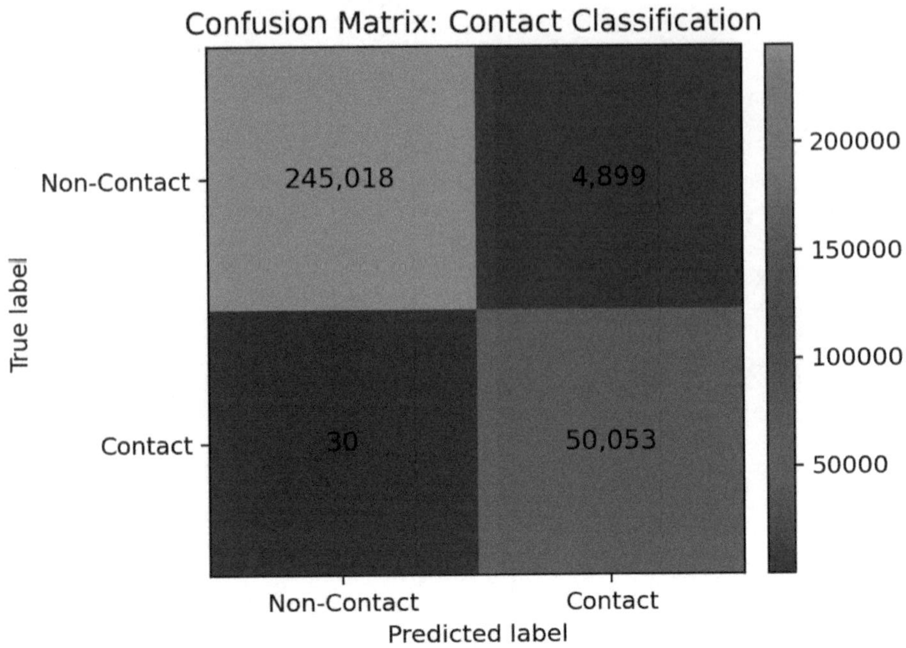

Fig. 5. Confusion Matrix for all timestep

rectly classified, while 4,899 are incorrectly predicted as contacts, corresponding to a modest false-positive rate. Crucially, the model exhibits an exceptionally low number of false negatives, missing only 30 true contacts out of more than 50,000. This results in a contact recall of 99.94%, indicating that nearly all physically relevant contacts are correctly identified. Precision for the contact class reaches 91.1%, reflecting that the majority of predicted contacts correspond to true particle interactions. The overall F1 score of approximately 95.4% highlights a favorable balance between precision and recall. While overall accuracy exceeds 98%, recall and F1 score provide more meaningful indicators of performance given the inherent imbalance in contact versus non-contact samples. From a physical modeling perspective, the observed error distribution is desirable. Missing a true contact leads directly to underestimation of force transmission, stress chains, and energy dissipation, whereas false-positive contacts typically correspond to weak or near-contact interactions that can be filtered during post-processing based on predicted force magnitude or contact probability thresholds. The model, therefore, prioritizes contact detection robustness, ensuring reliable downstream force prediction and stress reconstruction. These results confirm that the learned edge-level representation effectively captures the geometric and kinematic signatures of particle contacts, providing a reliable foundation for subsequent contact force regression.

Figure 6 compares the ground-truth and predicted spatial distributions of strong contacts (top 10% in F_t/F_n) at $t = 134$. The axial distribution shows that

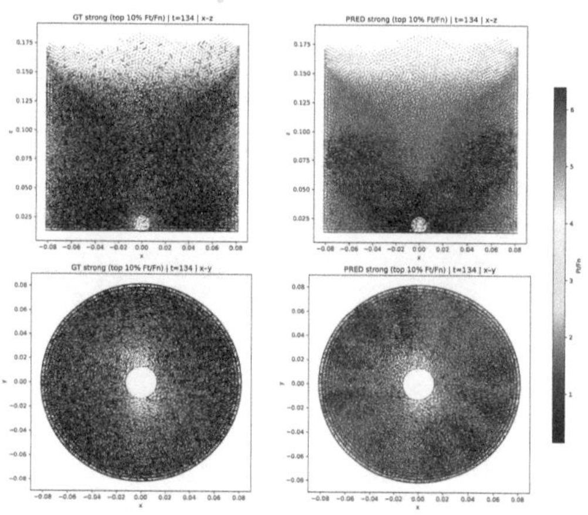

(a) Contact Network for DEM contact vs Predicted contact

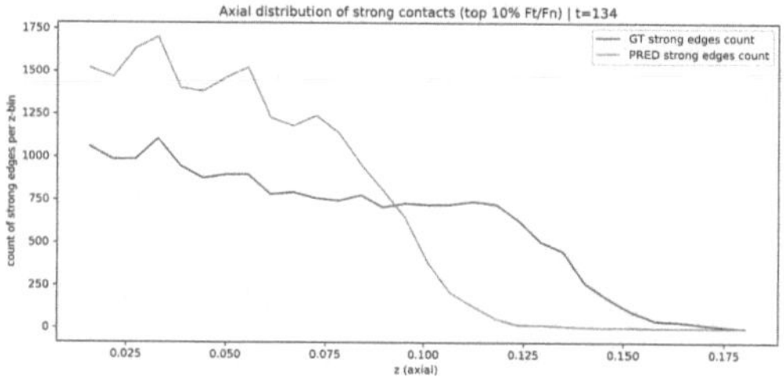

(b) axial distribution of strong edges

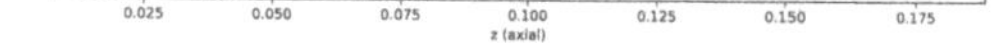

(c) mean distribution of contact

Fig. 6. True DEM simulation contact forces vs Predicted contact forces at single timestep t = 134 which at 6.70 s simulation time. (a) Axial and Radial distribution of contact forces in the mill and 10% strong contact force network for True and predicted contact, (b) Axial distribution of strong predicted and true contact (c) Axial distribution of normal and tangential force ratio for True and predicted contact

the model accurately captures the location and extent of the active grinding zone, reproducing the decay of strong contact density toward the free surface. Spatial projections in the x–z and x–y planes further demonstrate that the predicted force networks recover the dominant mesoscale structures observed in the ground truth, including vertically extended force chains and annular radial shear bands characteristic of stirred milling. While the predicted networks appear slightly sparser, this reflects a selective emphasis on mechanically dominant contacts rather than spurious weak interactions. Overall, the close agreement in both axial statistics and spatial topology indicates that the model successfully learns the underlying force-transmission mechanisms governing shear-driven grinding in stirred mills.

4 Conclusions

In this work, we presented an edge-aware graph neural network framework for predicting contact force networks in dense, shear-driven slurries within stirred media mills. By representing DEM snapshots as particle interaction graphs and jointly learning contact existence and contact forces at the edge level, the proposed model captures both local particle interactions and the mesoscale organization of force transmission that governs slurry rheology and grinding performance. The model accurately reproduces contact classification metrics, achieving high recall and F1 scores, and reliably predicts normal and tangential contact forces. Beyond pointwise accuracy, the GNN recovers physically meaningful force-network structures, including axial gradients, radial shear bands, anisotropic contact fabrics, and percolating force chains associated with shear thickening and jamming. These results demonstrate that the learned representations preserve the essential topology and mechanics of force transmission in dense suspensions, rather than merely fitting contact statistics. Importantly, the proposed framework enables rapid inference of contact force networks for large-scale systems without the need for computationally expensive DEM simulations. Despite being trained on a limited set of operating conditions, the model generalizes to new system states, highlighting its potential as a physics-consistent surrogate model. Future work will be aggregating predicted contact forces, bulk stress measures and rheological indicators will be explored and can be recovered in close agreement with DEM results, providing a direct link between particle-scale interactions and macroscopic mill behavior. This work demonstrates that edge-aware GNNs offer a powerful and efficient pathway for bridging the micro–macro divide in stirred milling. The proposed approach opens new opportunities for data-driven digital twins, rapid what-if analyses, and optimization of industrial fine grinding processes. Future work will extend the framework to a broader range of operating conditions, incorporate fluid–particle coupling effects, and integrate breakage and population balance models to further enhance predictive capability and practical relevance.

Acknowledgments. This work was supported by the Institute for Computational and Data Sciences (ICDS), which provided access to high-performance computing resources used to generate the training data.

Author Declarations. Conflict of Interest. The authors declare no conflicts of interest.

Conflict of Interest. The authors declare no conflicts of interest.

Author Contributions. Corlbert A. Ayuk: Conceptualization, formal analysis, and writing—original draft. Olumide Ogunmodimu: Supervision, review, and editing.

Data Availability. The data that support the findings of this study are available from the corresponding author upon reasonable request.

References

1. Edens, L.E., et al.: Shear stress dependence of force networks in 3D dense suspensions. Soft Matter **17**(32), 7476–7486 (2021)
2. Gameiro, M., Singh, A., Kondic, L., Mischaikow, K., Morris, J.F.: Interaction network analysis in shear thickening suspensions. Phys. Rev. Fluids **5**(3), 034307 (2020)
3. Küçüksönmez, E., Servantie, J.: Shear thinning and thickening in dispersions of spherical nanoparticles. Phys. Rev. E **102**(1), 012604 (2020)
4. Yan, H., Ding, J., Tao, C.: A physics-informed neural network solution for rheological modeling of cement slurries. Fluids **10**(7), 184 (2025)
5. Tapia, F., Ichihara, M., Pouliquen, O., Guazzelli, É.: Viscous to inertial transition in dense granular suspension. Phys. Rev. Lett. **129**(7), 078001 (2022)
6. Tapia, F., Hong, C.-W., Aussillous, P., Guazzelli, É.: Rheology of suspensions of non-brownian soft spheres across the jamming and viscous-to-inertial transitions. Phys. Rev. Lett. **133**(8), 088201 (2024)
7. Lin, N.Y., Ness, C., Cates, M.E., Sun, J., Cohen, I.: Tunable shear thickening in suspensions. Proc. Natl. Acad. Sci. **113**(39), 10774–10778 (2016)
8. Sinnott, M., Cleary, P.W., Morrison, R.D.: Slurry flow in a tower mill. Miner. Eng. **24**(2), 152–159 (2011)
9. Tordesillas, A., Walker, D.M., Lin, Q.: Force cycles and force chains. Phys. Rev. E-Stat. Nonlinear Soft Matter Phys. **81**(1), 011302 (2010)
10. Patino-Ramirez, F., O'Sullivan, C., Dini, D.: Percolating contacts network and force chains during interface shear in granular media. Granular Matter **25**(2), 31 (2023)
11. Essayah, A., Shire, T., Gao, Z.: The relationship between contact network and energy dissipation in granular materials. Granular Matter **24**(4), 100 (2022)
12. Song, Y., Shim, E.: Structure characterization of the clogging process of coarse fibrous filter media during solid particle loading with x-ray micro-computed tomography. Sep. Purif. Technol. **273**, 118980 (2021)
13. Cleary, P.W., Owen, P.J.: Using dem to understand scale-up for a hicom®mill. Miner. Eng. **92**, 86–109 (2016)
14. Sharma, S., Sharma, A., Singh, A.: Frictional contact network in dense suspension flow. arXiv preprint arXiv:2505.22747 (2025)
15. Aminimajd, A., Maia, J., Singh, A.: Robust prediction of frictional contact network in near-jamming suspensions employing deep graph neural networks. Phys. Fluids **37**(7) (2025)
16. Jiang, Y., Byrne, E., Glassey, J., Chen, X.: Integrating graph neural network-based surrogate modeling with inverse design for granular flows. Indu. Eng. Chem. Res. **63**(20), 9225–9235 (2024)

17. Mei, J., Ma, G., Cao, W., Wu, T., Zhou, W.: Graph neural network unveils the spatiotemporal evolution of structural defects in sheared granular materials. Int. J. Plast **184**, 104218 (2025)
18. Li, Z., Li, X., Zhang, H., Huang, D., Zhang, L.: The prediction of contact force networks in granular materials based on graph neural networks. J. Chem. Phys. **158**(5) (2023)
19. Qin, Y., DeWitt, S., Radhakrishnan, B., Biros, G.: GrainGNN: a dynamic graph neural network for predicting 3D grain microstructure. J. Comput. Phys. **510**, 113061 (2024)
20. Haeri, A., Holz, D., Skonieczny, K.: Subspace graph networks for real-time granular flow simulation with applications to machine-terrain interactions. Eng. Appl. Artif. Intell. **135**, 108765 (2024)
21. Majmudar, T.S., Behringer, R.P.: Contact force measurements and stress-induced anisotropy in granular materials. Nature **435**, 1079–1082 (2005)
22. Ostojic, S., Somfai, E., Nienhuis, B.: Scale invariance and universality of force networks in static granular matter. Nature **439**, 828–830 (2006)
23. Silbert, L.E., Grest, G.S., Landry, J.W.: Statistics of the contact network in frictional and frictionless granular packings. Phys. Rev. E **66**, 061303 (2002)
24. Dijksman, J.A., et al.: Characterizing granular networks using topological metrics. Phys. Rev. E **97**, 042903 (2018)
25. Cundall, P.A., Strack, O.D.L.: A discrete numerical model for granular assemblies. Géotechnique **29**, 47–65 (1979)
26. Makse, H.A., Kurchan, J.: Testing the thermodynamic approach to granular matter with a numerical model of a decisive experiment. Nature **415**, 614–617 (2002)
27. Plimpton, S.: Fast parallel algorithms for short-range molecular dynamics. J. Comput. Phys. **117**, 1–19 (1995)
28. Ringl, C., Urbassek, H.M.: A LAMMPS implementation of granular mechanics: inclusion of adhesive and microscopic friction forces. Comput. Phys. Commun. **183**, 986–992 (2012)
29. Tan, X., Guo, Y., Huang, D., Zhang, L.: A structural approach to vibrational properties ranging from crystals to disordered systems. Soft Matter **17**, 1330–1336 (2021)
30. Cheng, Z., Wang, J.: Estimation of contact forces of granular materials under uniaxial compression based on a machine learning model. Granular Matter **24**, 17 (2022)
31. Zhou, J., Long, S., Wang, Q., Dinsmore, A.D.: Measurement of forces inside a three-dimensional pile of frictionless droplets. Science **312**, 1631–1633 (2006)
32. Sanfratello, L., Fukushima, E., Behringer, R.P.: Using MR elastography to image the 3d force chain structure of a quasi-static granular assembly. Granular Matter **11**, 1–6 (2009)
33. Aminimajd, A., Maia, J., Singh, A.: Scalability of a graph neural network in accurate prediction of frictional contact networks in suspensions. Soft Matter **21**(15), 2826–2835 (2025)
34. Schoenholz, S.S.: Combining machine learning and physics to understand glassy systems. J. Phys.: Conf. Ser. **1036**, 012021 (2018)
35. Schoenholz, S.S., Cubuk, E.D., Sussman, D.M., Kaxiras, E., Liu, A.J.: A structural approach to relaxation in glassy liquids. Nat. Phys. **12**, 469–471 (2016)
36. Schoenholz, S.S., Cubuk, E.D., Kaxiras, E., Liu, A.J.: Relationship between local structure and relaxation in out-of-equilibrium glassy systems. PNAS **114**, 263–267 (2017)

37. Harrington, M., Liu, A.J., Durian, D.J.: Machine learning characterization of structural defects in amorphous packings of dimers and ellipses. Phys. Rev. E **99**, 022903 (2019)
38. Cubuk, E.D., et al.: Structure-property relationships from universal signatures of plasticity in disordered solids. Science **358**, 1033–1037 (2017)
39. Sharp, T.A., Thomas, S.L., Cubuk, E.D., Schoenholz, S.S., Srolovitz, D.J., Liu, A.J.: Machine learning determination of atomic dynamics at grain boundaries. PNAS **115**, 10943–10947 (2018)
40. Li, H., Jin, Y., Jiang, Y., Chen, J.Z.: Determining the nonequilibrium criticality of a gardner transition via a hybrid study of molecular simulations and machine learning. PNAS **118**, 2017392118 (2021)
41. Craven, G.T., Lubbers, N., Barros, K., Tretiak, S.: Machine learning approaches for structural and thermodynamic properties of a Lennard-Jones fluid. J. Chem. Phys. **153**, 104502 (2020)
42. Alkemade, R.M., Boattini, E., Filion, L., Smallenburg, F.: Comparing machine learning techniques for predicting glassy dynamics. J. Chem. Phys. **156**, 204503 (2022)
43. Tah, I., Ridout, S.A., Liu, A.J.: Fragility in glassy liquids: a structural approach based on machine learning. J. Chem. Phys. **157**, 124501 (2022)
44. Scarselli, F., Gori, M., Tsoi, A.C., Hagenbuchner, M., Monfardini, G.: The graph neural network model. IEEE Trans. Neural Networks **20**, 61–80 (2008)
45. Wu, Z., Pan, S., Chen, F., Long, G., Zhang, C., Philip, S.Y.: A comprehensive survey on graph neural networks. IEEE Trans. Neural Networks Learn. Syst. **32**, 4–24 (2020)
46. Sanchez-Gonzalez, A., Godwin, J., Pfaff, T., Ying, R., Leskovec, J., Battaglia, P.: Learning to simulate complex physics with graph networks. In: Proceedings of the ICML, pp. 8459–8468 (2020)
47. Bapst, V., Keck, T., Grabska-Barwińska, A., Donner, C., Cubuk, E.D., Schoenholz, S.S., et al.: Unveiling the predictive power of static structure in glassy systems. Nat. Phys. **16**, 448–454 (2020)
48. Mandal, R., Casert, C., Sollich, P.: Robust prediction of force chains in jammed solids using graph neural networks. Nat. Commun. **13**(1), 4424 (2022)
49. Fischer-Cripps, A.: The hertzian contact surface. J. Mater. Sci. **34**(1), 129–137 (1999)
50. Sinnott, M., Cleary, P.W., Morrison, R.: Analysis of stirred mill performance using dem simulation: part 1-media motion, energy consumption and collisional environment. Miner. Eng. **19**(15), 1537–1550 (2006)
51. Xu, K., Hu, W., Leskovec, J., Jegelka, S.: How powerful are graph neural networks? arXiv preprint arXiv:1810.00826 (2018)
52. Hamilton, W., Ying, Z., Leskovec, J.: Inductive representation learning on large graphs. In: Advances in Neural Information Processing Systems, p. 30 (2017)

Author Index

H. R. Arabnia et al. (Eds.): CSCI 2025, CCIS 2967, pp. 575–577, 2027.
https://doi.org/10.1007/978-3-032-28454-9

X

Y

Z

MIX
Papier | Fördert gute Waldnutzung
FSC® C083411

Kolibri 360 GmbH
Ferdinand-Jühlke-Straße 7
99095 Erfurt, Deutschland
produktsicherheit@kolibri360.de